BARRON'S

Regents Exams and Answers

Physics

The Physical Setting

ALBERT S. TARENDASH
Assistant Principal, Supervision (Retired),
Department of Chemistry and Physics
Stuyvesant High School, New York, New York

The section on how to answer Part C questions and the glossary were adapted from *Let's Review: Physics* by Miriam A. Lazar and Albert S. Tarendash, published by Kaplan, Inc., d/b/a Barron's Educational Series, 1996.

Acknowledgments

The 2014 Answers Explained were written by Gregory Guido.

The 2015–2019 Answers Explained were written by Miriam A. Lazar.

Published by Kaplan, Inc., d/b/a Barron's Educational Series
750 Third Avenue
New York, NY 10017
www.barronseduc.com

ISBN: 978-1-5062-5411-1

Printed in Canada

10 9 8 7 6 5 4 3 2 1

Kaplan, Inc., d/b/a Barron's Educational Series print books are available at special quantity discounts to use for sales promotions, employee premiums, or educational purposes. For more information or to purchase books, please call the Simon & Schuster special sales department at 866-506-1949.

Contents

Reference Tables for Physics   **45**

Using the Equations to Solve Physics Problems   **57**

Regents Examinations, Answers, and Self-Analysis Charts   **73**

Preface

A HELPFUL WORD TO THE STUDENT

As you are aware, the purpose of this book is to help you review for the New York State Regents examination in physics. You can also use it effectively to prepare for classroom, midterm, and final examinations. The book contains a number of special sections and other features designed to aid you in achieving good grades on your examinations. Included are the following:

- **How to Use This Book.** Your journey begins here. This section explains how to use the entire book as an effective test-preparation tool. It provides a method of identifying those areas you have mastered and those that will require additional work.
- **Test-Taking Techniques.** In this section, you will learn how to take the Regents examination effectively. Included are a list of materials you will need to bring to the examination and directions on how to fill out your answer sheet, how to read the questions, and how to deal with difficult items. In addition, it provides some suggestions on how to approach the extended-response questions found in Part C of the examination.
- **What to Expect on the Regents Examination in Physics.** This section explains the format, content, and grading of the examination.
- **New York State Physical Setting/Physics Core:** Topic Outline and Question Index. The Topic Outline provides a detailed description of the Regents Physics syllabus. The Question Index keys the Regents examination questions that appear in this book to the topic outline. The index will help you find questions on similar topics and will provide information about the areas that are stressed most heavily on the exam.
- **Glossary of Important Terms.** The glossary will help you understand the technical terms that may appear on the examination.

- *Reference Tables for Physics.* A significant part of the Regents examination requires you to be able to find and use the information contained in these reference tables. This section describes each table and the types of information that may be obtained from it.
- *Using the Equations to Solve Physics Problems.* This section provides examples that show you how to use each of the equations given in Reference Table *K*.
- *Recent Regents Examinations.* These questions will provide the bulk of your study and review. When you have answered all of the questions that appear in this book, you will be able to approach the Regents examination with confidence!
- *Explanation of Answers and Self-Analysis Chart for Each Examination.* The detailed answer explanations will help you to understand why certain choices are correct while others are incorrect. The self-analysis chart will pinpoint your strengths and weaknesses on each practice examination you take.

This book was written to provide you with the basic tools you need to perform well on the Regents Physics examination. If you follow its advice and prepare diligently, you will achieve your goal.

Best wishes for success!

Albert S. Tarendash

How to Use This Book

1. Read the section entitled *Test-Taking Techniques* to learn how to prepare properly for an examination and how to take it with maximum efficiency.
2. Read the section entitled *What to Expect on the Regents Examination in Physics* to familiarize yourself with the structure and contents of this examination.
3. Read the section entitled *Reference Tables for Physics* to familiarize yourself with the contents and use of these tables.
   Note: On occasion, the New York State Education Department may change the content or format of the Regents examination and/or the reference tables. Your classroom teacher is your best source of information about such changes.
4. Read the section entitled *Using the Equations to Solve Physics Problems.*
5. Take the first Regents examination in this book, answering *all* of the questions.
6. Refer to the *Glossary of Important Terms* to learn the meanings of words and terms you do not understand.
7. Check your answers, and then complete the self-analysis chart at the end of the examination to pinpoint your strengths and weaknesses.
8. Read the detailed explanation of *all of the questions,* paying closest attention to the questions you answered incorrectly. Occasionally, the *wrong choices* are explained, and these explanations may help you understand why you chose an incorrect answer.
9. When you have determined your areas of weakness, refer to the *New York State Regents Physics Core: Topic Outline and Question Index* to locate similar questions on other recent examinations. (You can also use this outline and index to determine which areas have been stressed in recent years.)
10. Repeat steps 5–9 for the other examinations, *with the exception of the most recent test.*

11. When you have completed your studying, but no more than 1 or 2 days before the actual examination, take the most recent examination in this book *under strict examination conditions.*

After you have checked your answers to this last examination, you will have a rough idea of how you will perform on the Regents examination you will take.

IMPORTANT NOTE ABOUT THE REFERENCE
TABLES AND EQUATIONS

In this book, the reference tables and equations beginning on page 45 have been *slightly modified* from the official New York State Reference Tables for Physics (2006 edition). The tables in this book are indexed by letter, and the equations are indexed by letter and number. Some of the tables have been moved in order to separate them from the equations, and each table is accompanied by a brief explanation. Otherwise, the material on these tables and the official tables is *identical.*

All of the explanations and answers in this book refer to the tables and equations *contained within this book.*

Test-Taking Techniques

The following pages contain several tips to help you achieve a good grade on the Physics Regents exam.

GENERAL HELPFUL TIPS

TIP 1

Be confident and prepared.

SUGGESTIONS

- Review previous tests.
- Use a clock or watch, and take previous exams at home under examination conditions (i.e., don't have the radio or television on).
- Get a review book. (One review book is Barron's *Let's Review: Physics*.)
- Talk over the answers to questions on these tests with someone else, such as another student in your class or someone at home.
- Finish all your homework assignments.
- Look over classroom exams that your teacher gave during the term.
- Take class notes carefully.
- Practice good study habits.
- Know that there is an answer for every question.
- Be aware that the people who made up the Regents exam want you to pass.
- Remember that thousands of students over the last few years have taken and passed a Physics Regents. You can pass too!
- Complete your study and review at least one day before the examination. Last-minute cramming does not help and may hurt your performance.

- On the night prior to the exam day: lay out all the things you will need, such as clothing, pens, and admission cards.
- Go to bed early; eat wisely.
- Bring the required materials to the examination. This generally means a pen, two sharpened pencils, and a good quality eraser. In addition, Parts A, B, and C may require the use of a ruler and a protractor. If your school does not supply a calculator, be certain to bring one to the examination. Some schools also require a signed Regents admission card for identification.
- Good advice: Assume your school will *not* supply you with any materials!
- Once you are in the exam room, arrange things, get comfortable, be relaxed, attend to personal needs (the bathroom).
- Keep your eyes on your own paper; do not let them wander over to anyone else's paper.
- Be polite in making any reasonable requests of the exam room proctor, such as changing your seat or having window shades raised or lowered.

TIP 2

Read test instructions carefully.

SUGGESTIONS

- Be familiar with the format of the examination.
- Know how the test will be graded.
- If your school supplies an electronic scoring sheet, be certain you are familiar with the additional directions for recording and changing answers.
- If you decide to change an answer, be certain that you erase your original response completely.
- Any stray marks on your answer sheet should be erased completely.
- Be familiar with the directions for Parts B and C. Answer each question completely. Explanations should be written as *whole sentences* and substitutions into equations *must include units*. Be certain that your answers are clearly labeled and well organized. Place a box around numerical answers. Be neat!
- Ask for assistance from the exam room proctor if you do not understand the directions.

TIP 3

Read each question carefully before you record your answer.

SUGGESTIONS

- Be sure you understand *what* the question is asking; circle key words and numbers.
- Try to recognize information that is *given* in the question.
- On extended response problems, check the blank answer sheet; it will provide clues to answering the questions.
- Will a physics formula help you find the answer to the question?
- Some choices in Parts A and B may look appealing yet be incorrect. (These traps are known as *distractors*.)
- Try to eliminate those choices that are *obviously* incorrect.

TIP 4

Budget your test time (3 hours).

SUGGESTIONS

- Bring a watch or clock to the test.
- The Regents examination is designed to be completed in 1½ to 2 hours.
- If you are absolutely uncertain of the answer to a question, mark your question booklet and move on to the next question.
- If you persist in trying to answer every difficult question *immediately*, you may find yourself rushing or unable to finish the remainder of the examination.
- When you have finished the examination, return to those unanswered questions.
- Good advice: If at all possible, reread the *entire* examination—and your responses—at least one more time. (This will help you eliminate those errors that result from misreading questions.)

TIP 5

Use your reasoning skills.

SUGGESTIONS

- Answer *all* questions.
- Relate (connect) the question to anything that you studied, wrote in your notebook, or heard your teacher say in class.
- Relate (connect) the question to any film, demonstration, or experiment you saw in class, any project you did, or to anything you may have learned from newspapers, magazines, or television.
- Look over the entire test to see whether one part of it can help you answer another part.

TIP 6

Use your reference tables and refer frequently to the "Equations for Physics."

SUGGESTIONS

- You should be familiar with the *content* of each table.
- Frequently, the answers to questions can be found from information contained within the table.
- The equations are grouped according to where they appear in the Core.
- The definition of each symbol in the equation is also provided.
- These equations will aid you in answering many of the questions on the examination.

TIP 7

Do not be afraid to guess on multiple-choice questions.

SUGGESTIONS

- In general, go with your first answer choice.
- Eliminate obvious incorrect choices.
- If still unsure of an answer, make an educated guess.
- There is no penalty for guessing; therefore, answer ALL questions. An omitted answer gets no credit.

TIP 8
Sign the declaration found on your answer sheet.

SUGGESTION

- Unless this declaration is signed, your paper cannot be scored.

SUMMARY OF TIPS

1. Be confident and prepared.
2. Read test instructions carefully.
3. Read each question carefully and read each choice before you record your answer.
4. Budget your test time (3 hours).
5. Use your reasoning skills.
6. Use your reference tables and refer frequently to the "Summary of Equations."
7. Don't be afraid to guess.
8. Sign the declaration found on your answer sheet.

HOW TO ANSWER PART C QUESTIONS

An *extended-response question* is an examination question that requires the test taker to do more than to choose among several responses or to fill in a blank. You may need to perform numerical calculations, draw and interpret graphs, and provide extended written responses to a question or problem.

Part C of the New York State Regents Examination in Physics contains free-response questions. This appendix is designed to provide you with a number of general guidelines for answering them.

SOLVING PROBLEMS INVOLVING NUMERICAL CALCULATIONS

To receive full credit you must:

- Provide the appropriate equation(s).
- Substitute values and units into the equation(s).
- Display the answer, with appropriate units and to the correct number of significant figures.
- If the answer is a vector quantity, include its direction.

Although SI units are used on the Regents examination, you are expected to have some familiarity with other metric units such as the gram and the kilometer.

You should write as legibly as possible. Teachers are human, and nothing irks them more than trying to decipher a careless, messy scrawl. It is also a good idea to identify your answer clearly, either by placing it in a box or by writing the word "answer" next to it.

A final word: If you provide the correct answer but do not show any work, you will not receive any credit for the problem!

The following is a sample problem and its model solution.

Problem

A 5.0-kilogram object has a velocity of 10 meters per second [east]. Calculate the momentum of this object.

Solution

$$\mathbf{p} = m\mathbf{v}$$
$$\mathbf{p} = (5.0 \text{ kg})(10 \text{ m/s [east]})$$

$$\boxed{\mathbf{p} = 50 \text{ kg.m/s [east]}}$$

GRAPHING EXPERIMENTAL DATA

To receive full credit you must:

- Label both axes with the appropriate variables and units.
- Divide the axes so that the data ranges fill the graph as nearly as possible.
- Plot all data points accurately.
- Draw a best-fit line carefully with a straightedge. The line should pass through the origin *only if the data warrant it.*
- If a part of the question requires that the slope be calculated, calculate the slope *from the line,* not from individual data points.

A graph should have a title, and the *independent variable* is usually drawn along the x-axis.

The following is a sample problem and its model solution.

Problem

The weights of various masses, measured on Planet X, are given in the table below.

Mass (kg)	Weight (N)
15	21
20	32
25	35
30	48
35	56

1. Draw a graph that illustrates these data.
2. Use the *graph* to calculate the acceleration due to gravity on Planet X.

Solution

1. The graph shown below incorporates the essential items that were listed in the table.

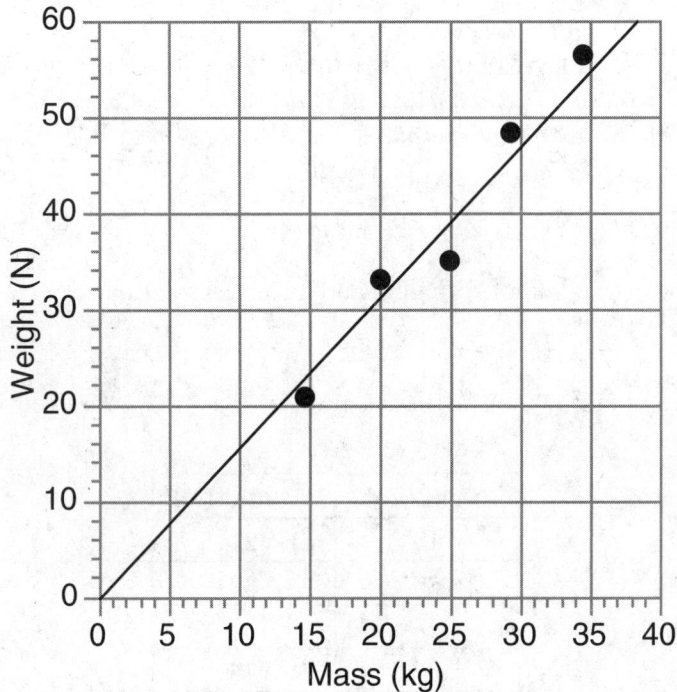

2. Since the magnitude of the gravitational acceleration can be determined by calculating the ratio of weight to mass ($g = W/m$), we can calculate the value of g from the slope of the graph.

We choose two points on the line; we do not use the data points themselves:

$$g = \frac{\Delta W}{\Delta m} = \frac{50 \text{ N} - 10 \text{ N}}{32 \text{ kg} - 6 \text{ kg}} = 1.5 \frac{\text{N}}{\text{kg}} = 1.5 \text{ m/s}^2$$

DRAWING DIAGRAMS

To receive full credit you must:

- Draw your diagrams neatly, and label them clearly.
- Draw vectors *to scale* and *in the correct direction*. If you are given a scale, you must draw your vectors to that scale.
- Bring a straightedge and a protractor with you so that you can draw neat, accurate diagrams.

WRITING AN EXTENDED-RESPONSE ANSWER

To receive full credit you must:

- Use complete, clear sentences that make sense to the reader.
- Use correct physics in your explanations.

A sample question and acceptable and unacceptable answers are given below.

Question

A student draws a graph of the potential difference across a conductor versus the current in the conductor. How can the resistance of the conductor be determined from the graph the student has drawn?

Acceptable Answers

- The resistance can be determined from the slope of the graph.
- The change in the potential difference divided by the change in current determines the resistance of the conductor.
- The slope of the straight line is the resistance of the conductor.

Unacceptable Answers

- The slope. (Incomplete sentence)
- The y-intercept of the graph determines the resistance. (Incorrect physics)

What to Expect on the Regents Examination in Physics

FORMAT OF THE PHYSICS EXAMINATION

The physics examination will be three hours long and will include four parts: A, B–1, B–2, and C. You should be prepared to answer multiple-choice questions as well as questions that require an extended written response.

In general, questions will fall into three categories:

Content questions will test your knowledge and understanding of the material contained within the New York State Physics Core. You may be asked to provide definitions of physical phenomena, interpret diagrams, and solve simple problems.

Skills questions will test your ability to apply, analyze, and evaluate the material contained within the Core. You may be asked to draw and/or interpret graphs and diagrams and solve problems of a more complex nature.

Applications questions will test your ability to apply your scientific knowledge and skills to real-world situations.

Note: The topic outline found on pages 18–30 contains the content, skills, and real-world applications that make up the New York State Physics Core.

Some of the questions on the examination will require use of the 2006 edition of the *Reference Tables for Physical Setting/Physics* (see pages 45–56).

You will be required to answer ALL of the questions on the Physical Setting/Physics Regents examination.

Analysis of a Recent Physical Setting/ Physics Regents Examination

Part	Question Format	Types of Questions	*Percent of Examination
A	Multiple-choice questions	Content questions	41
B–1	Multiple-choice questions	Content and skills questions	14
B–2	Multiple-choice and extended response questions	Content and skills questions	21
C	Extended response questions	Content, skills, and applications questions	24

*These percentages may vary slightly in future examinations.

The maximum *raw* score on the examination is 85 points. A teacher's chart will be provided for converting your *raw* score to a *scaled* score that has a maximum of 100 points. A sample conversion table taken from a recent Regents Physics Examination is shown on the next page. (Remember: Conversion tables can vary slightly from year to year.)

Sample Conversion Table

Raw Score	Scale Score	Raw Score	Scale Score	Raw Score	Scale Score	Raw Score	Scale Score
85	100	63	78	41	57	19	31
84	99	62	77	40	56	18	29
83	98	61	76	39	55	17	28
82	97	60	76	38	54	16	26
81	96	59	75	37	53	15	25
80	95	58	74	36	52	14	23
79	94	57	73	35	50	13	22
78	93	56	72	34	49	12	20
77	92	55	71	33	48	11	19
76	91	54	70	32	47	10	17
75	90	53	69	31	46	9	16
74	89	52	68	30	45	8	14
73	88	51	67	29	43	7	12
72	87	50	66	28	42	6	11
71	86	49	65	27	41	5	9
70	85	48	64	26	40	4	7
69	84	47	63	25	38	3	5
68	83	46	62	24	37	2	4
67	82	45	61	23	36	1	2
66	81	44	60	22	35	0	0
65	80	43	59	21	33		
64	79	42	58	20	32		

The table is used to convert the number of points you actually received on the examination (your "raw" score) to your final score on the examination (your "scaled" score). **Note that this table will change from one examination to another.**

TOPICS COVERED ON THE REGENTS EXAMINATION IN PHYSICS

All of the questions on the Physics examination will test major understandings, skills, and real-world applications drawn from following five subject areas:

M.   Math Skills
I.   Mechanics
II.   Energy
III.   Electricity and Magnetism
IV.   Waves
V.   Modern Physics

It is suggested that you read the *Topic Outline* found on pages 18–30 in order to learn the exact nature of the material that is subject to testing.

New York State Physical Setting/ Physics Core

TOPIC OUTLINE

The Topic Outline on pages 18–30 is taken from Appendices C and D of the New York State Physical Setting/Physics Core. All Regents physics examinations are based on this Core. The Topic Outline is divided into six sections:

M.   Math Skills
I.   Mechanics
II.   Energy
III.   Electricity and Magnetism
IV.   Waves
V.   Modern Physics

Each course area contains one or more of the following items:

- *Performance indicators*, that is, the major understandings that you must have mastered for the examination.
- *Process skills* that you need to be able to demonstrate during the examination.
- *Real-world applications* that relate physics concepts to the world around you.

Note: When an asterisk (*) is associated with a performance indicator, it means that you need to be able to solve problems using one or more of the equations given in the reference tables on pages 45–56.

M. Math Skills

Sequence	Process Skills (The student will be able to . . .)	Real-World Application
M.1	use algebraic and geometric representations to describe and compare data	use scaled diagrams to represent and manipulate vector quantities, represent physical quantities in graphical form, construct graphs of real-world data (scatter plots, line or curve of best fit), manipulate equations to solve for unknowns, use dimensional analysis to confirm algebraic solutions
M.2	use deductive reasoning to construct and evaluate arguments	interpret graphs to determine the mathematical relationship between the variables
M.3	apply algebraic and geometric concepts and skills in the solution of problems	explain the physical relevance of the properties of a graph of real-world data using slope, intercepts, and area under a curve

I. Mechanics

Sequence	Performance Indicators
I.1	Measured quantities can be classified as either vector or scalar.
I.2	An object in linear motion may travel with a constant velocity* or with acceleration*. *(Note: Testing of acceleration will be limited to cases in which acceleration is constant.)*
I.3	An object in free fall accelerates due to the force of gravity*. Friction and other forces cause the actual motion of a falling object to deviate from its theoretical motion. *(Note: Initial velocities of objects in free fall may be in any direction.)*
I.4	The resultant of two or more vectors, acting at any angle, is determined by vector addition.
I.5	A vector may be resolved into perpendicular components.*
I.6	The path of a projectile is the result of the simultaneous effect of the horizontal and vertical components of its motion; these components act independently.
I.7	A projectile's time of flight is dependent upon the vertical components of its motion.
I.8	The horizontal displacement of a projectile is dependent upon the horizontal component of its motion and its time of flight.
I.9	According to Newton's First Law, the inertia of an object is directly proportional to its mass. An object remains at rest or moves with constant velocity, unless acted upon by an unbalanced force.
I.10	When the net force on a system is zero, the system is in equilibrium.
I.11	According to Newton's Second Law, an unbalanced force causes a mass to accelerate*.
I.12	Weight is the gravitational force with which a planet attracts a mass.* The mass of an object is independent of the gravitational field in which it is located.

I. Mechanics (*Continued*)

I.13	Kinetic friction* is a force that opposes motion.
I.14	Centripetal force* is the net force which produces centripetal acceleration*. In uniform circular motion, the centripetal force is perpendicular to the tangential velocity.
I.15	The impulse* imparted to an object causes a change in its momentum*.
I.16	The elongation or compression of a spring depends upon the nature of the spring (its spring constant) and the magnitude of the applied force*.
I.17	According to Newton's Third Law, forces occur in action/reaction pairs. When one object exerts a force on a second, the second exerts a force on the first that is equal in magnitude and opposite in direction.
I.18	Momentum is conserved in a closed system.* (*Note: Testing will be limited to momentum in one dimension.*)
I.19	Gravitational forces are only attractive, whereas electrical and magnetic forces can be attractive or repulsive.
I.20	The inverse square law applies to electrical* and gravitational* fields produced by point sources.
I.21	Field strength* and direction are determined using a suitable test particle. (*Notes: 1) Calculations are limited to electrostatic and gravitational fields. 2) The gravitational field near the surface of Earth and the electrical field between two oppositely charged parallel plates are treated as uniform.*)

Is. Mechanics Skills

Sequence	Process Skills (The student will be able to . . .)	Real-World Application
Is.1	construct and interpret graphs of position, velocity, or acceleration versus time	Global Positioning Systems (GPS), track and field
Is.2	determine and interpret slopes and areas of motion graphs	mathematical slopes, calculus
Is.3	determine the acceleration due to gravity near the surface of the Earth	weights, bungee jumping, skydiving
Is.4	determine the resultant of two or more vectors graphically or algebraically	navigation (e.g., boats, planes)
Is.5	draw scaled force diagrams, using a ruler and a protractor	building design (stress analysis), cranes, picture hangers
Is.6	resolve a vector into perpendicular components: graphically and algebraically	push lawn mower, amusement park wave swing
Is.7	sketch the theoretical path of a projectile	tennis, soccer, golf, archery
Is.8	use vector diagrams to analyze mechanical systems (equilibrium and nonequilibrium)	cars, elevators, tightrope walker, apparent weightlessness (micro-gravity)
Is.9	verify Newton's Second Law for linear motion	space shuttle, cruise control
Is.10	determine the coefficient of friction for two surfaces	skidding on driving surfaces, ice skating, Teflon surfaces, sledding
Is.11	verify Newton's Second Law for uniform circular motion	amusement park rides (e.g., merry-go-rounds)
Is.12	verify conservation of momentum	car crashes, balls, bats
Is.13	determine a spring constant	car suspension systems, rubber bands, spring scales

II. Energy

Sequence	Performance Indicators
II.1	When work* is done on or by a system, there is a change in the total energy* of the system.
II.2	Work done against friction results in an increase in the internal energy of the system.
II.3	Power* is the time-rate at which work is done or energy is expended.
II.4	All energy transfers are governed by the law of conservation of energy.*
II.5	Energy may be converted among mechanical, electromagnetic, nuclear, and thermal forms.
II.6	Potential energy is the energy an object possesses by virtue of its position or condition. Types of potential energy are gravitational* and elastic*.
II.7	Kinetic energy* is the energy an object possesses by virtue of its motion.
II.8	In an ideal mechanical system, the sum of the macroscopic kinetic and potential energies (mechanical energy) is constant.*
II.9	In a nonideal mechanical system, as mechanical energy decreases there is a corresponding increase in other energies such as internal energy.*

IIs. Energy Skills

Sequence	Process Skills (The student will be able to . . .)	Real-World Application
IIs.1	describe and explain the exchange between potential energy, kinetic energy, and internal energy for simple mechanical systems, such as a pendulum, a roller coaster, a spring, a freely falling object	skiing, skateboarding
IIs.2	predict velocities, heights, and spring compressions based on energy conservation	diving board, trampoline
IIs.3	determine the energy stored in a spring	ballpoint pen, pop-up toys
IIs.4	observe and explain energy conversions in real-world situations	hydroelectric power, solar power, Sun, engines
IIs.5	recognize and describe conversions among different forms of energy in real or hypothetical devices such as a motor, a generator, a photocell, a battery	solar-powered calculator, electric fan, heat pumps, air conditioners, Peltier devices
IIs.6	compare the power developed when the same work is done at different rates	elevators, running versus walking up stairs, motorcycles versus tractor-trailers
IIs.7	determine the factors that affect the period of a pendulum	Pirate Ship and Sky Coaster (amusement park rides), grandfather clock, swing

III. Electricity and Magnetism

Sequence	Performance Indicators
III.1	Gravitational forces are only attractive, whereas electrical and magnetic forces can be attractive or repulsive.
III.2	The inverse square law applies to electrical* and gravitational* fields produced by point sources.
III.3	Energy may be stored in electric* or magnetic fields. This energy may be transferred through conductors or space and may be converted to other forms of energy.
III.4	The factors affecting resistance in a conductor are length, cross-sectional area, temperature, and resistivity.*
III.5	All materials display a range of conductivity. At constant temperature, common metallic conductors obey Ohm's Law*.
III.6	A circuit is a closed path in which a current* can exist. (*Note: Use conventional current.*)
III.7	Electrical power* and energy* can be determined for electric circuits.
III.8	Circuit components may be connected in series* or in parallel.* Schematic diagrams are used to represent circuits and circuit elements.
III.9	Moving electric charges produce magnetic fields. The relative motion between a conductor and a magnetic field may produce a potential difference in the conductor.

IIIs. Electricity and Magnetism Skills

Sequence	Process Skills (The student will be able to . . .)	Real-World Application
IIIs.1	measure current and voltage in a circuit	transformers, power supplies, battery testers, power meters, multi-meters
IIIs.2	use measurements to determine the resistance of a circuit element	dimmer switches, volume controls, temperature controls (potentiometers)
IIIs.3	interpret graphs of voltage versus current	power meters, sound-board meters
IIIs.4	measure and compare the resistance of conductors of various lengths and cross-sectional areas	toasters, hair dryers, power transmission lines
IIIs.5	construct simple series and parallel circuits	household wiring, jumper cables, fuses, and circuit breakers
IIIs.6	draw and interpret circuit diagrams which include voltmeters and ammeters	schematic plans
IIIs.7	predict the behavior of lightbulbs in series and parallel circuits	holiday lights, flash-lights
IIIs.8	map the magnetic field of a permanent magnet, indicating the direction of the field between the N (north-seeking) and S (south-seeking) poles	compass, magnets, magnetic storage media (e.g., floppy disks, hard drives, tapes)

IV. Waves

Sequence	Performance Indicators
IV.1	An oscillating system produces waves. The nature of the system determines the type of wave produced.
IV.2	Waves carry energy and information without transferring mass. This energy may be carried by pulses or periodic waves.
IV.3	Waves are categorized by the direction in which particles in a medium vibrate about an equilibrium position relative to the direction of propagation of the wave such as transverse and longitudinal waves.
IV.4	Mechanical waves require a material medium through which to travel.
IV.5	The model of a wave incorporates the characteristics of amplitude, wavelength*, frequency*, period*, wave speed*, and phase.
IV.6	Electromagnetic radiation exhibits wave characteristics. Electromagnetic waves can propagate through a vacuum.
IV.7	All frequencies of electromagnetic radiation travel at the same speed in a vacuum.*
IV.8	When a wave strikes a boundary between two media, reflection*, transmission, and absorption occur. A transmitted wave may be refracted.
IV.9	When a wave moves from one medium into another, the wave may refract due to a change in speed. The angle of refraction (measured with respect to the normal) depends on the angle of incidence and the properties of the media (indices of refraction).*
IV.10	The absolute index of refraction is inversely proportional to the speed of a wave.*

IV. Waves (*Continued*)

IV.11	When waves of a similar nature meet, the resulting interference may be explained using the Principle of Superposition. Standing waves are a special case of interference.
IV.12	Resonance occurs when energy is transferred to a system at its natural frequency.
IV.13	Diffraction occurs when waves pass by obstacles or through openings. The wavelength of the incident wave and the size of the obstacle or opening affect how the wave spreads out.
IV.14	When a wave source and an observer are in relative motion, the observed frequency of the waves traveling between them is shifted (Doppler effect).

IVs. Waves Skills

Sequence	Process Skills (The student will be able to . . .)	Real-World Application
IVs.1	compare the characteristics of two transverse waves such as amplitude, frequency, wavelength, speed, period, and phase	stadium waves, electromagnetic waves, S-waves (secondary earthquake waves)
IVs.2	draw wave forms with various characteristics	oscilloscopes
IVs.3	identify nodes and antinodes in standing waves	guitar string (vibrating stretched wire), pipe organ (vibrating air column)
IVs.4	differentiate between transverse and longitudinal waves	polarized sunglasses, liquid crystal displays (e.g., computer screens, watches, calculator), speakers, 3-D movies
IVs.5	determine the speed of sound in air	echoes
IVs.6	predict the superposition of two waves interfering constructively and destructively (indicating nodes, antinodes, and standing waves)	stereo speakers, surround sound, iridescence (e.g., butterfly wings, soap bubbles), Tacoma Narrows Bridge, beats, electronic tuners
IVs.7	observe, sketch, and interpret the behavior of wave fronts as they reflect, refract, and diffract	ocean waves, amusement park wave pools, harbor waves, pond ripples, ultrasonic cleaners (standing waves)
IVs.8	draw ray diagrams to represent the reflection and refraction of waves	barcode scanners, mirrors
IVs.9	determine empirically the index of refraction of a transparent medium	diamonds, spearfishing, lenses

V. Modern Physics

Sequence	Performance Indicators
V.1	States of matter and energy are restricted to discrete values (quantized).
V.2	Charge is quantized on two levels. On the atomic level, charge is restricted to the elementary charge (charge on an electron or proton). On the subnuclear level charge appears as fractional values of the elementary charge (quarks).
V.3	On the atomic level, energy is emitted or absorbed in discrete packets called photons.*
V.4	The energy of a photon is proportional to its frequency.*
V.5	On the atomic level, energy and matter exhibit the characteristics of both waves and particles.
V.6	Among other things, mass-energy and charge are conserved at all levels (from subnuclear to cosmic).
V.7	The Standard Model of Particle Physics has evolved from previous attempts to explain the nature of the atom and states that: 1. Atomic particles are composed of subnuclear particles. 2. The nucleus is a conglomeration of quarks which manifest themselves as protons and neutrons. 3. Each elementary particle has a corresponding antiparticle.
V.8	Behaviors and characteristics of matter, from the microscopic to the cosmic levels, are manifestations of its atomic structure. The macroscopic characteristics of matter, such as electrical and optical properties, are the result of microscopic interactions.
V.9	The total of the fundamental interactions is responsible for the appearance and behavior of the objects in the universe.
V.10	The fundamental source of all energy in the universe is the conversion of mass into energy.*

Vs. Modern Physics Skills

Sequence	Process Skills (The student will be able to . . .)	Real-World Application
Vs.1	interpret energy-level diagrams	black light posters, lasers
Vs.2	correlate spectral lines with energy-level diagram	neon lights, street lights

QUESTION INDEX

What follows is an index to the examination questions that are explained in this book. The questions are indexed according to the sequence numbers given in the core and in the skills section.

Some questions embrace more than one topic and/or skill. These questions are marked with a dagger (†).

SEQUENCE	JUNE 2013	JUNE 2014	JUNE 2015	JUNE 2016	JUNE 2017	JUNE 2018	JUNE 2019
M. Math Skills							
M.1	66, 70	†51, †52, †53, †54	16, 67, 73, 74, 75	†2, †5, 7–†8, †12–15, 17–18, 23, 30–33, 38–39, 41–43, †59–60, 68, †70–†73, 76–84	†4–†7, 9–†13, 15–17, 19, 21–22, 28–29, 32, 35, 39–43, 48, 49, †55–57, 60–61, †63–64, †66–68, 71–75, 77–82, 84, 85	†3, †4, †6, 9, †13, †14, 15, 19, 20, 26, 29, 30, 32, 36–39, 41–44, 50, 51, †58–†70, 76–78, 82, 84	†2, 7, †8, 10, †12–16, 23, 26, 32, 34, †36, 39, 41, 45, 46, 48, 49, 51–53, †57–75, 78–80
M.2	57	†41, †46	24, 43, 48	37, 40, †44–45	†44		11, 33, 37
M.3	58, 59	48, †64, †65	61, 62	46, 62, †65	3, 36, 38	53, †59, 71, 72	56
I. Mechanics							
I.1	1, 36	1, 36	1	1	1	1	1
I.2	3, 4, 72, 73	2, 7	2, 39, 40	31, 37	†4, †6	†3, †4, †6	23, 34
I.3		4	22		†7	7	3, 59, 60
I.4	2		3, 41			53	†2, 19
I.5		3	37, 41		2	5, 71, 92	7, †40
I.6	5		†38		37		†40
I.7	6		†38, 61, 62	56–58	†55	8	†40
I.8		47	†38		†55	5	†40
I.9	7	16	†4, 33	4	8, 32, 33	2	
I.10	9	44	†4	3	8		
I.11	12, 38, 40	8, 73, 74	†4	51, 52	43	40	41
I.12	10, 63, 64, 65	75, 77, 78	5, 11, 66	†2, †65	†5		5, 66, 67
I.13	8	76	†65, 68, 70	9, 53		11, 73, 74, 75	
I.14		38, 45	57, 58, 59		†12, 79, 80	12, 74, 76, 77, 79, 80	6, 42
I.15		9, 10	7	†59, 60, 61		9	61–65
I.16			6	†70, 71		32	†8
I.17	11	5	9	†5		10	18, 33
I.18	37, 54, 55		10	†8	11, 40, 41	39	
I.19							
I.20	47	71, 72	12			†14, 20	†36
I.21	13	6	11	25			
Is. Mechanics Skills							
Is.1		63, †64	60	63, 64			37, 55
Is.2		†65		62		45	
Is.3							
Is.4	71, 74, 75	†52, †53	3, 41, 69	67, 68, 69		52	2

SEQUENCE	JUNE 2013	JUNE 2014	JUNE 2015	JUNE 2016	JUNE 2017	JUNE 2018	JUNE 2019
Is.5	†51, 54						
Is.6			37	39		71, 72	7
Is.7		57					
Is.8		55, 56					
Is.9							
Is.10		79, 80					
Is.11							
Is.12							
Is.13					60	32	
II. Energy							
II.1	81		42, 48	†12, †13	19, 66, 70	54, 55	10
II.2	84					56	
II.3	39	58, 59	44, 51, 52	17, 18		15	†12
II.4	21	84	†15, †45				39
II.5	15, 27, 85	15	14	74			
II.6	14, 82, 83		†45	14	67, 68	†38, 60, 61	9, 39, 52, 53
II.7	45	81, 82	†45	15	77, 78	†38	39
II.8							11, 52, 53
II.9		42, 62	†15		39, 69		39
IIs. Energy Skills							
IIs.1		31					
IIs.2							39
IIs.3		19		72, †73			
IIs.4							51
IIs.5							
IIs.6							†12
IIs.7		50					
III. Electricity and Magnetism							
III.1	23	32	30, 71, 72		14, 30, 34	16, 34	4, 35
III.2	43	12, 33		49	15		
III.3	33, 41, 46, 56	13, 37, 39	16, 31	32	10, †13, 42, 48	†13, 18, 28, 37	†13, †14, 47, 68–70
III.4	†51, 52	14	17	45, 46	18, 47	17, 19, 66, 67, 68	†15
III.5			81, 82			43	73, 74
III.6	42	22	18	36, †44, 77, 78	63, 64, 73	33, †58, †59	
III.7	30, 32	40, 69, 70	83	38	21, 22, 56, 57, 74, 75	69, †70	17, 48, 75
III.8			19, 81, 82	33, 76, 77, 78, 79, 80	20, 71, 72, 76	33, 50	†16, 43, 71–72
III.9	28	11		10, 22			
IIIs. Electricity and Magnetism Skills							
IIIs.1	31						
IIIs.2	60, 61, 62	67, 68	13				
IIIs.3							84, 85

SEQUENCE	JUNE 2013	JUNE 2014	JUNE 2015	JUNE 2016	JUNE 2017	JUNE 2018	JUNE 2019
IIIs.4				42			
IIIs.5		66		33, 50, 75			54
IIIs.6					35, 71–76	33, 43	
IIIs.7			83, 84				
IIIs.8				11, 35	51		19
IV. Waves							
IV.1			26		26		
IV.2	16		20	20	23		
IV.3	19	61	26	6, 34	29	47	20
IV.4					25		20
IV.5	18, 22, 24, 29, 80	20, 21, 23, 35	20, 21, 24, 53, 54	21, 24	17	26, 35, 82, 83, 84, 85	21, 22, 78–80
IV.6	17		23				
IV.7		24	23	16			
IV.8	76, 77	18	34		9, 27		24
IV.9	49, 78, 79	49	8		27, 80, 82	23	25
IV.10	20	†41	46	23	28	24	26, †57, 58
IV.11	48		35	26, 49	45	27, 48	
IV.12	25	27	27	48	24	22	
IV.13			28	47		25	29
IV.14	26	26	32	19	50	49	28
IVs. Waves Skills							
IVs.1		34	25, 49		62	41	
IVs.2	53				58, 59		76, 77
IVs.3					46	81	27
IVs.4		17					
IVs.5							
IVs.6		25	35		65		50
IVs.7		60					29, 56
IVs.8			63		83	57	
IVs.9			64, †65	54, 55	84, 85	62, 63	
V. Modern Physics							
V.1	69		29, 76				
V.2	35	28	50		31, 52		†44, †38
V.3			77, 80	81, 82			
V.4	67, 68	†46, 83	47, 78, 79	27, 83, 84	44	29, 64, 65	
V.5							
V.6		85					
V.7	44			28, 29	52	42	31, †38, †44
V.8						30	
V.9		29			54		
V.10	50	30	55, 56	30	61	31	45
Vs. Modern Physics Skills							
Vs.1	34		29, 76	81	16		32, †40
Vs.2		43		85			46

Glossary of Important Terms

absolute index of refraction The ratio of the speed of light in a vacuum to the speed of light in a medium.

absorption spectrum A series of dark spectral lines or bands formed by the absorption of specific wavelengths of light by atoms or molecules.

acceleration The time rate of change in velocity. The SI unit is meters per second^2.

accuracy The agreement of a measured value with an accepted standard.

alpha particle A helium nucleus; a particle consisting of two protons and two neutrons.

alternating current An electric current that varies in magnitude and alternates in direction.

ammeter A device used to measure electric current. It is constructed by placing a low-resistance shunt across the coil of a galvanometer.

ampere (A) The SI unit of electric current, equivalent to the unit coulomb per second.

amplitude The maximum displacement in periodic phenomena such as wave motion, pendulum motion, and spring oscillation.

angle of incidence The angle made by the incident wave with the surface of a medium; the angle made by the incident ray with the normal to the surface of the medium.

angle of reflection The angle made by the reflected wave with the surface of a medium; the angle made by the reflected ray with the normal to the surface of the medium.

angle of refraction The angle made by the refracted wave with the surface of a medium; the angle made by the refracted ray with the normal to the surface of the medium.

anode The positive terminal of a DC source of potential difference.

antimatter One or more atoms composed entirely of antiparticles.

antinode The point or locus of points on an interference pattern (such as a standing wave or double slit pattern) that results in maximum constructive interference.

antiparticle The counterpart of a subatomic particle. An antiparticle has the same mass as its companion particle, but its electric charge is opposite in sign.

atomic number The number of protons in the nucleus of an atom. The atomic number defines the element.

Balmer series The visible-ultraviolet line spectrum of atomic hydrogen. It is the result of electrons falling from higher levels to the $n = 2$ state.

baryon A "heavy" particle, such as a proton or a neutron. Baryons are composed of three quarks. For example, a neutron is composed of an up quark and two down quarks (udd).

battery A combination of two or more electric cells.

beta (−) particle An electron formed in the nucleus by the disintegration of a neutron.

beta (+) particle A positron, the antiparticle of the electron, formed in the nucleus by the disintegration of a proton.

binding energy The energy equivalent of the mass defect of a nucleus.

cathode The negative terminal of a DC source of potential difference.

cathode ray tube A device for visualizing an electron beam. It consists of an evacuated tube with a source of electrons at one end and a fluorescent screen at the other end. The electron beam is controlled by electric and magnetic fields.

Celsius scale (°C) The temperature scale that fixes the (atmospheric) freezing point of water at 0° and the boiling point of water at 100°.

centripetal acceleration The acceleration that is directed along the radius and toward the center of a curved path in which an object is moving.

centripetal force The force that causes centripetal acceleration. It is responsible for changing an object's direction, not its speed.

chromatic aberration A lens defect in which different colors of light are focused at different points.

circuit A closed loop formed by a source of potential difference connected to one or more resistances.

coefficient of friction The ratio of the force of friction on an object to the normal force on it.

coherent light A series of light waves that have a fixed phase relationship; the type of light produced by a laser. Lasers produce beams of monochromatic coherent light.

component One of the two or more vectors into which a given vector may be resolved.

concurrent forces Two or more forces acting at the same point.

conductor A material that allows electrons to flow through it freely. Metals such as copper and silver are conductors.

constructive interference The combination of two in-phase wave disturbances to produce a single wave disturbance whose amplitude is the sum of the amplitudes of the individual disturbances.

converging lens A lens that focuses its transmitted light to a point. Generally, convex lenses are converging lenses.

convex lens A lens that is thicker in the middle than at the edges.

coulomb (C) The SI unit of electric charge, approximately equal to 6.25×10^{18} elementary charges.

Coulomb's law The electrostatic force between two point charges is directly proportional to the product of the charges and inversely proportional to the square of the distance between the charges.

critical angle The angle of incidence for which the corresponding angle of refraction is 90°.

cycle One complete repetition of the pattern in any periodic phenomenon.

de Broglie wavelength The wavelength of a matter wave.

destructive interference The combination of two out-of-phase wave disturbances to produce a single wave disturbance whose amplitude is the difference of the amplitudes of the individual disturbances.

diffraction The bending of a wave around a barrier.

diffuse reflection The reflection of parallel light rays by irregular surfaces.

direct current An electric current that flows in one direction only.

dispersion The separation of polychromatic light into its individual colors.

displacement A change of position in a specific direction.

Doppler effect An apparent change in frequency that results when a wave source and an observer are in relative motion with respect to each other.

elastic potential energy The energy stored in a spring when it is compressed or stretched.

electric current The time rate of flow of charged particles. The SI unit of electric current is the ampere (A).

electric field The region of space around a charged object that affects other charges.

electric field intensity The ratio of the force that an electric field exerts on a charge to the magnitude of the charge.

electric motor A device that converts electrical energy into mechanical energy.

electric potential The total work done by an electric field in bringing 1 coulomb of positive charge from infinity to a specific point. The potential is a positive number if the charge is repelled by the field and a negative number if the charge is attracted by the field. At infinity, the potential is taken to be zero. Electric potential is measured in volts.

electromagnetic force The fundamental force that governs the attraction or repulsion among charged particles, whether at rest or in motion.

electromagnetic induction The process by which the magnetic field and the mechanical energy are used to generate a potential difference.

electromagnetic radiation The propagation of electromagnetic waves in space.

electromagnetic spectrum The entire range of electromagnetic waves from the lowest to the highest frequencies.

electromagnetic wave A periodic wave, consisting of mutually perpendicular electric and magnetic fields, that is radiated away from the vicinity of an accelerating charge.

electromotive force The potential difference produced as a result of the conversion of other forms of energy into electrical energy.

electron A fundamental, negatively charged, subatomic particle; a lepton.

electron-volt (eV) A unit of energy equal to the work needed to move an elementary charge across a potential difference of 1 volt.

electroscope A device used to detect the presence of electric charges.

elementary charge The magnitude of charge present on a proton or an electron. An elementary charge is approximately equal to 1.6×10^{-19} coulomb.

emission spectrum A series of bright spectral lines or bands formed by the emission of certain wavelengths of light by excited atoms falling to lower energy states.

energy A quantity related to work.

equilibrant A single balancing force that maintains the static equilibrium of an object.

equivalent resistance A single resistance that can be substituted for a group of resistances in series or in parallel.

excited state A condition in which the energy of an atom is greater than its lowest energy state.

ferromagnetic Referring to a material, such as iron, that has the ability to strengthen greatly the magnetic field of a current-carrying coil.

field lines A series of lines used to represent the magnitude and direction of a field.

fission The process of splitting a heavy nucleus, such as uranium-235, into lighter fragments. Fission is accompanied by the release of large quantities of energy.

force A push or a pull on an object. If the force is unbalanced, an acceleration will result.

frame of reference A coordinate grid and a set of synchronized clocks that can be used to determine the position and time of an event.

free fall A motion in the Earth's gravitational field without regard to air resistance.

frequency The number of repetitions produced per unit time by periodic phenomena.

friction The force present as the result of contact between two surfaces. The direction of a frictional force is opposite to the direction of motion.

fundamental forces The four forces in nature responsible for all interactions among matter. (See also *electromagnetic force, gravitational force, strong force, weak force*.)

fusion (1) The process of uniting lighter nuclei, such as deuterium, into a heavier nucleus. Fusion is accompanied by the release of large quantities of energy. (2) In the study of heat and thermodynamics, a synonym for *melting*.

galvanometer A device, consisting of a coil-shaped wire placed between the opposite poles of a permanent magnet, that is used to detect small amounts of electric current.

gamma radiation Very high energy photons of electromagnetic radiation. Gamma photons have the highest frequencies in the electromagnetic spectrum.

geiger counter A device that detects charged nuclear particles.

generator A device that uses a magnetic field and mechanical energy to induce a source of electromotive force.

gravitational field The region of space around a mass that affects other masses.

gravitational field intensity The ratio of the force that a gravitational field exerts on a mass to the magnitude of the mass, numerically equal to the acceleration due to gravity.

gravitational force The fundamental universal attraction between masses.

gravitational potential energy The energy that an object acquires as a result of the work done in moving the object against a gravitational field.

ground An extremely large source or reservoir of electrons, which can supply or accept electrons as the need arises.

ground state The lowest energy state of an atom.

hadron Any particle that interacts through the strong force. Hadrons are classified as baryons or as mesons.

heat energy The energy that is transferred from warmer objects to cooler ones because of a temperature difference between them.

hertz (Hz) The SI unit of frequency, equivalent to the unit second^{-1}.

image An optical reproduction of an object.

impulse The product of the net force acting on an object and the time during which the force acts. The impulse delivered to an object is equal to its change in momentum. The direction of the impulse is the direction of the force. The SI unit of impulse is the newton • second, which is equivalent to the kilogram • meter per second.

incident ray A ray of a wave impinging on a surface.

incident wave A wave impinging on a surface.

induced current An electric current that is the result of an induced electromotive force.

induced emf A potential difference created when a magnetic field is interrupted over a time period.

induction (1) A method of charging a neutral object by using a charged object and a ground. The induced charge is always opposite to the charge on the charged object. (2) See *induced current* and *induced emf.*

inertia The property of matter that resists changes in motion. Mass is the quantitative measure of inertia.

instantaneous velocity The ratio of displacement to time at any given instant; the slope of a line tangent to a displacement-time graph at any given point.

insulator A material that is a very poor conductor because it has few conduction electrons. Wood and glass are examples of insulators.

interference pattern Regions of constructive and destructive interference that are present in a medium as a result of the combination of two or more waves.

internal energy The total kinetic and potential energy associated with the atoms and molecules of an object.

joule (J) The SI unit of work and energy, equivalent to the unit newton • meter.

kilogram (kg) The SI unit of mass; a fundamental unit.
kinetic energy The energy that an object possesses because of its motion.

laser An acronym for light amplification by the stimulated emission of radiation. A laser is a device that emits extremely intense, monochromatic, coherent light.
lepton Any particle that participates in the weak force: electron, muon, tau, and their neutrinos.
longitudinal wave A wave in which the disturbance is parallel to the direction of the wave's motion. Sound waves are longitudinal.

magnet Any material that aligns itself, when free to do so, in an approximate north-south direction. Magnets exert forces on one another and on charged particles in motion.
magnetic field The region of space around a magnet or charge in motion that exerts a force on magnets or other moving charges.
magnification The ratio of image size to object size.
mass (1) The measure of an object's ability to obey Newton's second law of motion. (2) The measure of an object's ability to obey Newton's law of universal gravitation. The SI unit of mass is the kilogram.
mass defect The mass lost by a nucleus when it is assembled from its nucleons. (See also *binding energy*.)
mass number The sum of the number of protons and neutrons in a nucleus; the number of nucleons the nucleus contains.
matter waves According to quantum theory, the waves associated with moving particles.
medium A material through which a disturbance, such as a wave, travels.
meson Any particle that is composed of a quark and an antiquark. For example, a π^+ meson is composed of an up quark and an antidown quark $\left(u\bar{d} \right)$.
momentum The product of mass and velocity. The direction of an object's momentum is the direction of its velocity. The SI unit of momentum is the kilogram • meter per second.

natural frequency A specific frequency with which an elastic body may vibrate if disturbed.
net force The unbalanced force present on an object; the accelerating force.
neutrino A lepton with no charge and questionable mass. It and its antiparticle are products of beta-decay reactions.

Newton (N) The SI unit of force, equivalent to the unit kilogram • meter per second^2.

Newton's first law of motion Objects remain in a state of uniform motion unless acted upon by an unbalanced force.

Newton's law of universal gravitation Any two bodies in the universe are attracted to each other with a force that is directly proportional to their masses and inversely proportional to the square of the distance between them.

Newton's second law of motion The unbalanced force on an object is equal to the product of its mass and acceleration.

Newton's third law of motion If object A exerts a force on object B, then object B exerts an equal and opposite force on object A.

node The point or locus of points on an interference pattern, such as a standing wave or double-slit pattern, that results in total destructive interference.

normal A line perpendicular to a surface.

normal force The force that keeps two surfaces in contact. If an object is on a *horizontal* surface, the normal force on the object is equal to its weight.

nucleon A proton or a neutron.

nucleus The dense, positively charged core of an atom.

ohm (Ω) The SI unit of electrical resistance, equivalent to the unit volt per ampere.

Ohm's law A relationship in which the ratio of the potential difference across certain conductors to the current in them is constant at constant temperature.

parallel circuit An electric circuit with more than one current path.

particle accelerator A device used to accelerate charged nuclear particles.

period The time for one complete repetition of a periodic phenomenon. The SI unit of period is the second.

periodic wave A regularly repeating series of waves.

phase (1) A form in which matter can exist, including liquid, solid, gas, and plasma. (2) In wave motion, the points on the wave that have specific time and space relationships.

photoelectric effect A phenomenon in which light causes electrons to be ejected from certain materials.

photon The fundamental particle of electromagnetic radiation.

Planck's constant A universal constant (h) relating the energy of a photon to its frequency; its approximate value is 6.62×10^{-34} joule • second.

point charge A charge with negligible physical dimensions.

polarization A process that produces transverse waves that vibrate in only one plane. Polarization is limited to transverse waves: light can be polarized; sound cannot.

polychromatic Referring to light waves of different colors (frequencies).

potential difference The ratio of the work required to move a test charge between two points in an electric field to the magnitude of the test charge. The unit of potential difference is the volt.

potential energy The energy that a system has because of its relative position or condition.

power The time rate at which work is done or energy is expended. The SI unit of power is the watt, which is equivalent to the unit joule per second.

precision The limit of the ability of a measuring device to reproduce a measurement.

pressure The force on a surface per unit area. The SI unit of pressure is the pascal (Pa).

proton A positively charged subatomic particle with a charge equal in magnitude to that of the electron; a baryon.

pulse A nonperiodic disturbance in a medium.

quantum A discrete quantity of energy.

quarks The particles of which protons, neutrons, baryons, and mesons are composed. Quarks carry a charge of either one-third or two-thirds of an elementary charge and come in six "flavors": top, bottom, up, down, charm, and strange. Each quark has a companion antiquark.

radioactive decay A spontaneous change in the nucleus of an atom.

radioactivity Changes in the nucleus of an atom that produce the emission of subatomic particles or photons.

ray A straight line indicating the direction of travel of a wave.

real image An image created by the actual convergence of light waves. Real images from single mirrors and single lenses are inverted and can be projected on a screen.

refraction The change in the direction of a wave when it passes obliquely from one medium to another in which it moves at different speed.

regular reflection The reflection of parallel light rays incident on a smooth plane surface.

resistance The opposition of a material to the flow of electrons through it; the ratio of potential difference to current.

resistivity A quantity that allows the resistance of substances to be compared. Numerically, it is the resistance of a 1-meter conductor with a cross-sectional area of 1 square meter. The SI unit of resistivity is the ohm • meter.

resistor A device that supplies resistance to a circuit.

resolution The process of determining the magnitude and direction of the components of a vector.

resonance The spontaneous vibration of an object at a frequency equal to that of the wave that initiates the resonant vibration.

resultant A vector sum.

satellite A body that revolves around a larger body as a result of a gravitational force.

scalar quantity A quantity, such as mass or work, that has magnitude but not direction.

series circuit A circuit with only one current path.

significant digits The digits that are part of any measurement.

solenoid A coil of wire wound as a helix. When a current is passed through the solenoid, it becomes an electromagnet.

speed The time rate of change of distance; the magnitude of velocity. The SI unit of speed is the meter per second.

spring constant The ratio of the force required to stretch or compress a spring to the magnitude of the stretch or compression.

standard model A model of matter and the fundamental interactions that govern it.

standing wave A wave pattern created by the continual interference of an incident wave with its reflected counterpart. The standing wave does not travel, but oscillates about an equilibrium position.

static equilibrium The condition of a body when a net force of zero is acting on it.

superconductor A material with no electrical resistance.

strong force The strongest of the four fundamental forces. It mediates interactions among certain nuclear particles.

temperature The "hotness" of an object, measured with respect to a chosen standard.

torque A force, applied perpendicularly to a designated line, that tends to produce rotational motion.

total internal reflection The reflection of a wave inside a relatively dense medium produced when the angle of the wave with the boundary exceeds the critical angle.

total mechanical energy The sum of the potential and kinetic energies of a mechanical system.

transverse wave A wave in which the disturbance is perpendicular to the direction of the wave's motion. Light waves are transverse.

uniform In the study of motion, a term that is equivalent to *constant*.

vector A representation of a vector quantity; an arrow in which the length represents the magnitude of the quantity and the arrowhead points in the direction of its orientation.

vector quantity A quantity, such as force or velocity, that has both magnitude and direction.

velocity The time rate of change of displacement.

virtual image An image formed by projecting diverging light behind a mirror or a lens.

volt (V) The SI unit of potential difference, equivalent to the unit joule per coulomb.

voltage Another term for *potential difference*.

voltmeter A device used to measure potential difference and constructed by placing a large resistor in series with the coil of a galvanometer.

watt (W) The SI unit of power, equivalent to the unit joule per second.

wave A series of periodic oscillations of a particle or a field both in time and in space.

wave front All points on a wave that are in phase with each other.

wavelength The length of one complete wave cycle.

weak force The fundamental force that arises in certain types of radioactive decay.

weight The gravitational force present on an object.

work The product of the force on an object and its displacement. The SI unit of work is the joule.

Reference Tables
for Physics

When you take the New York State Regents examination in physics, you will be provided with a set of reference tables to aid you in answering the questions on the examination. For those students not taking the examination, the following tables will provide a convenient reference when answering the questions and problems presented in this book. Each of these tables is given below along with a brief description of it. Please read the Important Note on page 2 concerning these tables.

A. LIST OF PHYSICAL CONSTANTS

List of Physical Constants		
Name	Symbol	Value
Universal gravitational constant	G	6.67×10^{-11} N•m^2/kg^2
Acceleration due to gravity	g	9.81 m/s^2
Speed of light in a vacuum	c	3.00×10^8 m/s
Speed of sound in air at STP		3.31×10^2 m/s
Mass of Earth		5.98×10^{24} kg
Mass of the Moon		7.35×10^{22} kg
Mean radius of Earth		6.37×10^6 m
Mean radius of the Moon		1.74×10^6 m
Mean distance—Earth to the Moon		3.84×10^8 m
Mean distance—Earth to the Sun		1.50×10^{11} m
Electrostatic constant	k	8.99×10^9 N•m^2/C^2
1 elementary charge	e	1.60×10^{-19} C
1 coulomb (C)		6.25×10^{18} elementary charges
1 electronvolt (eV)		1.60×10^{-19} J
Planck's constant	h	6.63×10^{-34} J•s
1 universal mass unit (u)		9.31×10^2 MeV
Rest mass of the electron	m_e	9.11×10^{-31} kg
Rest mass of the proton	m_p	1.67×10^{-27} kg
Rest mass of the neutron	m_n	1.67×10^{-27} kg

The most important physical constants and their symbols, where appropriate, are given in this table. The value of each constant is given to three significant digits along with its units.

For example, if you were asked to calculate the gravitational force between the Moon and Earth, you would use this table to find the masses of the Moon and Earth and the mean Earth–Moon distance.

B. PREFIXES FOR POWERS OF 10

Prefix	Symbol	Notation
tera	T	10^{12}
giga	G	10^{9}
mega	M	10^{6}
kilo	k	10^{3}
deci	d	10^{-1}
centi	c	10^{-2}
milli	m	10^{-3}
micro	μ	10^{-6}
nano	n	10^{-9}
pico	p	10^{-12}

The metric prefixes, along with their symbols and values, are given for numbers between 10^{-12} and 10^{12}.

For example, 10^{-1} meter is known as a *decimeter* (dm).

C. APPROXIMATE COEFFICIENTS OF FRICTION

Approximate Coefficients of Friction	Kinetic	Static
Rubber on concrete (dry)	0.68	0.90
Rubber on concrete (wet)	0.58	
Rubber on asphalt (dry)	0.67	0.85
Rubber on asphalt (wet)	0.53	
Rubber on ice	0.15	
Waxed ski on snow	0.05	0.14
Wood on wood	0.30	0.42
Steel on steel	0.57	0.74
Copper on steel	0.36	0.53
Teflon on Teflon	0.04	

The coefficients of friction (μ) for a number of pairs of surfaces are provided. The *kinetic* coefficients of friction are used when the surfaces are in relative motion. The *static* coefficients of friction are used when the surfaces are at rest with respect to each other.

D. THE ELECTROMAGNETIC SPECTRUM

The Electromagnetic Spectrum

The wavelength and frequency ranges for the principal types of electromagnetic radiation are given in this chart. Note that the type of radiation is principally determined by its *source* and not necessarily by its wavelength (or frequency). As a result, there is some overlap between certain types of radiation. In addition, the frequency ranges for visible light are provided at the bottom of the chart.

E. ABSOLUTE INDICES OF REFRACTION

Absolute Indices of Refraction ($f = 5.09 \times 10^{14}$ Hz)	
Air	1.00
Corn oil	1.47
Diamond	2.42
Ethyl alcohol	1.36
Glass, crown	1.52
Glass, flint	1.66
Glycerol	1.47
Lucite	1.50
Quartz, fused	1.46
Sodium chloride	1.54
Water	1.33
Zircon	1.92

The absolute index of refraction is defined as the ratio of the speed of light in a vacuum to the speed of light in a medium. (Monochromatic yellow light of 5.09×10^{14} Hz was used to compute these indices.)

The larger the index of refraction, the slower light travels in the medium. Therefore, according to this table, light travels slowest in a diamond and fastest in air. In a vacuum, the index of refraction would be exactly 1.

This table is useful for solving problems in which light is refracted as well as for comparing the speed of light in different media.

F. ENERGY LEVEL DIAGRAMS FOR HYDROGEN AND MERCURY

Energy Level Diagrams

A Few Energy Levels for the Mercury Atom

Each energy state is represented by a horizontal line. Its energy value, in eV, is given at the right. Note that these values are negative numbers that increase to a maximum value of 0.00. The lowest energy state is known as the *ground state*. Every other state is known as an *excited state*. At a value of 0.00 eV, the electron is no longer associated with the atom, a condition known as *ionization*.

At the left is the label corresponding to the energy state. For hydrogen, these states are integers known as *principal quantum numbers*. For mercury, the atom is much more complex and the various states are represented as letters.

In order to calculate the energy involved in a particular transition, one *subtracts* the final energy value from the initial value. If the difference is negative, then energy is released by the atom as a photon. If it is positive, then the energy is absorbed by the atom.

G. CLASSIFICATION OF MATTER

Classification of Matter

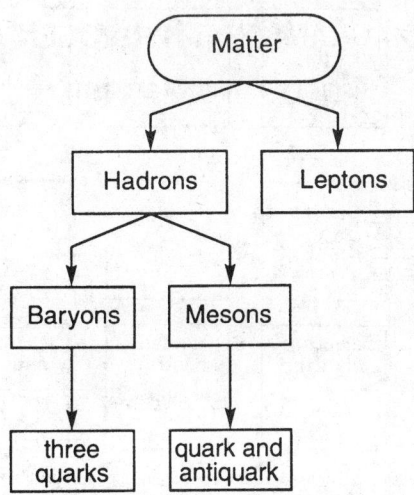

This chart classifies matter according to the *standard model*. For example, all nuclear particles are classified as either hadrons or leptons. Leptons (and their antiparticles) are not composed of smaller units and have a charge of ± 1 or 0. Hadrons are composed of smaller units and are ultimately composed of quarks and/or antiquarks.

H. PARTICLES OF THE STANDARD MODEL

Particles of the Standard Model

Quarks

Name	up	charm	top

Name
Symbol
Charge

up	charm	top
u	c	t
$+\frac{2}{3}e$	$+\frac{2}{3}e$	$+\frac{2}{3}e$

down	strange	bottom
d	s	b
$-\frac{1}{3}e$	$-\frac{1}{3}e$	$-\frac{1}{3}e$

Leptons

electron	muon	tau
e	μ	τ
$-1e$	$-1e$	$-1e$

electron neutrino	muon neutrino	tau neutrino
ν_e	ν_μ	ν_τ
0	0	0

Note: For each particle there is a corresponding antiparticle with a charge opposite that of its associated particle.

The names, symbols, and charges of the six quarks and six leptons are given in this chart. Electric charges are given in terms of the elementary charge (e). Each particle has a corresponding *antiparticle* that has an electric charge opposite that of the particle. For example, the antimuon ($\bar{\mu}$) has an electric charge of $+1e$.

I. CIRCUIT SYMBOLS

Circuit Symbols

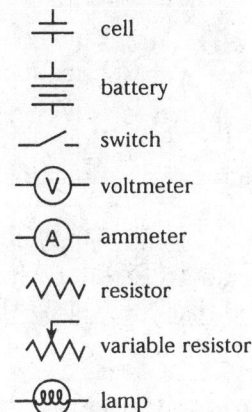

The symbols used in simple electric circuits are provided in this chart. It is useful in interpreting the elements that make up a particular type of circuit.

J. RESISTIVITIES AT 20°C

Resistivities at 20°C	
Material	Resistivity ($\Omega \cdot$m)
Aluminum	2.82×10^{-8}
Copper	1.72×10^{-8}
Gold	2.44×10^{-8}
Nichrome	$150. \times 10^{-8}$
Silver	1.59×10^{-8}
Tungsten	5.60×10^{-8}

The resistivities, in ohm • meters, are given for a number of conductors at 20°C. Resistivities are useful for comparing the relative abilities of various materials to conduct an electric current. They are also used to calculate the resistance of a conductor whose length and cross-sectional area are known.

K. EQUATIONS FOR PHYSICS

Reference Table K is divided into the following five areas:

- Mechanics: Equations ME1–ME23
- Electricity: Equations EL1–EL10
- Waves: Equations W1–W6
- Modern Physics: Equations MP1–MP3
- Geometry and Trigonometry: Equations GT1–GT4

In each area, the relevant equation is generally given on the left side of the page. The definition of the symbols used in each equation is given on the right side of the page. It is expected that you will be able to solve quantitative problems using any of the equations given below.

MECHANICS (EQUATIONS ME1–ME23)

ME1. $\bar{v} = \dfrac{d}{t}$

ME2. $a = \dfrac{\Delta v}{t}$

ME3. $v_f = v_i + at$

ME4. $d = v_i t + \frac{1}{2}at^2$

ME5. $v_f^2 = v_i^2 + 2ad$

ME6. $A_y = A \sin \theta$

ME7. $A_x = A \cos \theta$

ME8. $a = \dfrac{F_{net}}{m}$

ME9. $F_f = \mu F_N$

ME10. $F_g = \dfrac{Gm_1 m_2}{r^2}$

ME11. $g = \dfrac{F_g}{m}$

a = acceleration
a_c = centripetal acceleration
A = any vector quantity
d = displacement or distance
E_T = total energy
F = force
F_c = centripetal force
F_f = force of friction
F_g = weight or force due to gravity
F_N = normal force
F_{net} = net force
F_s = force on a spring
g = acceleration due to gravity or gravitational field strength
G = universal gravitational constant
h = height
J = impulse
k = spring constant

ME12. $p = mv$

KE = kinetic energy

m = mass

ME13. $p_{before} = p_{after}$

p = momentum

ME14. $J = F_{net}t = \Delta p$

P = power

ME15. $F_s = kx$

PE = potential energy

PE_s = potential energy stored in a spring

ME16. $PE_s = \frac{1}{2}kx^2$

Q = internal energy

ME17. $F_c = ma_c$

r = radius or distance between centers

t = time interval

ME18. $a_c = \dfrac{v^2}{r}$

v = velocity or speed

ME19. $\Delta PE = mg\Delta h$

$\bar{v}$ = average velocity or average speed

W = work

ME20. $KE = \frac{1}{2}mv^2$

x = change in spring length from the equilibrium position

ME21. $W = Fd = \Delta E_T$

Δ = change

ME22. $E_T = PE + KE + Q$

θ = angle

ME23. $p = \dfrac{W}{t} = \dfrac{Fd}{t} = F\bar{v}$

μ = coefficient of friction

Equations are provided for motion, vectors, Newton's laws, momentum and impulse, work, energy, and power.

ELECTRICITY (EQUATIONS EL1–EL10)

EL1. $F_e = \dfrac{kq_1q_2}{r^2}$

EL2. $E = \dfrac{F_e}{q}$

EL3. $V = \dfrac{W}{q}$

EL4. $I = \dfrac{\Delta q}{t}$

EL5. $R = \dfrac{V}{I}$

EL6. $R = \dfrac{\rho L}{A}$

EL7. $P = VI = I^2R = \dfrac{V^2}{R}$

EL8. $W = Pt = VIt = I^2Rt = \dfrac{V^2t}{R}$

A = cross-sectional area
E = electric field strength
F_e = electrostatic force
I = current
k = electrostatic constant
L = length of conductor
P = electrical power
q = charge
R = resistance
R_{eq} = equivalent resistance
r = distance between centers
t = time
V = potential difference
W = work (electrical energy)
Δ = change
ρ = resistivity

EL9. **Series Circuits**

EL9A. $I = I_1 = I_2 = I_3 = \ldots$

EL9B. $V = V_1 + V_2 + V_3 + \ldots$

EL9C. $R_{eq} = R_1 + R_2 + R_3 + \ldots$

EL10. **Parallel Circuits**

EL10A. $I = I_1 + I_2 + I_3 + \ldots$

EL10B. $V = V_1 = V_2 = V_3 = \ldots$

EL10C. $\dfrac{1}{R_{eq}} = \dfrac{1}{R_1} + \dfrac{1}{R_2} + \dfrac{1}{R_3} + \ldots$

Equations are provided for electrostatic forces, electric fields, potential difference, current, resistance, and power and energy in electric circuits. In addition, the current, potential difference, and resistance relationships are provided for series and parallel circuits.

WAVES (EQUATIONS W1–W6)

W1. $v = f\lambda$

W2. $T = \dfrac{1}{f}$

W3. $\theta_i = \theta_r$

W4. $n = \dfrac{c}{v}$

W5. $n_1 \sin \theta_1 = n_2 \sin \theta_2$

W6. $\dfrac{n_2}{n_1} = \dfrac{v_1}{v_2} = \dfrac{\lambda_1}{\lambda_2}$

c = speed of light in a vacuum
f = frequency
n = absolute index of refraction
T = period
v = velocity
λ = wavelength
θ = angle
θ_i = incident angle
θ_r = reflected angle

Equations are provided for all types of waves (relationships among speed, wavelength, frequency, and period) as well as for reflection and refraction.

MODERN PHYSICS (EQUATIONS MP1–MP3)

MP1. $E_{photon} = hf = \dfrac{hc}{\lambda}$

MP2. $E_{photon} = E_i - E_f$

MP3. $E = mc^2$

c = speed of light in a vacuum
E = energy
f = frequency
h = Planck's constant
m = mass
λ = wavelength

Three energy equations are included in this table. MP1 is the Planck equation for relating the energy of a photon to its frequency (or wavelength). MP2 is the energy of a photon that is emitted or absorbed as the result of a transition between two energy levels in an atom. MP3 is the Einstein equation that relates energy and mass.

GEOMETRY AND TRIGONOMETRY (EQUATIONS GT1–GT4)

GT1. **Rectangle**

$A = bh$

GT2. **Triangle**

$A = \frac{1}{2}bh$

GT3. **Circle**

GT3A. $A = \pi r^2$

GT3B. $C = 2\pi r$

GT4. **Right Triangle**

GT4A. $c^2 = a^2 + b^2$

GT4D. $\sin\theta = \frac{a}{c}$

GT4C. $\cos\theta = \frac{b}{c}$

GT4D. $\tan\theta = \frac{a}{b}$

A = area
b = base
C = circumference
h = height
r = radius

The geometric relationships for the areas of rectangles, triangles, and circles are given. In addition, the relationship for the circumference of a circle is provided. This table also gives the Pythagorean theorem and several trigonometric relationships for a right triangle.

Using the Equations to Solve Physics Problems

Note to the student: This section contains simple examples that use each of the equations given in *Reference Table K*. On the Regents examination in physics, you may be required to rearrange these equations or apply more than one equation in order to solve a problem.

MECHANICS

ME1. $\bar{v} = \dfrac{d}{t}$

EXAMPLE: It takes a car 0.20 hour to travel a distance of 12 kilometers.

$$\bar{v} = \frac{d}{t}$$
$$= \frac{12 \text{ km}}{0.20 \text{ h}}$$
$$= 60 \text{ km/h}$$

ME2. $a = \dfrac{\Delta v}{t}$

EXAMPLE: An object is uniformly accelerated from rest to a speed of 25 meters per second in 10 seconds.

$$\text{acceleration } a = \frac{(25 - 0) \text{ m/s}}{10 \text{ s}}$$
$$= 2.5 \text{ m/s}^2$$

ME3. $v_f = v_i + at$

    EXAMPLE: An object moving with an initial velocity of 3.0 meters per second accelerates for 15 seconds at 2.0 meters per second. What is the final velocity of the object?

$$v_f = 3.0 \text{ m/s} + \left(2.0 \text{ m/s}^2\right)(15 \text{ s})$$
$$= 33 \text{ m/s}$$

ME4. $d = v_i t + \dfrac{1}{2} at^2$

    EXAMPLE: A cart with an initial velocity of 6.00 meters per second rolls down an inclined plane with an acceleration of 4.00 meters per second squared. What distance will it travel in 3.00 seconds?

$$d = (6.00 \text{ m/s})(3.00 \text{ s}) + \frac{1}{2}\left(4.00 \text{ m/s}^2\right)(3.00 \text{ s})^2$$
$$= 36.0 \text{ m}$$

ME5. $v_f^2 = v_i^2 + 2ad$

    EXAMPLE: A car moving at a constant speed of 4.10 meters per second then accelerates uniformly at 3.20 meters per second squared. What will be its speed when it has traveled a distance of 40.0 meters?

$$v_f^2 = \left(4.10 \text{ m/s}^2\right) + (2)\left(3.20 \text{ m/s}^2\right)(40.0 \text{ m})$$
$$= 16.8 \text{ m}^2/\text{s}^2 + 256 \text{ m}^2/\text{s}^2$$
$$v_f = 16.5 \text{ m/s}$$

ME6. $A_y = A \sin \theta$

ME7. $A_x = A \cos \theta$

EXAMPLE: A projectile is fired at an angle of 30° to the horizontal with an initial velocity of 100 meters per second. Calculate the vertical and horizontal components of the projectile's initial velocity.

$$A_y = A \sin \theta$$

$$= \left(100 \; \frac{\text{m}}{\text{s}}\right)(0.500)$$

$$= 50.0 \; \frac{\text{m}}{\text{s}}$$

$$A_x = A \cos \theta$$

$$= \left(100 \; \frac{\text{m}}{\text{s}}\right)(0.866)$$

$$= 86.6 \; \frac{\text{m}}{\text{s}}$$

ME8. $a = \dfrac{F_{net}}{m}$

EXAMPLE: A 3.0-kilogram mass is being moved along a horizontal surface by a force of 6.0 newtons. If the surface is frictionless, what is the acceleration produced by the 6.0-newton force?

$$F_{net} = ma$$

$$6.0 \; \text{N} = (3.0 \; \text{kg})(a)$$

$$a = 2.0 \; \text{m/s}^2$$

ME9. $F_f = \mu F_N$

EXAMPLE: A 10-newton rubber object is dragged across a dry, horizontal concrete floor at constant speed. Calculate the force of friction on the object.

Since the object is on a horizontal surface, the normal force is equal to its weight (10 N). Use *Reference Table C* to obtain the *kinetic* coefficient of friction between rubber and dry concrete (0.68).

$$F_f = (0.68)(10 \; \text{N})$$

$$= 6.8 \; \text{N}$$

ME10. $F_g = \dfrac{Gm_1 m_2}{r^2}$

EXAMPLE: Two masses of 10.0 kg and 1.0 kg, respectively, are located 1.0 meter apart. How large a gravitational force does each mass exert on the other?

$$F_g = \left(6.67 \times 10^{-11} \text{N} \bullet \text{m}^2/\text{kg}^2\right)(10 \text{ kg})\dfrac{1.0 \text{ kg}}{(1.0 \text{ m})^2}$$

$$= 6.7 \times 10^{-10} \text{ N}$$

ME11. $g = \dfrac{F_g}{m}$

EXAMPLE: The acceleration due to gravity on planet A is 20 meters per second squared. On this planet what is the gravitational force on an object whose mass is 2.0 kilograms?

$$F_g = mg$$

$$= (2.0 \text{ kg})\left(20 \text{ m/s}^2\right)$$

$$= 40 \text{ N}$$

ME12. $p = mv$

EXAMPLE: A cart whose mass is 20 kg moves with a velocity of 5.0 meters per second eastward. What is its momentum?

$$p = mv$$

$$= (20 \text{ kg})(5.0 \text{ m/s})$$

$$= 1.0 \times 10^2 \text{ kg} \bullet \text{m/s in the direction of the velocity}$$

ME13. $p_{before} = p_{after}$

EXAMPLE: A 5.0-kilogram object traveling at 5.0 meters per second east collides with a 2.0-kilogram object traveling at 2.0 meters per second east. After the collision, the 2.0-kilogram object is moving at 4.0 meters per second east. What is the velocity of the 5.0-kilogram object? [Neglect friction.]

Designate east by using a positive sign.

$$(5.0 \text{ kg})(+5.0 \text{ m/s}) + (2.0 \text{ kg})(+2.0 \text{ m/s}) = (5.0 \text{ kg})(v) + (2.0 \text{ kg})(+4.0 \text{ m/s})$$
$$v = +4.2 \text{ m/s (east)}$$

ME14. $J = Ft = \Delta p$

EXAMPLE: A spring exerts a force of 50.0 newtons on a cart located on a frictionless plane. The cart has a mass of 2.0 kilograms and the force acts for 0.20 second.

$$J = \text{Impulse } Ft = (50.0 \text{ N})(0.20 \text{ s})$$
$$= 10 \text{ N} \bullet \text{s}$$

The change of momentum of the cart is also 10 N • s, and that is the same thing as 10 kilogram-meters per second.

$$\Delta p = 10 \text{ kg} \bullet \text{m/s}$$

ME15. $F_s = kx$

EXAMPLE: A block is suspended from a spring which has a spring constant, k, of 200 newtons per meter. What is the force on the spring if the spring is stretched 0.20 meter?

$$F_s = kx$$
$$= (200 \text{ N/m})(0.20 \text{ m})$$
$$= 40.0$$

ME16. $PE_s = \frac{1}{2}kx^2$

EXAMPLE: A block is suspended from a spring and as a result the spring is stretched 0.20 meter. The spring constant is 200 newtons per meter. How much potential energy is stored in the spring?

$$PE_s = \frac{1}{2}kx^2$$
$$= \frac{1}{2}(200 \text{ N/m})(0.20 \text{ m})^2$$
$$= 8.0 \text{ J}$$

ME17. $Fc = ma_c$

ME18. $a_c = \dfrac{v^2}{r}$

EXAMPLE: An object moves around a circle whose radius is 2.0 meters. The constant speed of the object is 6.0 meters per second and the mass of the object is 0.20 kilogram.

$$\text{centripetal acceleration } a_c = \frac{v^2}{r}$$

$$= \frac{(6.0 \text{ m/s})^2}{2.0 \text{ m}}$$

$$= 18. \text{ m/s}^2$$

$$\text{centripetal force } F_c = ma_c \left(= \frac{mv^2}{r} \right)$$

$$= (0.20 \text{ kg})(18 \text{ m/s}^2)$$

$$= 3.6 \text{ N}$$

ME19. $\Delta PE = mg\Delta h$

EXAMPLE: A 3.0-kilogram mass is raised 4.0 meters from a surface. Calculate its gain in gravitational potential energy.

$$\Delta PE = (3.0 \text{ kg})(9.8 \text{ m/s}^2)(4.0 \text{ m})$$

$$= 1.2 \times 10^2 \text{ J}$$

ME20. $KE = \dfrac{1}{2}mv^2$

EXAMPLE: A car with a mass of 1000 kilograms travels with a speed of 20 meters per second.

$$E_k = \frac{1}{2}(1000 \text{ kg})(20 \text{ m/s})^2$$

$$= 200{,}000 \text{ J}$$

ME21. $W = Fd = \Delta E_T$

EXAMPLE: If a horizontal force of 30 newtons is used to push an object 40 meters along a horizontal surface, the work done on the object

$$= Fd$$
$$= (30 \text{ N})(40 \text{ m})$$
$$= 1200 \text{ J}$$

Note: 1 newton-meter = 1 joule

ME22. $E_T = PE + KE + Q$

EXAMPLE: As an object, initially at rest, slides down a plane, it loses 25 joules of potential energy. If the object gains 18 joules of kinetic energy, what is the change in the internal energy of the object? How is the internal energy change used?

At the top of the incline, the total energy of the object is present as potential energy and internal energy. At the bottom of the incline, the lost potential energy is converted to kinetic energy and additional internal energy. The total energy of the object remains constant throughout the slide. Modify equation ME22 to read:

$$\Delta E_T = \Delta PE + \Delta KE + \Delta Q$$
$$\Delta E_T = 0 \text{ J}$$
$$= (-25 \text{ J}) + (+18 \text{ J}) + \Delta Q$$
$$\Delta Q = +7 \text{ J}$$

The change in the internal energy is used to do work against the friction between the object and the incline.

ME23. $P = \dfrac{W}{t} = \dfrac{Fd}{t} = F\bar{v}$

Power is the rate of doing work.

EXAMPLE: An object whose mass is 3.0 kilograms is moved at constant speed by a force of 5.0 newtons. If the object is moved 40 meters in 20 seconds, then the required power $= \dfrac{\text{work}}{\text{time}}$.

But the work which is done is equal to the product of the applied force and the distance that the object is moved:

$$W = Fd$$
$$= (5.0 \text{ N})(40 \text{ m})$$
$$= 2.0 \times 10^2 \text{ joules}$$

$$\text{power} = \frac{\text{work done}}{\text{time required}}$$
$$= \frac{200 \text{ joules}}{20 \text{ s}}$$
$$= 10 \text{ watts.}$$

$$\text{Also, power} = \frac{\text{force} \times \text{distance moved}}{\text{time required}}$$
$$= \frac{5.0 \text{ N} \times 40 \text{ m}}{20 \text{ s}}$$
$$= 10 \text{ watts}$$

ELECTRICITY

EL1. $\dfrac{F = kq_1q_2}{r^2}$

EXAMPLE: Two small charged spheres are 3.0 meters apart, and each sphere has a charge of 2.0×10^{-6} coulomb. Calculate the magnitude of the force exerted by one sphere on the other.

$$F = \frac{(9.00 \times 10^9 \text{ N m}^2/\text{C}^2)(2.0 \times 10^{-6} \text{ C})^2}{(3.0 \text{ m})^2}$$

$$= 4.0 \times 10^{-3} \text{ N}$$

EL2. $E = \dfrac{F_e}{q}$

EXAMPLE: When a charge of 0.040 coulomb is placed at a point in the electric field, the force on the charge is 100 newtons. What is the magnitude of the electric field at that point?

$$E = \frac{100 \text{ N}}{0.040 \text{ C}}$$
$$= 2500 \text{ N/C}$$

EL3. $V = \dfrac{W}{q}$

EXAMPLE: The work required to move a charge of 0.04 coulomb from one point to another in an electric field is 200 joules.

$$\text{potential difference} = \frac{200 \text{ J}}{0.04 \text{ C}}$$
$$= 5000 \text{ V}$$

EL4. $I = \dfrac{\Delta q}{t}$

EXAMPLE: Calculate the current in a conductor if 130 coulombs of charge is transferred in 20.0 seconds.

$$I = \frac{130 \text{ C}}{20.0 \text{ s}}$$
$$= 6.50 \text{ A}$$

EL5. $R = \dfrac{V}{I}$

EXAMPLE: A resistor has a resistance of 10 ohms and a current in it of 0.50 ampere. What is the potential difference across the resistor?

$$V = (0.50 \text{ A})(10 \ \Omega)$$
$$= 5.0 \text{ V}$$

EL6. $R = \dfrac{\rho L}{A}$

EXAMPLE: Calculate the resistance of a bar of silver if its length is 2.00 meters and its cross-sectional area is 0.0100 meter².

Use *Reference Table J* to obtain the resistivity of silver (1.59 × 10⁻⁸ Ω · m).

$$R = \left(1.59 \times 10^{-8}\ \Omega \bullet m\right)\left(\dfrac{2.00\ m}{0.0100\ m^2}\right)$$
$$= 3.18 \times 10^{-6}\ \Omega$$

EL7. $P = VI = I^2 R = \dfrac{V^2}{R}$

a. $P = VI$

EXAMPLE: The potential difference across a resistor is 15 volts and the current in it is 0.50 ampere.

$$P = (15\ V)(0.50\ A)$$
$$= 7.5\ W$$

b. $P = I^2 R$

EXAMPLE: The current in a resistor is 3.0 ampere, and its resistance is 10 ohms.

$$P = (30\ A)^2 (10\ \Omega)$$
$$= 90\ W$$

c. $P = \dfrac{V^2}{R}$

EXAMPLE: The potential difference across a resistor is 100 volts, and its resistance is 50.0 ohms.

$$P = \dfrac{(100\ V)^2}{50.0\ \Omega}$$
$$= 200\ W$$

EL8. $W = Pt = VIt = I^2Rt = \dfrac{V^2t}{R}$

a. $W = Pt$

EXAMPLE: An electric toaster rated at 1500 watts is used for 40 seconds. How much energy does it use?

$$W = (1500 \text{ J})(40 \text{ s})$$
$$= 6.0 \times 10^4 \text{ J}$$

b. $W = VIt$

EXAMPLE: The potential difference across a resistor is 15 volts and the current in it is 0.50 ampere. How much energy is used by the resistor in 5.0 minutes when operating at the rated power?

$$W = (15 \text{ V})(0.50 \text{ A})(300 \text{ s})$$
$$= 2.3 \times 10^3 \text{ J}$$

c. $W = I^2Rt$

EXAMPLE: The current in a resistor is 3.0 ampere, and its resistance is 10. ohms. How much energy is used by the resistor in 5.0 minutes?

$$W = (3.0 \text{ A})^2 (10 \text{ } \Omega)(300 \text{ s})$$
$$= 2.7 \times 10^4 \text{ J}$$

EL9A. Series Circuit: $I = I_1 = I_2 = I_3 = \ldots$

EXAMPLE: In a series circuit containing three resistors having resistances of 5, 10, and 15 ohms, respectively, we know that the current in the 15-ohm resistor is 0.5 ampere. This tells us that the current is also 0.5 ampere in the 5-ohm and 10-ohm resistors as well as in the generator connected to the circuit.

EL9B.   Series Circuit: $V = V_1 + V_2 + V_3 + ...$

EXAMPLE: In the circuit described in the previous question, if we know that the voltages across the three resistors are 2.5, 5.0, and 7.5 volts, respectively, then the voltage supplied by the source is the sum of these three values, namely 15 volts.

EL9C.   Series Circuit: $R_{eq} = R_1 + R_2 + R_3 + ...$

EXAMPLE: In the circuit described in the previous two questions, we can add the individual resistances to get the total resistance.

$$\text{total resistance} = (5 + 10 + 15)\ \Omega$$
$$= 30\ \Omega$$

EL10A.   Parallel Circuit: $I = I_1 + I_2 + I_3 + ...$

EXAMPLE: A parallel circuit has two branches, a 10-ohm resistor connected parallel to a 20-ohm resistor. If the current in the 10-ohm resistor is 6.0 amperes, and the current in the 20-ohm resistor is 3.0 amperes, how much current is supplied by the source?

$$I = 6.0\ \text{A} + 3.0\ \text{A}$$
$$= 9.0\ \text{A}$$

EL10B.   Parallel Circuit: $V = V_1 = V_2 = V_3 = ...$

In a parallel circuit the potential difference of the source is equal to the potential difference across each of the branches.

EXAMPLE: If the potential difference across the 20-ohm resistor is 60 volts, what is the potential difference supplied by the source?

$$V = V_1$$
$$= 60\ \text{V}$$

EL10C. Parallel Circuit: $\dfrac{1}{R_{eq}} = \dfrac{1}{R_1} + \dfrac{1}{R_2} + \dfrac{1}{R_3} + \ldots$

In a parallel circuit the reciprocal of the total resistance (or equivalent resistance) is equal to the sum of the reciprocals of the individual resistances.

EXAMPLE: In the circuit described in the previous two questions, what is the total resistance of the circuit?

$$\frac{1}{R_{eq}} = \frac{1}{10\ \Omega} + \frac{1}{20\ \Omega}$$
$$\frac{1}{R_{eq}} = \frac{3}{20\ \Omega}$$
$$R_{eq} = 6.7\ \Omega$$

WAVES

W1. $v = f\lambda$

EXAMPLE: The frequency of a wave is 2.0 hertz, and its wavelength is 0.030 meter.

$$v = (2.0\ \text{Hz})(0.030\ \text{m})$$
$$= 0.060\ \text{m/s}$$

W2. $T = \dfrac{1}{f}$

EXAMPLE: If the frequency of a wave is 0.25 cycle per second, what is the period of the wave?

$$T = \frac{1}{f}$$
$$= \frac{1}{0.25\ \text{Hz}}$$
$$= 4.0\ \text{s}$$

W3. $\theta_i = \theta_r$

EXAMPLE: When a ray of monochromatic light strikes a plane mirror, the incident ray makes an angle of 30° with the surface of the mirror. Calculate the angle of reflection for the reflected ray.

Angles of incidence and reflection are measured with respect to a *normal* drawn to the surface of the mirror. Therefore, the angle of incidence is the *complement* of 30°, that is, 60°.

$$\theta_r = \theta_i = 60°$$

W4. $n = \dfrac{c}{v}$

EXAMPLE: A ray of monochromatic light travels from a vacuum into a material whose index of refraction is 1.4. Calculate the speed of light in this material.

$$v = \frac{3.0 \times 10^8 \text{ m/s}}{1.4}$$

$$= 2.1 \times 10^8 \text{ m/s}$$

W5. $n_1 \sin \theta_1 = n_2 \sin \theta_2$

EXAMPLE: If the angle of incidence in air is 30°, what is the sine of the angle of refraction in crown glass?

$$1.00 \ \sin 30° = 1.52 \ \sin \theta_2$$

$$\sin \theta_2 = \frac{0.50}{1.52}$$

$$\sin \theta_2 = 0.33$$

W6. $\dfrac{n_2}{n_1} = \dfrac{v_1}{v_2} = \dfrac{\lambda_1}{\lambda_2}$

EXAMPLE: If the speed of light in a medium is 1.5×10^8 meters per second, what is the index of refraction of the medium?

MODERN PHYSICS

MP1. $E_{photon} = hf = \dfrac{hc}{\lambda}$

a. **EXAMPLE:** If the frequency of a monochromatic light beam is 1.2×10^{15} cycles per second, what is the energy of each of its photons?

$$E = \left(6.6 \times 10^{-34} \text{ J} \bullet \text{s}\right)\left(1.2 \times 10^{15} \text{ Hz}\right)$$
$$= 7.9 \times 10^{-19} \text{ J}$$

b. **EXAMPLE:** Calculate the energy of a photon of monochromatic orange light whose wavelength is 6.0×10^{-7} m.

$$E_{photon} = \left(6.63 \times 10^{-34} \text{ J} \bullet \text{s}\right)\left(3.00 \times 10^{8} \text{ m/s}\right)/6.0 \times 10^{-7} \text{ m}$$
$$= 3.3 \times 10^{-19} \text{ J}$$

MP2. $E_{photon} = E_i - E_f$

EXAMPLE: A hydrogen atom jumps from an energy level of -0.85 eV to an energy level of -13.60 eV. What is the energy of the photon emitted as a result of this transition?

$$E_{photon} = (-0.85 \text{ eV}) - (-13.60 \text{ eV})$$
$$= 12.75 \text{ eV}$$

MP3. $E = mc^2$

EXAMPLE: What is the energy equivalence, in joules, of 1.0×10^{-3} kilogram of matter?

$$E = \left(1.0 \times 10^{-3} \text{ kg}\right)\left(3.0 \times 10^{8} \text{ m/s}\right)$$
$$= 9.0 \times 10^{13} \text{ J}$$

Regents Examinations, Answers, and Self-Analysis Charts

IMPORTANT NOTE ABOUT THE REFERENCE
TABLES AND EQUATIONS

In this book, the reference tables and equations beginning on page 45 have been *slightly modified* from the official *New York State Reference Tables for Physics (2006 edition)*. The tables in this book are indexed by letter, and the equations are indexed by letter and number. Some of the tables have been moved in order to separate them from the equations, and each table is accompanied by a brief explanation. Otherwise, the material on these tables and the official tables is *identical*.

All of the explanations and answers in this book refer to the tables and equations *contained within this book*.

Examination June 2013

Physics: The Physical Setting

PART A

Answer all questions in this part.

Directions (1–35): For *each* statement or question, select the *number* of the word or expression that, of those given, best completes the statement or answers the question. Some questions may require the use of the *2006 Edition Reference Tables for Physical Setting/Physics*. Record your answers in the spaces provided.

1 Which term identifies a scalar quantity?

  (1) displacement   (3) velocity

  (2) momentum   (4) time   1 _____

2 Two 20-newton forces act concurrently on an object. What angle between these forces will produce a resultant force with the greatest magnitude?

  (1) 0°   (3) 90°

  (2) 45°   (4) 180°   2 _____

3 A car traveling west in a straight line on a highway decreases its speed from 30.0 meters per second to 23.0 meters per second in 2.00 seconds. The car's average acceleration during this time interval is

  (1) 3.5 m/s^2 east   (3) 13 m/s^2 east

  (2) 3.5 m/s^2 west   (4) 13 m/s^2 west   3 _____

4 In a race, a runner traveled 12 meters in 4.0 seconds as she accelerated uniformly from rest. The magnitude of the acceleration of the runner was

(1) 0.25 m/s^2 (3) 3.0 m/s^2

(2) 1.5 m/s^2 (4) 48 m/s^2 4 _____

5 A projectile is launched at an angle above the ground. The horizontal component of the projectile's velocity, v_x, is initially 40 meters per second. The vertical component of the projectile's velocity, v_y, is initially 30 meters per second. What are the components of the projectile's velocity after 2.0 seconds of flight? [Neglect friction.]

(1) $v_x = 40 \text{ m/s and } v_y = 10 \text{ m/s}$

(2) $v_x = 40 \text{ m/s and } v_y = 30 \text{ m/s}$

(3) $v_x = 20 \text{ m/s and } v_y = 10 \text{ m/s}$

(4) $v_x = 20 \text{ m/s and } v_y = 30 \text{ m/s}$ 5 _____

6 A ball is thrown with an initial speed of 10 meters per second. At what angle above the horizontal should the ball be thrown to reach the greatest height?

(1) 0°   (3) 45°

(2) 30°   (4) 90°   6 _____

7 Which object has the greatest inertia?

(1) a 0.010-kg bullet traveling at 90 m/s

(2) a 30-kg child traveling at 10 m/s on her bike

(3) a 490-kg elephant walking with a speed of 1.0 m/s

(4) a 1500-kg car at rest in a parking lot   7 _____

8 An 8.0-newton wooden block slides across a horizontal wooden floor at constant velocity. What is the magnitude of the force of kinetic friction between the block and the floor?

(1) 2.4 N   (3) 8.0 N

(2) 3.4 N   (4) 27 N   8 _____

9 Which situation represents a person in equilibrium?

  (1) a child gaining speed while sliding down a slide
  (2) a woman accelerating upward in an elevator
  (3) a man standing still on a bathroom scale
  (4) a teenager driving around a corner in his car   9 ____

10 A rock is thrown straight up into the air. At the highest point of the rock's path, the magnitude of the net force acting on the rock is

  (1) less than the magnitude of the rock's weight, but greater than zero
  (2) greater than the magnitude of the rock's weight
  (3) the same as the magnitude of the rock's weight
  (4) zero   10 ____

11 The diagram below shows a compressed spring between two carts initially at rest on a horizontal, frictionless surface. Cart A has a mass of 2 kilograms and cart B has a mass of 1 kilogram. A string holds the carts together.

The string is cut and the carts move apart. Compared to the magnitude of the force the spring exerts on cart A, the magnitude of the force the spring exerts on cart B is

  (1) the same
  (2) half as great
  (3) twice as great
  (4) four times as great   11 ____

12 An 8.0-newton block is accelerating down a frictionless ramp inclined at 15° to the horizontal, as shown in the diagram below.

What is the magnitude of the net force causing the block's acceleration?

(1) 0 N   (3) 7.7 N
(2) 2.1 N   (4) 8.0 N   12 _____

13 At a certain location, a gravitational force with a magnitude of 350 newtons acts on a 70-kilogram astronaut. What is the magnitude of the gravitational field strength at this location?

(1) 0.20 kg/N   (3) 9.8 m/s^2
(2) 5.0 N/kg   (4) 25 000 N • kg   13 _____

14 A spring gains 2.34 joules of elastic potential energy as it is compressed 0.250 meter from its equilibrium position. What is the spring constant of this spring?

(1) 9.36 N/m   (3) 37.4 N/m
(2) 18.7 N/m   (4) 74.9 N/m   14 _____

15 When a teacher shines light on a photocell attached to a fan, the blades of the fan turn. The brighter the light shining on the photocell, the faster the blades turn. Which energy conversion is illustrated by this demonstration?

(1) light → thermal → mechanical
(2) light → nuclear → thermal
(3) light → electrical → mechanical
(4) light → mechanical → chemical   15 _____

16 Which statement describes a characteristic common to all electromagnetic waves and mechanical waves?

(1) Both types of waves travel at the same speed.
(2) Both types of waves require a material medium for
propagation.
(3) Both types of waves propagate in a vacuum.
(4) Both types of waves transfer energy.   16 _____

17 An electromagnetic wave is produced by charged particles vibrating at a rate of 3.9×10^8 vibrations per second. The electromagnetic wave is classified as

(1) a radio wave   (3) an x ray
(2) an infrared wave   (4) visible light   17 _____

18 The energy of a sound wave is most closely related to the wave's

(1) frequency   (3) wavelength
(2) amplitude   (4) speed   18 _____

19 A sound wave traveling eastward through air causes the air molecules to

(1) vibrate east and west
(2) vibrate north and south
(3) move eastward, only
(4) move northward, only   19 _____

20 What is the speed of light ($f = 5.09 \times 10^{14}$ Hz) in ethyl alcohol?

(1) 4.53×10^{-9} m/s
(2) 2.43×10^2 m/s
(3) 1.24×10^8 m/s
(4) 2.21×10^8 m/s   20 _____

21 In the diagram below, an ideal pendulum released from position A swings freely to position B.

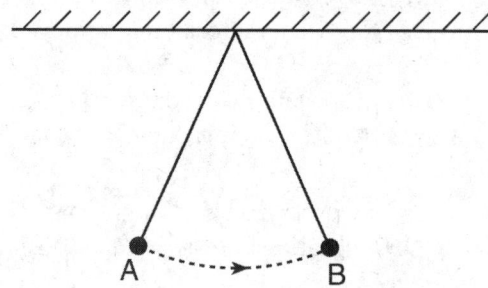

As the pendulum swings from A to B, its total mechanical energy

(1) decreases, then increases
(2) increases, only
(3) increases, then decreases
(4) remains the same

21 _____

22 The diagram below represents a periodic wave.

Which two points on the wave are out of phase?

(1) A and C (3) C and E
(2) B and F (4) D and G

22 _____

23 A dry plastic rod is rubbed with wool cloth and then held near a thin stream of water from a faucet. The path of the stream of water is changed, as represented in the diagram below.

Which force causes the path of the stream of water to change due to the plastic rod?

(1) nuclear   (3) electrostatic
(2) magnetic   (4) gravitational   23 ____

24 A distance of 1.0×10^{-2} meter separates successive crests of a periodic wave produced in a shallow tank of water. If a crest passes a point in the tank every 4.0×10^{-1} second, what is the speed of this wave?

(1) 2.5×10^{-4} m/s   (3) 2.5×10^{-2} m/s
(2) 4.0×10^{-3} m/s   (4) 4.0×10^{-1} m/s   24 ____

25 One vibrating 256-hertz tuning fork transfers energy to another 256-hertz tuning fork, causing the second tuning fork to vibrate. This phenomenon is an example of

(1) diffraction   (3) refraction
(2) reflection   (4) resonance   25 ____

26 Sound waves are produced by the horn of a truck that is approaching a stationary observer. Compared to the sound waves detected by the driver of the truck, the sound waves detected by the observer have a greater

(1) wavelength   (3) period
(2) frequency   (4) speed   26 ____

27 The electronvolt is a unit of

   (1) energy
   (2) charge
   (3) electric field strength
   (4) electric potential difference   27 _____

28 Which particle would produce a magnetic field?

   (1) a neutral particle moving in a straight line
   (2) a neutral particle moving in a circle
   (3) a stationary charged particle
   (4) a moving charged particle   28 _____

29 A physics student takes her pulse and determines that her heart
    beats periodically 60 times in 60 seconds. The period of her heart-
    beat is

   (1) 1 Hz   (3) 1 s
   (2) 60 Hz   (4) 60 s   29 _____

30 Moving 4.0 coulombs of charge through a circuit requires 48 joules
    of electric energy. What is the potential difference across this
    circuit?

   (1) 190 V   (3) 12 V
   (2) 48 V   (4) 4.0 V   30 _____

31 The diagram below shows currents in a segment of an electric circuit.

    What is the reading of ammeter A?

   (1) 1 A   (3) 9 A
   (2) 5 A   (4) 15 A   31 _____

32 An electric dryer consumes 6.0×10^6 joules of electrical energy when operating at 220 volts for 1.8×10^3 seconds. During operation, the dryer draws a current of

(1) 10 A   (3) 9.0×10^2 A

(2) 15 A   (4) 3.3×10^3 A   32 _____

33 Which net charge could be found on an object?

(1) $+4.80 \times 10^{-19}$ C

(2) $+2.40 \times 10^{-19}$ C

(3) -2.40×10^{-19} C

(4) -5.60×10^{-19} C   33 _____

34 A photon is emitted as the electron in a hydrogen atom drops from the $n = 5$ energy level directly to the $n = 3$ energy level. What is the energy of the emitted photon?

(1) 0.85 eV   (3) 1.51 eV

(2) 0.97 eV   (4) 2.05 eV   34 _____

35 In a process called pair production, an energetic gamma ray is converted into an electron and a positron. It is *not* possible for a gamma ray to be converted into two electrons because

(1) charge must be conserved

(2) momentum must be conserved

(3) mass-energy must be conserved

(4) baryon number must be conserved   35 _____

PART B–1
Answer all questions in this part.

Directions (36–50): For *each* statement or question, select the *number* of the word or expression that, of those given, best completes the statement or answers the question. Some questions may require the use of the *2006 Edition Reference Tables for Physical Setting/Physics*. Record your answers in the spaces provided.

36 The approximate length of an unsharpened No. 2 pencil is

(1) 2.0×10^{-2} m   (3) 2.0×10^{0} m

(2) 2.0×10^{-1} m   (4) 2.0×10^{1} m   36 _____

37 The diagram below shows an 8.0-kilogram cart moving to the right at 4.0 meters per second about to make a head-on collision with a 4.0-kilogram cart moving to the left at 6.0 meters per second.

Frictionless, horizontal surface

After the collision, the 4.0-kilogram cart moves to the right at 3.0 meters per second. What is the velocity of the 8.0-kilogram cart after the collision?

(1) 0.50 m/s left   (3) 5.5 m/s left

(2) 0.50 m/s right   (4) 5.5 m/s right   37 _____

38 Four forces act concurrently on a block on a horizontal surface as shown in the diagram below.

As a result of these forces, the block

(1) moves at constant speed to the right
(2) moves at constant speed to the left
(3) accelerates to the right
(4) accelerates to the left   38 _____

39 If a motor lifts a 400-kilogram mass a vertical distance of 10 meters in 8.0 seconds, the *minimum* power generated by the motor is

(1) 3.2×10^2 W   (3) 4.9×10^3 W
(2) 5.0×10^2 W   (4) 3.2×10^4 W   39 _____

40 A 4.0-kilogram object is accelerated at 3.0 meters per second^2 north by an unbalanced force. The same unbalanced force acting on a 2.0-kilogram object will accelerate this object toward the north at

(1) 12 m/s^2 (3) 3.0 m/s^2
(2) 6.0 m/s^2 (4) 1.5 m/s^2 40 _____

41 An electron is located in an electric field of magnitude 600 newtons per coulomb. What is the magnitude of the electrostatic force acting on the electron?

(1) 3.75×10^{21} N   (3) 9.60×10^{-17} N
(2) 6.00×10^2 N   (4) 2.67×10^{-22} N   41 _____

42 The current in a wire is 4.0 amperes. The time required for 2.5×10^{19} electrons to pass a certain point in the wire is

(1) 1.0 s   (3) 0.50 s

(2) 0.25 s   (4) 4.0 s   42 _____

43 When two point charges of magnitude q_1 and q_2 are separated by a distance, r, the magnitude of the electrostatic force between them is F. What would be the magnitude of the electrostatic force between point charges 2_{q1} and 4_{q2} when separated by a distance of $2r$?

(1) F (3) $16F$

(2) $2F$ (4) $4F$ 43 _____

44 The composition of a meson with a charge of -1 elementary charge could be

(1) $s\bar{c}$ (3) $u\bar{b}$

(2) $d\,s\,s$ (4) $\overline{ucd}$ 44 _____

45 Which graph represents the relationship between the kinetic energy and the speed of a freely falling object?

(1)

(3)

(2)

(4)   45 _____

46 Which diagram represents the electric field between two oppositely charged conducting spheres?

(1)   (3)

(2)   (4)   46 _____

47 Which graph represents the relationship between the magnitude of the gravitational force, F_g, between two masses and the distance, r, between the centers of the masses?

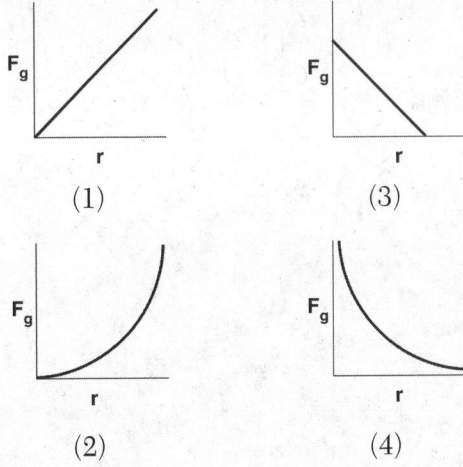

(1)   (3)

(2)   (4)   47 _____

48 The diagram below shows two waves traveling toward each other at equal speed in a uniform medium.

When both waves are in the region between points *A* and *B*, they will undergo

(1) diffraction
(2) the Doppler effect
(3) destructive interference
(4) constructive interference

48 _____

49 The diagram below shows a series of straight wave fronts produced
in a shallow tank of water approaching a small opening in a barrier.

Wave fronts

Which diagram represents the appearance of the wave fronts after
passing through the opening in the barrier?

49 _____

50 The graph below represents the relationship between energy and the equivalent mass from which it can be converted.

The slope of this graph represents

(1) c (3) g

(2) c^2 (4) g^2

50 _____

PART B–2
Answer all questions in this part.

Directions (51–65): Record your answers on the answer sheet provided after the questions. Some questions may require the use of the *2006 Edition Reference Tables for Physical Setting/Physics*.

IMPORTANT NOTE:

- 2-point calculation problems appear as two linked questions in the test, for example, 51–52.
- In the first question, 1 point is allowed for showing the equation *and* for substituting the correct values with their units.
- In the second question, 1 point is allowed for a correct answer (number *and* unit) or for an answer that is consistent with your work in the first question.

51–52 A 25.0-meter length of platinum wire with a cross-sectional area of 3.50×10^{-6} meter^2 has a resistance of 0.757 ohm at 20°C. Calculate the resistivity of the wire. [Show all work, including the equation and substitution with units.] [2]

53 The diagram below represents a periodic wave moving along a rope.

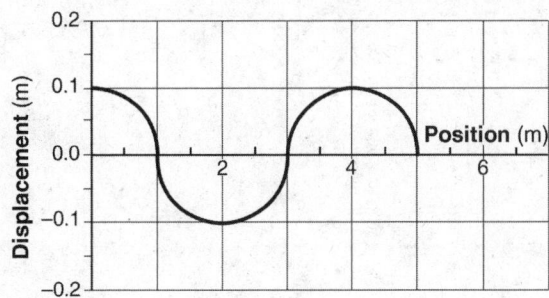

On the grid *on the answer sheet*, draw *at least one* full wave with the same amplitude and half the wavelength of the given wave. [1]

54–55 A baseball bat exerts an average force of 600 newtons east on a ball, imparting an impulse of 3.6 newton • seconds east to the ball. Calculate the amount of time the baseball bat is in contact with the ball. [Show all work, including the equation and substitution with units.] [2]

56 The diagram below shows the north pole of one bar magnet located near the south pole of another bar magnet.

On the diagram *on the answer sheet*, draw *three* magnetic field lines in the region between the magnets. [1]

Base your answers to questions 57 through 59 on the information and graph below.

The graph below shows the relationship between speed and elapsed time for a car moving in a straight line.

Speed vs. Time

57 Determine the magnitude of the acceleration of the car. [1]

58–59 Calculate the total distance the car traveled during the time interval 4.0 seconds to 8.0 seconds. [Show all work, including the equation and substitution with units.] [2]

Base your answers to questions 60 through 62 on the information below.

A 20-ohm resistor, R_1, and a resistor of unknown resistance, R_2, are connected in parallel to a 30-volt source, as shown in the circuit diagram below. An ammeter in the circuit reads 2.0 amperes.

60 Determine the equivalent resistance of the circuit. [1]

61–62 Calculate the resistance of resistor R_2. [Show all work, including the equation and substitution with units.] [2]

Base your answers to questions 63 through 65 on the information below.

A 28-gram rubber stopper is attached to a string and whirled clockwise in a horizontal circle with a radius of 0.80 meter. The diagram *on the answer sheet* represents the motion of the rubber stopper. The stopper maintains a constant speed of 2.5 meters per second.

63–64 Calculate the magnitude of the centripetal acceleration of the stopper. [Show all work, including the equation and substitution with units.] [2]

65 On the diagram *on the answer sheet*, draw an arrow showing the direction of the centripetal force acting on the stopper when it is at the position shown. [1]

PART C
Answer all questions in this part.

Directions (66–85): Record your answers on the answer sheet provided after the questions. Some questions may require the use of the *2006 Edition Reference Tables for Physical Setting/Physics.*

Base your answers to questions 66 through 69 on the information below.

Auroras over the polar regions of Earth are caused by collisions between charged particles from the Sun and atoms in Earth's atmosphere. The charged particles give energy to the atoms, exciting them from their lowest available energy level, the ground state, to higher energy levels, excited states. Most atoms return to their ground state within 10 nanoseconds.

In the higher regions of Earth's atmosphere, where there are fewer interatom collisions, a few of the atoms remain in excited states for longer times. For example, oxygen atoms remain in an excited state for up to 1.0 second. These atoms account for the greenish and red glows of the auroras. As these oxygen atoms return to their ground state, they emit green photons ($f = 5.38 \times 10^{14}$ Hz) and red photons ($f = 4.76 \times 10^{14}$ Hz). These emissions last long enough to produce the changing aurora phenomenon.

IMPORTANT NOTE:

- 2-point calculation problems appear as two linked questions in the test, for example, 67–68.

- In the first question, 1 point is allowed for showing the equation *and* for substituting the correct values with their units.

- In the second question, 1 point is allowed for a correct answer (number *and* unit) or for an answer that is consistent with your work in the first question.

---

66 What is the order of magnitude of the time, in seconds, that most atoms spend in an excited state? [1]

67–68 Calculate the energy of a photon, in joules, that accounts for the red glow of the aurora. [Show all work, including the equation and substitution with units.] [2]

69 Explain what is meant by an atom being in its ground state. [1]

Base your answers to questions 70 through 75 on the information below.

A girl rides her bicycle 1.40 kilometers west, 0.70 kilometer south, and 0.30 kilometer east in 12 minutes. The vector diagram *on the answer sheet* represents the girl's first two displacements in sequence from point *P*. The scale used in the diagram is 1.0 centimeter = 0.20 kilometer.

70–71 On the vector diagram *on the answer sheet*, using a ruler and a protractor, construct the following vectors:

- Starting at the arrowhead of the second displacement vector, draw a vector to represent the 0.30 kilometer east displacement. Label the vector with its magnitude. [1]
- Draw the vector representing the resultant displacement of the girl for the entire bicycle trip and label the vector *R*. [1]

72–73 Calculate the girl's average speed for the entire bicycle trip. [Show all work, including the equation and substitution with units.] [2]

74 Determine the magnitude of the girl's resultant displacement for the entire bicycle trip, in kilometers. [1]

75 Determine the measure of the angle, in degrees, between the resultant and the 1.40-kilometer displacement vector. [1]

Base your answers to questions 76 through 80 on the information below.

A light ray with a frequency of 5.09×10^{14} hertz traveling in water has an angle of incidence of 35° on a water-air interface. At the interface, part of the ray is reflected from the interface and part of the ray is refracted as it enters the air.

76 What is the angle of reflection of the light ray at the interface? [1]

77 On the diagram *on the answer sheet*, using a protractor and a straightedge, draw the reflected ray. [1]

78–79 Calculate the angle of refraction of the light ray as it enters the air. [Show all work, including the equation and substitution with units.] [2]

80 Identify *one* characteristic of this light ray that is the same in *both* the water and the air. [1]

Base your answers to questions 81 through 85 on the information and diagram below.

A 30.4-newton force is used to slide a 40.0-newton crate a distance of 6.00 meters at constant speed along an incline to a vertical height of 3.00 meters.

81 Determine the total work done by the 30.4-newton force in sliding the crate along the incline. [1]

82–83 Calculate the total increase in the gravitational potential energy of the crate after it has slid 6.00 meters along the incline. [Show all work, including the equation and substitution with units.] [2]

84 State what happens to the kinetic energy of the crate as it slides along the incline. [1]

85 State what happens to the internal energy of the crate as it slides along the incline. [1]

Answer Sheet
June 2013
Physics: The Physical Setting

PART B–2

51–52

53

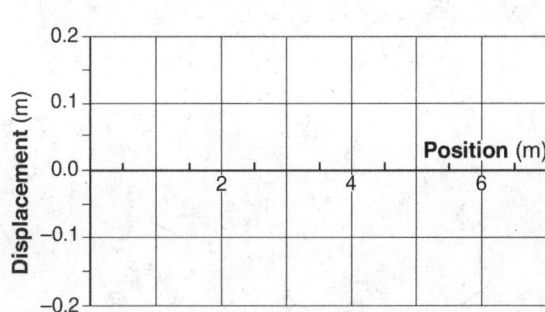

54–55

56

| N | | S |

57 _____ m/s^2

58–59

60 _____ Ω

61–62

63–64

65

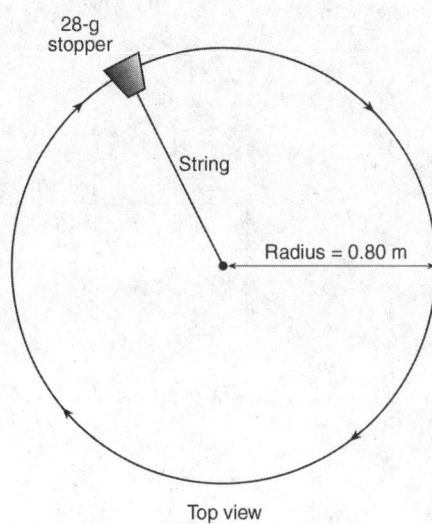

28-g
stopper

String

Radius = 0.80 m

Top view

PART C

66 _____

67–68

69 _____

_____

70–71

Scale
1.0 cm = 0.20 km

1.40 km

0.70 km

P

72–73

74 _____ **km**

75 _____ °

76 _____ °

77

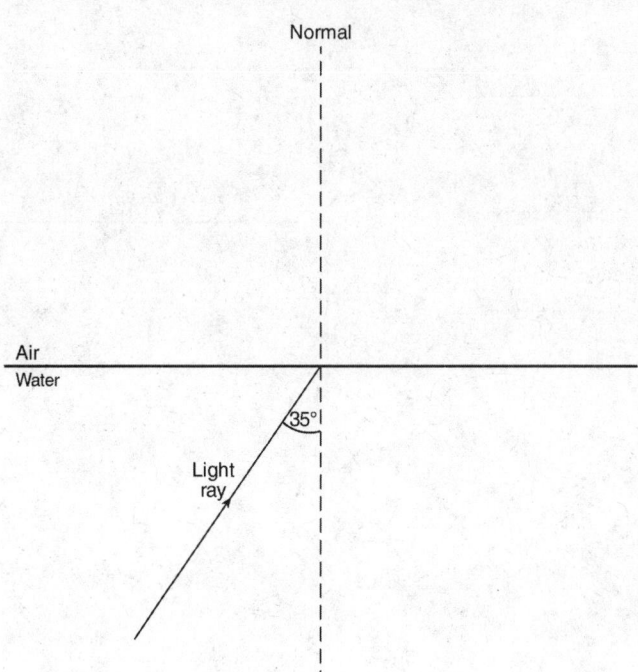

78–79

80 _____

81 _____**J**

82–83

84 _____

85 _____

Answers
June 2013
Physics: The Physical Setting

Answer Key

PART A

1. 4	8. 1	15. 3	22. 4	29. 3
2. 1	9. 3	16. 4	23. 3	30. 3
3. 1	10. 3	17. 1	24. 3	31. 3
4. 2	11. 1	18. 2	25. 4	32. 2
5. 1	12. 2	19. 1	26. 2	33. 1
6. 4	13. 2	20. 4	27. 1	34. 2
7. 4	14. 4	21. 4	28. 4	35. 1

PART B–1

36. 2	39. 3	42. 1	45. 3	48. 4
37. 1	40. 2	43. 2	46. 3	49. 1
38. 4	41. 3	44. 1	47. 4	50. 2

PART B–2 and **PART C.** *See* **Answers Explained**

Answers Explained

PART A

1 A scalar quantity has magnitude but not direction. Time is never associated with a direction and is therefore a scalar quantity.

WRONG CHOICES EXPLAINED:
(1), (2), (3) Displacement, momentum, and velocity are all associated with both magnitude and direction. They are vector quantities.

2 When two forces act at the same point, they produce the maximum resultant force (40.0 newtons) when the angle between them is 0°.

WRONG CHOICE EXPLAINED:
(4) When the angle between the two forces is 180°, they will produce the minimum resultant force (0.0 newton).

3 Assume that east is positive and west is negative. Use *Equation ME2* on *Reference Table K*:

$$\text{Solution:} \quad a = \frac{\Delta v}{t}$$

$$= \frac{-23.0 \text{ m/s} - (-30.0 \text{ m/s})}{2.00 \text{ s}}$$

$$= +3.5 \text{ m/s}^2 \text{ or } 3.5 \text{ m/s}^2 \text{ east}$$

4 Use *Equation ME4* on *Reference Table K*:

$$\text{Solution:} \quad d = v_i t + \frac{1}{2}at^2$$

$$12 \text{ m} = (0.0 \text{ m/s})(4.0 \text{ s}) + \frac{1}{2}a(4.0 \text{ s})^2$$

$$a = 1.5 \text{ m/s}^2$$

5 When a projectile is in motion, the horizontal component of its velocity, v_x, does not change. So v_x remains 40 m/s. The vertical component, however, does change because of the influence of gravity. In order to find the final value of v_y, use *Equation ME3* on *Reference Table K*:

$$\text{Solution:} \quad v_f = v_i + at$$
$$= 30 \text{ m/s} + \left(-9.8 \text{ m/s}^2\right)\left(2.0 \text{ s}\right)$$
$$\approx 10 \text{ s}$$

6 The vertical component of the ball's velocity determines the time that the ball will remain in the air and, therefore, the height to which it will rise. The vertical component is greatest when the angle with the horizontal is 90°.

WRONG CHOICE EXPLAINED:
(3) If thrown at 45°, the ball will travel the greatest horizontal distance.

7 Inertia is determined solely by an object's mass. The 1500-kilogram car has the greatest inertia.

8 In order to calculate the force of kinetic friction between the wooden block and the wood floor, we must know the coefficient of kinetic friction and the normal force on the block. According to *Reference Table C*, the coefficient of kinetic friction between wood and wood is 0.30. Since the wood is on a horizontal surface, its weight (8.0 newtons) is equal to the normal force on the block. Use *Equation ME9* on *Reference Table K*:

$$\text{Solution:} \quad F_f = \mu F_N$$
$$= \left(0.30\right)\left(8.0 \text{ N}\right)$$
$$= 2.4 \text{ N}$$

9 An object in equilibrium has no net force acting on it. Therefore, according to Newton's second law, the object must either be at rest or be traveling at constant speed in a straight line. Of the choices given, only choice (3), a man standing still on a bathroom scale, meets these criteria.

10 At its highest point, the rock is at rest. The net force acting on it is the gravitational force, which is the same as the object's weight.

11 The spring is the means by which cart A interacts with cart B. Therefore, according to Newton's third law, the force exerted by the spring on cart A must be equal in magnitude (and opposite in direction) to the force exerted by the spring on cart B.

12 The force acting on the block parallel to the ramp ($F_{||}$) is the component of the block's weight parallel to the incline of the ramp. This component is equal to the product of the block's weight (mg) and the sine of the angle of the ramp ($\sin \theta$). Use the following equation:

$$\text{Solution:} \qquad F_{||} = mg\sin\theta$$
$$= (8.0 \text{ N})\sin 15°$$
$$= 2.1 \text{ N}$$

13 Use *Equation ME11* on *Reference Table K*:

$$\text{Solution:} \qquad g = \frac{F_g}{m}$$
$$= \frac{350 \text{ N}}{70 \text{ kg}}$$
$$= 5.0 \text{ N/kg}$$

14 Use *Equation ME16* on *Reference Table K*:

$$\text{Solution:} \qquad PE_s = \frac{1}{2}kx^2$$
$$k = \frac{2\,PE_s}{x^2}$$
$$= \frac{2(2.34 \text{ J})}{(0.250 \text{ m})^2}$$
$$= 74.9 \text{ N/m}$$

15 A photocell changes light energy into electrical energy and produces an electric current in the circuit containing the fan. The motor in the fan changes the electrical energy into the mechanical energy of the rotating fan blades.

16 All waves transfer energy without transferring mass.

WRONG CHOICES EXPLAINED:

(1) The speed of an electromagnetic wave is much faster than the speed of a mechanical wave.

(2) Only mechanical waves require a material medium for propagation.

(3) Only electromagnetic waves propagate in a vacuum.

17 The vibrating particles produce an electromagnetic wave whose frequency is 3.9×10^8 hertz. See *Reference Table D*. The electromagnetic wave is in the portion of the electromagnetic spectrum known as radio waves.

18 The energy of any mechanical wave depends on the wave's amplitude.

19 Sound is a longitudinal wave in which the air particles vibrate parallel to the wave's direction of motion. Therefore, the air molecules will vibrate east and west.

WRONG CHOICE EXPLAINED:

(2) The particles in this wave are moving perpendicular to the wave's direction of motion. This is the characteristic of a transverse wave.

20 See *Reference Tables A and E*. The speed of light (c) is 3.00×10^8 meters per second and the absolute index of refraction of ethyl alcohol is 1.36. Now use *Equation W4* on *Reference Table K*:

$$\text{Solution:} \quad n = \frac{c}{v}$$

$$v = \frac{c}{n}$$

$$= \frac{3.00 \times 10^8 \text{ m/s}}{1.36}$$

$$= 2.21 \times 10^8 \text{ m/s}$$

21 The total mechanical energy of the pendulum is the sum of its kinetic and potential energies. In an ideal pendulum (no friction), kinetic energy is converted to potential energy and vice versa. However, the total mechanical energy does *not* change.

22 When two points on a periodic wave are a whole number of wavelengths apart $(0\lambda, 1\lambda, 2\lambda, 3\lambda, \ldots)$, the two points are in phase with each other. When two points on a periodic wave are an odd number of half-wavelengths apart $\left(\frac{1}{2}\lambda, \frac{3}{2}\lambda, \frac{5}{2}\lambda, \frac{7}{2}, \ldots\right)$, the two points are out of phase with each other. Points D and G are $\frac{3}{2}\lambda$ apart and are, therefore, out of phase.

WRONG CHOICES EXPLAINED:

(1), (3) Points A and C and points C and E are each 1 wavelength apart and are therefore in phase.

(2) Points B and F are 2 wavelengths apart and are therefore in phase.

23 When the plastic rod is rubbed with wool, the rod acquires a negative electric charge. The charged rod causes the stream of water to deviate because it places an electrostatic force on the positive ends of the water molecules.

24 The wave crest travels a distance of 1.0×10^{-2} meter in 4.0×10^{-1} second. Use *Equation ME1* on *Reference Table K*:

$$\text{Solution:} \quad v = \frac{d}{t}$$

$$= \frac{1.0 \times 10^{-2}\ \text{m}}{4.0 \times 10^{-1}\text{s}}$$

$$= 2.5 \times 10^{-2}\ \text{m/s}$$

An alternate way of solving this problem is to recognize that the wavelength (λ) is 1.0×10^{-2} meter and that the frequency (f) is $1/(4.0 \times 10^{-1})$ hertz. Then use *Equation W1* on *Reference Table K*.

25 Sound resonance is the phenomenon by which a vibrating tuning fork can cause another tuning fork with the same frequency to vibrate. This phenomenon can also be observed in vibrating air columns and vibrating strings.

WRONG CHOICES EXPLAINED:

(1) Diffraction is the bending of a wave around a barrier.

(2) Reflection is the bouncing back of a wave when the wave strikes the boundary of two different media.

(3) Refraction is the bending of a wave entering a second medium because the wave speed is different in each of the two media.

26 When a sound source and an observer are in relative motion, the apparent frequency heard by the observer will be different from the frequency emitted by the source. If the source and the observer are moving toward each other, the apparent frequency perceived by the observer will be greater than the emitted frequency. This phenomenon is known as the Doppler effect.

27 The electronvolt is a unit of energy. As shown on *Reference Table A*, 1 electronvolt is equivalent to 1.60×10^{-19} joule.

28 Magnetic fields are produced by charged particles in motion.

29 The period is the time for 1 heartbeat. Since 60 heartbeats take 60 seconds, 1 heartbeat will take 1 second.

30 See *Equation EL3* on *Reference Table K*:

$$\text{Solution:} \quad V = \frac{W}{q}$$
$$= \frac{48 \text{ J}}{4.0 \text{ C}}$$
$$= 12 \text{V}$$

31 Current is the flow of charge in a circuit. Since charge is conserved in the circuit, the sum of the currents entering the junction (the place where the two wires cross) must equal the sum of the currents leaving the junction. A total of 12 amperes enter. So a total of 12 amperes must leave the junction. Therefore, the amount of current passing through the ammeter must be 9 amperes.

32 Use *Equation EL8* on *Reference Table K*:

$$\text{Solution:} \quad W = VIt$$
$$I = \frac{W}{Vt}$$
$$= \frac{6.0 \times 10^6 \text{ J}}{(220 \text{ V})(1.8 \times 10^3 \text{ s})}$$
$$= 15 \text{ A}$$

33 All electric charges must be whole-number multiples of the elementary charge. See *Reference Table* A; $e = 1.60 \times 10^{-19}$ C. This relationship is given by the equation $q = ne$, where n is the number of elementary charges comprising the charge q. Of the choices given, only choice (1), $+4.80 \times 10^{-19}$ C, is a whole-number multiple of the elementary charge. (In this case, $q = +3.00e$.)

WRONG CHOICES EXPLAINED:
(2) $q = +1.500e$
(3) $q = -1.500e$
(4) $q = -3.5e$

34 Use the hydrogen energy level diagram on *Reference Table* F:

$$E_{photon} = E_3 - E_5$$
$$= (-1.51 \text{ eV}) - (-0.54 \text{ eV})$$
$$= -0.97 \text{ eV}$$

The negative sign means that the photon was emitted.

35 A gamma ray photon has no electric charge. Since electric charge must be conserved, the production of an electron, which has a negative charge, must be accompanied by the production of a positive particle.

PART B–1

36 An unsharpened pencil is about 10 inches long. Since 1 inch is approximately equal to 2 centimeters, the length of the pencil is approximately 20 centimeters or 2×10^{-1} meter.

WRONG CHOICES EXPLAINED:
(1) 2×10^{-2} meter is approximately equal to 1 inch.
(3) 2 meters is approximately equal to 80 inches.
(4) 20 meters is approximately equal to 60 feet.

37 This question involves the conservation of (linear) momentum. Use *Equations ME12* and *ME13* on *Reference Table K*:

Solution: $p = mv$

$$p_{before} = p_{after}$$

$$\left(8.0 \text{ kg}\right)\left(+4.0 \text{ m/s}\right) + \left(4.0 \text{ kg}\right)\left(-6.0 \text{ m/s}\right)$$

$$= \left(8.0 \text{ kg}\right)v + \left(4.0 \text{ kg}\right)\left(3.0 \text{ m/s}\right)$$

$$v = \frac{\left(8.0 \text{ kg}\right)\left(+4.0 \text{ m/s}\right) + \left(4.0 \text{ kg}\right)\left(-6.0 \text{ m/s}\right) - \left(4.0 \text{ kg}\right)\left(3.0 \text{ m/s}\right)}{8.0 \text{ kg}}$$

$$= -0.5 \text{ m/s}$$

This means the 8.0-kilogram cart is moving to the left at 0.50 meter per second.

38 We need only consider the forces in the x-direction since the vertical forces add to 0 newton. The two horizontal forces produce a net force of 20 newtons to the left. Therefore, the block will accelerate to the left.

39 The force needed to lift the mass is equal to its weight (mg). The power generated by the motor is the work done in lifting the mass (mgh) divided by the time needed to lift it. Use *Equation ME23* on *Reference Table K*:

Solution: $P = \dfrac{W}{t}$

$$= \frac{Fd}{t}$$

$$= \frac{mgh}{t}$$

$$= \frac{\left(400 \text{ kg}\right)\left(9.81 \text{ m/s}^2\right)\left(10 \text{ m}\right)}{8.00 \text{ s}}$$

$$= 4.9 \times 10^3 \text{ W}$$

40 The net force needed to accelerate the 4.0-kilogram object at 3.0 meters per second^2 is determined by using Newton's second law (*Equation ME8* on *Reference Table K*):

$$F_{net} = ma$$
$$= (4.0 \text{ kg})(3.0 \text{ m/s}^2)$$
$$= 12 \text{ N}$$

Now apply Newton's second law to the 2.0-kilogram mass:

Solution: $a = \dfrac{F_{net}}{m}$

$$= \dfrac{12 \text{ N}}{2.0 \text{ kg}}$$

$$= 6.0 \text{ m/s}^2$$

41 The electron has a charge whose magnitude is equal to the magnitude of the elementary charge. See *Reference Table A* and use *Equation EL2* on *Reference Table K*:

Solution: $E = \dfrac{F_e}{q}$

$$F_e = qE$$
$$= (1.60 \times 10^{-19} \text{ C})(600 \text{ N/C})$$
$$= 9.60 \times 10^{-17} \text{ N}$$

42 See *Reference Table A* for the magnitude of the charge on the electron, and use *Equation EL4* on *Reference Table K*:

Solution: $I = \dfrac{\Delta q}{t}$

$$t = \dfrac{\Delta q}{I}$$

$$= \dfrac{(2.5 \times 10^{19} \text{ electrons})\left(\dfrac{1.6 \times 10^{-19} \text{C}}{1 \text{ electron}}\right)}{4.0 \text{ A}}$$

$$= 1.0 \text{ s}$$

43 Use Coulomb's law (*Equation EL1* on *Reference Table K*):

$$\text{Solution:} \quad F_e = \frac{kq_1q_2}{r^2}$$

$$F'_e = \frac{k(2q_1)(4q_2)}{(2r)^2}$$

$$= \frac{8}{4} \cdot \frac{kq_1q_2}{r^2}$$

$$= 2F$$

44 See *Reference Tables G* and *H*. A meson is composed of a quark and an antiquark. An antiquark always has a charge opposite to that of its companion quark. The combination $s\bar{c}$ is a meson that has a charge of $\left(-\frac{1}{3}e\right) + \left(-\frac{2}{3}e\right) = -1e$ as required.

WRONG CHOICES EXPLAINED:
  (2), (4) These particles are not mesons.
  (3) The meson $u\bar{b}$ has a charge of $\left(+\frac{2}{3}e\right) + \left(+\frac{1}{3}e\right) = +1e$

45 Refer to *Equation ME20* on *Reference Table K*. Kinetic increases with the square of the speed. A graph of kinetic energy versus speed will produce an upward curve, known as a parabola.

46 Electric field lines always point away from positive charges and toward negative charges. Of the choices given, only the diagram shown in choice (3) meets these criteria.

47 Refer to *Equation ME10* on *Reference Table K*. The gravitational force between two masses varies inversely as the square of the distance between the masses. As the distance decreases, the force increases rapidly. As the distance increases, the force decreases rapidly. Neither the force nor the distance can ever equal zero. Only the graph shown in choice (4) meets all of these criteria.

48 When the waves meet in region *AB*, they will coincide at every point. In other words, the waves will be completely in phase. This is the condition for constructive interference.

49 When the straight wave fronts reach the barrier, they diffract. In other words, they bend around the barrier, causing circular wave fronts to emerge. The circular wave fronts have the same wavelength (1.4 centimeters) as the straight wave fronts incident on the barrier. This is illustrated in choice (1).

50 Refer to *Equation MP3* on *Reference Table K*. According to this relationship, energy is directly proportional to the equivalent mass. A graph of energy versus equivalent mass will yield a straight line that passes through the origin and has a positive slope equal to c^2.

PART B–2

51–52 Use *Equation EL6* on *Reference Table K*:

$$\text{Solution:} \qquad R = \frac{\rho L}{A}$$

$$\rho = \frac{RA}{L}$$

$$= \frac{(0.757 \ \Omega)(3.50 \times 10^{-6} \, \text{m}^2)}{25.0 \ \text{m}}$$

[1 point]

52 Solution: $\rho = 1.06 \times 10^{-7} \ \Omega \bullet \text{m}$

[1 point]

53 The original periodic wave illustrated in the question has a wavelength of 4 meters. It consists of the right half of a crest (from 0 m to 1 m), a full trough (from 1 m to 3 m), and the left half of a trough (from 3 m to 4 m). The amplitude of the wave is 0.1 meter.

In order to draw a wave with the same amplitude (0.1 m) and half the wavelength (2 m), you must draw the right half of a crest from 0 m to 0.5 m, a full trough from 0.5 m to 1.5 m, and the left half of a crest from 1.5 m to 2 m. See the diagram below.

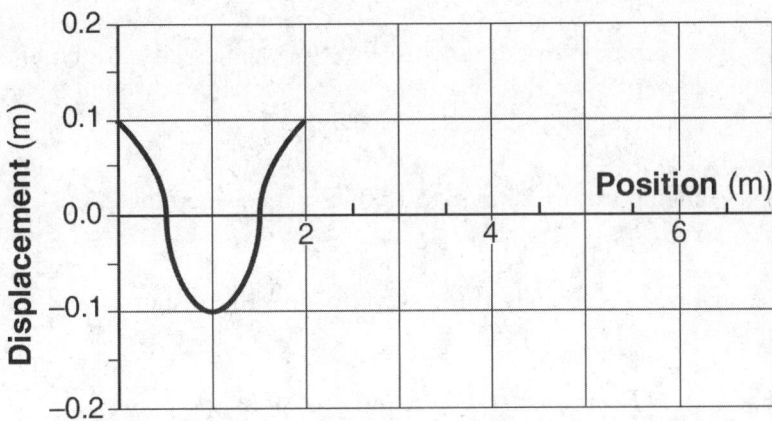

[1 point]

54–55 Use *Equation ME14* on *Reference Table K*:

Solution: $J = F_{net}t$

$$t = \frac{J}{F_{net}}$$

$$= \frac{3.6 \text{ N} \cdot \text{s}}{600 \text{ N}}$$

[1 point]

55 Solution: $t = 0.0060$ s or 6.0×10^{-3} s

[1 point]

56 Magnetic field lines are directed from the north pole to the south pole. Refer to the diagrams below. (Either one is acceptable for credit.)

[1 point]

57 Refer to the graph given with this question. The acceleration of the car is determined from the slope of the graph. For example, choose the coordinates (0.0 s, 0.0 m/s) and (8.0 s, 10.0 m/s):

$$\text{Solution:} \quad a = \frac{\Delta v}{\Delta t}$$

$$= \frac{10.0 \text{ m/s} - 0.0 \text{ m/s}}{8.0 \text{ s} - 0.0 \text{ s}}$$

$$= 1.25 \text{ m/s}^2$$

Answers ranging from 1.20 m/s^2 to 1.30 m/s^2 receive credit.

[1 point]

58–59 There are actually three ways to find the distance, and any one is acceptable for credit.

Method 1: Use *Equation ME4* on *Reference Table K*:

$$\text{Solution:} \quad d = v_i t + \frac{1}{2} a t^2$$

$$= (5.0 \text{ m/s})(4.0 \text{ s}) + \frac{1}{2}\left(1.25 \text{ m/s}^2\right)(4.0 \text{ s})^2$$

Method 2: Use *Equation ME1* on *Reference Table K*. The average speed between 4.0 s and 8.0 s is found by adding the speeds at these two times (5.0 m/s and 10.0 m/s) and then dividing by 2, which is 7.5 m/s:

$$\text{Solution:} \quad \bar{v} = \frac{d}{t}$$

$$d = \bar{v} t$$

$$= (7.5 \text{ m/s})(4.0 \text{ s})$$

Method 3: Find the area under the graph between 4.0 s and 8.0 s by breaking the resultant trapezoid into a right triangle and a rectangle. The area of the rectangle is 20 m and the area of the right triangle is 10 m.

[1 point]

59 Solution: $d = 30$ m

[1 point]

60 Use *Equation EL5* on *Reference Table K*:

$$\text{Solution:} \quad R_{eq} = \frac{V}{I}$$

$$= \frac{30 \text{ V}}{2.0 \text{ A}}$$

$$= 15 \; \Omega$$

[1 point]

61–62 Use *Equation EL10C* on *Reference Table K*:

$$\text{Solution:} \quad \frac{1}{R_{eq}} = \frac{1}{R_1} + \frac{1}{R_2}$$

$$\frac{1}{R_2} = \frac{1}{R_{eq}} - \frac{1}{R_1}$$

$$= \frac{1}{15 \; \Omega} - \frac{1}{20 \; .\Omega}$$

$$= \frac{1}{60 \; \Omega}$$

[1 point]

62 Solution: $R_{eq} = 60 \; \Omega$

[1 point]

63–64 Refer to the diagram on the answer sheet. The radius of the circular path is 0.80 meter, and the speed of the stopper is 2.5 meters per second. Use *Equation ME18* on *Reference Table K*:

$$\text{Solution:} \quad a_c = \frac{v^2}{r}$$

$$= \frac{(2.5 \text{ m/s})^2}{0.80 \text{ m}}$$

[1 point]

64 Solution: $a_c = 7.8 \text{ m/s}^2$

[1 point]

65 Refer to the diagram below, which shows an appropriate drawing that is acceptable for credit. The arrow pointing to the center of the circle represents the centripetal force vector.

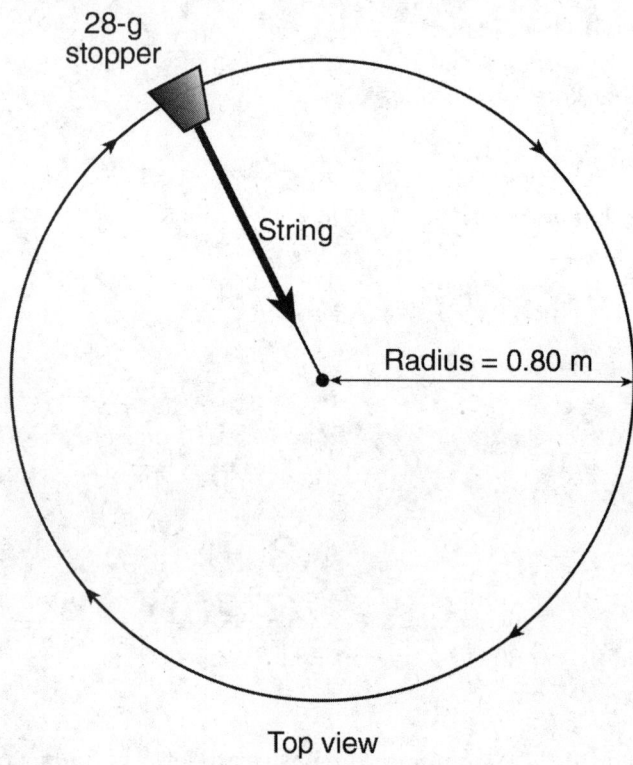

28-g
stopper

String

Radius = 0.80 m

Top view

[1 point]

PART C

66 The order of magnitude of a measurement is the measurement stated to the nearest power of 10 or the power itself. For example, the order of magnitude for 10 nanoseconds is 10^{-8} or -8.

This particular question, however, stirred a great deal of controversy due to its wording. Therefore, any of the following answers were also acceptable:

- ≤ -8
- $\leq 10^{-9}$
- < -8
- $< 10^{-9}$
- any whole number less than -8
- any exponent less than -8

[1 point]

67–68 Use *Reference Table A* and *Equation MP1* on *Reference Table K*:

$$\text{Solution:} \quad E_{photon} = hf$$
$$= \left(6.63 \times 10^{-34} \text{ J} \cdot \text{s}\right)\left(4.76 \times 10^{14} \text{ Hz}\right)$$

[1 point]

68 Solution: $E_{photon} = 3.16 \times 10^{-19}$ J

[1 point]

69 The ground state is the lowest available energy level that an atom can have. The ground state is the most stable energy state of an atom. (Either statement is acceptable for credit.)

[1 point]

70–71 Refer to the diagram below, which applies to questions 70 and 71.

In order to draw the 0.30-kilometer displacement, use the conversion factor provided in the question and employ the factor-label method:

$$0.30 \text{ km}\left(\frac{1.0 \text{ cm}}{0.20 \text{ km}}\right) = 1.5 \text{ cm}$$

Now draw a 1.5-centimeter arrow pointing to the right. Label it "0.30 km," as shown in the diagram.

The range of arrow lengths acceptable for credit is 1.3 cm to 1.7 cm.

[1 point]

71 Draw the resultant vector from point P to the end of the 0.3-kilometer vector. Label the resultant as R, as shown in the diagram above.

[1 point]

72–73 The girl's average speed for the entire trip is the total distance she covers divided by the total elapsed time. Use *Equation ME1* on *Reference Table K*:

Solution: $\bar{v} = \dfrac{d}{t}$

$= \dfrac{1.40 \text{ km} + 0.70 \text{ km} + 0.30 \text{ km}}{12 \text{ min}}$

$= \dfrac{2.4 \text{ km}}{12 \text{ min}}$

or

$\bar{v} = \dfrac{d}{t}$

$= \dfrac{1400 \text{ m} + 700 \text{ m} + 300 \text{ m}}{12 \text{min}\left(\dfrac{60 \text{ s}}{1 \text{ min}}\right)}$

$= \dfrac{2400 \text{ m}}{720 \text{ s}}$

[1 point]

73 Solution: $\bar{v} = 0.20$ km/min or $\bar{v} = 3.3$ m/s   [1 point]

74 In order to calculate the magnitude of the resultant displacement vector, use a ruler to measure the length of the arrow marked R. The length is 6.6 centimeters. Now apply the factor-label method using the conversion factor given:

Solution: 6.6 cm $\left(\dfrac{0.20 \text{ km}}{1 \text{ cm}}\right) = 1.3$ km

Answers ranging from 1.1 km to 1.5 km received credit.   [1 point]

75 Measure the angle between the resultant vector and the 1.4-kilometer displacement with a protractor. The correct angle is 32°.
Answers ranging from 30° to 34° receive credit.   [1 point]

76 The angle of reflection of any wave at a boundary separating two media is always equal to the angle of incidence. Therefore, the angle of reflection of the light ray at the air-water interface is 35°.

[1 point]

77 Refer to the diagram below:

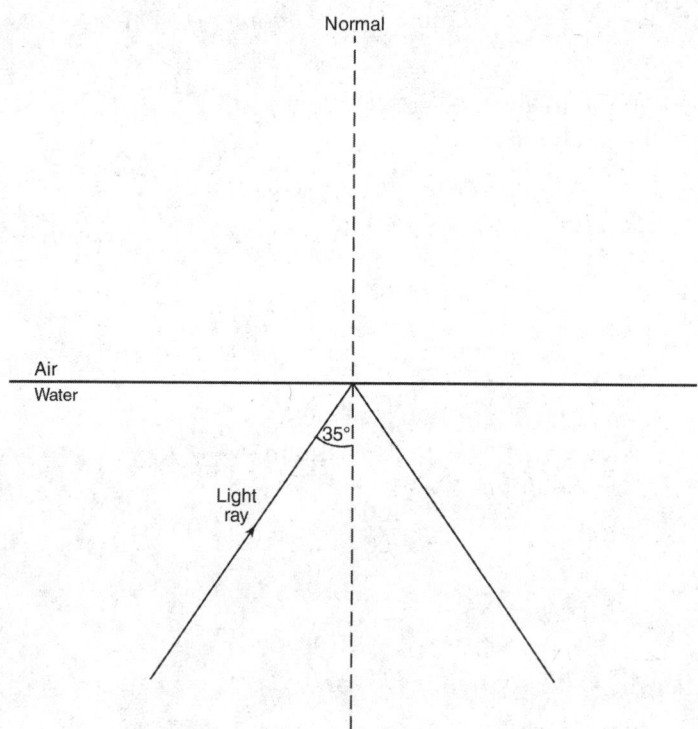

Use a protractor and a straight edge to draw the reflected ray in the water at an angle of 35° to the normal.

[1 point]

78–79 Use *Reference Table E* and *Equation W5* on *Reference Table K*:

$$\text{Solution:} \quad n_{\text{water}} \ \sin\theta_{\text{water}} = n_{\text{air}} \ \sin\theta_{\text{air}}$$

$$\sin\theta_{\text{air}} = \frac{n_{\text{water}} \ \sin\theta_{\text{water}}}{n_{\text{air}}}$$

$$= \frac{(1.33)(\sin 35°)}{1.00}$$

$$= 0.763$$

$$\theta_{\text{air}} = \sin^{-1}(0.763)$$

[1 point]

79 Solution: $\theta_{air} = 49.7°$

Either 49° or 50° is acceptable for credit.   [1 point]

80 Any one of the following characteristics is the same in both water and air
and is acceptable for credit:

- frequency
- period
- phase
- color
- the transverse nature of the light wave

[1 point]

81 Use *Equation ME21* on *Reference Table K*:

$$\begin{aligned} \text{Solution:} \quad W &= Fd \\ &= (30.4\ N)(6.00\ m) \\ &= 182\ J \end{aligned}$$

[1 point]

82–83 Use *Equation ME19* on *Reference Table K*:

$$\begin{aligned} \text{Solution:} \quad \Delta PE &= mgh \\ &= (40.0\ N)(3.00\ m) \end{aligned}$$

Remember:

- The value 40.0 newtons is the weight of the object (mg) and not its mass.
- The vertical distance must be used in the calculation.

[1 point]

83 Solution: $\Delta PE = 120\ J$ [1 point]

84 Since the speed remains constant, the kinetic energy of the crate remains
constant.

[1 point]

85 The work done on the crate is used to counteract the friction between the
crate and the incline. As a result, the internal energy of the crate increases.

[1 point]

Topic	Question Numbers (total)	Wrong Answers (x)	Grade
Math Skills	57–59, 66, 70: (5)		$\dfrac{100(5-x)}{5} = \%$
Mechanics	1–13, 36–38, 40, 47, 54, 55, 63–65, 71–75: (28)		$\dfrac{100(28-x)}{28} = \%$
Energy	14, 15, 21, 27, 39, 45, 81–85: (11)		$\dfrac{100(11-x)}{11} = \%$
Electricity/ Magnetism	23, 28, 30–33, 41–43, 46, 51, 52, 56, 60–62: (16)		$\dfrac{100(16-x)}{16} = \%$
Waves	16–20, 22, 24–26, 29, 48, 49, 53, 76–80: (18)		$\dfrac{100(18-x)}{18} = \%$
Modern Physics	34, 35, 44, 50, 67–69: (7)		$\dfrac{100(7-x)}{7} = \%$

Examination
June 2014

Physics: The Physical Setting

PART A
Answer all questions in this part.

Directions (1–35): For *each* statement or question, select the *number* of the word or expression that, of those given, best completes the statement or answers the question. Some questions may require the use of the *2006 Edition Reference Tables for Physical Setting/Physics*. Record your answers in the spaces provided.

1 Which quantity is scalar?

(1) mass   (3) momentum

(2) force   (4) acceleration   1 _____

2 What is the final speed of an object that starts from rest and accelerates uniformly at 4.0 meters per second^2 over a distance of 8.0 meters?

(1) 8.0 m/s   (3) 32 m/s

(2) 16 m/s   (4) 64 m/s   2 _____

3 The components of a 15-meters-per-second velocity at an angle of 60° above the horizontal are

(1) 7.5 m/s vertical and 13 m/s horizontal

(2) 13 m/s vertical and 7.5 m/s horizontal

(3) 6.0 m/s vertical and 9.0 m/s horizontal

(4) 9.0 m/s vertical and 6.0 m/s horizontal   3 _____

4 What is the time required for an object starting from rest to fall freely 500 meters near Earth's surface?

(1) 51.0 s   (3) 10.1 s
(2) 25.5 s   (4) 7.14 s   4 _____

5 A baseball bat exerts a force of magnitude F on a ball. If the mass of the bat is three times the mass of the ball, the magnitude of the force of the ball on the bat is

(1) F (3) $3F$
(2) $2F$ (4) $F/3$ 5 _____

6 A 2.0-kilogram mass is located 3.0 meters above the surface of Earth. What is the magnitude of Earth's gravitational field strength at this location?

(1) 4.9 N/kg   (3) 9.8 N/kg
(2) 2.0 N/kg   (4) 20 N/kg   6 _____

7 A truck, initially traveling at a speed of 22 meters per second, increases speed at a constant rate of 2.4 meters per second^2 for 3.2 seconds. What is the total distance traveled by the truck during this 3.2-second time interval?

(1) 12 m   (3) 70 m
(2) 58 m   (4) 83 m   7 _____

8 A 750-newton person stands in an elevator that is accelerating downward. The upward force of the elevator floor on the person must be

(1) equal to 0 N   (3) equal to 750 N
(2) less than 750 N   (4) greater than 750 N   8 _____

9 A 3.0-kilogram object is acted upon by an impulse having a magnitude of 15 newton • seconds. What is the magnitude of the object's change in momentum due to this impulse?

(1) 5.0 kg • m/s   (3) 3.0 kg • m/s
(2) 15 kg • m/s   (4) 45 kg • m/s   9 _____

10 An air bag is used to safely decrease the momentum of a driver in a car accident. The air bag reduces the magnitude of the force acting on the driver by

   (1) increasing the length of time the force acts on the driver
   (2) decreasing the distance over which the force acts on the driver
   (3) increasing the rate of acceleration of the driver
   (4) decreasing the mass of the driver   10 _____

11 An electron moving at constant speed produces

   (1) a magnetic field, only
   (2) an electric field, only
   (3) both a magnetic and an electric field
   (4) neither a magnetic nor an electric field   11 _____

12 A beam of electrons passes through an electric field where the magnitude of the electric field strength is 3.00×10^3 newtons per coulomb. What is the magnitude of the electrostatic force exerted by the electric field on each electron in the beam?

   (1) 5.33×10^{-23} N   (3) 3.00×10^3 N
   (2) 4.80×10^{-16} N   (4) 1.88×10^{22} N   12 _____

13 How much work is required to move 3.0 coulombs of electric charge a distance of 0.010 meter through a potential difference of 9.0 volts?

   (1) 2.7×10^3 J   (3) 3.0 J
   (2) 27 J   (4) 3.0×10^{-2} J   13 _____

14 What is the resistance of a 20.0-meter-long tungsten rod with a cross-sectional area of 1.00×10^{-4} meter^2 at 20°C?

   (1) 2.80×10^{-5} Ω   (3) 89.3 Ω
   (2) 1.12×10^{-2} Ω   (4) 112 Ω   14 _____

15 Two pieces of flint rock produce a visible spark when they are struck together. During this process, mechanical energy is converted into

  (1) nuclear energy and electromagnetic energy
  (2) internal energy and nuclear energy
  (3) electromagnetic energy and internal energy
  (4) elastic potential energy and nuclear energy　　　　　15 _____

16 A 15-kilogram cart is at rest on a horizontal surface. A 5-kilogram box is placed in the cart. Compared to the mass and inertia of the cart, the cart-box system has

  (1) more mass and more inertia
  (2) more mass and the same inertia
  (3) the same mass and more inertia
  (4) less mass and more inertia　　　　　16 _____

17 Transverse waves are to radio waves as longitudinal waves are to

  (1) light waves　　　　　　　　(3) ultraviolet waves
  (2) microwaves　　　　　　　　(4) sound waves　　　　　17 _____

18 As a monochromatic light ray passes from air into water, two characteristics of the ray that will *not* change are

  (1) wavelength and period
  (2) frequency and period
  (3) wavelength and speed
  (4) frequency and speed　　　　　18 _____

19 When a mass is placed on a spring with a spring constant of 60.0 newtons per meter, the spring is compressed 0.500 meter. How much energy is stored in the spring?

  (1) 60.0 J　　　　　　　　(3) 15.0 J
  (2) 30.0 J　　　　　　　　(4) 7.50 J　　　　　19 _____

20 A boy pushes his sister on a swing. What is the frequency of oscillation of his sister on the swing if the boy counts 90 complete swings in 300 seconds?

(1) 0.30 Hz

(2) 2.0 Hz

(3) 1.5 Hz

(4) 18 Hz

20 _____

21 What is the period of a sound wave having a frequency of 340 hertz?

(1) 3.40×10^2 s

(2) 1.02×10^0 s

(3) 9.73×10^{-1} s

(4) 2.94×10^{-3} s

21 _____

22 An MP3 player draws a current of 0.120 ampere from a 3.00-volt battery. What is the total charge that passes through the player in 900 seconds?

(1) 324 C

(2) 108 C

(3) 5.40 C

(4) 1.80 C

22 _____

23 A beam of light has a wavelength of 4.5×10^{-7} meter in a vacuum. The frequency of this light is

(1) 1.5×10^{-15} Hz

(2) 4.5×10^{-7} Hz

(3) 1.4×10^2 Hz

(4) 6.7×10^{14} Hz

23 _____

24 When x-ray radiation and infrared radiation are traveling in a vacuum, they have the same

(1) speed

(2) frequency

(3) wavelength

(4) energy per photon

24 _____

25 The diagram below represents two identical pulses approaching each other in a uniform medium.

As the pulses meet and are superposed, the maximum displacement of the medium is

(1) −6 cm   (3) 3 cm
(2) 0 cm   (4) 6 cm   25 _____

26 As a car approaches a pedestrian crossing the road, the driver blows the horn. Compared to the sound wave emitted by the horn, the sound wave detected by the pedestrian has a

(1) higher frequency and a lower pitch
(2) higher frequency and a higher pitch
(3) lower frequency and a higher pitch
(4) lower frequency and a lower pitch   26 _____

27 When air is blown across the top of an open water bottle, air molecules in the bottle vibrate at a particular frequency and sound is produced. This phenomenon is called

(1) diffraction   (3) resonance
(2) refraction   (4) the Doppler effect   27 _____

28 An antibaryon composed of two antiup quarks and one antidown quark would have a charge of

(1) +1e   (3) 0e
(2) −1e   (4) −3e   28 _____

29 Which force is responsible for producing a stable nucleus by opposing the electrostatic force of repulsion between protons?

(1) strong   (3) frictional

(2) weak   (4) gravitational   29 _____

30 What is the total energy released when 9.11×10^{-31} kilogram of mass is converted into energy?

(1) 2.73×10^{-22} J   (3) 9.11×10^{-31} J

(2) 8.20×10^{-14} J   (4) 1.01×10^{-47} J   30 _____

31 A shopping cart slows as it moves along a level floor. Which statement describes the energies of the cart?

(1) The kinetic energy increases and the gravitational potential energy remains the same.

(2) The kinetic energy increases and the gravitational potential energy decreases.

(3) The kinetic energy decreases and the gravitational potential energy remains the same.

(4) The kinetic energy decreases and the gravitational potential energy increases.   31 _____

32 Two identically sized metal spheres, A and B, are on insulating stands, as shown in the diagram below. Sphere A possesses an excess of 6.3×10^{10} electrons and sphere B is neutral.

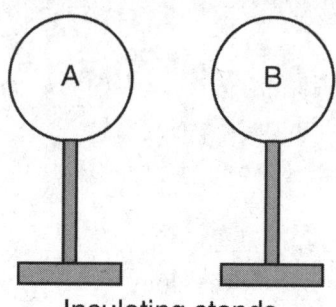

Insulating stands

Which diagram best represents the charge distribution on sphere B?

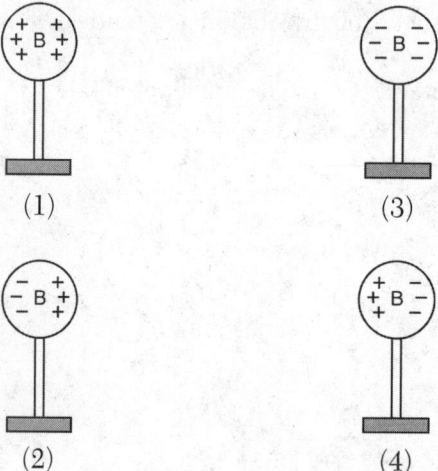

(1)   (3)

(2)   (4)

32 ____

33 Two points, A and B, are located within the electric field produced by a -3.0 nanocoulomb charge. Point A is 0.10 meter to the left of the charge and point B is 0.20 meter to the right of the charge, as shown in the diagram below.

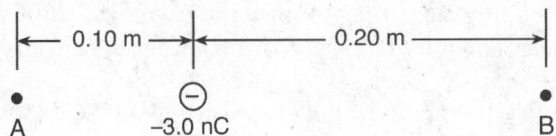

Compared to the magnitude of the electric field strength at point A, the magnitude of the electric field strength at point B is

(1) half as great   (3) one-fourth as great
(2) twice as great   (4) four times as great   33 _____

34 The diagram below represents two waves, A and B, traveling through the same uniform medium.

Which characteristic is the same for both waves?

(1) amplitude   (3) period
(2) frequency   (4) wavelength   34 _____

35 The diagram below shows a periodic wave.

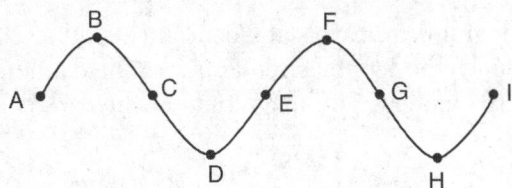

Which two points on the wave are 180° out of phase?

(1) A and C (3) F and G
(2) B and E (4) D and H 35 _____

PART B-1
Answer all questions in this part.

Directions (36–50): For *each* statement or question, select the *number* of the word or expression that, of those given, best completes the statement or answers the question. Some questions may require the use of the *2006 Edition Reference Tables for Physical Setting/Physics*. Record your answers in the spaces provided.

36 The height of a 30-story building is approximately

(1) 10^0 m   (3) 10^2 m

(2) 10^1 m   (4) 10^3 m   36 _____

37 Two identically sized metal spheres on insulating stands are positioned as shown below. The charge on sphere A is -4.0×10^{-6} coulomb and the charge on sphere B is -8.0×10^{-6} coulomb.

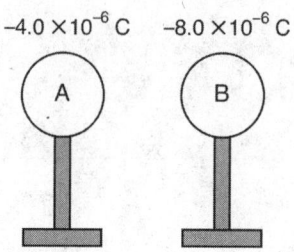

The two spheres are touched together and then separated. The total number of excess electrons on sphere A after the separation is

(1) 2.5×10^{13} (3) 5.0×10^{13}

(2) 3.8×10^{13} (4) 7.5×10^{13} 37 _____

38 A 1.0×10^3-kilogram car travels at a constant speed of 20 meters per second around a horizontal circular track. The diameter of the track is 1.0×10^2 meters. The magnitude of the car's centripetal acceleration is

(1) 0.20 m/s^2 (3) 8.0 m/s^2

(2) 2.0 m/s^2 (4) 4.0 m/s^2 38 _____

39 Which combination of units can be used to express electrical energy?

(1) $\dfrac{\text{volt}}{\text{coulomb}}$

(2) $\dfrac{\text{coulomb}}{\text{volt}}$

(3) volt • coulomb

(4) volt • coulomb • second

39 _____

40 The total amount of electrical energy used by a 315-watt television during 30.0 minutes of operation is

(1) 5.67×10^5 J

(2) 9.45×10^3 J

(3) 1.05×10^1 J

(4) 1.75×10^{-1} J

40 _____

41 Which graph best represents the relationship between the absolute index of refraction and the speed of light ($f = 5.09 \times 10^{14}$ Hz) in various media?

(1)

(3)

(2)

(4)

41 _____

42 A 25-gram paper cup falls from rest off the edge of a tabletop 0.90 meter above the floor. If the cup has 0.20 joule of kinetic energy when it hits the floor, what is the total amount of energy converted into internal (thermal) energy during the cup's fall?

(1) 0.02 J

(2) 0.22 J

(3) 2.2 J

(4) 220 J

42 _____

43 Which electron transition between the energy levels of hydrogen causes the emission of a photon of visible light?

(1) $n = 6$ to $n = 5$ (3) $n = 5$ to $n = 2$

(2) $n = 5$ to $n = 6$ (4) $n = 2$ to $n = 5$ 43 _____

44 Which graph best represents an object in equilibrium moving in a straight line?

(1)

(3)

(2)

(4)   44 _____

45 A body, B, is moving at constant speed in a horizontal circular path around point P. Which diagram shows the direction of the velocity (v) and the direction of the centripetal force (F_c) acting on the body?

(1)

(3)

(2)

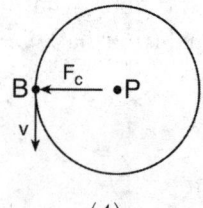

(4)   45 _____

46 Which graph best represents the relationship between photon energy and photon wavelength?

(1)

(3)

(2)

(4)

46 _____

47 Which combination of initial horizontal velocity (v_H) and initial vertical velocity (v_v) results in the greatest horizontal range for a projectile over level ground? [Neglect friction.]

v_v = 6 m/s

v_H = 8 m/s

(1)

v_v = 6 m/s

v_H = 6 m/s

(3)

v_v = 8 m/s

v_H = 6 m/s

(2)

v_v = 8 m/s

v_H = 8 m/s

(4)

47 _____

48 Which graph represents the greatest amount of work?

(1)

(3)

(2)

(4)

48 _____

49 When a ray of light traveling in water reaches a boundary with air, part of the light ray is reflected and part is refracted. Which ray diagram best represents the paths of the reflected and refracted light rays?

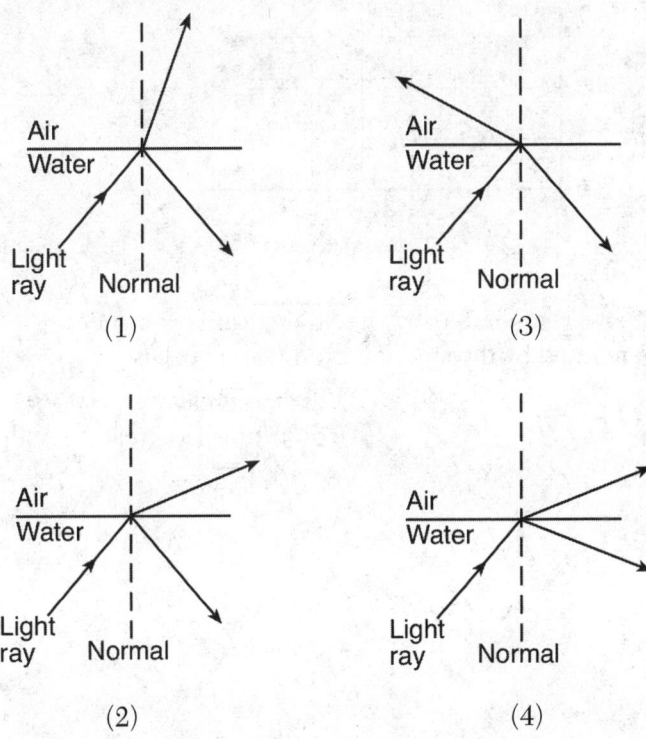

(1)

(2)

(3)

(4)

49 _____

50 The graph below represents the work done against gravity by a student as she walks up a flight of stairs at constant speed.

Compared to the power generated by the student after 2.0 seconds, the power generated by the student after 4.0 seconds is

(1) the same   (3) half as great

(2) twice as great   (4) four times as great   50 _____

PART B–2
Answer all questions in this part.

Directions (51–65): Record your answers on the answer sheet provided after the questions. Some questions may require the use of the *2006 Edition Reference Tables for Physical Setting/Physics*.

IMPORTANT NOTE:

- 2-point calculation problems appear as two linked questions in the test, for example, 55–56.
- In the first question, 1 point is allowed for showing the equation *and* for substituting the correct values with their units.
- In the second question, 1 point is allowed for a correct answer (number *and* unit) or for an answer that is consistent with your work in the first question.

Base your answers to questions 51 through 54 on the information below and the scaled vector diagram on the answer sheet and on your knowledge of physics.

Two forces, a 60-newton force east and an 80-newton force north, act concurrently on an object located at point P, as shown.

51 Using a ruler, determine the scale used in the vector diagram. [1]

52 Draw the resultant force vector to scale on the diagram *on the answer sheet*. Label the vector "*R*." [1]

53 Determine the magnitude of the resultant force, *R*. [1]

54 Determine the measure of the angle, in degrees, between north and the resultant force, *R*. [1]

55–56 A 3.00-newton force causes a spring to stretch 60.0 centimeters. Calculate the spring constant of this spring. [Show all work, including the equation and substitution with units.] [2]

57 A 7.28-kilogram bowling ball traveling 8.50 meters per second east collides head-on with a 5.45 kilogram bowling ball traveling 10.0 meters per second west. Determine the magnitude of the total momentum of the two-ball system after the collision. [1]

58–59 Calculate the average power required to lift a 490-newton object a vertical distance of 2.0 meters in 10 seconds. [Show all work, including the equation and substitution with units.] [2]

60 The diagram *on the answer sheet* shows wave fronts approaching an opening in a barrier. The size of the opening is approximately equal to one-half the wavelength of the waves. On the diagram *on the answer sheet*, draw the shape of *at least three* of the wave fronts after they have passed through this opening. [1]

61 The diagram *on the answer sheet* shows a mechanical transverse wave traveling to the right in a medium. Point A represents a particle in the medium. Draw an arrow originating at point A to indicate the initial direction that the particle will move as the wave continues to travel to the right in the medium. [1]

62 Regardless of the method used to generate electrical energy, the amount of energy provided by the source is always greater than the amount of electrical energy produced. Explain why there is a difference between the amount of energy provided by the source and the amount of electrical energy produced. [1]

[1 point]

Base your answers to questions 63 through 65 on the graph below, which represents the relationship between velocity and time for a car moving along a straight line, and your knowledge of physics.

63 Determine the magnitude of the average velocity of the car from $t = 6.0$ seconds to $t = 10$ seconds. [1]

64 Determine the magnitude of the car's acceleration during the first 6.0 seconds. [1]

65 Identify the physical quantity represented by the shaded area on the graph. [1]

PART C
Answer all questions in this part.

Directions (66–85): Record your answers on the answer sheet provided after the questions. Some questions may require the use of the *2006 Edition Reference Tables for Physical Setting/Physics.*

IMPORTANT NOTE:

- 2-point calculation problems appear as two linked questions in the test, for example, 69–70.
- In the first question, 1 point is allowed for showing the equation *and* for substituting the correct values with their units.
- In the second question, 1 point is allowed for a correct answer (number and unit) or for an answer that is consistent with your work in the first question.

Base your answers to questions 66 through 70 on the information below and on your knowledge of physics.

A student constructed a series circuit consisting of a 12.0-volt battery, a 10.0-ohm lamp, and a resistor. The circuit does not contain a voltmeter or an ammeter. When the circuit is operating, the total current through the circuit is 0.50 ampere.

66 In the space *on the answer sheet*, draw a diagram of the series circuit constructed to operate the lamp, using symbols from the *Reference Tables for Physical Setting/Physics*. [1]

67 Determine the equivalent resistance of the circuit. [1]

68 Determine the resistance of the resistor. [1]

69–70 Calculate the power consumed by the lamp. [Show all work, including the equation and substitution with the units.] [2]

Base your answers to questions 71 through 75 on the information below and on your knowledge of physics.

Pluto orbits the Sun at an average distance of 5.91×10^{12} meters. Pluto's diameter is 2.30×10^6 meters and its mass is 1.31×10^{22} kilograms.

Charon orbits Pluto with their centers separated by a distance of 1.96×10^7 meters. Charon has a diameter of 1.21×10^6 meters and a mass of 1.55×10^{21} kilograms.

71–72 Calculate the magnitude of the gravitational force of attraction that Pluto exerts on Charon. [Show all work, including the equation and substitution with units.] [2]

73–74 Calculate the magnitude of the acceleration of Charon toward Pluto. [Show all work, including the equation and substitution with units.] [2]

75 State the reason why the magnitude of the Sun's gravitational force on Pluto is greater than the magnitude of the Sun's gravitational force on Charon. [1]

Base your answers to questions 76 through 80 on the information below and on your knowledge of physics.

A horizontal 20-newton force is applied to a 5.0-kilogram box to push it across a rough, horizontal floor at a constant velocity of 3.0 meters per second to the right.

76 Determine the magnitude of the force of friction acting on the box. [1]

77–78 Calculate the weight of the box. [Show all work, including the equation and substitution with units.] [2]

79–80 Calculate the coefficient of kinetic friction between the box and the floor. [Show all work, including the equation and substitution with units] [2]

Base your answers to questions 81 through 85 on the information below and on your knowledge of physics.

An electron traveling with a speed of 2.50×10^6 meters per second collides with a photon having a frequency of 1.00×10^{16} hertz. After the collision, the photon has 3.18×10^{-18} joule of energy.

81–82 Calculate the original kinetic energy of the electron. [Show all work, including the equation and substitution with units.] [2]

  83 Determine the energy in joules of the photon before the collision. [1]

  84 Determine the energy lost by the photon during the collision. [1]

  85 Name *two* physical quantities conserved in the collision. [1]

Answer Sheet
June 2014
Physics: The Physical Setting

PART B–2

51 1.0 cm = _____ **N**

52

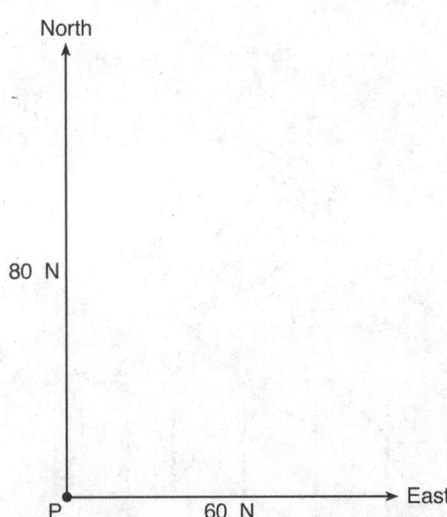

53 _____ **N**

54 _____ °

55–56

57 _____ **kg • m/s**

58–59

60

61

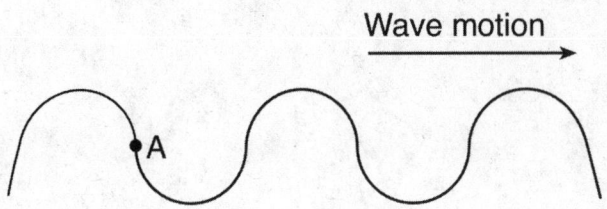

62 _____

_____

63 _____ **m/s**

64 _____ **m/s^2**

65 _____

PART C

66

67 _____ Ω

68 _____ Ω

69–70

71–72

73–74

75 _____

_____

76 _____ **N**

77–78

79–80

81–82

83 _____**J**

84 _____**J**

85 _____ and _____

Answers
June 2014
Physics: The Physical Setting

Answer Key

PART A

1. 1	8. 2	15. 3	22. 2	29. 1
2. 1	9. 2	16. 1	23. 4	30. 2
3. 2	10. 1	17. 4	24. 1	31. 3
4. 3	11. 3	18. 2	25. 4	32. 4
5. 1	12. 2	19. 4	26. 2	33. 3
6. 3	13. 2	20. 1	27. 3	34. 1
7. 4	14. 2	21. 4	28. 2	35. 1

PART B–1

36. 3	39. 3	42. 1	45. 3	48. 2
37. 2	40. 1	43. 3	46. 1	49. 2
38. 3	41. 2	44. 4	47. 4	50. 1

PART B–2 and **PART C**. *See* **Answers Explained**.

Answers Explained

PART A

1 A scalar quantity has magnitude (a measured value) but not direction. The mass of an object has magnitude only (for example, 50 kilograms) but not direction. Therefore, mass is a scalar quantity.

WRONG CHOICES EXPLAINED:
(2), (3), (4) Force, momentum, and acceleration all have direction associated with them. For example, a force can be 20. newtons north. Since these quantities have both magnitude and direction, they are vector quantities.

2 Use *Equation ME5* on *Reference Table K*:

Solution: $v^2_f = v_i^2 + 2ad$

$$= (0.0 \text{ m/s})^2 + 2(4.0 \text{ m/s}^2)(8.0 \text{ m})$$

$$= 64 \text{ m}^2/\text{s}^2$$

$$v_f = 8.0 \text{ m/s}$$

3 Use *Equations ME6* and *ME7* on *Reference Table K*:

Solution: $A_y = A \sin \theta$

$$= (15 \text{ m/s}) \sin \ 60°$$

$$= (15 \text{ m/s})(0.8660)$$

$$= 13 \text{ m/s}$$

(vertical component)

$$A_x = A \cos \theta$$

$$= (15 \text{ m/s}) \cos \ 60°$$

$$= (15 \text{ m/s})(0.5)$$

$$= 7.5 \text{ m/s}$$

(horizontal component)

4 Since it is released from rest, the initial speed, v_i, is 0.0 meters per second. The acceleration due to gravity is 9.81 meters per second squared (see *Reference Table A*). Use *Equation ME4* on *Reference Table K*:

Solution:
$$d = v_i t + \frac{1}{2}at^2$$

$$d = \frac{1}{2}at^2$$

$$t = \sqrt{\frac{2d}{a}}$$

$$= \sqrt{\frac{2(500 \text{ m})}{9.81 \text{ m/s}^2}}$$

$$= 10.1 \text{ s}$$

5 Newton's third law states that if object *A* exerts a force on object *B*, then object *B* exerts an equal and opposite force on object *A*. An alternate way of stating this is that for every action, there is an equal and opposite reaction. Hence these forces are called an "action-reaction pair." So the magnitude of the force that the ball exerts on the bat is equal to the force that the bat exerts on the ball.

WRONG CHOICES EXPLAINED:
(2), (3), (4) The relative masses of the objects do not affect the forces described.

6 Earth's gravitational field strength is symbolized by the letter *g*. See the definition of symbols on *Reference Table K*. The letter *g* is also the symbol for the acceleration due to gravity, which near the surface of Earth is 9.81 m/s^2 (see *Reference Table A*). Since the question states that the mass is 3.0 meters above the surface of Earth, this location is considered to be near the surface. To 2 significant digits, the gravitational field strength can be rounded to 9.8. Note that N/kg are equivalent units to m/s^2.

WRONG CHOICES EXPLAINED:
(1), (4) In the absence of air resistance, all objects accelerate at the same rate when they fall. So the gravitational field strength does not depend on the mass of the falling object.

7 Use *Equation ME4* on *Reference Table K*:

Solution:
$$d = v_i t + \frac{1}{2} at^2$$
$$= (22 \text{ m/s})(3.2 \text{ s}) + \frac{1}{2}(2.4 \text{ m/s}^2)(3.2 \text{ s})^2$$
$$= 83 \text{ m}$$

8 According to Newton's second law, when the forces acting on an object are unbalanced, the object will accelerate in the direction of the greater force. To accelerate downward, the downward force must be greater than the upward force. In this case, the downward force is the gravitational force (the person's weight) of 750 newtons. So the upward force (the normal force coming from the floor of the elevator) must be less than 750 newtons.

9 An impulse is a force acting for a certain amount of time. An impulse acting on an object causes the object to change its momentum. Use *Equation ME14* on *Reference Table K*. The impulse is numerically equal to the object's change in momentum ($J = \Delta p$). If the impulse is 15 newton • seconds, then the change in momentum is 15 kg • m/s. Note that kg • m/s are equivalent units to newton • seconds.

WRONG CHOICES EXPLAINED:
(1), (4) Since impulse and change in momentum are equal, the mass of the object is irrelevant.

10 Use *Equation ME14* on *Reference Table K*. The product of the force acting on an object and the time that force acts equals the object's change in momentum. The change in momentum of a driver in a car accident will be the same whether he or she hits an air bag or the steering wheel. In either case, the driver goes from moving very fast to being stopped. If the driver hits the steering wheel he or she stops very quickly since the wheel is hard. So a large force must be applied. If the driver hits an air bag, he or she stops more slowly since the air bag compresses. Thus a smaller force is applied, doing less damage to the driver. Look at the equation to see this mathematically:

$$\Delta p_{\text{hitting steering wheel}} = \Delta p_{\text{hitting air bag}}$$
$$F_{net} \times t = F_{net} \times t$$
$$(\text{large force}) \times (\text{small time}) = (\text{small force}) \times (\text{large time})$$

11 All charged particles produce an electric field. Magnetic fields are produced by electric charges that are moving. Thus, a moving electron produces both types of fields.

12 Use *Equation EL2* on *Reference Table K*. Find the elementary charge (the charge on an electron or a proton) from *Reference Table A*:

Solution:

$$E = \frac{F_e}{q}$$

$$F_e = Eq$$

$$= \left(3.00 \times 10^3 \text{ N/C}\right)\left(1.60 \times 10^{-19} \text{ C}\right)$$

$$= 4.80 \times 10^{-16} \text{ N}$$

13 The distance is irrelevant. Use *Equation EL3* on *Reference Table K*:

Solution:

$$V = \frac{W}{q}$$

$$W = qV$$

$$= (3.0\text{C})(9.0 \text{ V})$$

$$= 27 \text{ J}$$

14 Use *Equation EL6* on *Reference Table K*. The resistivity (ρ) of tungsten can be found on *Reference Table J*:

Solution:

$$R = \frac{\rho L}{A}$$

$$= \frac{\left(5.60 \times 10^{-8} \text{ } \Omega \cdot \text{m}\right)\left(20.0 \text{ m}\right)}{\left(1.00 \times 10^{-4} \text{ m}^2\right)}$$

$$= 1.12 \times 10^{-2} \text{ } \Omega$$

15 Mechanical energy (kinetic and potential energy) is converted into electromagnetic energy and internal energy when the two pieces of rock are struck together since a visible spark is produced. Remember that visible light is a type of electromagnetic radiation. (See *Reference Table D.*) In addition, the rock pieces get warmer (increase in internal energy) as a result of the collision.

WRONG CHOICES EXPLAINED:

(1), (2), (4) Nuclear energy is not released since no atom has had its nucleus split. Visible light results instead from electrons transitioning between energy levels. Electrons are not in the nucleus.

16 Inertia is the property of an object that resists a change in motion. The measure of an object's inertia is its mass. Therefore, the more mass an object has the more inertia it has.

17 All electromagnetic waves are transverse. As shown on *Reference Table D* radio waves are a type of electromagnetic wave. So radio waves are transverse Sound waves are longitudinal and mechanical.

WRONG CHOICES EXPLAINED:

(1), (2), (3) Light waves, microwaves, and ultraviolet waves are all on the electromagnetic spectrum as shown on *Reference Table D*. So they are all classified as transverse waves.

18 As a light wave passes from air into water, it slows down. However, its frequency remains the same. Since frequency and period are reciprocals of each other as shown on *Equation W2* on *Reference Table K*, the period also remains the same

WRONG CHOICES EXPLAINED:

(1), (3) Speed and wavelength are directly proportional, as shown by *Equation W1* on *Reference Table K*. Since the speed decreases, the wavelength decreases as well.

19 Use *Equation ME16* on *Reference Table K*:

Solution:

$$PE_s = \frac{1}{2}kx^2$$

$$= \frac{1}{2}(60.0 \text{ N/m})(0.500 \text{ m})^2$$

$$= 7.50 \text{ J}$$

WRONG CHOICE EXPLAINED:
(3) Did you forget to square the 0.500 m? Beware the square!

20 Frequency is a measure of the number of cycles (oscillations) per second:

Solution:
$$f = \frac{\text{number of swings}}{\text{number of seconds}}$$
$$= \frac{90 \text{ swings}}{300 \text{ seconds}}$$
$$= 0.30 \text{ Hz}$$

21 Use *Equation W2* on *Reference Table K*:

Solution:
$$T = \frac{1}{f}$$
$$= \frac{1}{340 \text{ Hz}}$$
$$= 2.94 \times 10^{-3} \text{ s}$$

22 The voltage is irrelevant. Use *Equation EL4* on *Reference Table K*:

Solution:
$$I = \frac{\Delta q}{t}$$
$$\Delta q = I \cdot t$$
$$= (0.120 \text{ A})(900 \text{ s})$$
$$= 108 \text{ C}$$

23 Use *Equation W1* on *Reference Table K* and see *Reference Table A*. The speed of all electromagnetic waves in a vacuum, including light, is 3.00×10^8 meters per second:

Solution:
$$v = f\lambda$$
$$f = \frac{v}{\lambda}$$
$$= \frac{3.00 \times 10^8 \text{ m/s}}{4.5 \times 10^{-7} \text{ m}}$$
$$= 6.7 \times 10^{14} \text{ Hz}$$

24 See *Reference Table A* and *Reference Table D*. Since the speed of all electromagnetic waves in a vacuum is 3.00×10^8 meters per second, x-ray and infrared radiation (waves) have the same speed. They are both types of electromagnetic waves.

WRONG CHOICES EXPLAINED:

(2), (3) You can also see on *Reference Table D* that x-ray radiation and infrared radiation have different frequencies and wavelengths.

(4) The energy per photon depends on the frequency (or wavelength) of the photon. See *Equation MP1* on *Reference Table K*.

25 Each pulse has an amplitude (height or maximum displacement) of 3 centimeters. When the two pulses meet, the amplitude of the resultant pulse will be the sum of the two individual amplitudes, which is 6 centimeters. In other words, the medium will be displaced twice as much when the waves overlap. The waves will be superposed.

26 When a source of sound (the horn) and an observer (the pedestrian) are moving relative to each other, the apparent frequency heard by the observer will be different than the actual frequency emitted by the source. This is known as the Doppler effect. When the source and the observer are moving toward each other the apparent frequency heard by the observer will be greater than the actual frequency emitted by the source. Additionally, the term *pitch* is a common term that means frequency. Hence, a high pitch sound has a high frequency.

27 All systems have one or more frequencies at which they naturally tend to vibrate. Think of a guitar string or a child on a swing. If they are forced to vibrate at one of these natural frequencies by some external driving force, the amount they vibrate is large. In this case, blowing across the top of the bottle forces the air molecules in the bottle to vibrate enough to produce a sound. This phenomenon is known as *resonance*.

WRONG CHOICES EXPLAINED:

(1) Diffraction is the bending of a wave as it passes through a small opening or around a barrier.

(2) Refraction is the change in direction of a wave when it crosses a boundary between two different media at an angle if the speed of the wave is different in the two media.

(4) The Doppler effect is the apparent change in frequency of a wave due to the relative motion between the source of the wave and the observer of the wave.

28 See the note that is written below *Reference Table H*. The charge on any antiparticle is the opposite charge of the associated particle. Each antiup quark has a charge of $-\frac{2}{3}$e, and the antidown quark has a charge of $+\frac{1}{3}$e. Add these together to find the total charge on the antibaryon:

Solution: $\left(-\frac{2}{3}e\right) + \left(-\frac{2}{3}e\right) + \left(+\frac{1}{3}e\right) = -1e$

29 The strong force is an attractive force that governs the interactions of particles in the nucleus.

WRONG CHOICES EXPLAINED:
(2) The weak force governs certain types of radioactive decay.
(3) Frictional force is a type of electromagnetic force.
(4) The gravitational force governs the interactions among particles that have mass.

30 Use *Equation MP3* on *Reference Table K*. Find the speed of light in a vacuum (c) on *Reference Table A*:

Solution: $E = mc^2$

$$= \left(9.11 \times 10^{-31} \text{kg}\right)\left(3.00 \times 10^8 \text{ m/s}\right)^2$$

$$= 8.20 \times 10^{-14} \text{ J}$$

WRONG CHOICE EXPLAINED:
(1) Did you forget to square the speed of light? Beware the square!

31 As shown in *Equation ME20* on *Reference Table K*, kinetic energy is a measure of the energy of motion. This means that the faster an object travels, the more kinetic energy it has. As shown in *Equation ME19* on *Reference Table K*, gravitational potential energy is a measure of the energy of position. This means that the higher an object is above the ground, the more gravitational potential energy it has. Since the shopping cart is slowing down, its kinetic energy decreases. Since the shopping cart is on a level floor, its gravitational potential energy remains the same.

WRONG CHOICE EXPLAINED:
(4) Did you think that whenever an object's kinetic energy decreases, its gravitational potential energy increases (like when a ball is thrown up into the air)? That is not always the case!

32 Since sphere B is neutral, it has an equal number of protons and electrons. Because like charges repel each other, the electrons on sphere A will repel the electrons on sphere B to the side farthest away from sphere A.

WRONG CHOICES EXPLAINED:
(1), (3) Neither of these show a neutral sphere.

33 Use *Equations EL1 and EL2* on *Reference Table K*. Let $q_1 = -3.0$ nC, and let q_2 be a test charge placed at points A and B.

$$F_e = \frac{kq_1q_2}{r^2} \text{ and } E = \frac{F_e}{q_2}$$

$$E = \frac{\frac{kq_1q_2}{r^2}}{q_2}$$

$$= \frac{kq_1}{r^2}$$

Since k and q_1 remain the same and the distance is doubled, proportional reasoning can be used:

Solution: $$E = \frac{(1)(1)}{(2)^2}$$

$$= \frac{1}{4}$$

34 Amplitude is the maximum displacement of the wave from equilibrium, which is the maximum height of the wave. By looking at the diagram, the height of each wave appears to be the same.

WRONG CHOICES EXPLAINED:
(2), (3) Since frequency and period are reciprocals of each other, see *Equation W2* on *Reference Table K*, these are either both the same or both different.
(4) Wavelength is the length of the wave and can be measured from one crest to the next crest. By looking at the diagram, the wavelength of wave B appears to be shorter than the wavelength of wave A.

35 Two points on a wave are out of phase by 180° if the horizontal distance between them is an odd number of half-wavelengths. Points *A* and *C* are one half-wavelength (½ λ) apart from each other.

WRONG CHOICES EXPLAINED:
(2) *B* and *E* are ¾ λ apart, so they are out of phase by 270°.
(3) *F* and *G* are ¼ λ apart, so they are out of phase by 90°.
(4) *D* and *H* are a whole wavelength apart (1λ), so they are in phase.

PART B–1

36 Each story of a building is approximately 3 meters. How would you know that? Take a look at the room you're in right now. Imagine how many meter sticks would be needed to go from floor to ceiling—about 3 of them. A 30-story building would then be about 90 meters high. This is approximately 100 meters, which is 10^2 meters or roughly the size of a football field. Of the choices, this seems the most reasonable.

WRONG CHOICES EXPLAINED:
(1) 10^0 m is 1 meter, which is clearly too small.
(2) 10^1 m is 10 meters. This is a better answer than choice (1) but is still too small. This would be a the height of a 3-story building.
(4) 10^3 m is 1,000 m or 1 kilometer, which is about $\frac{2}{3}$ of a mile. This measurement is much too large.

37 Since the spheres are identical, the final charge on each sphere is the average of the two initial charges. The electrons are shared equally between the two spheres after the spheres are touched together. This charge has to be converted by the factor-label method into the number of elementary charges (electrons) it represents. Use the conversion factor found on *Reference Table A*:

Solution:
$$\text{Average charge} = \frac{\left(-4.0 \times 10^{-6}\,\text{C}\right) + \left(-8.0 \times 10^{-6}\,\text{C}\right)}{2}$$

$$= -6.0 \times 10^{-6}\,\text{C}$$

$$\text{Number of electrons} = \left(\frac{-6.0 \times 10^{-6}\,\text{C}}{1}\right)\left(\frac{1e}{-1.60 \times 10^{-19}\,\text{C}}\right)$$

$$\text{Number of electrons} = 3.8 \times 10^{13}\,e$$

38 Use *Equation ME18* on *Reference Table K*. The mass of the car is irrelevant. Be sure to divide the diameter by 2 to get the radius:

Solution:
$$a_c = \frac{v^2}{r}$$
$$= \frac{(20 \text{ m/s})^2}{(0.5 \times 10^2 \text{ m})}$$
$$= 8.0 \text{ m/s}^2$$

WRONG CHOICES EXPLAINED:

(1) You probably forgot to square the velocity and also to divide the diameter by 2.

(4) You probably forgot to divide the diameter by 2.

39 Use the equation for electrical energy found at *Equation EL3* on *Reference Table K*. Substitute the appropriate units without including any numbers:

Solution:
$$V = \frac{W}{q}$$
$$W = Vq$$
$$= (\text{V})(C)$$
$$= \text{volt} \bullet \text{coulomb}$$

40 Use *Equation EL8* on *Reference Table K*. Be sure to convert minutes into seconds by using the factor-label method:

Solution:
$$t = \left(\frac{30.0 \text{ minutes}}{1}\right)\left(\frac{60 \text{ seconds}}{1.0 \text{ minute}}\right)$$
$$t = 1,800 \text{ seconds}$$
$$W = Pt$$
$$= (315 \text{ W})(1,800 \text{ s})$$
$$= 5.67 \times 10^5 \text{ J}$$

WRONG CHOICE EXPLAINED:

(2) You probably forgot to convert minutes into seconds.

41 Use *Equation W4* on *Reference Table K*. The absolute index of refraction of a substance is inversely proportional to the speed of light in that substance. Only choice (2) shows a graph in which the index decreases as the speed increases.

42 According to the conservation of energy principle, the total energy (E_T) of a closed system at one instant is the same as its total energy at another instant. In other words, $E_T = E_T$. Use *Equation ME22* and *Equation ME19* on *Reference Table K* and use *Reference Table A*. Be sure to convert grams to kilograms:

Solution:

$$E_T = E_T$$
$$E_T = PE + KE + Q$$
$$PE + KE + Q = PE + KE + Q$$
$$mg\Delta h + 0 + 0 = 0 + KE + Q$$
$$(0.025 \text{ kg})(9.81 \text{ m/s}^2)(0.90 \text{ m}) = (0.20 \text{ J}) + Q$$
$$0.22 \text{ J} = (0.20 \text{ J}) + Q$$
$$Q = 0.02 \text{ J}$$

WRONG CHOICE EXPLAINED:
(4) You probably forgot to convert grams into kilograms.

43 Photons are emitted (as opposed to absorbed) only when an electron transitions from a higher energy level to a lower one. This reduces the possible answer choices to only (1) or (3). Check both of these choices to see if the emitted photon is visible, which involves a multistep process:

First determine the energy of the photon in choice (3), when the photon drops from energy level 5 to 2. Use *Equation MP2* on *Reference Table K* and the energy level diagram for hydrogen on *Reference Table F*:

$$
\begin{aligned}
E_{photon} &= E_i - E_f \\
&= E_5 - E_2 \\
&= (-0.54 \text{ eV}) - (-3.40 \text{ eV}) \\
&= 2.86 \text{ eV}
\end{aligned}
$$

Then convert this energy from electronvolts into joules by the factor-label method. Use the conversion factor found in *Reference Table A*:

$$E_{photon} = \left(\frac{2.86 \text{ eV}}{1}\right)\left(\frac{1.60 \times 10^{-19} \text{ J}}{1 \text{ eV}}\right)$$

$$= 4.58 \times 10^{-19} \text{ J}$$

Next determine the frequency of this photon. Use *Equation MP1* on *Reference Table K* and Planck's constant (h) found on *Reference Table A*:

$$E_{photon} = hf$$

$$f = \frac{E_{photon}}{h}$$

$$= \frac{4.58 \times 10^{-19} \text{ J}}{6.63 \times 10^{-34} \text{ J} \cdot \text{s}}$$

$$= 6.91 \times 10^{14} \text{ Hz}$$

Finally check to see if this frequency is in the range of visible light by looking at the electromagnetic spectrum found in *Reference Table D*. Light with this frequency is in the range of violet light. Whew! If you had determined the energy of the photon in choice (1) instead, you would have found that its frequency was not in the visible light range. At that point, you would not have had to redo the equations using choice (3). You could simply have chosen choice (3) because you would have eliminated all other answer choices.

44 An object in equilibrium is either at rest or moving with a constant velocity (constant speed in a straight line). Since this object is moving in a straight line, it must have a constant speed. *Equation ME20* on *Reference Table K* shows that if the speed is constant, so is the kinetic energy.

WRONG CHOICES EXPLAINED:

(1) This graph shows an object that is accelerating, not moving at a constant speed.

(2) This graph shows an object whose velocity (speed) is decreasing.

(3) This graph shows an object whose velocity (speed) is increasing. As shown in *Equation ME12* on *Reference Table K*, momentum is the product of mass and velocity.

45 When an object travels at constant speed in a horizontal circle, the force causing it to move in a circle (centripetal force) is always pointed toward the center of the circle. The word "centripetal" means *center-seeking*. If it were not for this force, according to Newton's first law, the object would continue traveling in a straight line in the direction it was already moving, that is, tangent to the circle.

46 Use *Equation MP1* on *Reference Table K*. The energy of a photon is inversely proportional to its wavelength. Only choice (1) shows a graph in which the photon energy decreases as the wavelength increases.

47 The greatest horizontal range for a projectile, in the absence of air resistance, is when the angle of launch is at 45° above the horizontal. See *Equations ME6 and ME7* on *Reference Table K*. This means that both the horizontal and vertical components of the initial velocity are equal.

48 Use *Equation ME21* on *Reference Table K*. The total work done is the area underneath a force-distance graph. By looking at the graphs, you can see that if you shade in the area underneath each graph for the first 0.5 meter, the greatest shading would be under graph (2). Alternatively, you could calculate the area underneath each graph.

49 When a ray of light travels from a substance with a higher index of refraction to a substance with a lower index of refraction, see *Reference Table E*, the light ray bends away from the normal. Additionally, the reflected ray must reflect at the same angle that the incident ray makes with the normal. This is called the law of reflection. See *Equation W3* on *Reference Table K*. Choice (2) shows both of these behaviors.

WRONG CHOICES EXPLAINED:
(1) This diagram shows the refracted ray bending toward the normal, not away from it.
(3) This diagram shows the refracted ray crossing to the other side of the normal.
(4) This diagram shows the angle of reflection not equal to the angle of incidence.

50 Use *Equation ME23* on *Reference Table K*. The power generated by the student is equal to the slope of the graph. Since the slope is the same at all points, the power generated is the same as well.

PART B–2

51 Using a ruler, measure the length of either vector. Then use the factor-label method to convert the answer to newtons. For example, the east vector is marked 60. N and measures 6.0 centimeters:

Solution: $\left(\dfrac{1.0 \text{ cm}}{1}\right)\left(\dfrac{60 \text{ N}}{6.0 \text{ cm}}\right) = 10 \text{ N}$

Answers ranging from 9.0 N to 11 N receive credit.

[1 point]

52 Refer to the diagram below. Be sure to draw an arrowhead at the end of the resultant vector or you will not receive credit.

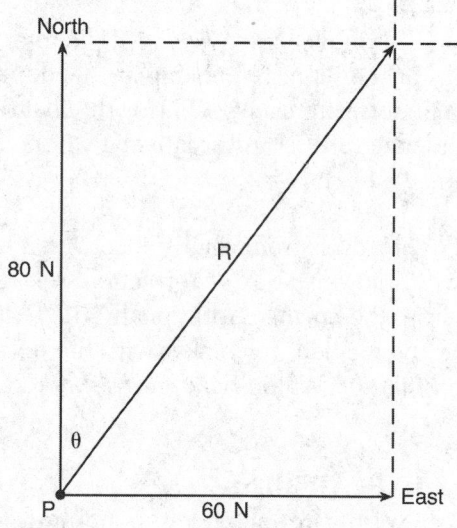

[1 point]

53 This question can be done in one of two ways, either by measuring it or by calculating it.

Method 1: Measure the resultant vector with a ruler, and use your scale to determine its magnitude. The vector measures 10.0 centimeters:

Solution: $\left(\dfrac{10.0 \text{ cm}}{1}\right)\left(\dfrac{10 \text{ N}}{1.0 \text{ cm}}\right) = 100 \text{ N}$

Method 2: Use the Pythagorean theorem found in *Equation GT4A* on *Reference Table K* to calculate the magnitude of the resultant vector:

Solution: $c^2 = a^2 + b^2$

$= (60 \text{ N})^2 + (80 \text{ N})^2$

$c = 100 \text{ N}$

Answers ranging from 97 N to 103 N receive credit.

[1 point]

54 This question can be done in one of two ways, either by measuring it or by calculating it.

Method 1: Measure the angle with a protractor.

Method 2: Use one of the trigonometric formulas found in *Equation GT4B, Equation GT4C*, or *Equation GT4D* on *Reference Table K* to calculate the angle:

Solution: $\sin \theta = \dfrac{a}{c}$

$= \dfrac{60 \text{ N}}{100 \text{ N}}$

$\theta = 37°$

Answers ranging from 35° to 39° receive credit.

[1 point]

55–56 Use *Equation ME15* on *Reference Table K*. Converting centimeters into meters is optional:

Solution: $F_s = kx$ $F_s = kx$

$k = \dfrac{F_s}{x}$ or $k = \dfrac{F_s}{x}$

$= \dfrac{3.00 \text{ N}}{0.600 \text{ m}}$ $= \dfrac{3.00 \text{ N}}{60.0 \text{ cm}}$

[1 point]

56 Use the equation you found in question 55.

   Solution: $k = 5.00$ N/m or $k = 0.0500$ N/cm

[1 point]

57 Use *Equation ME12* and *Equation ME13* on *Reference Table K*. Assume
east is the positive direction. Be sure to set negative any velocity that is directed
west:

   Solution:

$$p_{before} = p_{after}$$
$$p_{1initial} + p_{2initial} = p_{after}$$
$$m_1 v_{1i} + m_2 v_{2i} = p_{after}$$
$$(7.28 \text{ kg})(8.50 \text{ m/s}) + (5.45 \text{ kg})(-10.0 \text{ m/s}) = p_{after}$$
$$61.88 \text{ kg} \cdot \text{m/s} + (-54.5 \text{kg} \cdot \text{m/s}) = p_{after}$$
$$p_{after} = 7.38 \text{kg} \cdot \text{m/s}$$

Credit is allowed for 7.3 kg • m/s or 7.4 kg • m/s.

[1 point]

58–59 Use *Equation ME23* on *Reference Table K*:

   Solution: $$P = \frac{Fd}{t}$$
$$\frac{(490 \text{ N})(2.0 \text{ m})}{10 \text{ s}}$$

[1 point]

59 Use the equation you found in question 57.

   Solution: $P = 98$ W

[1 point]

60 Refer to the diagram below, which shows an appropriate drawing that will receive credit.

[1 point]

61 Refer to the diagram below, which shows an appropriate drawing that will receive credit.

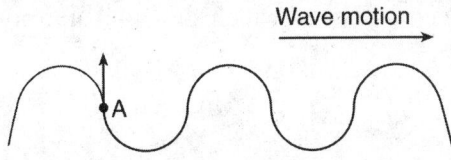

[1 point]

62 Acceptable answers include but are not limited to:

• Energy is needed to overcome friction.
• Energy is converted into internal (thermal) energy in the moving parts.
• Energy is converted into sound.

[1 point]

63 Since the velocity of the car is constant from 6.0 seconds to 10 seconds, the answer may be read directly off the graph.

   Solution: 15 m/s

[1 point]

64 Use *Equation ME2* on *Reference Table K*:

Solution: $a = \dfrac{\Delta v}{t}$

$= \dfrac{(15 \text{ m/s}) - (0.0 \text{ m/s})}{6.0 \text{ s}}$

$= 2.5 \text{ m/s}^2$

[1 point]

65 The area underneath a velocity-time graph measures how far an object has traveled in a given time. Acceptable answers include but are not limited to:

- displacement
- distance
- how far the car traveled

[1 point]

PART C

66 Refer to the diagram below, which shows an appropriate drawing that will receive credit.

[1 point]

67 Use *Equation EL5* on *Reference Table K*. Since you are calculating the total resistance (equivalent resistance) of the circuit, use the total potential difference and the total current:

Solution: $R = \dfrac{V}{I}$

$= \dfrac{12.0 \text{ V}}{0.50 \text{ A}}$

$= 24 \text{ }\Omega$

[1 point]

68 Use *Equation EL9C* on *Reference Table K:*

Solution: $R_{eq} = R_1 + R_2$

$R_2 = R_{eq} - R_1$

$R_2 = 24\ \Omega - 10.0\ \Omega$

$R_2 = 14\ \Omega$

[1 point]

69–70   Use *Equation EL7* on *Reference Table K:*

Solution: $P = I^2R$

$= (0.50\ \text{A})^2 (10.0\ \Omega)$

[1 point]

70 Use the equation you found in question 69.

Solution: $P = 2.5\ \text{W}$

[1 point]

71–72   Use *Equation ME10* on *Reference Table K.* The universal gravitational constant (G) is found on *Reference Table A.* The diameters of Pluto and Charon are irrelevant:

Solution:

$$F_g = \frac{Gm_1m_2}{r^2}$$

$$= \frac{\left(6.67 \times 10^{-11}\ \text{N} \cdot \text{m}^2/\text{kg}^2\right)\left(1.31 \times 10^{22}\ \text{kg}\right)\left(1.55 \times 10^{21}\ \text{kg}\right)}{\left(1.96 \times 10^7\ \text{m}\right)^2}$$

[1 point]

72 Use the equation you found in question 71.

Solution: $F_g = 3.53 \times 10^{18}\ \text{N}$

[1 point]

73–74 Use *Equation ME8* on *Reference Table K.* Use the mass of Charon since it is the object that is accelerating:

Solution:
$$a = \frac{F_{net}}{m}$$
$$= \frac{3.53 \times 10^{18}\,\text{N}}{1.55 \times 10^{21}\,\text{kg}}$$

[1 point]

74 Use the equation you found in question 73.

Solution: $a = 2.28 \times 10^{-3}$ m/s^2 or $a = 2.27 \times 10^{-3}$ m/s^2

Credit is given for either answer as long as it is consistent with the equation you found in question 73.

[1 point]

75 The gravitational force of attraction between two objects depends on the mass of both objects. The gravitational force from the Sun is greater on Pluto than on Charon because the mass of Pluto is greater than the mass of Charon.

[1 point]

76 According to Newton's first law, since the box is moving at a constant velocity, the forces acting on it are balanced. Therefore, the force of friction acting to the left is equal to the applied force pushing the box to the right:

Solution: 20 N

[1 point]

77–78 Use *Equation ME11* on *Reference Table K.* The acceleration due to gravity (g) is found on *Reference Table A.*

Solution:
$$g = \frac{F_g}{m}$$
$$F_g = mg$$
$$F_g = (5.0\ \text{kg})(9.81\ \text{m/s}^2)$$

[1 point]

78 Use the equation you found in question 77.

Solution: $F_g = 49$ N

[1 point]

79–80 Use *Equation ME9* on *Reference Table* K. The normal force, *FN*, which is the force that the floor exerts upward on the box, is equal and opposite to the weight of the box:

Solution: $F_f = \mu F_N$

$$\mu = \frac{F_f}{F_N}$$

$$= \frac{20 \text{ N}}{49 \text{ N}}$$

[1 point]

80 Use the equation you found in question 79.

Solution: $\mu = 0.41$ or 0.40.

Note that the answer has no units.

[1 point]

81–82 Use *Equation ME20* on *Reference Table* K. The mass of the electron is found on *Reference Table A*:

Solution: $KE = \frac{1}{2}mv^2$

$$= \frac{1}{2}\left(9.11 \times 10^{-31} \text{ kg}\right)\left(2.50 \times 10^6 \text{ m/s}\right)^2$$

[1 point]

82 Use the equation you found in question 81.

Solution: $KE = 2.85 \; 10^{-18}$ J

[1 point]

83 Use *Equation MP1* on *Reference Table K*. The value of Planck's constant (h) is found on *Reference Table A*:

Solution: $E_{photon} = hf$

$$= \left(6.63 \times 10^{-34} \text{ J} \bullet \text{s}\right)\left(1.00 \times 10^{16} \text{ Hz}\right)$$

$$= 6.63 \times 10^{-18} \text{ J}$$

[1 point]

84 The energy lost by the photon during the collision is the difference between its initial and final energies.

Solution: $E_{lost} = E_i - E_f$

$$= \left(6.63 \times 10^{-18} \text{ J}\right) - \left(3.18 \times 10^{-18} \text{ J}\right)$$

$$= 3.45 \times 10^{-18} \text{ J}$$

[1 point]

85 1 point is given for *two* acceptable quantities. Acceptable answers include but are not limited to:

- mass
- charge
- momentum
- energy

[1 point]

Topic	Question Numbers (total)	Wrong Answers (x)	Grade
Math Skills	41, 46, 48, 51–54, 64, 65: (9)		$\dfrac{100\,(9 - x)}{9} = \%$
Mechanics	1–10, 16, 36, 38, 44, 45, 47, 55–57, 63, 71–80: (30)		$\dfrac{100\,(30 - x)}{30} = \%$
Energy	15, 19, 31, 42, 50, 58, 59, 62, 81, 82, 84: (11)		$\dfrac{100\,(11 - x)}{11} = \%$
Electricity/ Magnetism	11–14, 22, 32, 33, 37, 39, 40, 66–70: (15)		$\dfrac{100\,(15 - x)}{15} = \%$
Waves	17, 18, 20, 21, 23–27, 34, 35, 49, 60, 61: (14)		$\dfrac{100\,(14 - x)}{14} = \%$
Modern Physics	28–30, 43, 83, 85: (6)		$\dfrac{100\,(6 - x)}{6} = \%$

Examination
June 2015
Physics: The Physical Setting

PART A
Answer all questions in this part.

Directions (1–35): For *each* statement or question, select the *number* of the word or expression that, of those given, best completes the statement or answers the question. Some questions may require the use of the *2006 Edition Reference Tables for Physical Setting/Physics.* Record your answers in the spaces provided.

1 Which quantities are scalar?
  (1) speed and work
  (2) velocity and force
  (3) distance and acceleration
  (4) momentum and power   1 ____

2 A 3.00-kilogram mass is thrown vertically upward with an initial speed of 9.80 meters per second. What is the maximum height this object will reach? [Neglect friction.]
  (1) 1.00 m   (3) 9.80 m
  (2) 4.90 m   (4) 19.6 m   2 ____

3 An airplane traveling north at 220 meters per second encounters a 50-meters-per-second crosswind from west to east, as represented in the diagram below.

220. m/s

50.0 m/s

What is the resultant speed of the plane?

(1) 170 m/s   (3) 226 m/s

(2) 214 m/s   (4) 270 m/s   3 _____

4 A 160-kilogram space vehicle is traveling along a straight line at a constant speed of 800 meters per second. The magnitude of the net force on the space vehicle is

(1) 0 N   (3) 8.00×10^2 N

(2) 1.60×10^2 N   (4) 1.28×10^5 N   4 _____

5 A student throws a 5.0-newton ball straight up. What is the net force on the ball at its maximum height?

(1) 0.0 N   (3) 5.0 N, down

(2) 5.0 N, up   (4) 9.8 N, down   5 _____

6 A vertical spring has a spring constant of 100 newtons per meter. When an object is attached to the bottom of the spring, the spring changes from its unstretched length of 0.50 meter to a length of 0.65 meter. The magnitude of the weight of the attached object is

(1) 1.1 N   (3) 50 N
(2) 15 N   (4) 65 N   6 ____

7 A 1.5-kilogram cart initially moves at 2.0 meters per second. It is brought to rest by a constant net force in 0.30 second. What is the magnitude of the net force?

(1) 0.40 N   (3) 10 N
(2) 0.90 N   (4) 15 N   7 ____

8 Which characteristic of a light wave must increase as the light wave passes from glass into air?

(1) amplitude   (3) period
(2) frequency   (4) wavelength   8 ____

9 As a 5.0×10^2-newton basketball player jumps from the floor up toward the basket, the magnitude of the force of her feet on the floor is 1.0×10^3 newtons. As she jumps, the magnitude of the force of the floor on her feet is

(1) 5.0×10^2 N   (3) 1.5×10^3 N
(2) 1.0×10^3 N   (4) 5.0×10^5 N   9 ____

10 A 0.0600-kilogram ball traveling at 60.0 meters per second hits a concrete wall. What speed must a 0.0100-kilogram bullet have in order to hit the wall with the same magnitude of momentum as the ball?

(1) 3.60 m/s   (3) 360 m/s
(2) 6.00 m/s   (4) 600 m/s   10 ____

11 The Hubble telescope's orbit is 5.6×10^5 meters above Earth's surface. The telescope has a mass of 1.1×10^4 kilograms. Earth exerts a gravitational force of 9.1×10^4 newtons on the telescope. The magnitude of Earth's gravitational field strength at this location is

(1) 1.5×10^{-20} N/kg   (3) 8.3 N/kg

(2) 0.12 N/kg   (4) 9.8 N/kg   11 _____

12 When two point charges are a distance d apart, the magnitude of the electrostatic force between them is F. If the distance between the point charges is increased to $3d$, the magnitude of the electrostatic force between the two charges will be

(1) $\frac{1}{9}F$ (3) $2F$

(2) $\frac{1}{3}F$ (4) $4F$ 12 _____

13 A radio operating at 3.0 volts and a constant temperature draws a current of 1.8×10^{-4} ampere. What is the resistance of the radio circuit?

(1) $1.7 \times 10^4 \, \Omega$ (3) $5.4 \times 10^{-4} \, \Omega$

(2) $3.0 \times 10^1 \, \Omega$ (4) $6.0 \times 10^{-5} \, \Omega$ 13 _____

14 Which energy transformation occurs in an operating electric motor?

(1) electrical → mechanical

(2) mechanical → electrical

(3) chemical → electrical

(4) electrical → chemical   14 _____

15 A block slides across a rough, horizontal tabletop. As the block comes to rest, there is an increase in the block-tabletop system's

(1) gravitational potential energy

(2) elastic potential energy

(3) kinetic energy

(4) internal (thermal) energy   15 _____

16 How much work is required to move an electron through a potential difference of 3.00 volts?

(1) 5.33×10^{-20} J   (3) 3.00 J

(2) 4.80×10^{-19} J   (4) 1.88×10^{19} J   16 ____

17 During a laboratory experiment, a student finds that at 20° Celsius, a 6.0-meter length of copper wire has a resistance of 1.3 ohms. The cross-sectional area of this wire is

(1) 7.9×10^{-8} m^2 (3) 4.6×10^{0} m^2

(2) 1.1×10^{-7} m^2 (4) 1.3×10^{7} m^2 17 ____

18 A net charge of 5.0 coulombs passes a point on a conductor in 0.050 second. The average current is

(1) 8.0×10^{-8} A   (3) 2.5×10^{-1} A

(2) 1.0×10^{-2} A   (4) 1.0×10^{2} A   18 ____

19 If several resistors are connected in series in an electric circuit, the potential difference across each resistor

(1) varies directly with its resistance
(2) varies inversely with its resistance
(3) varies inversely with the square of its resistance
(4) is independent of its resistance   19 ____

20 The amplitude of a sound wave is most closely related to the sound's

(1) speed   (3) loudness

(2) wavelength   (4) pitch   20 ____

21 A duck floating on a lake oscillates up and down 5.0 times during a 10.-second interval as a periodic wave passes by. What is the frequency of the duck's oscillations?

(1) 0.10 Hz   (3) 2.0 Hz

(2) 0.50 Hz   (4) 50.0 Hz   21 ____

22 Which diagram best represents the position of a ball, at equal time intervals, as it falls freely from rest near Earth's surface?

(1)   (2)   (3)   (4)   22 _____

23 A gamma ray and a microwave traveling in a vacuum have the same

(1) frequency   (3) speed

(2) period   (4) wavelength   23 _____

24 A student produces a wave in a long spring by vibrating its end. As the frequency of the vibration is doubled, the wavelength in the spring is

(1) quartered   (3) unchanged

(2) halved   (4) doubled   24 _____

25 Which two points on the wave shown in the diagram below are in phase with each other?

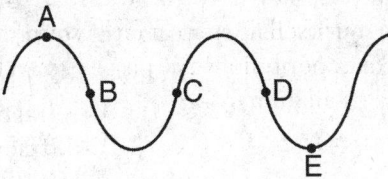

(1) *A* and *B*   (3) *B* and *C*

(2) *A* and *E*   (4) *B* and *D*   25 _____

26 As a longitudinal wave moves through a medium, the particles of the medium

(1) vibrate parallel to the direction of the wave's propagation
(2) vibrate perpendicular to the direction of the wave's propagation
(3) are transferred in the direction of the wave's motion, only
(4) are stationary   26 _____

27 Wind blowing across suspended power lines may cause the power lines to vibrate at their natural frequency. This often produces audible sound waves. This phenomenon, often called an Aeolian harp, is an example of

(1) diffraction   (3) refraction
(2) the Doppler effect   (4) resonance   27 _____

28 A student listens to music from a speaker in an adjoining room, as represented in the diagram below.

She notices that she does not have to be directly in front of the doorway to hear the music. This spreading of sound waves beyond the doorway is an example of

(1) the Doppler effect   (3) refraction
(2) resonance   (4) diffraction   28 _____

29 What is the minimum energy required to ionize a hydrogen atom in the $n = 3$ state?

(1) 0.00 eV   (3) 1.51 eV

(2) 0.66 eV   (4) 12.09 eV   29 _____

Base your answers to questions 30 and 31 on the diagram below and on your knowledge of physics. The diagram represents two small, charged, identical metal spheres, A and B, that are separated by a distance of 2.0 meters.

$\longleftarrow$ 2.0 m $\longrightarrow$

(A)   (B)

-1.6×10^{-6} C $+1.0 \times 10^{-6}$ C

30 What is the magnitude of the electrostatic force exerted by sphere A on sphere B?

(1) 7.2×10^{-3} N   (3) 8.0×10^{-13} N

(2) 3.6×10^{-3} N   (4) 4.0×10^{-13} N   30 _____

31 If the two spheres were touched together and then separated, the charge on sphere A would be

(1) -3.0×10^{-7} C   (3) -1.3×10^{-6} C

(2) -6.0×10^{-7} C   (4) -2.6×10^{-6} C   31 _____

32 The horn of a moving vehicle produces a sound of constant frequency. Two stationary observers, A and C, and the vehicle's driver, B, positioned as represented in the diagram below, hear the sound of the horn.

Compared to the frequency of the sound of the horn heard by driver B, the frequency heard by observer A is

(1) lower and the frequency heard by observer C is lower
(2) lower and the frequency heard by observer C is higher
(3) higher and the frequency heard by observer C is lower
(4) higher and the frequency heard by observer C is higher   32 _____

33 A different force is applied to each of four different blocks on a frictionless, horizontal surface. In which diagram does the block have the greatest inertia 2.0 seconds after starting from rest?

33 _____

34 The diagram below shows a ray of monochromatic light incident
on a boundary between air and glass.

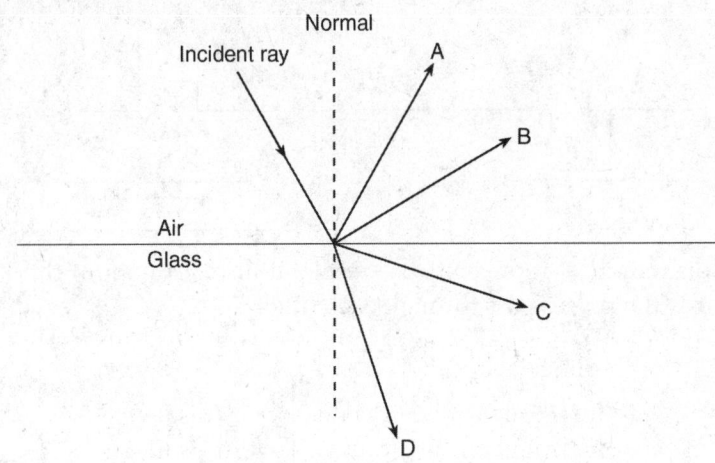

Which ray best represents the path of the reflected light ray?

(1) *A*   (3) *C*

(2) *B*   (4) *D*   34 _____

35 Two pulses approach each other in the same medium. The diagram below represents the displacements caused by each pulse.

Which diagram best represents the resultant displacement of the medium as the pulses pass through each other?

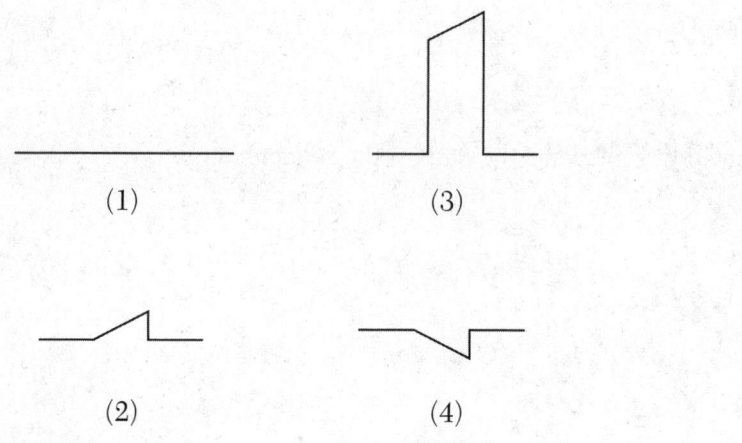

(1)   (3)

(2)   (4)   35 _____

PART B–1

Answer all questions in this part.

Directions (36–50): For *each* statement or question, select the *number* of the word or expression that, of those given, best completes the statement or answers the question. Some questions may require the use of the *2006 Edition Reference Tables for Physical Setting/Physics.* Record your answers in the spaces provided.

36 The diameter of an automobile tire is closest to

    (1) 10^{-2} m   (3) 10^{1} m

    (2) 10^{0} m   (4) 10^{2} m   36 _____

37 The vector diagram below represents the velocity of a car traveling 24 meters per second 35° east of north.

What is the magnitude of the component of the car's velocity that is directed eastward?

    (1) 14 m/s   (3) 29 m/s

    (2) 20. m/s   (4) 42 m/s   37 _____

38 Without air resistance, a kicked ball would reach a maximum
    height of 6.7 meters and land 38 meters away. With air resistance,
    the ball would travel

    (1) 6.7 m vertically and more than 38 m horizontally
    (2) 38 m horizontally and less than 6.7 m vertically
    (3) more than 6.7 m vertically and less than 38 m horizontally
    (4) less than 38 m horizontally and less than 6.7 m vertically   38 ____

39 A car is moving with a constant speed of 20 meters per second.
    What total distance does the car travel in 2.0 minutes?

    (1) 10 m   (3) 1200 m
    (2) 40 m   (4) 2400 m   39 ____

40 A car, initially traveling at 15 meters per second north, accelerates
    to 25 meters per second north in 4.0 seconds. The magnitude of
    the average acceleration is

    (1) 2.5 m/s^2 (3) 10 m/s^2
    (2) 6.3 m/s^2 (4) 20 m/s^2 40 ____

41 An object is in equilibrium. Which force vector diagram could rep-
    resent the force(s) acting on the object?

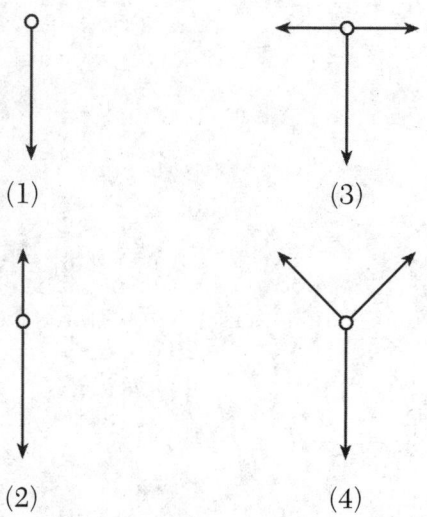

(1)   (3)

(2)   (4)   41 ____

42 Which combination of fundamental units can be used to express the amount of work done on an object?

(1) kg • m/s   (3) kg • m^2/s^2
(2) kg • m/s^2 (4) kg • m^2/s^3 42 _____

43 Which graph best represents the relationship between the potential energy stored in a spring and the change in the spring's length from its equilibrium position?

(1)

(3)

(2)   (4)   43 _____

44 An electric motor has a rating of 4.0×10^2 watts. How much time will it take for this motor to lift a 50.-kilogram mass a vertical distance of 8.0 meters? [Assume 100% efficiency.]

(1) 0.98 s   (3) 98 s
(2) 9.8 s   (4) 980 s   44 _____

45 A compressed spring in a toy is used to launch a 5.00-gram ball. If the ball leaves the toy with an initial horizontal speed of 5.00 meters per second, the minimum amount of potential energy stored in the compressed spring was

(1) 0.0125 J   (3) 0.0625 J
(2) 0.0250 J   (4) 0.125 J   45 _____

46 A ray of yellow light ($f = 5.09 \times 10^{14}$ Hz) travels at a speed of 2.04×10^8 meters per second in

   (1) ethyl alcohol   (3) lucite

   (2) water   (4) glycerol   46 ____

47 A blue-light photon has a wavelength of 4.80×10^{-7} meter. What is the energy of the photon?

   (1) 1.86×10^{22} J   (3) 4.14×10^{-19} J

   (2) 1.44×10^2 J   (4) 3.18×10^{-26} J   47 ____

48 The graph below represents the relationship between the force exerted on an elevator and the distance the elevator is lifted.

How much total work is done by the force in lifting the elevator from 0.0 m to 9.0 m?

   (1) 9.0×10^4 J   (3) 1.5×10^5 J

   (2) 1.2×10^5 J   (4) 1.8×10^5 J   48 ____

49 The diagram below shows waves A and B in the same medium.

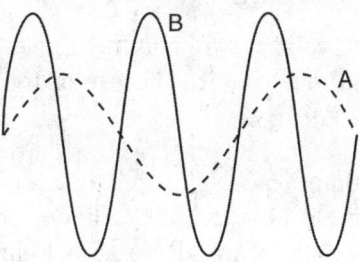

Compared to wave A, wave B has

(1) twice the amplitude and twice the wavelength
(2) twice the amplitude and half the wavelength
(3) the same amplitude and half the wavelength
(4) half the amplitude and the same wavelength   49 _____

50 What is the quark composition of a proton?

(1) uud   (3) csb
(2) udd   (4) uds   50 _____

PART B–2
Answer all questions in this part.

Directions (51–65): Record your answers on the answer sheet provided after the questions. Some questions may require the use of the *2006 Edition Reference Tables for Physical Setting/Physics.*

51–52 Calculate the minimum power output of an electric motor that lifts a 1.30×10^4-newton elevator car vertically upward at a constant speed of 1.50 meters per second. [Show all work, including the equation and substitution with units.] [2]

53–54 A microwave oven emits a microwave with a wavelength of 2.00×10^{-2} meter in air. Calculate the frequency of the microwave. [Show all work, including the equation and substitution with units.] [2]

55–56 Calculate the energy equivalent in joules of the mass of a proton. [Show all work, including the equation and substitution with units.] [2]

Base your answers to questions 57 through 59 on the information and diagram below and on your knowledge of physics.

A 1.5×10^3-kilogram car is driven at a constant speed of 12 meters per second counterclockwise around a horizontal circular track having a radius of 50. meters, as represented below.

Direction of car's motion

◆—50. m—▶

Track, as Viewed from Above

57 On the diagram *on the answer sheet*, draw an arrow to indicate the direction of the velocity of the car when it is at the position shown. Start the arrow on the car. [1]

58–59 Calculate the magnitude of the centripetal acceleration of the car. [Show all work, including the equation and substitution with units.] [2]

___

Base your answers to questions 60 through 62 on the information below and on your knowledge of physics.

A football is thrown at an angle of 30° above the horizontal. The magnitude of the horizontal component of the ball's initial velocity is 13.0 meters per second. The magnitude of the vertical component of the ball's initial velocity is 7.5 meters per second. [Neglect friction.]

60 On the axes *on the answer sheet*, draw a graph representing the relationship between the horizontal displacement of the football and the time the football is in the air. [1]

61–62 The football is caught at the same height from which it is thrown. Calculate the total time the football was in the air. [Show all work, including the equation and substitution with units.] [2]

Base your answers to questions 63 through 65 on the information and diagram below and on your knowledge of physics.

A ray of light ($f = 5.09 \times 10^{14}$ Hz) traveling through a block of an unknown material, passes at an angle of incidence of 30° into air, as shown in the diagram below.

63 Use a protractor to determine the angle of refraction of the light ray as it passes from the unknown material into air. [1]

64–65 Calculate the index of refraction of the unknown material. [Show all work including the equation and substitution with units.] [2]

PART C
Answer all questions in this part.

Directions (66–85): Record your answers on the answer sheet provided after the questions. Some questions may require the use of the *2006 Edition Reference Tables for Physical Setting/Physics.*

Base your answers to questions 66 through 70 on the information below and on your knowledge of physics.

The diagram below represents a 4.0-newton force applied to a 0.200-kilogram copper block sliding to the right on a horizontal steel table.

66 Determine the weight of the block. [1]

67–68 Calculate the magnitude of the force of friction acting on the moving block. [Show all work, including the equation and substitution with units.] [2]

69 Determine the magnitude of the net force acting on the moving block. [1]

70 Describe what happens to the magnitude of the velocity of the block as the block slides across the table. [1]

Base your answers to questions 71 through 75 on the information and diagram below and on your knowledge of physics.

Two conducting parallel plates 5.0×10^{-3} meter apart are charged with a 12-volt potential difference. An electron is located midway between the plates. The magnitude of the electrostatic force on the electron is 3.8×10^{-16} newton.

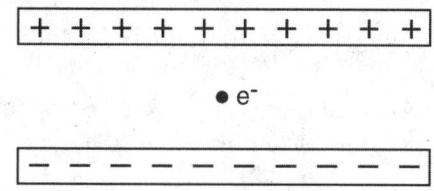

71 On the diagram *on the answer sheet*, draw *at least three* field lines to represent the direction of the electric field in the space between the charged plates. [1]

72 Identify the direction of the electrostatic force that the electric field exerts on the electron. [1]

73–74 Calculate the magnitude of the electric field strength between the plates in newtons per coulomb. [Show all work, including the equation and substitution with units.] [2]

75 Describe what happens to the magnitude of the net electrostatic force on the electron as the electron is moved toward the positive plate. [1]

Base your answers to questions 76 through 80 on the information below and on your knowledge of physics.

An electron in a mercury atom changes from energy level b to a higher energy level when the atom absorbs a single photon with an energy of 3.06 electron-volts.

76 Determine the letter that identifies the energy level to which the electron jumped when the mercury atom absorbed the photon. [1]

77 Determine the energy of the photon, in joules. [1]

78–79 Calculate the frequency of the photon. [Show all work, including the equation and substitution with units.] [2]

80 Classify the photon as one of the types of electromagnetic radiation listed in the electromagnetic spectrum. [1]

Base your answers to questions 81 through 85 on the information and circuit diagram below and on your knowledge of physics.

Three lamps are connected in parallel to a 120-volt source of potential difference, as represented below.

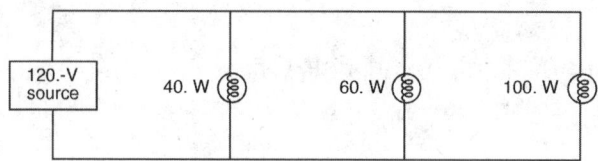

81–82 Calculate the resistance of the 40-watt lamp. [Show all work, including the equation and substitution with units.] [2]

83 Describe what change, if any, would occur in the power dissipated by the 100-watt lamp if the 60-watt lamp were to burn out. [1]

84 Describe what change, if any, would occur in the equivalent resistance of the circuit if the 60-watt lamp were to burn out. [1]

85 The circuit is disassembled. The same three lamps are then connected in series with each other and the source. Compare the equivalent resistance of this series circuit to the equivalent resistance of the parallel circuit. [1]

Answer Sheet
June 2015
Physics: The Physical Setting

PART B–2

51–52

53–54

55–56

57

Track, as Viewed from Above

58–59

60

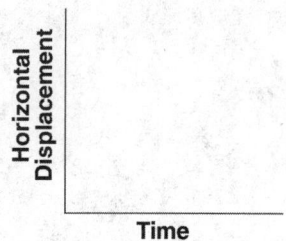

61–62

63 _____ °

64–65

PART C

66 _____ N

67–68

69 _____ N

70 _____

_____

71

72 _____

73–74

75 _____

_____

76 _____

77 _____**J**

78–79

80 _____

81–82

83 _____

_____

84 _____

_____

85 _____

_____

Answers
June 2015
Physics: The Physical Setting

Answer Key

PART A

1. 1	8. 4	15. 4	22. 2	29. 3
2. 2	9. 2	16. 2	23. 3	30. 2
3. 3	10. 3	17. 1	24. 2	31. 1
4. 1	11. 3	18. 4	25. 4	32. 2
5. 3	12. 1	19. 1	26. 1	33. 4
6. 2	13. 1	20. 3	27. 4	34. 1
7. 3	14. 1	21. 2	28. 4	35. 2

PART B–1

36. 2	39. 4	42. 3	45. 3	48. 3
37. 1	40. 1	43. 1	46. 4	49. 2
38. 4	41. 4	44. 2	47. 3	50. 1

PART B–2 and PART C. *See* **Answers Explained**.

PART A

1 Both speed and work are scalar quantities.

WRONG CHOICES EXPLAINED:
(2) Both velocity and force are vector quantities.
(3) Distance is a scalar quantity while acceleration is a vector quantity.
(4) Momentum is a vector quantity while power is a scalar quantity.

2 When dealing with a word problem requiring a mathematical solution, it is most helpful to translate the words into which quantities/variables are explicitly given and which constants or other variables can be deduced from the situation or words. We also write down the variable that needs to be solved for or found. Then it is much easier to determine which formula is required and then use it to solve the problem.

Given:	Find:	Solution:
$m = 3.00$ kg	$d_{max} = ?$	$v_f^2 = v_i^2 + 2ad$
$v_i = 9.80$ m/s		$d = \dfrac{v_f^2 - v_i^2}{2a}$
$a = -9.81$ m/s^2		$d = \dfrac{(0 \text{ m/s})^2 - (9.80 \text{ m/s})^2}{2(-9.81 \text{ m/s}^2)}$
$v_f = 0$ m/s		$d = 4.90$ m

3 To solve this problem we use the technique of vector addition. We recognize that if we place the vectors head to tail, the resultant vector would be the hypotenuse of a right triangle and we can use the Pythagorean theorem to solve algebraically for the resultant vector:

$$a^2 + b^2 = c^2$$
$$c = \sqrt{a^2 + b^2}$$
$$c = \sqrt{(220 \text{ m/s})^2 + (50 \text{ m/s})^2}$$
$$c = 226 \text{ m/s}$$

4 Because the vehicle is traveling in a straight line at constant speed, according to Newton's First Law, the net force on the vehicle would be equal to 0 N.

5 The only force acting on the ball is the force due to gravity, in this case the weight of the ball, which is 5.0 N.

6

Given:	Find:	Solution:
$k = 100$ N/m	$F = ?$	$F = (100 \text{ N/m})(0.15 \text{ m})$
$x = 0.65 \text{ m} - 0.50 \text{ m}$		$F = 15$ N

7

Given:

$m = 1.5$ kg

$v_i = 2.0$ m/s

$v_f = 0$ m/s

$t = 0.30$ s

Find:

$F = ?$

Solution:

$$F_{net}\Delta t = \Delta p$$

$$F_{net}\Delta t = m\Delta v$$

$$F_{net}\Delta t = m(v_f - v_i)$$

$$F_{net} = \frac{m\left(v_f - v_i\right)}{\Delta t}$$

$$F_{net} = \frac{(1.5 \text{ kg})(0 \text{ m/s} - 2.0 \text{ m/s})}{(0.30 \text{ s})}$$

$$F_{net} = 10 \text{ N}$$

8 When a light wave passes from glass into air, the speed of the light wave increases. Velocity of wave is equal to the product of the frequency of the wave times its wavelength ($v = f\lambda$). Since the frequency does not change, the wavelength changes in proportion to the change in velocity. Since the velocity increases, the wavelength must also increase.

WRONG CHOICES EXPLAINED:
  (1) Amplitude is related to the energy carried by the wave.
  (2) Frequency of the wave does not change.
  (3) Period is the reciprocal of frequency and, since frequency does not change, neither would the period.

9 According to Newton's Third Law, if object A exerts a force on object B, then object B exerts an equal and opposite force back on object A. In this case, the basketball player's feet exert a 1.0×10^3 N force on the floor; therefore, the floor exerts a 1.0×10^3 N force on her feet.

10

Given:	Find:	Solution
$m_{ball} = 0.0600$ kg	$v_{bullet} = ?$	$p_{ball} = p_{bullet}$
$v_{ball} = 60.0$ m/s		$m_{ball} \bullet v_{ball} = m_{bullet} \bullet v_{bullet}$
$m_{bullet} = 0.0100$ kg		$v_{bullet} = \dfrac{m_{ball} \bullet v_{ball}}{m_{bullet}}$

$$v_{bullet} = \frac{(0.0600 \text{ kg}) (60 \text{ m/s})}{(0.0100 \text{ kg})}$$

$$v_{bullet} = 360 \text{ m/s}$$

11

Given:	Find:	Solution:
$m = 1.1 \times 10^4$ kg	$g = ?$	$g = \dfrac{F_g}{m}$
$F_g = 9.1 \times 10^4$ N		$g = \dfrac{9.1 \times 10^4 \text{ N}}{1.1 \times 10^4 \text{ kg}}$

$$g = 8.3 \text{ N/kg}$$

12

$$Fe = \frac{kq_1q_2}{r^2}$$

Electrostatic force is inversely proportional to the square of the distance between the centers of the point charges. If the electrostatic force is equal to F if $r = d$, then:

$$F = \frac{kq_1q_2}{d^2}$$

If $r = 3d$, then:

$$F_e = \frac{kq_1q_2}{(3d)^2}$$

$$F_e = \frac{kq_1q_2}{9d^2}$$

$$F_e = \left(\frac{1}{9}\right)\left(\frac{kq_1q_2}{d^2}\right)$$

$$F_e = \frac{1}{9}F$$

13

Given:

$V = 3.0\,\text{V}$

$I = 1.8 \times 10^{-4}\,\text{A}$

Find:

$R = ?$

Solution:

$V = IR$

$R = \frac{V}{I}$

$$R = \frac{3.0\text{ V}}{1.8 \times 10^{-4}\,\text{A}}$$

$$R = 1.7 \times 10^4\ \Omega$$

14 An electric motor converts electrical energy into mechanical energy.

WRONG CHOICES EXPLAINED:
(1) A generator converts mechanical into electrical energy.
(2) A battery is an example of converting chemical to electrical energy.
(3) Charging a battery would be an example of converting electrical into chemical energy.

15 The law of conservation of energy states that the total energy of an isolated system remains constant: $E_{Ti} = E_{Tf}$

$$PE_i + KE_i + Q_i = PE_f + KE_f + Q_f.$$

If the block comes to rest, then the kinetic energy of the block decreases to zero. Neither the block nor the table changes in height; therefore, there is no change in potential energy. The force of friction between the table and the block which causes the block to come to rest results in an increase in the heat (thermal) energy of the system.

WRONG CHOICES EXPLAINED:
(1) No change in height relative to earth surface so no change in gravitational potential energy
(2) No springs in the system, so no change in elastic potential energy
(3) Since the block came to rest and velocity decreased, there is a decrease in kinetic energy, not an increase.

16

Given:	Find:	Solution:
$q = 1.6 \times 10^{-19}$ J	$W = ?$	$V = \dfrac{W}{q}$
$V = 3.00$ V		$W = Vq$
		$W = (3.00 \text{ V})(1.6 \times 10^{-19} \text{ J})$
		$W = 4.80 \times 10^{-19}$ J

17

Given:	Find:	Solution:
$T = 20°C$	$A = ?$	$R = \dfrac{\rho L}{A}$
Copper wire		$A = \dfrac{\rho L}{R}$
$\rho = 1.72 \times 10^{-8}\ \Omega\ \text{m}$		$A = \dfrac{\left(1.72 \times 10^{-8}\Omega\text{m}\right)(6.0\ \text{m})}{1.3\ \Omega}$
$L = 6.0\ \text{m}$		$A = 7.9 \times 10^{-8}\ \text{m}^2$
$R = 1.3\ \Omega$		

18

Given:	Find:	Solution:
$q = 5.0\ \text{C}$	$I = ?$	$I = \dfrac{q}{t}$
$t = 0.050\ \text{s}$		$I = \dfrac{5.0\ \text{C}}{0.050\ \text{s}}$
		$I = 1.0 \times 10^2\ \text{A}$

19 Resistors that are connected in series (all in a row) have the identical amount of current flow through each of them. The relationship between potential difference, current, and resistance is calculated using the relationship $V = IR$. If I remains constant, potential difference is directly proportional to resistance.

20 Amplitude is related to the amount of energy in a wave. In a sound wave, amplitude corresponds to loudness.

21 Frequency is defined as the amount of times a motion repeats itself in a given time period. In this instance

$$\frac{5.0}{10\ \text{s}} = 0.50\ \text{Hz}$$

22 As a ball falls freely from rest, it is acted upon by a gravitational force downward, which results in its speed increasing from rest at a rate of 9.81 meters per second. Therefore, with each increasing second, the ball will fall a greater distance during each time second interval.

WRONG CHOICES EXPLAINED:
(1) This diagram represents movement at constant velocity as the distance for each time interval is identical.

(3) This diagram represents movement with decreasing velocity as the distance traveled for each time interval decreases.

(4) This diagram represents movement with decreasing then increasing velocity as the distance traveled for each time interval decreases then increases.

23 Gamma rays and microwaves are both electromagnetic waves and thus travel at the same speed, the speed of light in a vacuum.

WRONG CHOICES EXPLAINED:
(1), (2), (4) Gamma rays and microwaves travel at the same speed but have different frequencies and thus different periods (reciprocal of frequency) and different wavelengths [wavelength and frequency are inversely proportional if velocity is constant $(v = f\lambda)$].

24 If velocity is constant, wavelength and frequency are inversely proportional $(v = f\lambda)$. If frequency is doubled, wavelength is halved.

$$f\lambda = (2f)\left(\frac{1}{2}\lambda\right)$$

25 Points on a wave that are at equal displacements from their rest position and are experiencing identical movements, i.e., toward or away from the rest position, are said to be in phase. Points B and D fit this definition.

WRONG CHOICES EXPLAINED:
(2) A and E are completely out of phase.

(1), (3) A and B and B and C are partially out of phase.

26 As a longitudinal wave travels through a medium, the particles of the medium vibrate in simple harmonic motion in a direction parallel to the direction of motion of the wave.

WRONG CHOICES EXPLAINED:
(2) This motion is characteristic for a transverse wave.
(3) The energy of the wave is transferred in the direction of motion.

27 Resonance is the spontaneous vibration of an object at a frequency equal to that of the wave that initiates that resonant vibration. The resulting standing wave created has amplitudes large enough that sound can be heard.

WRONG CHOICES EXPLAINED:
(1) Diffraction is the bending of a wave around a barrier.
(2) The Doppler effect is the apparent change in frequency that results when a wave source and an observer are in relative motion with respect to each other.
(3) Refraction is the change in direction of a wave when it passes obliquely from one medium into another in which it moves at a different speed.

28 Diffraction is the bending of a wave around a barrier.

WRONG CHOICES EXPLAINED:
(1) The Doppler effect is the apparent change in frequency that results when a wave source and an observer are in relative motion with respect to each other.
(2) Resonance is the spontaneous vibration of an object at a frequency equal to that of the wave that initiates that resonant vibration.
(3) Refraction is the change in direction of a wave when it passes obliquely from one medium into another in which it moves at a different speed.

29 Refer to the *Energy Level Diagram for Hydrogen* in *Reference Table F*. At $n = 3$, the energy level of the hydrogen atom is equal to -1.51 eV. Therefore, 1.51 eV is the minimum amount of energy required to ionize the hydrogen atom, i.e., disassociate the electron from the atom by bringing the energy level to zero.

30

Given: Find: Solution:

$q_A = -1.6 \times 10^{-6}$ C $F_e = ?$ $F_e = \dfrac{kq_1 q_2}{r^2}$

$q_B = 1.0 \times 10^{-6}$ C $F_e =$

$$\frac{\left(8.99 \times 10^9 \text{ N m}^2/\text{C}^2\right)\left(-1.6 \times 10^{-6} \text{ C}\right)(1.0 \times 10^{-6} \text{ C})}{(2.0 \text{ m})^2}$$

$r = 2.0$ m $F_e = 3.6 \times 10^{-3}$ N

$k = 8.99 \times 10^9$ Nm^2/C^2

31 If the two spheres were touched together there would be a redistribution of the total charge of the 2 spheres such that each sphere would have the same net charge. We add the charges on all the spheres and divide by the total number of spheres:

$$\frac{\left(-1.6 \times 10^{-6} \text{ C}\right) + \left(1.0 \times 10^{-6} \text{ C}\right)}{2} = -3.0 \times 10^{-7} \text{C}$$

32 This scenario is an example of the Doppler effect. The vehicle is moving away from observer A and moving toward observer C. The apparent pitch or frequency is decreased as the sound source moves away from and increased as the sound source moves toward an observer.

WRONG CHOICES EXPLAINED:

(1) The sound source would have to be traveling away from both observers for this to be true.

(3) The vehicle would have to be moving toward observer A and away from observer C for this to be true.

(4) The sound source would have to be traveling toward both observers for this to be true.

33 Mass is the quantitative measure of inertia. The block in diagram 4 has the greatest mass of all the 4 blocks, 20 kg, and thus has the greatest inertia.

34 When light rays are reflected from a surface, the angle that the incident ray makes with the normal to the surface is equal to the angle that the reflected ray makes with the normal to the surface. Ray A meets this definition.

WRONG CHOICES EXPLAINED:

(2) Ray B has an angle of reflection greater than the angle of incidence.

(3), (4) Rays C and D represent refracted rays.

35 These two displacements are in opposite directions and when they meet will result in destructive interference. The resulting displacement is the algebraic sum of the displacements of the individual pulses. The pulse on the right is the virtual opposite to the pulse on the left with a small addition. When they meet, the rectangular portion of each will cancel out and all that will be left is the small top triangular area above the center. The diagram in choice (2) accurately represents the resultant displacement at the instant the pulses pass through each other.

WRONG CHOICES EXPLAINED:

(1) The two pulses would have to be equal in magnitude and opposite in direction for this diagram to be the resultant.

(3) The two pulses would have to have displacements in the same direction for this diagram to be the resultant.

(4) The pulse on the left would have had to have been displaced up instead of down and the pulse on the right displaced down instead of up for this diagram to be the resultant.

PART B–1

36 This is an order of magnitude question. 10 to the zero power is equivalent to 1. One meter is approximately equal to 3.3 feet, which is a reasonable approximation for the diameter of a tire.

WRONG CHOICES EXPLAINED:

(1) 10 centimeters is not a reasonable approximation of the diameter of a tire.

(3) 10 meters (approx. 33 ft) is not a reasonable approximation of the diameter of a tire.

(4) 100 meters (approx. 330 ft) is not a reasonable approximation of the diameter of a tire.

37

Given:

$v = 24$ m/s

$\theta = 35°$ E of N

Find:

$v_{east} = ?$

Solution:

$v_{east} = v \sin \theta$

$v_{east} = (24 \text{ m/s})(\sin 35°)$

$v_{east} = 14$ m/s

38 Air resistance would have the effect of slowing the motion or velocity of the ball. As a result the horizontal and vertical components of the velocity would be decreased. The ball would not reach as great a height. It would therefore have a shorter amount of time in the air or hang time and thus not travel as far horizontally.

39

Given:

$v = 20$ m/s

$t = 2.0$ min • $(60 \text{ s}/1 \text{ min}) = 120$ s

Find:

$d = ?$

Solution:

$v = \dfrac{d}{t}$

$d = vt$

$d = (20 \text{ m/s})(120 \text{ s})$

$d = 2{,}400$ m

40

Given:

$v_i = 15$ m/s

$v_f = 25$ m/s

$t = 4.0$ s

Find:

$a = ?$

Solution:

$a = \dfrac{\Delta v}{t}$

$a = \dfrac{v_f - v_i}{t}$

$a = \dfrac{(25 \text{ m/s} - 15 \text{ m/s})}{4.0 \text{ s}}$

$a = 2.5$ m/s^2

41 In diagram 4 the horizontal components of the generally upward vector forces could negate each other and the vertical components could negate the downward force vector.

WRONG CHOICES EXPLAINED:
(1) This diagram indicates a net force in the downward direction.
(2) In this diagram, the magnitude of the upward force vector is smaller than the downward vector, which would result in a net force downward.
(3) In this diagram, the 2 horizontal vectors look equal in magnitude so, being that they are opposite in direction, would result in a net force of zero in the horizontal direction. However, there is no force to counteract the downward vector; therefore, there would be a net force in the downward direction.

42

$$W = Fd$$
$$F = ma$$

Therefore $W = mad$

$$W = (kg)(m/s^2)(m)$$
$$W = kg \bullet m^2/s^2$$

WRONG CHOICES EXPLAINED:
(1) unit for momentum
(2) equivalent to newtons, unit for force
(4) equivalent to joules per second or watts, unit for power

43 The potential energy stored in a spring is proportional to the square of the displacement of the spring from its equilibrium position.

$$PE_s = \frac{1}{2}kx^2$$

The graph in choice (1) illustrates an exponential relationship.

WRONG CHOICES EXPLAINED:
(2) This graph illustrates a direct proportionality.
(3) This graph illustrates an inverse relationship.
(4) This graph illustrates an inversely proportional relationship.

44

Given:	Find:	Solution:
$P = 4.0 \times 10^2$ W	$t = ?$	$P = \dfrac{Fd}{t}$
$m = 50$ kg		$t = \dfrac{Fd}{P}$ $F_g = mg$
$d = 8.0$ m		$t = \dfrac{mgd}{P}$
$g = 9.81$ m/s^2		$t = \dfrac{(50.\ \text{kg})(9.81\ \text{m/s}^2)(8.0\ \text{m})}{(4.0 \times 10^2\ \text{W})}$
		$t = 9.8$ s

45

Given:	Find:	Solution:
$m_{ball} = 5.00$ g $\bullet$ (1 kg/1,000 g)		
$= 5.00 \times 10^{-3}$ kg	$PE_s = ?$	$PE_s = KE_{ball}$
$v_{ball} = 5.00$ m/s		$PE_s = \dfrac{1}{2} m_{ball} v_{ball}{}^2$
		$PE_s =$
		$\dfrac{1}{2}(5.00 \times 10^{-3}\ \text{kg})(5.00\ \text{m/s})^2$
		$PE_s = 0.0625$ J

46 Refer to the *Absolute Indices of Refraction* table in *Reference Table E*.

Given: Find: Solution:

$v = 2.04 \times 10^8$ m/s $n = ?$ $n = \dfrac{c}{v}$

$c = 3.00 \times 10^8$ m/s $n = \dfrac{3.00 \times 10^8 \text{ m/s}}{2.04 \times 10^8 \text{ m/s}}$

$n = 1.47$

An absolute index of refraction of 1.47 corresponds to corn oil or glycerol.

WRONG CHOICES EXPLAINED:
(1) has an index of refraction of 1.36.
(2) has an index of refraction of 1.33.
(3) has an index of refraction of 1.50.

47

Given: Find: Solution:

$\lambda = 4.80 \times 10^{-7}$ m $E_{photon} = ?$ $E_{photon} = \dfrac{hc}{\lambda}$

$$E_{photon} = \dfrac{\left(6.63 \times 10^{-34} \text{ J} \bullet \text{s}\right)\left(3.00 \times 10^8 \text{ m/s}\right)}{\left(4.80 \times 10^{-7} \text{m}\right)}$$

$$E_{photon} = 4.14 \times 10^{-19} \text{ J}$$

48

Given: Find: Solution:

$F_1 = 1.0 \times 10^4$ N $W_{total} = ?$ $W_{total} = W_1 + W_2$

$d_1 = 3.0$ m $W_{total} = F_1 d_1 + F_2 d_2$

$F_2 = 2.0 \times 10^4$ N $W_{total} =$

$(1.0 \times 10^4 \text{ N})(3.0 \text{ m}) + (2.0 \times 10^4 \text{ N})(6.0 \text{ m})$

$d_2 = 6.0$ m $W_{total} = 3.0 \times 10^4 \text{ Nm} + 1.2 \times 10^5$
Nm

$W_{total} = 1.5 \times 10^5 \text{ J}$

49 Wave *B* has a greater amplitude than wave *A*, i.e., a greater displacement from rest position, so we can eliminate choices (3) and (4). Wave *B* has a smaller, or shorter wavelength than wave *A* so we can eliminate choice (1).

50 A proton is made up of up and down quarks and has a net charge of +1. Refer to the *Particles of the Standard Model* chart in *Reference Table H*. The correct combination of up and down quarks to yield a net charge of +1 would be uud.

WRONG CHOICES EXPLAINED:
(2) This would yield a net charge of 0 and is the correct combination for a neutron.
(3) This would yield a net charge of 0.
(4) This would yield a net charge of 0.

PART B–2

51–52

Given:	Find:	Solution:
$F = 1.30 \times 10^4$ N	$P = ?$	$P = F\bar{v}$
$\bar{v} = 1.50$ m/s		$P = (1.30 \times 10^4 \text{ N})(1.50 \text{ m/s})$
		$P = 1.95 \times 10^4$ W

51 One credit is awarded for the correct equation and substitution of values with units.

52 One credit is awarded for the correct answer with units or for an answer, with units, that is consistent with the answer to question 51.
Note: Students will not be penalized more than 1 credit for errors in units in questions 51 and 52.

53–54

Given: Find: Solution:

$\lambda = 2.00 \times 10^{-2}$ m $f = ?$ $c = f\lambda$

$c = 3.0 \times 10^8$ m/s $f = \dfrac{c}{\lambda}$

$$f = \frac{3.0 \times 10^8 \text{ m/s}}{2.00 \times 10^{-2} \text{m}}$$

$$f = 1.50 \times 10^{10} \text{ Hz}$$

53 One credit is awarded for the correct equation and substitution of values with units.

54 One credit is awarded for the correct answer with units or for an answer, with units, that is consistent with the answer to question 53.
Note: Students will not be penalized more than 1 credit for errors in units in questions 53 and 54.

55–56

Given: Find: Solution:

$m = 1.67 \times 10^{-27}$ kg $E = ?$ $E = mc^2$

$c = 3.0 \times 10^8$ m/s $E = (1.67 \times 10^{-27} \text{ kg})(3.0 \times 10^8 \text{ m/s})^2$

                                            $E = 1.50 \times 10^{-10}$ J

Or

Given: Find: Solution:

$m = 1$ u $= 9.31 \times 10^2$ eV) $E = ?$ $E = (9.31 \times 10^8 \text{ eV})(1.60 \times 10^{-19} \text{ J/eV})$

MeV $= 9.31 \times 10^8$ eV

1 eV $= 1.60 \times 10^{-19}$ J $E = 1.50 \times 10^{-10}$ J

55 One credit is awarded for the correct equation and substitution of values with units.

56 One credit is awarded for the correct answer with units or for an answer, with units, that is consistent with the answer to question 55.

Note: Students will not be penalized more than 1 credit for errors in units in questions 55 and 56.

Track, as Viewed from Above

57 One credit is awarded for an arrow drawn from the car with the arrowhead pointed to the left.

58–59

Given:	Find:	Solution:
$m = 1.5 \times 10^3 \text{ kg}$	$a_c = ?$	$a_c = \dfrac{v^2}{r}$
$v = 12 \text{ m/s}$		$a_c = \dfrac{(12 \text{ m/s})^2}{50 \text{ m}}$
$r = 50 \text{ M}$		$a_c = 2.9 \text{ m/s}^2$

58 One credit is awarded for the correct equation and substitution of values with units.

59 One credit is awarded for the correct answer with units or for an answer, with units, that is consistent with the answer to question 58.

Note: Students will not be penalized more than 1 credit for errors in units in questions 58 and 59.

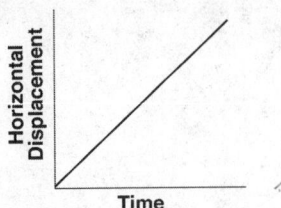

60 Velocity in the horizontal direction is constant, therefore displacement is directly proportional to time elapsed.

One credit is awarded for a straight line with a positive slope.

Note: The line may be sketched in. It must approximate a straight line, but need not pass through the origin.

61–62

Given:	Find:	Solution:
$v_{ix} = 13.0$ m/s	$t_{total} = ?$	$a = \dfrac{\Delta v}{t}$
$v_{iy} = 7.5$ m/s		$t = \dfrac{\Delta v}{a}$
$d_{yi} = d_{yf} = 0$		$t = \dfrac{v_f - v_i}{a}$
$a = -9.81$ m/s^2		$t = \dfrac{(-7.5 \text{ m/s}) - (7.5 \text{ m/s})}{(-9.81 \text{ m/s}^2)}$
$v_{fy} = -7.5$ m/s		$t = 1.5$ s

61 One credit is awarded for the correct equation and substitution of values with units.

62 One credit is awarded for the correct answer with units or for an answer, with units, that is consistent with the answer to question 61.

Note: Students will not be penalized more than 1 credit for errors in units in questions 61 and 62.

63 One credit is awarded for an answer of $56° \pm 2°$.

64–65

Given: Find: Solution:

$\theta_{unknown} = 30°$ $n_{unknown} = 1.00$ $n_1 \sin \theta_1 = n_2 \sin \theta_2$

$\theta_{air} = 56°$ $n_{unknown} \sin \theta_{unknown} = n_{air} \sin \theta_{air}$

$n_{air} = 1.00$ $n_{unknown} = \dfrac{n_{air} \sin \theta_{air}}{\sin \theta_{unknown}}$

$$n_{unknown} = \frac{(1.00)(\sin 56°)}{\sin 30°}$$

$$n_{unknown} = 1.7 \text{ or } 1.6$$

64 One credit is awarded for the correct equation and substitution of values with units.

65 One credit is awarded for the correct answer with units or for an answer, with units, that is consistent with the answer to question 64.

Note: Students will not be penalized more than 1 credit for errors in units in questions 64 and 65.

PART C

66

Given:	Find:	Solution:

$m = 0.200$ kg $F_w = ?$ $F_w = mg$

$g = 9.81$ m/s^2

$F_w = (0.200$ kg$)(9.81$ m/s$^2)$

$F_w = 1.96$ N

One credit is awarded for an answer of 1.96 N or 2.0 N or 1.9 N.

67–68

Given:	Find:	Solution:

$\mu_k = 0.36$ $F_f = ?$ $F_f = \mu_k F_N$

$F_w = 1.96$ N

$F_w = F_N$

$F_f = (0.36)(1.96$ N$)$

$F_f = 0.71$ N

67 One credit is awarded for the correct equation and substitution of values with units.

68 One credit is awarded for the correct answer with units (0.71 N or 0.70 N) or for an answer, with units, that is consistent with the answer to question 67.
Note: Students will not be penalized more than 1 credit for errors in units in questions 67 and 68.

69

Given:	Find:	Solution:

$F_f = 0.71$ N $F_{net} = ?$ $F_{net} = F_{applied} - F_f$

$F_{applied} = 4.0$ N

$F_{net} = (4.0$ N$) - (0.71$ N$)$

$F_{net} = 3.3$ N

One credit is awarded for an answer of 3.3 N or for an answer that is consistent with the answer to question 68.

70 Given that there is a net force acting on the block, the velocity of the block would increase as the block slides across the table.

One credit is awarded for an acceptable answer as above or for an answer that is consistent with the answer to question 69.

71 One credit is awarded for at least 3 arrows pointing away from the positive plate and toward the negative plate.

Note: Use of a straightedge is not necessary to draw the field lines. Field lines near the edge of the plate may be curved.

72 The direction of the electrostatic force on the electron would be toward the positive plate and away from the negative plate, i.e., opposite the direction of the field.

One credit is awarded for an answer consistent with the above explanation.

Given: Find: Solution:

$d = 5.0 \times 10^{-3}$ m $E = ?$ $E = \dfrac{V}{d}$ or $E = \dfrac{F_e}{q}$

$V = 12$ V $E = \dfrac{12 \text{ V}}{5.0 \times 10^{-3}\,\text{m}}$

                                                    $E = \dfrac{3.8 \times 10^{-16} \text{ N}}{1.6 \times 10^{-19} \text{ C}}$

$q = 1.6 \times 10^{-19}$ C $E = 2.4 \times 10^3$ V/m

                                                    $E = 2.4 \times 10^3$ N/C

$F = 3.8 \times 10^{-16}$ N $E = 2.4 \times 10^3$ N/C

73–74 One credit is awarded for the correct equation and substitution of values with units.

74 One credit is awarded for a correct answer with units or for an answer, with units, that is consistent with the answer to question 73.
Note: Students will not be penalized more than 1 credit for errors in units in questions 73 and 74.

75 The magnitude of the net electrostatic force on the electron remains constant. One credit is awarded for an answer consistent with the explanation above.

76 Refer to *Reference Table F*. The difference in energy between level b and level f is equal to 3.06 eV.
One credit is awarded for f or energy level f.

77

Given:

$E_{photon} = 3.06$ eV

1 eV $= 1.6 \times 10^{-19}$ J

Find:

E_{photon} in joules

Solution:

3.06 eV$(1.6 \times 10^{-19}$ J/eV$) = 4.90 \times 10^{-19}$ J

One credit is awarded for an answer of 4.90×10^{-19} J or 4.89×10^{-19} J.

78–79

Given:

$E_{photon} = 4.90 \times 10^{-19}$ J

$h = 6.63 \times 10^{-34}$ J/s

Find:

$f = ?$

Solution:

$E_{photon} = hf$

$f = \dfrac{E_{photon}}{h}$

$f = \dfrac{4.90 \times 10^{-19} \text{J}}{6.63 \times 10^{-34} \text{J/s}}$

$f = 7.39 \times 10^{14}$ Hz

78 One credit is awarded for the correct equation and substitution of values with units.

79 One credit is awarded for a correct answer with units or for an answer, with units, that is consistent with the answer to question 78.
Note: Students will not be penalized more than 1 credit for errors in units in questions 78 and 79.

80 Refer to the *Electromagnetic Spectrum* diagram in *Reference Table D.* An $f = 7.39 \times 10^{14}$ Hz is within the visible light spectrum, specifically visible violet light.
One credit is awarded for an answer of visible light or violet light or for an answer that is consistent with the answer to question 79.

81–82

Given: Find: Solution:

$V = 120$ V $R = ?$ $P = \dfrac{V^2}{R}$

$P = 40$ W $R = \dfrac{V^2}{P}$

$$R = \dfrac{(120 \text{ V})^2}{40 \text{ W}}$$
$$R = 360 \ \Omega$$

81 One credit is awarded for the correct equation and substitution of values with units.

82 One credit is awarded for a correct answer with units or for an answer, with units, that is consistent with the answer to question 81.
Note: Students will not be penalized more than 1 credit for errors in units in questions 81 and 82.

83 The voltage across the lamp and its resistance are unchanged regardless of whether the other lamps are functional or not; therefore, there would be no change in the power dissipated by the 100 W lamp.
One credit is awarded for an answer indicating that there is no change in the power dissipated.

84 The total resistance in a parallel circuit would increase if a resistor is removed from the circuit. Therefore, if the 60 W lamp were to burn out, that would be akin to removing a resistor from a parallel circuit and the total resistance in the circuit would increase.

One credit is awarded for an answer indicating that the equivalent resistance would increase.

85 In a series circuit, the total resistance of the circuit is equal to the sum of the individual resistances. Therefore, the total resistance in a series circuit would be greater than the total resistance of the resistors connected in parallel.

One credit is awarded for an answer indicating that the equivalent resistance of the series circuit would be greater than the equivalent resistance of the parallel circuit.

Topic	Question Numbers (total)	Wrong Answers (x)	Grade
Math Skills	2, 3, 6, 7, 10, 13, 16–18, 21, 30, 31, 36, 37, 39, 40, 44–48, 53–56, 58–60, 64–68, 72–79, 81, 82: (43)		$\dfrac{100\,(43 - x)}{43} = \%$
Mechanics	1–7, 9–12, 22, 33, 38–41, 57–59, 61, 62, 68, 70: (24)		$\dfrac{100\,(24 - x)}{24} = \%$
Energy	14, 15, 42, 44, 45, 51, 52: (7)		$\dfrac{100\,(7 - x)}{7} = \%$
Electricity/ Magnetism	13, 19, 83, 84: (4)		$\dfrac{100\,(4 - x)}{4} = \%$
Waves	8, 20, 21, 23–28, 32, 34, 35, 46, 53, 54: (15)		$\dfrac{100\,(15 - x)}{15} = \%$
Modern Physics	29, 47, 50, 55, 56, 76–79 (9)		$\dfrac{100\,(9 - x)}{9} = \%$

Examination June 2016

Physics: The Physical Setting

PART A

Answer all questions in this part.

Directions (1–35): For *each* statement or question, select the *number* of the word or expression that, of those given, best completes the statement or answers the question. Some questions may require the use of the *2006 Edition Reference Tables for Physical Setting/Physics.* Record your answers in the spaces provided.

1 Which quantity is a vector?

  (1) power   (3) speed

  (2) kinetic energy   (4) weight   1 _____

2 A 65.0-kilogram astronaut weighs 638 newtons at the surface of Earth. What is the mass of the astronaut at the surface of the Moon, where the acceleration due to gravity is 1.62 meters per second squared?

  (1) 10.7 kg   (3) 105 N

  (2) 65.0 kg   (4) 638 N   2 _____

3 When the sum of all the forces acting on a block on an inclined plane is zero, the block

  (1) must be at rest

  (2) must be accelerating

  (3) may be slowing down

  (4) may be moving at constant speed   3 _____

4 The greatest increase in the inertia of an object would be produced by increasing the

   (1) mass of the object from 1.0 kg to 2.0 kg
   (2) net force applied to the object from 1.0 N to 2.0 N
   (3) time that a net force is applied to the object from 1.0 s to 2.0 s
   (4) speed of the object from 1.0 m/s to 2.0 m/s   4 _____

5 A 100-kilogram cart accelerates at 0.50 meter per second squared west as a horse exerts a force of 60. newtons west on the cart. What is the magnitude of the force that the cart exerts on the horse?

   (1) 10 N   (3) 60 N
   (2) 50 N   (4) 110 N   5 _____

6 Sound waves are described as

   (1) mechanical and transverse
   (2) mechanical and longitudinal
   (3) electromagnetic and transverse
   (4) electromagnetic and longitudinal   6 _____

7 An electrical force of 8.0×10^{-5} newton exists between two point charges, $q1$ and $q2$. If the distance between the charges is doubled, the new electrical force between the charges will be

   (1) 1.6×10^{-4} N   (3) 3.2×10^{-4} N
   (2) 2.0×10^{-5} N   (4) 4.0×10^{-5} N   7 _____

8 A blue lab cart is traveling west on a track when it collides with and sticks to a red lab cart traveling east. The magnitude of the momentum of the blue cart before the collision is 2.0 kilogram • meters per second, and the magnitude of the momentum of the red cart before the collision is 3.0 kilogram • meters per second. The magnitude of the total momentum of the two carts after the collision is

   (1) 1.0 kg • m/s   (3) 3.0 kg • m/s
   (2) 2.0 kg • m/s   (4) 5.0 kg • m/s   8 _____

9 The diagram below represents the path of a thrown ball through the air.

Ground

Which arrow best represents the direction in which friction acts on the ball at point *P*?

   (1)   (2)   (3)   (4)   9 _____

10 A magnetic field would be produced by a beam of

(1) x rays   (3) protons
(2) gamma rays   (4) neutrons   10 _____

11 The diagram below represents the electric field in the region of two small charged spheres, *A* and *B*.

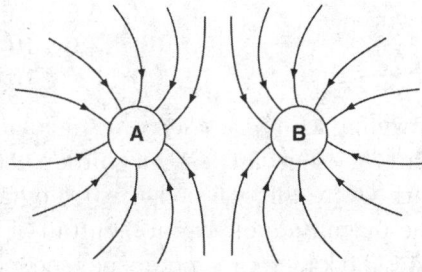

What is the sign of the net charge on *A* and *B*?

(1) *A* is positive and *B* is positive.
(2) *A* is positive and *B* is negative.
(3) *A* is negative and *B* is negative.
(4) *A* is negative and *B* is positive.   11 _____

12 A horizontal force of 20 newtons eastward causes a 10-kilogram box to have a displacement of 5 meters eastward. The total work done on the box by the 20-newton force is

  (1) 40 J   (3) 200 J
  (2) 100 J   (4) 1000 J   12 _____

13 A block initially at rest on a horizontal, frictionless surface is accelerated by a constant horizontal force of 5.0 newtons. If 15 joules of work is done on the block by this force while accelerating it, the kinetic energy of the block increases by

  (1) 3.0 J   (3) 20 J
  (2) 15 J   (4) 75 J   13 _____

14 Two objects, A and B, are held one meter above the horizontal ground. The mass of B is twice as great as the mass of A. If PE is the gravitational potential energy of A relative to the ground, then the gravitational potential energy of B relative to the ground is

  (1) PE (3) $\dfrac{PE}{2}$
  (2) $2PE$ (4) $4PE$ 14 _____

15 What is the kinetic energy of a 55-kilogram skier traveling at 9.0 meters per second?

  (1) 2.5×10^2 J   (3) 2.2×10^3 J
  (2) 5.0×10^2 J   (4) 4.9×10^3 J   15 _____

16 A 5.09×10^{14}-hertz electromagnetic wave is traveling through a transparent medium. The main factor that determines the speed of this wave is the

  (1) nature of the medium
  (2) amplitude of the wave
  (3) phase of the wave
  (4) distance traveled through the medium   16 _____

17 A motor does a total of 480 joules of work in 5.0 seconds to lift a 12-kilogram block to the top of a ramp. The average power developed by the motor is

(1) 8.0 W   (3) 96 W

(2) 40. W   (4) 2400 W   17 _____

18 A 5.8×10^4-watt elevator motor can lift a total weight of 2.1×10^4 newtons with a maximum constant speed of

(1) 0.28 m/s   (3) 2.8 m/s

(2) 0.36 m/s   (4) 3.6 m/s   18 _____

19 A stationary police officer directs radio waves emitted by a radar gun at a vehicle moving toward the officer. Compared to the emitted radio waves, the radio waves reflected from the vehicle and received by the radar gun have a

(1) longer wavelength   (3) longer period

(2) higher speed   (4) higher frequency   19 _____

20 A light wave strikes the Moon and reflects toward Earth. As the light wave travels from the Moon toward Earth, the wave carries

(1) energy, only

(2) matter, only

(3) both energy and matter

(4) neither energy nor matter   20 _____

21 The time required to produce one cycle of a wave is known as the wave's

(1) amplitude   (3) period

(2) frequency   (4) wavelength   21 _____

22 A magnetic compass is placed near an insulated copper wire. When the wire is connected to a battery and a current is created, the compass needle moves and changes its position. Which is the best explanation for the production of a force that causes the needle to move?

  (1) The copper wire magnetizes the compass needle and exerts the force on the compass needle.

  (2) The compass needle magnetizes the copper wire and exerts the force on the compass needle.

  (3) The insulation on the wire becomes charged, which exerts the force on the compass needle.

  (4) The current in the wire produces a magnetic field that exerts the force on the compass needle.   22 _____

23 A beam of monochromatic light ($f = 5.09 \times 10^{14}$ Hz) has a wavelength of 589 nanometers in air. What is the wavelength of this light in Lucite?

  (1) 150 nm   (3) 589 nm

  (2) 393 nm   (4) 884 nm   23 _____

24 If the amplitude of a sound wave is increased, there is an increase in the sound's

  (1) loudness   (3) velocity

  (2) pitch   (4) wavelength   24 _____

25 In the diagram below, point P is located in the electric field between two oppositely charged parallel plates.

Compared to the magnitude and direction of the electrostatic force on an electron placed at point P, the electrostatic force on a proton placed at point P has

(1) the same magnitude and the same direction
(2) the same magnitude, but the opposite direction
(3) a greater magnitude, but the same direction
(4) a greater magnitude and the opposite direction   25 _____

26 The effect produced when two or more sound waves pass through the same point simultaneously is called

(1) interference   (3) refraction
(2) diffraction   (4) resonance   26 _____

27 A gamma ray photon and a microwave photon are traveling in a vacuum. Compared to the wavelength and energy of the gamma ray photon, the microwave photon has a

(1) shorter wavelength and less energy
(2) shorter wavelength and more energy
(3) longer wavelength and less energy
(4) longer wavelength and more energy   27 _____

28 According to the Standard Model of Particle Physics, a neutrino is a type of

(1) lepton   (3) meson
(2) photon   (4) baryon   28 _____

29 Which combination of quarks produces a neutral baryon?

(1) cts   (3) uds

(2) dsb   (4) uct   29 _____

30 When 2.0×10^{-16} kilogram of matter is converted into energy, how much energy is released?

(1) 1.8×10^{-1} J   (3) 6.0×10^{-32} J

(2) 1.8×10^{1} J   (4) 6.0×10^{-8} J   30 _____

31 A ball is hit straight up with an initial speed of 28 meters per second. What is the speed of the ball 2.2 seconds after it is hit? [Neglect friction.]

(1) 4.3 m/s   (3) 22 m/s

(2) 6.4 m/s   (4) 28 m/s   31 _____

32 A particle with a charge of 3.00 elementary charges moves through a potential difference of 4.50 volts. What is the change in electrical potential energy of the particle?

(1) 1.07×10^{-19} eV   (3) 1.50 eV

(2) 2.16×10^{-18} eV   (4) 13.5 eV   32 _____

33 Which circuit has the largest equivalent resistance?

(1)   (3)

(2)   (4)

                                                            33 _____

34 A transverse wave is moving toward the right in a uniform medium. Point X represents a particle of the uniform medium. Which diagram represents the direction of the motion of particle X at the instant shown?

(1) ←——X   Direction of Wave Motion

(2) X——→   Direction of Wave Motion

(3) ↓X   Direction of Wave Motion

(4) X↓   Direction of Wave Motion

34 _____

35 Which diagram represents magnetic field lines between two north magnetic poles?

(1)   (3)

(2)   (4)

35 _____

PART B–1
Answer all questions in this part.

Directions (36–50): For *each* statement or question, select the *number* of the word or expression that, of those given, best completes the statement or answers the question. Some questions may require the use of the *2006 Edition Reference Tables for Physical Setting/Physics*. Record your answers in the spaces provided.

36 Which measurement is closest to 1×10^{-2} meter?

    (1) diameter of an atom
    (2) width of a student's finger
    (3) length of a football field
    (4) height of a schoolteacher   36 _____

37 Which graph represents the relationship between the speed of a freely falling object and the time of fall of the object near Earth's surface?

                                                                     37 _____

38 A hair dryer with a resistance of 9.6 ohms operates at 120 volts for 2.5 minutes. The total electrical energy used by the dryer during this time interval is

    (1) 2.9×10^3 J   (3) 1.7×10^5 J
    (2) 3.8×10^3 J   (4) 2.3×10^5 J   38 _____

39 A box weighing 46 newtons rests on an incline that makes an angle of 25° with the horizontal. What is the magnitude of the component of the box's weight perpendicular to the incline?

(1) 19 N   (3) 42 N
(2) 21 N   (4) 46 N   39 _____

40 Which graph represents the motion of an object traveling with a positive velocity and a negative acceleration?

(1)   (3)

(2)   (4)   40 _____

41 Car A, moving in a straight line at a constant speed of 20. meters per second, is initially 200 meters behind car B, moving in the same straight line at a constant speed of 15 meters per second. How far must car A travel from this initial position before it catches up with car B?

(1) 200 m   (3) 800 m
(2) 400 m   (4) 1000 m   41 _____

42 A 2700-ohm resistor in an electric circuit draws a current of 2.4 milliamperes. The total charge that passes through the resistor in 15 seconds is

(1) 1.6×10^{-4} C   (3) 1.6×10^{-1} C
(2) 3.6×10^{-2} C   (4) 3.6×10^{1} C   42 _____

43 A 1000–kilogram car traveling 20.0 meters per second east experiences an impulse of 2000 newton • seconds west. What is the final velocity of the car after the impulse has been applied?

  (1) 18.0 m/s east   (3) 20.5 m/s west

  (2) 19.5 m/s east   (4) 22.0 m/s west   43 _____

44 Which graph represents the relationship between the potential difference applied to a copper wire and the resulting current in the wire at constant temperature?

(1)

(3)

(2)

(4)   44 _____

45 A tungsten wire has resistance R at 20°C. A second tungsten wire at 20°C has twice the length and half the cross-sectional area of the first wire. In terms of R, the resistance of the second wire is

  (1) $\dfrac{R}{2}$ (3) $2R$

  (2) R (4) $4R$ 45 _____

46 After an incandescent lamp is turned on, the temperature of its filament rapidly increases from room temperature to its operating temperature. As the temperature of the filament increases, what happens to the resistance of the filament and the current through the filament?

  (1) The resistance increases and the current decreases.

  (2) The resistance increases and the current increases.

  (3) The resistance decreases and the current decreases.

  (4) The resistance decreases and the current increases.   46 _____

47 Parallel wave fronts are incident on an opening in a barrier. Which diagram shows the configuration of wave fronts and barrier opening that will result in the greatest diffraction of the waves passing through the opening? [Assume all diagrams are drawn to the same scale.]

(1)

(3)

(2)

(4)

47 _____

48 A singer demonstrated that she could shatter a crystal glass by singing a note with a wavelength of 0.320 meter in air at STP. What was the natural frequency of the glass?

(1) 9.67×10^{-4} Hz

(3) 1.03×10^{3} Hz

(2) 1.05×10^{2} Hz

(4) 9.38×10^{8} Hz

48 _____

49 The diagram below represents a standing wave in a string.

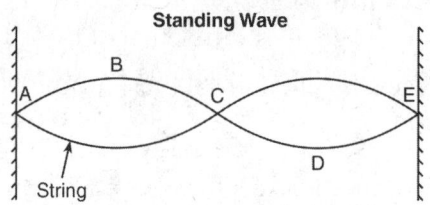

Maximum constructive interference occurs at the

(1) antinodes A, C, and E
(2) nodes A, C, and E
(3) antinodes B and D
(4) nodes B and D

49 _____

50 Which circuit diagram represents voltmeter V connected correctly to measure the potential difference across resistor R_2?

50 _____

PART B–2
Answer all questions in this part.

Directions (51–65): Record your answers on the answer sheet provided after the questions. Some questions may require the use of the *2006 Edition Reference Tables for Physical Setting/Physics.*

Base your answers to questions 51 through 53 on the information and diagram below and on your knowledge of physics.

As represented in the diagram below, a constant 15-newton force, F, is applied to a 2.5-kilogram box, accelerating the box to the right at 2.0 meters per second squared across a rough horizontal surface.

51–52 Calculate the magnitude of the net force acting on the box. [Show all work, including the equation and substitution with units.] [2]

53 Determine the magnitude of the force of friction on the box. [1]

Base your answers to questions 54 and 55 on the information and diagram below and on your knowledge of physics.

A ray of light ($f = 5.09 \times 10^{14}$ Hz) is traveling through a mineral sample that is submerged in water. The ray refracts as it enters the water, as shown in the diagram below.

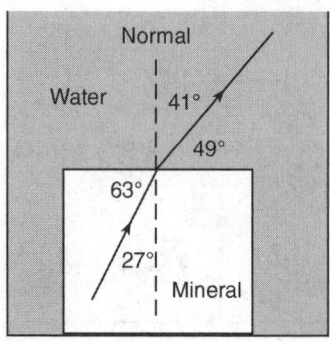

54–55 Calculate the absolute index of refraction of the mineral. [Show all work, including the equation and substitution with units.] [2]

Base your answers to questions 56 through 58 on the information below and on your knowledge of physics.

A ball is rolled twice across the same level laboratory table and allowed to roll off the table and strike the floor. In each trial, the time it takes the ball to travel from the edge of the table to the floor is accurately measured. [Neglect friction.]

56–57 In trial A, the ball is traveling at 2.50 meters per second when it reaches the edge of the table. The ball strikes the floor 0.391 second after rolling off the edge of the table. Calculate the height of the table. [Show all work, including the equation and substitution with units.] [2]

58 In trial B, the ball is traveling at 5.00 meters per second when it reaches the edge of the table. Compare the time it took the ball to reach the floor in trial B to the time it took the ball to reach the floor in trial A. [1]

Base your answers to questions 59 through 61 on the information and diagram below and on your knowledge of physics.

A toy airplane flies clockwise at a constant speed in a horizontal circle of radius 8.0 meters. The magnitude of the acceleration of the airplane is 25 meters per second squared. The diagram shows the path of the airplane as it travels around the circle.

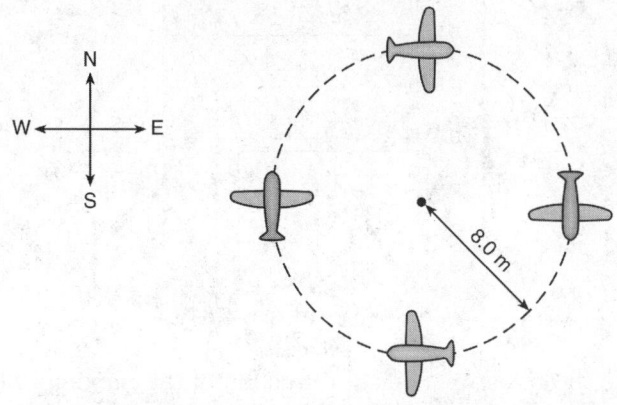

59–60 Calculate the speed of the airplane. [Show all work, including the equation and substitution with units.] [2]

61 State the direction of the velocity of the airplane at the instant the acceleration of the airplane is southward. [1]

Base your answers to questions 62 through 64 on the information and graph below and on your knowledge of physics.

The graph below represents the speed of a marble rolling down a straight incline as a function of time.

Speed vs. Time

62 What quantity is represented by the slope of the graph? [1]

63–64 Calculate the distance the marble travels during the first 3.0 seconds [Show all work, including the equation and substitution with units.] [2]

65 The graph below represents the relationship between weight and mass for objects on the surface of planet X.

Weight vs. Mass on Planet X

Determine the acceleration due to gravity on the surface of planet X. [1]

PART C

Answer all questions in this part.

Directions (66–85): Record your answers on the sheet provided in the back. Some questions may require the use of the *2006 Edition Reference Tables for Physical Setting/Physics*.

Base your answers to questions 66 through 69 on the information and vector diagram below and on your knowledge of physics.

A hiker starts at point *P* and walks 2.0 kilometers due east and then 1.4 kilometers due north. The vectors in the diagram below represent these two displacements.

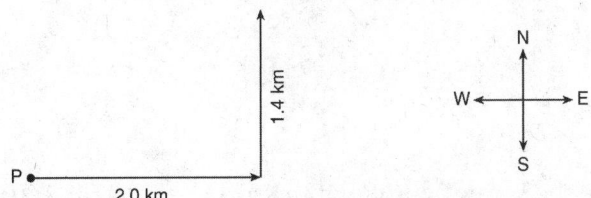

66 Using a metric ruler, determine the scale used in the vector diagram. [1]

67 On the diagram *on your answer sheet*, use a ruler to construct the vector representing the hiker's resultant displacement. [1]

68 Determine the magnitude of the hiker's resultant displacement. [1]

69 Using a protractor, determine the angle between east and the hiker's resultant displacement. [1]

Base your answers to questions 70 through 74 on the information and diagram below and on your knowledge of physics.

A jack-in-the-box is a toy in which a figure in an open box is pushed down, compressing a spring. The lid of the box is then closed. When the box is opened, the figure is pushed up by the spring. The spring in the toy is compressed 0.070 meter by using a downward force of 12.0 newtons.

70–71 Calculate the spring constant of the spring. [Show all work, including the equation and substitution with units.] [2]

72–73 Calculate the total amount of elastic potential energy stored in the spring when it is compressed. [Show all work, including the equation and substitution with units.] [2]

74 Identify *one* form of energy to which the elastic potential energy of the spring is converted when the figure is pushed up by the spring. [1]

Base your answers to questions 75 through 80 on the information below and on your knowledge of physics.

A 12-volt battery causes 0.60 ampere to flow through a circuit that contains a lamp and a resistor connected in parallel. The lamp is operating at 6.0 watts.

75 Using the circuit symbols shown on the *Reference Tables for Physical Setting/Physics*, draw a diagram of the circuit in the space provided *on your answer sheet*. [1]

76–77 Calculate the current through the lamp. [Show all work, including the equation and substitution with units.] [2]

78 Determine the current in the resistor. [1]

79–80 Calculate the resistance of the resistor. [Show all work, including the equation and substitution with units.] [2]

_____

Base your answers to questions 81 through 85 on the information below and on your knowledge of physics.

The Great Nebula in the constellation Orion consists primarily of excited hydrogen gas. The electrons in the atoms of excited hydrogen have been raised to higher energy levels. When these atoms release energy, a frequent electron transition is from the excited $n = 3$ energy level to the $n = 2$ energy level, which gives the nebula one of its characteristic colors.

81 Determine the energy, in electronvolts, of an emitted photon when an electron transition from $n = 3$ to $n = 2$ occurs. [1]

82 Determine the energy of this emitted photon in joules. [1]

83–84 Calculate the frequency of the emitted photon. [Show all work, including the equation and substitution with units.] [2]

85 Identify the color of light associated with this photon. [1]

Answer Sheet
June 2016
Physics: The Physical Setting

PART B–2

51–52

53 _____ N

54–55

56–57

58 _____

_____

59–60

61 _____

62 _____

63–64

5 _____ **m/s^2**

PART C

66 1.0 cm = _____ **km**

67

68 _____ **km**

69 _____ °

70–71

72–73

74 _____

75

76–77

78 _____ **A**

79–80

81 _____ **eV**

82 _____ **J**

83–84

85 _____

Answers
June 2016
Physics: The Physical Setting

Answer Key

PART A

1. 4	8. 1	15. 3	22. 4	29. 3
2. 2	9. 2	16. 1	23. 2	30. 2
3. 4	10. 3	17. 3	24. 1	31. 2
4. 1	11. 3	18. 3	25. 2	32. 4
5. 3	12. 2	19. 4	26. 1	33. 3
6. 2	13. 2	20. 1	27. 3	34. 3
7. 2	14. 2	21. 3	28. 1	35. 4

PART B–1

36. 2	39. 3	42. 2	45. 4	48. 3
37. 4	40. 2	43. 1	46. 1	49. 3
38. 4	41. 3	44. 1	47. 2	50. 3

PART B–2 and **PART C**. *See* **Answers Explained.**

Answers Explained

PART A

1 Weight is a vector quantity, it has magnitude and direction dependent on the direction of acceleration due to gravity.

WRONG CHOICES EXPLAINED:
(1) power, (2) kinetic energy, and (3) speed are all scalar quantities with only magnitude and no direction.

2 The mass of the astronaut on the surface of the Moon is the same as the mass of the astronaut on the surface of the Earth, 65.0 kg. Mass is constant and does not change. Weight is dependent on the force of gravity.

3 If the sum of all forces on an object is equal to zero (i.e., there is no net force), then according to Newton's First Law of Motion, the object is either at rest or moving at constant speed, no acceleration.

WRONG CHOICES EXPLAINED:
Choice (1) is incorrect because it states that the block *must* be at rest. Had it said that the block *may* be at rest, it would have been a correct choice. Choices (2) and (3) are incorrect because, in order to be accelerating or decelerating, there must be a net force (i.e., the sum of all forces acting on the object are not zero).

4 Mass is the quantitative measure of inertia; therefore, increasing mass increases inertia.

WRONG CHOICES EXPLAINED:
(2) Increasing net force would increase acceleration
(3) Increasing time that a net force is applied results in a greater impulse
(4) Increasing the speed of the object would increase its kinetic energy

5 According to Newton's Third Law of Motion, if object A exerts a force on object B, then object B exerts an equal but opposite force on object A.

6 Sound waves are longitudinal mechanical waves.

7 Refer to the *Mechanics* section of *Reference Table K*:

$$F_e = \frac{kq_1q_2}{r^2}$$

Electrostatic force is inversely proportional to the square of the distance between the centers of the point charges. If the electrostatic force is equal to 8.0×10^{-5} and if $r = d$, then

$$8.0 \times 10^{-5} = \frac{kq_1q_2}{d^2}$$

If $r = 2d$, then

$$F_e = \frac{kq_1q_2}{(2d)^2}$$

$$F_e = \frac{kq_1q_2}{4d^2}$$

$$F_e = \left(\frac{1}{4}\right)\left(\frac{kq_1q_2}{d^2}\right)$$

$$F_e = \frac{1}{4}8.0 \times 10^{-5}$$

$$F_e = 2.0 \times 10^{-5}$$

8 When dealing with a word problem requiring a mathematical solution, it is most helpful to translate the words into what quantities/variables are explicitly given and what constants or other variables can be deduced from the situation or words. We also write down the variable that needs to be solved for or found. Then it is much easier to determine which formula is required (refer to the appropriate section of *Reference Table K*), and then use it to solve the problem. It is also easier, and there is less chance of making an error, if we first solve the equation algebraically for the unknown and then substitute the given variables with units into the equation in order to calculate the answer.

Given:	Find:	Solution:

Given:

$p_{blue\ i} = 2.0$ kg • m/s west

$p_{red\ i} = 3.0$ kg • m/s east

Find:

$p_{total\ f} = ?$

Solution:

$p_{total\ i\ +} = p_{total\ f}$

$p_{blue\ i\ +}p_{red\ i} = p_{total\ f}$

2.0 kg • m/s west +

3.0 kg • m/s east $= p_{total\ f}$

1.0 kg • m/s east $= p_{total\ f}$

9 The direction of a frictional force is opposite to the direction of motion. At point *P*, the ball is moving diagonally upward to the right; therefore, friction acts in a downward direction to the left.

10 A magnetic field can be generated by a moving charged particle like a proton.

WRONG CHOICES EXPLAINED:
   Choices (1), (2), and (4) are incorrect because all these particles have no charge.

11 Field lines represent the path that a very small positive charge takes while in the field. The arrows on the field lines indicate the direction. The field lines in the diagram point toward *A*; therefore, *A* must be negative. The lines point toward *B* as well; therefore, *B* must be negative as well.

12 Write out the problem. Refer to the *Mechanics* section of *Reference Table K*.

Given:

$F = 20$ N

$m_{box} = 10$ kg

$d = 5$ m

Find:

$W = ?$

Solution:

$W = Fd$

$W = (20\ N)(5\ m)$

$W = 100$ Nm $= 100$ J

13 Work done on an object increases its kinetic energy. $W = \Delta KE$. Since 15 J of work is done, the kinetic energy increases by 15 J.

14 Refer to the *Mechanics* section of *Reference Table K*:

$$PE = \Delta mgh$$

Potential energy is proportional to mass. If the mass of A is equal to m, then

$$\Delta PE_A = \Delta mg\Delta h$$

If mass $B = 2m$, then

$$\Delta PE_B = (2m)g\Delta h$$
$$\Delta PE_B = (2mg\Delta h)$$
$$\Delta PE_B = 2\Delta PE_A$$

15 Write out the problem. Refer to the *Mechanics* section of *Reference Table K*:

Given:	Find:	Solution:
$m = 55 \text{ kg}$	$KE = ?$	$KE = \frac{1}{2}mv^2$
$v = 9.0 \text{ m/s}$		$KE = \frac{1}{2}(55 \text{ kg})(9.0 \text{ m/s})^2$
		$KE = 2.2 \times 10^3 \frac{kgm^2}{s^2}$
		$KE = 2.2 \times 10^3 \text{ J}$

16 The speed of an electromagnetic wave is less in a material medium than in a vacuum and is dependent on the nature of the medium and the frequency of the light.

17 Write out the problem. Refer to the *Mechanics* section of *Reference Table K*:

Given: Find: Solution:

$W = 480$ J $P = ?$ $P = \dfrac{W}{t}$

$t = 5.0$ s $P = \dfrac{480 \text{ J}}{5.0 \text{ s}}$

$m = 12$ kg $P = 96\dfrac{\text{J}}{\text{s}} = 96$ W

18 Write out the problem. Refer to the *Mechanics* section of *Reference Table K*:

Given: Find: Solution:

$P = 5.8 \times 10^4$ W $\bar{v} = ?$ $P = F\bar{v}$

$Fw = 2.1 \times 10^4$ N $\bar{v} = \dfrac{P}{F}$

$$\bar{v} = \dfrac{5.8 \times 10^4 \text{ W}}{2.1 \times 10^4 \text{ N}}$$

$$\bar{v} = 2.8\dfrac{\text{W}}{\text{N}} = 2.8\,\dfrac{\text{J/s}}{\text{N}} = 2.8\dfrac{\text{Nm/s}}{\text{N}} = 2.8 \text{ m/s}$$

19 This scenario is an example of the Doppler effect. The Doppler effect is the apparent change in frequency that results when a wave source and an observer are in relative motion with respect to each other. In this case, the vehicle is moving toward the stationary police officer. The apparent pitch or frequency is decreased as the sound source moves away from and increased as the sound source moves toward a stationary observer.

20 Waves transfer energy without the transfer of mass.

21 The period is the time required for periodic motion to repeat itself. Choice (3) is correct.

WRONG CHOICES EXPLAINED:
(1) Amplitude is the maximum displacement of any particle in a medium relative to its rest position.
(2) Frequency is the number of repetitions produced per unit time in periodic phenomena.
(4) Wavelength is the length of one complete wave cycle.

22 Moving electric charges produce magnetic fields. The current in the wire generates a magnetic field. The magnetic field exerts a force on the compass needle, which is itself a magnet.

23 Write out the problem. Refer to the *Modern Physics* section of *Reference Table K* and to *Reference Table E*:

Given:

$\lambda_{air} = 589$ nm

$n_{air} = 1.00$

$n_{Lucite} = 1.50$

Find:

$\lambda_{Lucite} = ?$

Solution:

$$\frac{n_2}{n_1} = \frac{\lambda_1}{\lambda_2}$$

$$\frac{n_{Lucite}}{n_{air}} = \frac{\lambda_{air}}{\lambda_{Lucite}}$$

$$\lambda_{Lucite} = \frac{(\lambda_{air})(n_{air})}{n_{Lucite}}$$

$$\lambda_{Lucite} = \frac{(589 \text{ nm})(1.00)}{1.50}$$

$$\lambda_{Lucite} = 393 \text{ nm}$$

WRONG CHOICES EXPLAINED:
If you selected choice (4), you substituted into the equation incorrectly and/or multiplied by 1.50 instead of dividing by 1.50

24 Amplitude is a measure of the energy transmitted by a wave and in sound waves this corresponds to loudness. Choice (1) is correct.

WRONG CHOICES EXPLAINED:
(2) Pitch in sound waves corresponds to frequency.
(3) Velocity of a sound wave is dependent on the properties of the medium.
(4) Wavelength is inversely proportional to the frequency of a wave.

25 Refer to the *Electricity* section in *Reference Table K*:

$$E = \frac{F_e}{q}$$

Protons and electrons have equal but opposite charges; therefore, the magnitude of the force on a electron will be the same as the magnitude on the proton, but it will be in the opposite direction.

26 The effect produced when two or more sound waves pass through the same point simultaneously is called interference. Choice (1) is correct.

WRONG CHOICES EXPLAINED:
(2) Diffraction is the bending of a wave around a barrier.
(3) Refraction is the change in direction of a wave when it passes obliquely from one medium into another in which it moves at a different speed.
(4) Resonance is the spontaneous vibration of an object at a frequency equal to that of the wave that initiates that resonant vibration.

27 Refer to the *Electromagnetic Spectrum* table in *Reference Table D*: Microwave photons have longer wavelengths than gamma ray photons. Refer to the *Modern Physics* section in *Reference Table K*:

$$E_{photon} = \frac{hc}{\lambda}$$

Therefore, microwave photons have less energy than gamma ray photons.

28 Refer to the *Particles of the Standard Model* chart in *Reference Table H*. The neutrino is a type of lepton.

29 Refer to the *Particles of the Standard Model* chart in the *Reference Table H*: A combination of an up, down, and strange quarks will result in a neutral baryon. Choice (3) is correct.

WRONG CHOICES EXPLAINED:

(1) A combination of a charm, top, and strange quarks will result in a particle with a charge of +1.

(2) A combination of a down, strange, and bottom quarks will result in a particle with a charge of −1.

(4) A combination of an up, charm, and top quark will result in a particle with a charge of +2.

30 Write out the problem. Refer to the *Modern Physics* section of *Reference Table K*.

Given:	Find:	Solution:
$m = 2.0 \times 10^{-16}$ kg	$E = ?$	$E = mc^2$
$c = 3.0 \times 10^8$ m/s		$E = (2.0 \times 10^{-16} \text{ kg})(3.0 \times 10^8 \text{ m/s})^2$
		$E = 1.8 \times 10^1$ J

31 Write out the problem. Refer to the *Modern Physics* section of *Reference Table K*:

Given:	Find:	Solution:
$v_i = 28$ m/s	$v_f = ?$	$v_f = v_i + at$
$a = -9.81$ m/s^2		$v_f = \left(28\frac{m}{s}\right) + \left(-9.81\frac{m}{s^2}\right)(2.2 \text{ s})$
$t = 2.2$ s		$v_f = \left(28\frac{m}{s}\right) + \left(-21.582\frac{m}{s}\right)$
		$v_f = 6.4\frac{m}{s}$

32 Refer to the *List of Physical Constants* table (*Reference Table A*): One elementary charge is equivalent to 1.60×10^{-19} C. Write out the problem. Refer to the *Electricity* section of the *Reference Table K*:

Given: Find: Solution:

$q = 4.8 \times 10^{-19}$ C $W = ?$ $V = \dfrac{W}{q}$

$V = 4.50$ V $W = Vq$

$$W = (4.50 \text{ V})(4.8 \times 10^{-19} \text{ C})$$

$$W = 2.16 \times 10^{-18} \text{ J}$$

Refer to the *List of Physical Constants* table (*Reference Table A*): 1 electronvolt (eV) is equivalent to 1.60×10^{-19} J.

$$W = (2.16 \times 10^{-18} \text{J})\left(\frac{1 \text{ eV}}{1.60 \times 10^{-19} \text{ J}}\right) = 13.5 \text{ eV}$$

WRONG CHOICE EXPLAINED:
If you selected choice (2), you forgot to convert joules to electronvolts.

33 In a series circuit, the total equivalent resistance of the circuit is equal to the sum of all the individual resistances. By contrast, in a parallel circuit, the equivalent resistance is less than any single resistance in the circuit.
We can eliminate choices (2) and (4) as they contain the same resistances as choices (1) and (3), respectively, but in parallel versus series circuits, and, thus, their total equivalent resistances will be less.
Choice (3) is correct as this circuit has an equivalent resistance of 10 Ω, whereas choice (1) only has an equivalent resistance of 8 Ω.

34 In a transverse wave, the particles of the medium vibrate or exhibit simple harmonic motion (SHM) about a rest position perpendicular to the direction of motion of the wave. This eliminates choices (1) and (2). Because the wave shown is moving toward the right of the page, at the instant shown, point *X* is moving upward along the crest of the wave. Choice (3) is correct.

35 By convention, magnetic field lines point away from north and towards the south. Since all of the diagrams shown are of 2 north poles of magnets, only the diagram in choice 4 shows the field lines pointing away from both north poles. Choice (4) is correct.

PART B–1

36 This is an order of magnitude question. Ten to the –2 power is equivalent to one hundredth. One hundredth of a meter is equal to 1 cm, which is a reasonable approximation for the width of a very small student's finger. One centimeter is not a reasonable approximation of the diameter of an atom, the length of a football field, or the height of a schoolteacher.

37 The graph in choice (4) illustrates a direct proportional relationship between speed and time. Choice (4) is the correct answer as an object in free fall experiences a constant acceleration due to gravity and speed increases 9.81 m/s every second.

WRONG CHOICES EXPLAINED:
    (1) This graph illustrates constant speed—no change in speed over time.
    (2) This graph illustrates an inverse proportional relationship between speed and time.
    (3) This graph illustrates an exponential relationship between speed and time (i.e., not constant acceleration, but increasing acceleration).

38 Write out the problem. Refer to the *Electricity* section of the *Reference Table K*:

Given:	Find:	Solution:
$R = 9.6\ \Omega$	$W = ?$	$W = \dfrac{V^2 t}{R}$
$V = 120\ \text{V}$		$W = \dfrac{(120\ \text{V})^2 (150\ \text{s})}{9.6\ \Omega}$
$t = 2.5\ \text{min} = 150\ \text{s}$		$W = 2.3 \times 10^5\ \text{J}$

WRONG CHOICE EXPLAINED:
    If you selected choice (2), you forgot to convert 2.5 min into 150 seconds before substituting into the equation.

39 Write out the problem. Refer to the *Mechanics* section of the *Reference Table K*:

Given:	Find:	Solution:
$F_W = 46$ N	$F_\perp = ?$	$F_\perp = F_W \cos \theta$
$\theta = 25°$ to the horizontal		$F_\perp = (46$ N$) \cos(25°)$
		$F_\perp = 42$ N

WRONG CHOICE EXPLAINED:

If you selected choice (1), you used sin instead of cos or you found the cos of the complementary angle 65°. This answer is equivalent to the component of the weight parallel to the surface of the inclined plane.

40 This graph shows an exponential proportional relationship between position and time; there is a positive velocity with a negative acceleration. Position is increasing over time, but the rate at which it is increasing decreases as time goes by.

WRONG CHOICES EXPLAINED:

(1) This graph shows a directly proportional relationship between position and time. Constant positive velocity (i.e., no acceleration).

(3) This graph shows a directly proportional relationship between position and time. Constant negative velocity (i.e., no acceleration).

(4) This graph shows an exponential proportional relationship between position and time, but there is a negative velocity with a negative acceleration.

41 Create a Data Table to compare position of Car *A* to Car *B* over time.

Time (s)	Position of Car *A* (m)	Position of Car *B* (m)
0	0	200
10	200	350
20	400	500
30	600	650
40	800	800

42 Write out the problem. Refer to the *Electricity* section of *Reference Table K*:

Given:	Find:	Solution:
$R = 2700\ \Omega$	$q = ?$	$I = \dfrac{q}{t}$
$I = 2.4\text{ mA} = 2.4 \times 10^{-3}\text{ A}$		$q = It$
$t = 15\text{ s}$		$q = (2.4 \times 10^{-3}\text{ A})(15\text{ s})$
		$q = 3.6 \times 10^{-2}\text{ C}$

WRONG CHOICE EXPLAINED:
If you selected choice (4), you did not convert milliamperes to amperes before substituting into the equation.

43 Write out the problem. Refer to the *Mechanics* section of *Reference Table K*:

Given:
$m = 1000\text{ kg}$
$v_i = 20.0\text{ m/s east}$
$J = 2000\text{ N·s west}$

Find:
$v_f = ?$

Solution:
$$J = \Delta p$$
$$J = m\Delta v$$
$$J = m(v_f - v_i)$$
$$J = mv_f - mv_i$$
$$mv_f = J + mv_i$$
$$v_f = \frac{J + mv_i}{m}$$
$$v_f = \frac{(-2000\text{Ns}) + (1000\text{ kg})\left(20.0\frac{\text{m}}{\text{s}}\right)}{1000\text{ kg}}$$
$$v_f = +18.0\frac{\text{m}}{\text{s}} = 18\text{ m/s east}$$

WRONG CHOICE EXPLAINED:
(4) The impulse is applied in the opposite direction as the initial velocity. You did not account for this by making either the velocity or the impulse negative and as a result obtained this value.

44 Refer to the *Electricity* section of *Reference Table K*: At constant temperature

$$R = \frac{\rho L}{A} \qquad R = \frac{V}{I}$$

Resistance is constant and potential difference is directly proportional to the current. The graph in choice (1) correctly represents a direct proportional relationship.

WRONG CHOICES EXPLAINED:
(2) This graph represents an exponentially proportional relationship.
(3) This graph shows that decreased potential difference results in increased current.
(4) This graph represents an inversely proportional relationship.

45 Refer to the *Electricity* section of *Reference Table K*: If the length of the first wire equals L, and its cross-sectional area equals A at 20°C, then

$$R = \frac{\rho L}{A}$$

If the length of the second wire is $2L$, and the cross-sectional area is ½A, then

$$R_2 = \frac{\rho(2L)}{(0.5A)}$$

$$R_2 = \frac{2(\rho L)}{0.5(A)}$$

$$R_2 = \frac{4(\rho L)}{(A)} = 4R$$

46 For most metallic conductors, as the temperature increases, resistance increases.
Refer to the *Electricity* section of *Reference Table K*:

$$R = \frac{V}{I} \qquad I = \frac{V}{R}$$

Current is inversely proportional to resistance. The greater the resistance, the smaller the current.

47 Diffraction is the bending of a wave around an obstacle. The longer the wavelength of the wave, the larger the amount of diffraction. The greatest bending or diffraction occurs when the width of the gap or opening is approximately the same size as the wavelength of the wave. The diagram in choice (2) accurately represents these criteria.

48 Refer to the *List of Physical Constants* table in *Reference Table A*. Write out the problem. Refer to the *Waves and Optics* section of *Reference Table K*:

Given:	Find:	Solution:
$\lambda = 0.320$ m	$f = ?$	$v = f\lambda$
$v = 3.31 \times 10^2$ m/s		$f = \dfrac{v}{\lambda}$

$$f = \frac{3.31 \times 10^2 \, \text{m/s}}{0.320 \, \text{m}}$$

$$f = 1.03 \times 10^3 \, \text{Hz}$$

49 The points on a standing wave that appear to stand still are called nodes and are the result of total destructive interference. The areas of maximum constructive interference, alternating crest/troughs are the antinodes: in this diagram, points B and D.

50 A voltmeter measures the potential difference across 2 points in a circuit. If we want to measure the potential difference across resistor R_2, then the voltmeter needs to be connected to both ends of resistor R_2. This is correctly represented in choice (3).

PART B–2

51–52 Write out the problem. Refer to the *Mechanics* section of *Reference Table K*:

Given:	Find:	Solution:
$m = 2.5$ kg	$F_{net} = ?$	$a = \dfrac{F_{net}}{m}$
$a = 2.0$ m/s^2		$F_{net} = ma$
		$F_{net} = (2.5 \text{ kg})(2.0 \text{ m/s}^2)$

51 One credit is awarded for the correct equation and substitution of values with units.

52 One credit is awarded for the correct answer with units or for an answer, with units, that is consistent with the answer to question 51.
Note: Students will not be penalized more than 1 credit for errors in units in questions 51 and 52.

53 The net force is the sum of all the applied forces. The force of friction is always in the opposite direction of motion so we subtract the friction force from the applied force to get the net force.

$$F_{net} = F_{applied} - F_f$$
$$F_f = F_{applied} - F_{net}$$
$$F_f = 15 \text{ N} - 5.0 \text{ N}$$
$$F_f = 10 \text{ N}$$

One credit is awarded for the correct answer or for an answer that is consistent with the difference between 15 N and the student's answer to question 52.

54–55 Write out the problem. Refer to the *Waves and Optics* section and the *Absolute Indices of Refraction* table (*Reference Table E*):

Given:	Find:	Solution:
$\theta_{mineral} = 27°$	$n_{mineral} = ?$	$n_1 \sin \theta_1 = n_2 \sin \theta_2$
$\theta_{water} = 41°$		$n_{mineral} \sin \theta_{mineral} = n_{water} \sin \theta_{water}$
$n_{water} = 1.33$		$n_{mineral} = \dfrac{n_{water} \sin \theta_{water}}{\sin \theta_{mineral}}$
		$n_{mineral} = \dfrac{(1.33)(\sin\ 41°)}{(\sin\ 27°)}$
		$n_{mineral} = 1.9$

54 One credit is awarded for the correct equation and substitution of values with units.

55 One credit is awarded for the correct answer or for an answer that is consistent with the answer to question 54.
Note: Students will not be penalized more than 1 credit for errors in units in questions 54 and 55.

56–57 Write out the problem. Refer to the *Mechanics* section of *Reference Table K*:

Given:	Find:	Solution:
$v_{ix} = 2.50$ m/s	$d_{table} = ?$	$d_y = v_{iy}t + \dfrac{1}{2}a_yt^2$
$t = 0.391$ s		$d_y = \left(0\dfrac{m}{s}\right)(0.391\ s) + \dfrac{1}{2}\left(9.81\dfrac{m}{s^2}\right)(0.391\ s)^2$
$v_{iy} = 0$ m/s		$d = 0$ m $+ 0.749881305$
$a_y = 9.81$ m/s^2		$d_y = 0.750$ m

56 One credit is awarded for the correct equation and substitution of values with units.

57 One credit is awarded for the correct answer with units or for an answer, with units, that is consistent with the answer to question 56.

Note: Students will not be penalized more than 1 credit for errors in units in questions 56 and 57.

58 Because we are dealing with a level table, the initial speed of the ball given in both cases is the horizontal speed. The initial vertical speed of the ball in both cases is 0 m/s. Since the table does not change height and the ball accelerates due to gravity the same way in both trials, the time it takes the ball to hit the ground is the same in both trials. What would be different is how far the ball would travel in the horizontal direction—the distance away from the table or the range.

One credit is awarded for stating that the measured times are the same.

59–60 Write out the problem. Refer to the *Mechanics* section of *Reference Table K*:

Given:	Find:	Solution:
$r = 8.0$ m	$v = ?$	$a_c = \dfrac{v^2}{r}$
$a_c = 25$ m/s^2		$v^2 = a_c r$
		$v = \sqrt{a_c r}$
		$v = \sqrt{\left(25\dfrac{\text{m}}{\text{s}^2}\right)(8.0\ \text{m})}$
		$v = \sqrt{200\dfrac{\text{m}^2}{\text{s}^2}}$
		$v = 14$ m/s

59 One credit is awarded for the correct equation and substitution of values with units.

60 One credit is awarded for the correct answer with units or for an answer, with units, that is consistent with the answer to question 59.

Note: Students will not be penalized more than 1 credit for errors in units in questions 58 and 59.

61 The centripetal acceleration is directed exactly southward when the plane is at the top of the circle or the most northward point of the circle. The direction of velocity for an object traveling in circular motion is to the circle. In this instance, directly to the right or east.

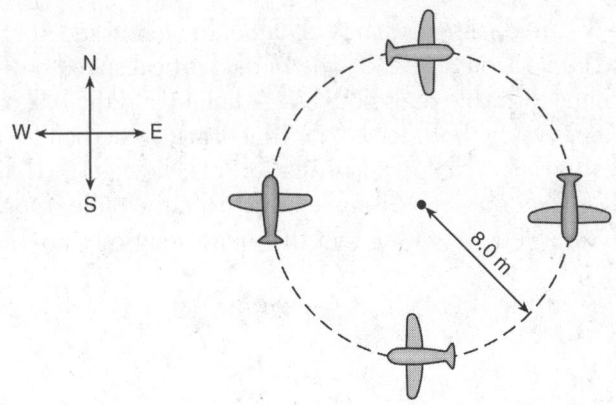

One credit is awarded for indicating/stating that the direction of the airplane's velocity is east or to the right.

62 The slope of a line is defined as the change in y over the change in x. In this graph, y is speed and x is time; thus, the slope on a speed versus time graph is equal to the change in speed over the change in time, which is equal to the acceleration.

$$slope = \frac{\Delta y}{\Delta x} = \frac{\Delta v}{\Delta t} = a$$

One credit is awarded for an answer of acceleration *or* rate of change of velocity (speed).

63–64   Write out the problem. Refer to the *Mechanics* section of *Reference Table K*:

Given: Find: Solution:

$t = 3.0$ s $d = ?$ $\bar{v} = \dfrac{d}{t}$

$v_i = 0$ m/s $d = \bar{v}t$

$v_f = 4$ m/s $d = \left(\dfrac{v_f + v_i}{2}\right)t$

$$d = \left(\dfrac{0 \text{ m/s} + 4 \text{ m/s}}{2}\right)(3.0 \text{ s})$$

$$d = 6.0 \text{ m}$$

        or

$$d = v_i t + \frac{1}{2}at^2$$

$$a = \frac{v}{t}$$

$$d = v_i t + \frac{1}{2}\left(\frac{v}{t}\right)t^2$$

$$a = \left(0\frac{\text{m}}{\text{s}}\right)(3.0 \text{ s}) + \frac{1}{2}\left(4.0 \ \frac{\text{m}}{\text{s}} - 0\frac{\text{m}}{\text{s}}\right)(3.0 \text{ s})$$

$$d = 6.0 \text{ m}$$

   or

The area under the curve in a speed-versus-time graph is equal to the displacement.

Refer to the *Geometry and Trigonometry* section of *Reference Table K*:

$$A = \tfrac{1}{2}bh$$

$$A = \tfrac{1}{2}(3.0 \text{ s})(4.0 \text{ m/s})$$

$$A = 6.0 \text{ m}$$

63   One credit is awarded for a correct equation and substitution of values with units.

64 One credit is awarded for the correct answer with units or for an answer, with units, that is consistent with the answer to question 63.

Note: Students will not be penalized more than 1 credit for errors in units in questions 63 and 64.

65 Refer to the *Mechanics* section of *Reference Table K*:

$$g = \frac{F_g}{m}$$

The slope of a line is defined as the change in y over the change in x. In this graph, y is weight or the force due to gravity and x is mass; thus, the slope on a weight-versus-mass graph is equal to the change in weight over the change in mass, which is equal to the gravitational acceleration.

$$slope = \frac{\Delta y}{\Delta x} = \frac{\Delta F_g}{m} = g$$

$$g = \frac{\Delta F_g}{\Delta m} = \frac{(6.0\ N - 0\ N)}{(2.0\ kg - 0\ kg)} = 3.0\frac{N}{kg} = 3.0\frac{m}{s^2}$$

One credit is awarded for an answer of 3.0 m/s^2 ± 0.25 m/s^2.

PART C

66 If we measure the 2.0-km line using a metric ruler, we find the line to measure approximately 5.0 cm long. 2.0 km/5.0 cm equals 0.4 km/cm.
One credit is awarded for an answer of 0.40 km ± 0.02 km.

67

One credit is awarded for constructing an arrow starting at point P and ending with an arrowhead at the top of the 1.4-km vector or for a vector 6.1 cm ± 0.2 cm long at an angle of 35° north of the 2.0-km vector.

68 If we measure the vector drawn with a metric ruler, we measure approximately 6.0 cm, and according to our scale 1.0 cm = 0.4 km, this translates (6.0 cm × 0.4k m/cm) to 2.4 km.

Alternatively, we use the technique of vector addition. We recognize that the vectors form a right triangle and the resultant vector would be the hypotenuse of a right triangle. We can use the Pythagorean theorem to solve algebraically for the resultant vector.

Refer to the *Geometry and Trigonometry* section of *Reference Table K*:

$$a^2 + b^2 = c^2$$
$$c = \sqrt{a^2 + b^2}$$
$$c = \sqrt{(2.0 \text{ km})^2 + (1.4 \text{ km})^2}$$
$$c = 2.4 \text{ km}$$

One credit is awarded for an answer of 2.4 ± 0.2 km or for an answer that is the product of the length of the resultant (in centimeters) drawn in question 67 and the response to question 66.

69 Assuming the vector was drawn correctly in question 67, a protractor properly placed at point *P* will yield an answer of 35°.

One credit is awarded for an answer of 35° ± 2° or for an answer that is the angle between the 2.0-km vector and the response to question 67.

70–71 Write out the problem. Refer to the *Mechanics* section of *Reference Table K*:

Given:	Find:	Solution:
$x = 0.070$ m	$k = ?$	$F_s = kx$
$F_s = 12.0$ N		$k = \dfrac{F_s}{x}$
		$k = \dfrac{12.0 \text{ N}}{0.070 \text{ m}}$
		$k = 170$ N/m

70 One credit is awarded for a correct equation and substitution of values with units.

71 One credit is awarded for the correct answer with units or for an answer, with units, that is consistent with the answer to question 70.
Note: Students will not be penalized more than 1 credit for errors in units in questions 70 and 71.

72–73 Write out the problem. Refer to the *Mechanics* section of *Reference Table K*:

Given:	Find:	Solution:
$k = 170$ N/m	$PE_s = ?$	$PE_s = \frac{1}{2}kx^2$
$x = 0.070$ m		$PE_s = \frac{1}{2}\left(170\frac{\text{N}}{\text{m}}\right)(0.070 \text{ m})^2$
		$PE_s = 0.42$ Nm $= 0.42$ J

72 One credit is awarded for a correct equation and substitution of values with units.

73 One credit is awarded for the correct answer with units or for an answer, with units, that is consistent with the answer to question 72.
Note: Students will not be penalized more than 1 credit for errors in units in questions 72 and 73.

74 One credit is awarded for an acceptable response. Acceptable responses include, but are not limited to,

- kinetic energy (energy that an object possesses because of its motion)
- sound (as the lid opens and it pops out there would be sounds)
- internal energy (thermal energy—the total kinetic and potential energy associated with the atoms and molecules of an object)
- gravitational potential energy (the energy that an object acquires as a result of the work done in moving the object against a gravitational force)

75 Refer to the *Circuit Symbols* section in *Reference Table I*.

One credit is awarded for a correct circuit diagram.

Note: Credit is allowed if the diagram uses a cell symbol instead of a battery symbol.

76–77 Write out the problem. Refer to the *Electricity* section of *Reference Table K*:

Given:	Find:	Solution:
$V = 12.0$ V	$I_{lamp} = ?$	$P_{lamp} = VI_{lamp}$
$I_{total} = 0.60$ A		$I = \dfrac{P}{V}$
$P_{lamp} = 6.0$ W		$I = \dfrac{6.0 \text{ W}}{12.0 \text{ V}}$
		$I = 0.50$ A

76 One credit is awarded for a correct equation and substitution of values with units.

77 One credit is awarded for the correct answer with units or for an answer, with units, that is consistent with the answer to question 76.

Note: Students will not be penalized more than 1 credit for errors in units in questions 76 and 77.

78 This is a parallel circuit. The total current is equal to the sum of the currents in the individual pathways.

Given: Find: Solution:

$I_{total} = 0.60$ A $I_{resistor} = ?$ $I_{total} = I_{lamp} + I_{resistor}$

$I_{lamp} = 0.50$ A $I_{resistor} = I_{total} - I_{lamp}$

$I_{resistor} = 0.60$ A $- 0.50$ A

$I_{resistor} = 0.10$ A

One credit is awarded for an answer of 0.10 A or for an answer that is the difference between the response to question 77 and 0.60 A or for an answer that is consistent with the response to question 75.

79–80 Write out the problem. Refer to the *Electricity* section of *Reference Table K*:

Given: Find: Solution:

$V = 12.0$ V $R_{resistor} = ?$ $V = IR$

$I_{resistor} = 0.10$ A $R = \dfrac{V}{I}$

$R = \dfrac{12.0 \text{ V}}{0.10 \text{ A}}$

$R = 120 \ \Omega$

79 One credit is awarded for a correct equation and substitution of values with units or for an answer that is consistent with the response given in question 78.

80 One credit is awarded for the correct answer with units or for an answer, with units, that is consistent with the answer to question 79.
Note: Students will not be penalized more than 1 credit for errors in units in questions 79 and 80.

81 Refer to the *Energy Level Diagram for Hydrogen* in *Reference Table F*. Find the difference in energy between level 3 and level 2.

$$(-1.51 - (-3.40) = 1.89 \text{ eV})$$

One credit is awarded for answer of 1.89 eV.

82 Write out the problem. Refer to *Reference Table A*:

Given:	Find:	Solution:
$E_{photon} = 1.89$ eV	E_{photon} in Joules	1.89 eV$(1.6 \times 10^{-19}$ J/eV) =
1 eV $= 1.6 \times 10^{-19}$ J		3.02×10^{-19} J

One credit is awarded for an answer of 3.02×10^{-19} J or for an answer that is the product of the response to question 81 and 1.6×10^{-19} J/eV

83–84 Write out the problem. Refer to the *Modern Physics* section of *Reference Table K*.
Refer to *Reference Table A*:

Given:	Find:	Solution:
$E_{photon} = 3.02 \times 10^{-19}$ J	$f = ?$	$E_{photon} = hf$
$h = 6.63 \times 10^{-34}$ Js		$f = \dfrac{E_{photon}}{h}$
		$f = \dfrac{3.02 \times 10^{-19} \text{J}}{6.63 \times 10^{-34} \text{Js}}$
		$f = 4.56 \times 10^{14}$ Hz

83 One credit is awarded for the correct equation and substitution of values with units or for an answer with units that is consistent with the response to question 82.

84 One credit is awarded for a correct answer with units or for an answer, with units, that is consistent with the answer to question 83.
Note: Students will not be penalized more than 1 credit for errors in units in questions 83 and 84.

85 Refer to the *Electromagnetic Spectrum* diagram in *Reference Table D*. A $f = 4.56 \times 10^{14}$ Hz is within the visible light spectrum, specifically visible red light.
One credit is awarded for an answer of red light or for an answer that is consistent with the answer to question 84.

Topic	Question Numbers (total)	Wrong Answers (x)	Grade
Math Skills	2, 5, 7, 8, 12, 13, 14, 15, 17, 18, 23, 30, 31, 32, 33, 36, 37, 38, 39, 40, 41, 42, 43, 44, 45, 46, 48, 51, 52, 53, 54, 55, 56, 57, 59, 60, 62, 63, 64, 65, 66, 68, 70, 71, 72, 73, 76, 77, 78, 79, 80, 81, 82, 83, 84: (55)		$\dfrac{100(55-x)}{55} = \%$
Mechanics	1–5, 8, 9, 25, 31, 37, 39, 51–53, 56–64, 65, 67–69, 70, 71: (29)		$\dfrac{100(29-x)}{29} = \%$
Energy	12–15, 17, 18, 72, 73, 74: (9)		$\dfrac{100(9-x)}{9} = \%$
Electricity/ Magnetism	7, 10, 11, 22, 32, 33, 35, 36, 38, 42, 44, 45, 46, 50, 75, 76, 77, 78, 80: (19)		$\dfrac{100(19-x)}{19} = \%$
Waves	6, 16, 19, 20, 21, 24, 23, 26, 34, 47, 48, 49: (12)		$\dfrac{100(12-x)}{12} = \%$
Modern Physics	27, 28, 29, 30, 81, 82, 83, 84, 85: (9)		$\dfrac{100(9-x)}{9} = \%$

Examination
June 2017
Physics: The Physical Setting

PART A
Answer all questions in this part.

Directions (1–35): For *each* statement or question, select the *number* of the word or expression that, of those given, best completes the statement or answers the question. Some questions may require the use of the *2006 Edition Reference Tables for Physical Setting/Physics*. Record your answers in the spaces provided.

1 A unit used for a vector quantity is

(1) watt
(2) newton
(3) kilogram
(4) second

1 ____

2 A displacement vector with a magnitude of 20 meters could have perpendicular components with magnitudes of

(1) 10 m and 10 m
(2) 12 m and 8.0 m
(3) 12 m and 16 m
(4) 16 m and 8.0 m

2 ____

3 A hiker travels 1.0 kilometer south, turns and travels 3.0 kilometers west, and then turns and travels 3.0 kilometers north. What is the total distance traveled by the hiker?

(1) 3.2 km
(2) 3.6 km
(3) 5.0 km
(4) 7.0 km

3 ____

4 A car with an initial velocity of 16.0 meters per second east slows uniformly to 6.0 meters per second east in 4.0 seconds. What is the acceleration of the car during this 4.0-second interval?

(1) 2.5 m/s^2 west
(2) 2.5 m/s^2 east
(3) 4.0 m/s^2 west
(4) 4.0 m/s^2 east

4 ____

5 On the surface of planet X, a body with a mass of 10 kilograms weighs 40 newtons. The magnitude of the acceleration due to gravity on the surface of planet X is

(1) 4.0×10^3 m/s^2 (3) 9.8 m/s^2

(2) 4.0×10^2 m/s^2 (4) 4.0 m/s^2 5 _____

6 A car traveling in a straight line at an initial speed of 8.0 meters per second accelerates uniformly to a speed of 14 meters per second over a distance of 44 meters. What is the magnitude of the acceleration of the car?

(1) 0.41 m/s^2 (3) 3.0 m/s^2

(2) 1.5 m/s^2 (4) 2.2 m/s^2 6 _____

7 An object starts from rest and falls freely for 40 meters near the surface of planet P. If the time of fall is 4.0 seconds, what is the magnitude of the acceleration due to gravity on planet P?

(1) 0 m/s^2 (3) 5.0 m/s^2

(2) 1.3 m/s^2 (4) 10 m/s^2 7 _____

8 If a block is in equilibrium, the magnitude of the block's acceleration is

(1) zero   (3) increasing

(2) decreasing   (4) constant, but not zero   8 _____

9 The diagram below shows a light ray striking a plane mirror.

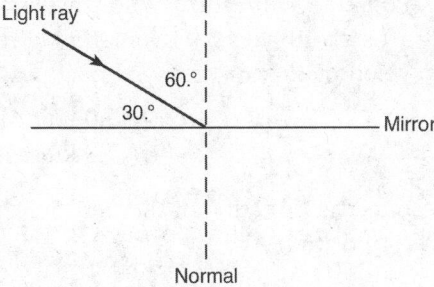

What is the angle of reflection?

(1) $30°$ (3) $90°$

(2) $60°$ (4) $120°$ 9 _____

10 An electric field exerts an electrostatic force of magnitude 1.5×10^{-14} newton on an electron within the field. What is the magnitude of the electric field strength at the location of the electron?

(1) 2.4×10^{-33} N/C   (3) 9.4×10^{4} N/C

(2) 1.1×10^{-5} N/C   (4) 1.6×10^{16} N/C   10 _____

11 A 7.0-kilogram cart, A, and a 3.0-kilogram cart, B, are initially held together at rest on a horizontal, frictionless surface. When a compressed spring attached to one of the carts is released, the carts are pushed apart. After the spring is released, the speed of cart B is 6.0 meters per second, as represented in the diagram below.

Frictionless surface

What is the speed of cart A after the spring is released?

(1) 14 m/s   (3) 3.0 m/s

(2) 6.0 m/s   (4) 2.6 m/s   11 _____

12 An electron in a magnetic field travels at constant speed in the circular path represented in the diagram below.

Which arrow represents the direction of the net force acting on the electron when the electron is at position A?

(1)   (2)   (3)   (4)   12 _____

13 The potential difference between two points, A and B, in an electric field is 2.00 volts. The energy required to move a charge of 8.00×10^{-19} coulomb from point A to point B is

(1) 4.00×10^{-19} J (3) 6.25×10^{17} J
(2) 1.60×10^{-18} J (4) 2.50×10^{18} J   13 _____

14 Which statement describes the gravitational force and the electrostatic force between two charged particles?

(1) The gravitational force may be either attractive or repulsive, whereas the electrostatic force must be attractive.
(2) The gravitational force must be attractive, whereas the electrostatic force may be either attractive or repulsive.
(3) Both forces may be either attractive or repulsive.
(4) Both forces must be attractive.   14 _____

15 An electrostatic force exists between two $+3.20 \times 10^{-19}$-coulomb point charges separated by a distance of 0.030 meter. As the distance between the two point charges is *decreased*, the electrostatic force of

(1) attraction between the two charges decreases
(2) attraction between the two charges increases
(3) repulsion between the two charges decreases
(4) repulsion between the two charges increases   15 _____

16 What is the energy of the photon emitted when an electron in a mercury atom drops from energy level f to energy level b?

(1) 8.42 eV (3) 3.06 eV
(2) 5.74 eV (4) 2.68 eV   16 _____

17 An observer counts 4 complete water waves passing by the end of a dock every 10 seconds. What is the frequency of the waves?

(1) 0.40 Hz (3) 40 Hz
(2) 2.5 Hz (4) 4.0 Hz   17 _____

18 Copper is a metal commonly used for electrical wiring in houses. Which metal conducts electricity better than copper at 20°C?

(1) aluminum   (3) nichrome

(2) gold   (4) silver   18 _____

19 A motor does 20 joules of work on a block, accelerating the block vertically upward. Neglecting friction, if the gravitational potential energy of the block increases by 15 joules, its kinetic energy

(1) decreases by 5 J   (3) decreases by 35 J

(2) increases by 5 J   (4) increases by 35 J   19 _____

20 When only one lightbulb blows out, an entire string of decorative lights goes out. The lights in this string must be connected in

(1) parallel with one current pathway

(2) parallel with multiple current pathways

(3) series with one current pathway

(4) series with multiple current pathways   20 _____

21 An electric toaster is rated 1200 watts at 120 volts. What is the total electrical energy used to operate the toaster for 30 seconds?

(1) 1.8×10^3 J   (3) 1.8×10^4 J

(2) 3.6×10^3 J   (4) 3.6×10^4 J   21 _____

22 What is the rate at which work is done in lifting a 35-kilogram object vertically at a constant speed of 5.0 meters per second?

(1) 1700 W   (3) 180 W

(2) 340 W   (4) 7.0 W   22 _____

23 When a wave travels through a medium, the wave transfers

(1) mass, only

(2) energy, only

(3) both mass and energy

(4) neither mass nor energy   23 _____

24 Glass may shatter when exposed to sound of a particular frequency. This phenomenon is an example of

(1) refraction   (3) resonance
(2) diffraction   (4) the Doppler effect   24 _____

25 Which waves require a material medium for transmission?

(1) light waves   (3) sound waves
(2) radio waves   (4) microwaves   25 _____

26 Which type of oscillation would most likely produce an electromagnetic wave?

(1) a vibrating tuning fork
(2) a washing machine agitator at work
(3) a swinging pendulum
(4) an electron traveling back and forth in a wire   26 _____

27 If monochromatic light passes from water into air with an angle of incidence of 35°, which characteristic of the light will remain the same?

(1) frequency   (3) speed
(2) wavelength   (4) direction   27 _____

28 The absolute index of refraction of medium Y is twice as great as the absolute index of refraction of medium X. As a light ray travels from medium X into medium Y, the speed of the light ray is

(1) halved   (3) quartered
(2) doubled   (4) quadrupled   28 _____

29 The diagram below shows a transverse wave moving toward the right along a rope.

At the instant shown, point *P* on the rope is moving toward the

(1) bottom of the page   (3) left

(2) top of the page   (4) right   29 _____

30 When an isolated conductor is placed in the vicinity of a positive charge, the conductor is attracted to the charge. The charge of the conductor

(1) must be positive

(2) must be negative

(3) could be neutral or positive

(4) could be neutral or negative   30 _____

31 The quarks that compose a baryon may have charges of

(1) $+\frac{2}{3}e, +\frac{2}{3}e,$ and $-\frac{1}{3}e$

(2) $+\frac{1}{3}e, +\frac{1}{3}e,$ and $-\frac{2}{3}e$

(3) $-1e, -1e,$ and 0

(4) $+\frac{2}{3}e, +\frac{2}{3}e,$ and 0 31 _____

32 A rubber block weighing 60 newtons is resting on a horizontal surface of dry asphalt. What is the magnitude of the minimum force needed to start the rubber block moving across the dry asphalt?

(1) 32 N   (3) 51 N

(2) 40 N   (4) 60 N   32 _____

33 The data table below lists the mass and speed of four different objects.

Object	Mass (kg)	Speed (m/s)
A	2.0	6.0
B	4.0	5.0
C	6.0	4.0
D	8.0	2.0

Which object has the greatest inertia?

(1) *A*　　　　　　　　　　　　(3) *C*
(2) *B*　　　　　　　　　　　　(4) *D*　　　　　33 _____

34 The electroscope shown in the diagram below is made completely of metal and consists of a knob, a stem, and leaves. A positively charged rod is brought near the knob of the electroscope and then removed.

The motion of the leaves results from electrons moving from the

(1) leaves to the knob, only
(2) knob to the leaves, only
(3) leaves to the knob and then back to the leaves
(4) knob to the leaves and then back to the knob　　　　　34 _____

35 Which circuit diagram represents the correct way to measure the current in a resistor?

(1)   (3)

(2)   (4)

35 _____

PART B-1
Answer all questions in this part.

Directions (36–50): For *each* statement or question, select the *number* of the word or expression that, of those given, best completes the statement or answers the question. Some questions may require the use of the *2006 Edition Reference Tables for Physical Setting/Physics*. Record your answers in the spaces provided.

36 The height of a typical kitchen table is approximately

  (1) 10^{-2} m   (3) 10^{1} m

  (2) 10^{0} m   (4) 10^{2} m   36 _____

37 A ball is thrown with a velocity of 35 meters per second at an angle of 30° above the horizontal. Which quantity has a magnitude of zero when the ball is at the highest point in its trajectory?

  (1) the acceleration of the ball

  (2) the momentum of the ball

  (3) the horizontal component of the ball's velocity

  (4) the vertical component of the ball's velocity   37 _____

38 The graph below represents the relationship between velocity and time of travel for a toy car moving in a straight line.

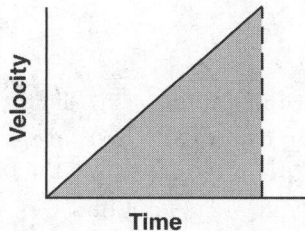

The shaded area under the line represents the toy car's

  (1) displacement   (3) acceleration

  (2) momentum   (4) speed   38 _____

39 A spring stores 10 joules of elastic potential energy when it is compressed 0.20 meter. What is the spring constant of the spring?

  (1) 5.0×10^{1} N/m   (3) 2.5×10^{2} N/m

  (2) 1.0×10^{2} N/m   (4) 5.0×10^{2} N/m   39 _____

Base your answers to questions 40 and 41 on the information below and on your knowledge of physics.

A cannonball with a mass of 1.0 kilogram is fired horizontally from a 500-kilogram cannon, initially at rest, on a horizontal, frictionless surface. The cannonball is acted on by an average force of 8.0×10^3 newtons for 1.0×10^{-1} second.

40 What is the magnitude of the change in momentum of the cannon-ball during firing?

(1) 0 kg • m/s   (3) 8.0×10^3 kg • m/s

(2) 8.0×10^2 kg • m/s   (4) 8.0×10^4 kg • m/s   40 _____

41 What is the magnitude of the average net force acting on the cannon?

(1) 1.6 N   (3) 8.0×10^3 N

(2) 16 N   (4) 4.0×10^6 N   41 _____

42 A metal sphere, X, has an initial net charge of -6×10^{-6} coulomb and an identical sphere, Y, has an initial net charge of $+2 \times 10^{-6}$ coulomb. The spheres touch each other and then separate. What is the net charge on sphere X after the spheres have separated?

(1) 0 C   (3) -4×10^{-6} C

(2) -2×10^{-6} C   (4) -6×10^{-6} C   42 _____

43 A constant eastward horizontal force of 70 newtons is applied to a 20-kilogram crate moving toward the east on a level floor. If the frictional force on the crate has a magnitude of 10 newtons, what is the magnitude of the crate's acceleration?

(1) 0.50 m/s^2 (3) 3.0 m/s^2

(2) 3.5 m/s^2 (4) 4.0 m/s^2 43 _____

44 Which graph represents the relationship between the energy of photons and the wavelengths of photons in a vacuum?

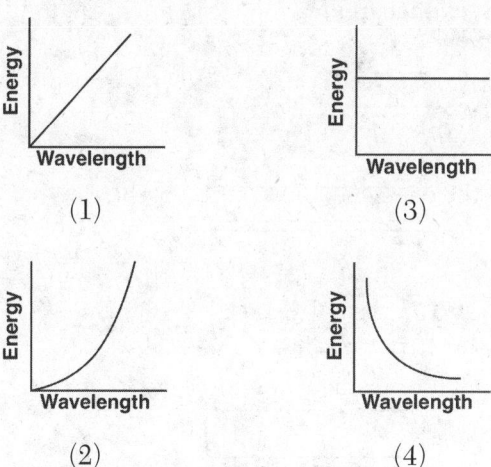

(1)   (3)

(2)   (4)   44 ____

Base your answers to questions 45 and 46 on the information and diagram below and on your knowledge of physics.

One end of a long spring is attached to a wall. A student vibrates the other end of the spring vertically, creating a wave that moves to the wall and reflects back toward the student, resulting in a standing wave in the spring, as represented in the diagram.

45 What is the phase difference between the incident wave and the reflected wave at point P?

(1) 0°   (3) 180°
(2) 90°   (4) 270°   45 ____

46 What is the total number of antinodes on the standing wave in the diagram?

(1) 6   (3) 3
(2) 2   (4) 4   46 ____

47 The diagrams below represent four pieces of copper wire at 20°C. For each piece of wire, ℓ represents a unit of length and A represents a unit of cross-sectional area.

The piece of wire that has the greatest resistance is

(1) wire 1   (3) wire 3
(2) wire 2   (4) wire 4   47 _____

Base your answers to questions 48 and 49 on the diagram below, which represents two charged, identical metal spheres, and on your knowledge of physics.

48 The number of excess elementary charges on sphere A is

(1) 6.4×10^{-25} (3) 2.5×10^{13}
(2) 6.4×10^{-19} (4) 5.0×10^{13} 48 _____

49 What is the magnitude of the electric force between the two spheres?

(1) 3.0×10^{-12} N   (3) 2.7×10^{-2} N
(2) 1.0×10^{-6} N   (4) 5.4×10^{-2} N   49 _____

50 The diagram below represents the wave fronts produced by a point source moving to the right in a uniform medium. Observers are located at points A and B.

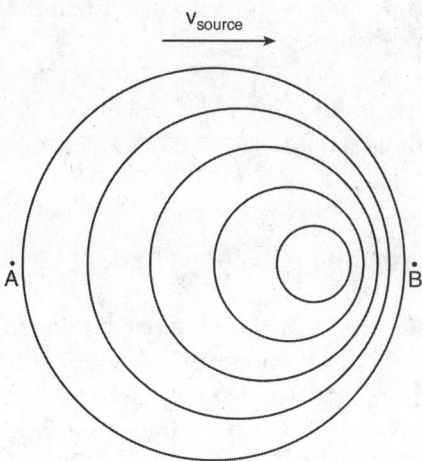

Compared to the wave frequency and wavelength observed at point A, the wave observed at point B has a

(1) higher frequency and a shorter wavelength
(2) higher frequency and a longer wavelength
(3) lower frequency and a shorter wavelength
(4) lower frequency and a longer wavelength   50 _____

PART B–2
Answer all questions in this part.

Directions (51–65): Record your answers on the answer sheet provided after the questions. Some questions may require the use of the *2006 Edition Reference Tables for Physical Setting/Physics.*

51 On the diagram *on your answer sheet*, sketch *at least four* magnetic field lines of force around a bar magnet. [Include arrows to show the direction of each field line.] [1]

Base your answers to questions 52 through 54 on the information below and on your knowledge of physics.

Tritium is a radioactive form of the element hydrogen. A tritium nucleus is composed of one proton and two neutrons. When a tritium nucleus decays, it emits a beta particle (an electron) and an antineutrino to create a stable form of helium. During beta decay, a neutron is spontaneously transformed into a proton, an electron, and an antineutrino.

52 What is the total number of quarks in a tritium nucleus? [1]

53 What is the total charge, in elementary charges, of a proton, an electron, and an antineutrino? [1]

54 What fundamental interaction is responsible for binding together the protons and neutrons in a helium nucleus? [1]

55 The diagram below represents a ball projected horizontally from a cliff at a speed of 10 meters per second. The ball travels the path shown and lands at time *t* and distance *d* from the base of the cliff. [Neglect friction.]

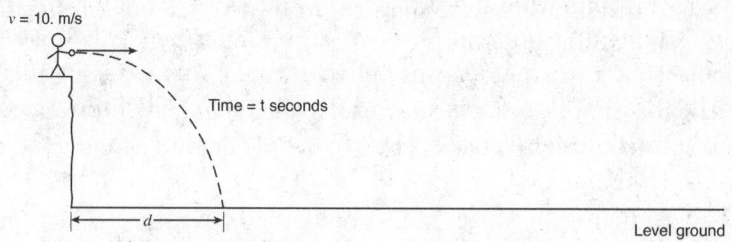

A second, identical ball is projected horizontally from the cliff at 20 meters per second. Determine the distance the second ball lands from the base of the cliff in terms of *d*. [1]

56–57 An operating television set draws 0.71 ampere of current when connected to a 120-volt outlet. Calculate the time it takes the television to consume 3.0×10^5 joules of electric energy. [Show all work, including the equation and substitution with units.] [2]

58–59 On the centimeter grid *on your answer sheet*, draw *at least one* cycle of a periodic transverse wave with an amplitude of 2.0 centimeters and a wavelength of 6.0 centimeters. [2]

60 The diagram below represents a 35-newton block hanging from a vertical spring, causing the spring to elongate from its original length.

Determine the spring constant of the spring. [1]

61 Determine the amount of matter, in kilograms, that must be converted to energy to yield 1.0 gigajoule. [1]

62 Thunder results from the expansion of air as lightning passes through it. The distance between an observer and a lightning strike may be determined if the time that elapses between the observer seeing the lightning and hearing the thunder is known. Explain why the lightning strike is seen before the thunder is heard. [1]

63–64 A bolt of lightning transfers 28 coulombs of charge through an electric potential difference of 3.2×10^7 volts between a cloud and the ground in 1.5×10^{-3} second. Calculate the average electric current between the cloud and the ground during this transfer of charge. [Show all work, including the equation and substitution with units.] [2]

65 The diagram below represents two pulses traveling toward each other in a uniform medium.

On the grid *on your answer sheet*, draw the resultant displacement of the medium when both pulses are located between points *A* and *B*. [1]

PART C
Answer all questions in this part.

Directions (66–85): Record your answers on the sheet provided in the back. Some questions may require the use of the 2006 *Edition Reference Tables for Physical Setting/Physics.*

Base your answers to questions 66 through 70 on the information and diagram below and on your knowledge of physics.

As represented in the diagram, a ski area rope-tow pulls a 72.0-kilogram skier from the bottom to the top of a 40.0-meter-high hill. The rope-tow exerts a force of magnitude 158 newtons to move the skier a total distance of 230.0 meters up the side of the hill at constant speed.

(Not to scale)

66 Determine the total amount of work done by the rope on the skier. [1]

67–68 Calculate the total amount of gravitational potential energy gained by the skier while moving up the hill. [Show all work, including the equation and substitution with units.] [2]

69 Describe what happens to the internal energy of the skier-hill system as the skier is pulled up the hill. [1]

70 Describe what happens to the total mechanical energy of the skier-hill system as the skier is pulled up the hill. [1]

Base your answers to questions 71 through 76 on the diagram and information below and on your knowledge of physics.

A 15-ohm resistor, .30-ohm resistor, and an ammeter are connected as shown with a 60-volt battery.

71–72 Calculate the equivalent resistance of R_1 and R_2. [Show all work, including the equation and substitution with units.] [2]

73 Determine the current measured by the ammeter. [1]

74–75 Calculate the rate at which the battery supplies energy to the circuit. [Show all work, including the equation and substitution with units.] [2]

76 If another resistor were added in parallel to the original circuit, what effect would this have on the current through resistor R_1? [1]

Base your answers to questions 77 through 80 on the information below and on your knowledge of physics.

A gas-powered model airplane has a mass of 2.50 kilograms. A student exerts a force on a cord to keep the airplane flying around her at a constant speed of 18.0 meters per second in a horizontal, circular path with a radius of 25.0 meters.

77–78 Calculate the kinetic energy of the moving airplane. [Show all work, including the equation and substitution with units.] [2]

79–80 Calculate the magnitude of the centripetal force exerted on the airplane to keep it moving in this circular path. [Show all work, including the equation and substitution with units.] [2]

Base your answers to questions 81 through 85 on the information and diagram below and on your knowledge of physics.

A ray of light with a frequency of 5.09×10^{14} hertz traveling in medium X is refracted at point P. The angle of refraction is 90°, as represented in the diagram.

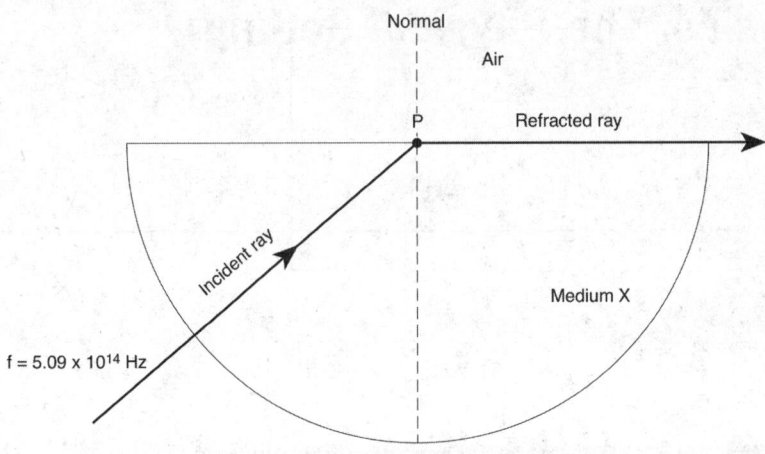

81–82 Calculate the wavelength of the light ray in air. [Show all work, including the equation and substitution with units.] [2]

   83 Measure the angle of incidence for the light ray incident at point P and record the value *on your answer sheet*. [1]

84–85 Calculate the absolute index of refraction for medium X. [Show all work, including the equation and substitution with units.] [2]

Answer Sheet
June 2017

Physics: The Physical Setting

PART B–2

51

52 _____ **quarks**

53 _____ **e**

54 _____

55 _____

56–57

58–59

60 _____ **N/m**

61 _____ **kg**

62 _____

63–64

65

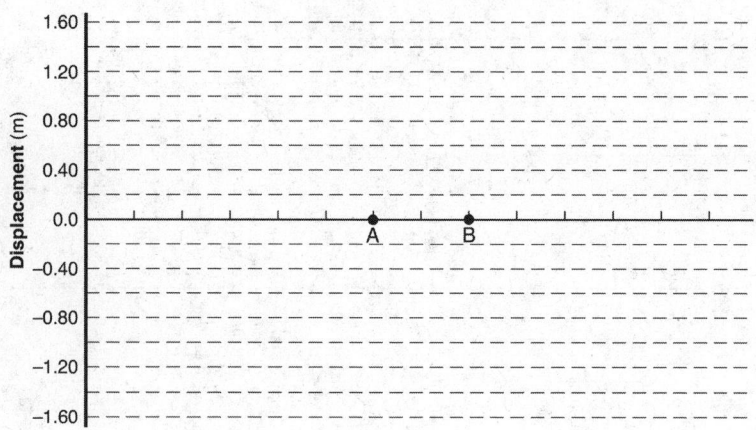

PART C

66 _____**J**

67–68

69 _____

70 _____

71–72

73 _____ **A**

74–75

76 _____

77–78

9–80

1–82

3 _____ °

4–85

Answers
June 2017

Physics: The Physical Setting

Answer Key

PART A

1. 2	8. 1	15. 4	22. 1	29. 2
2. 3	9. 2	16. 3	23. 2	30. 4
3. 4	10. 3	17. 1	24. 3	31. 1
4. 1	11. 4	18. 4	25. 3	32. 3
5. 4	12. 3	19. 2	26. 4	33. 4
6. 2	13. 2	20. 3	27. 1	34. 3
7. 3	14. 2	21. 4	28. 1	35. 1

PART B–1

36. 2	39. 4	42. 2	45. 3	48. 3
37. 4	40. 2	43. 3	46. 3	49. 3
38. 1	41. 3	44. 4	47. 2	50. 1

PART B–2 and **PART C**. *See* **Answers Explained**.

Answers Explained

PART A

1 The newton is the unit of measure for force, which is a vector quantity. Force is a vector quantity because it has both magnitude and direction.

WRONG CHOICES EXPLAINED:
(1), (3), and (4) Watt, kilogram, and second are all units representing scalar quantities. They represent power, mass, and time, respectively. These scalar quantities have magnitude but they do not have direction.

2 A displacement vector is formed by two perpendicular vectors. The three vectors form a right triangle. In this question, the resultant displacement vector measures 20 m. Any triangle whose sides are in a 3 : 4 : 5 ratio is a right triangle. The vectors 12 m, 16 m, and 20. m are in a 3 : 4 : 5 ratio, so they can be the perpendicular components and the given displacement vector.

WRONG CHOICES EXPLAINED:
(1), (2), (4) None of these answer choices can form a right triangle where the longest side measures 20 m.

3 To solve for total distance traveled, add all the individual distances traveled regardless of direction:

$$\text{Solution: } 1.0\ \text{km} + 3.0\ \text{km} + 3.0\ \text{km} = 7.0\ \text{km}$$

WRONG CHOICES EXPLAINED:
(1) This is a nonsense distracter.
(2) This is the total displacement, not the total distance traveled. Displacement is the change of position. In other words, it is the difference between the final position and the initial position.
(3) This answer adds the displacement traveled in the north-south direction and the distance traveled west. It is not the total distance traveled.

4 Refer to the *Mechanics* section of *Reference Table K.*

Given: $v_i = 16.0$ m/s

                 $v_f = 6.0$ m/s

                 $t = 4.0$ s

Find: $a = ?$

Solution: $a = \dfrac{\Delta v}{t}$

$$= \frac{v_f - v_i}{t}$$

$$= \frac{(6.0 \text{ m/s}) - (16.0 \text{ m/s})}{4.0 \text{ s}}$$

$$= -2.5 \text{ m/s}^2$$

The negative acceleration (deceleration) indicates that the direction of acceleration is opposite to the direction of motion, which is why the car slows down. So the acceleration can be described either as –2.5 m/s² east or as 2.5 m/s² west.

5 Remember that weight is the force due to gravity. Refer to the *Mechanics* section of *Reference Table K.*

Given: $m = 10$ kg

               $F_g = 40$ N

Find: $g = ?$

Solution: $g = \dfrac{F_g}{m}$

$$= \frac{40 \text{ N}}{10 \text{ kg}}$$

$$= 4.0 \text{ m/s}^2$$

6 Refer to the *Mechanics* section of *Reference Table K*. Rearrange the equation to solve for acceleration before substituting the values with units:

Given: $v_i = 8.0$ m/s

$v_f = 14$ m/s

Find: $d = 44$ m

$a = ?$

Solution: $v_f^2 = v_i^2 + 2ad$

$$a = \frac{v_f^2 - v_i^2}{2d}$$

$$= \frac{(14 \text{ m/s})^2 - (8.0 \text{ m/s})^2}{(2)(44 \text{ m})}$$

$$= 1.5 \text{ m/s}^2$$

7 Refer to the *Mechanics* section of *Reference Table K*. Rearrange the equation to solve for acceleration:

Given: $v_i = 0$ m/s

$d = 40$ m

$t = 4.0$ s

Find: $a = ?$

Solution: $d = v_i t + \frac{1}{2} a t^2$

$$a = \frac{2(d - v_i t)}{t^2}$$

$$= \frac{2(40 \text{ m} - (0 \text{ m/s})(4.0 \text{ s}))}{(4.0 \text{ s})^2}$$

$$= 5.0 \text{ m/s}^2$$

8 Acceleration is the result of an unbalanced force acting on an object. When an object is in equilibrium, the sum of the net forces acting on the object equals zero. Therefore, when a block is in equilibrium, the magnitude of the block's acceleration is zero.

WRONG CHOICES EXPLAINED:

(2), (3), and (4) When a block is not in equilibrium, there is a net force acting on the object. Therefore, the block could experience an acceleration that is decreasing, increasing, or constant but not zero.

9 Refer to the *Waves* section of *Reference Table K*. The angle of incidence is defined as the angle between the incoming light ray and the normal (the perpendicular). The angle of reflection is the angle of the outgoing light ray and the normal (the perpendicular):

$$Given: \qquad \theta_i = 60°$$

$$Find: \qquad \theta_r = ?$$

$$Solution: \qquad \theta_i = \theta_r$$

$$\theta_r = 60°$$

Remember that the angle of incidence equals the angle of reflection when light strikes a plane mirror.

WRONG CHOICES EXPLAINED:

(1) The angle of incidence is defined as the angle that the ray makes with the normal. The angle the ray makes with the surface of the mirror, 30°, is not equal to the angle of incidence and therefore is not the angle of reflection.

(3) Since the incoming light ray is not striking the surface at 90°, the angle of reflection does not equal 90°

(4) The angle of reflection can only be an acute angle (less than 90°). An angle of 120° is too large to be an angle of reflection.

10 Refer to the *Mechanics* section of *Reference Table K*. Remember that the charge on an electron is the elementary charge, *e*:

Given: $F_e = 1.5 \times 10^{-14}$ N

             $q = e = 1.6 \times 10^{-19}$ C

Find: $E = ?$

Solution: $E = \dfrac{F_e}{q}$

                 $= \dfrac{1.5 \times 10^{-14} \text{ N}}{1.6 \times 10^{-19} \text{ C}}$

                 $= 9.375 \times 10^4$ N/C

                 $= 9.4 \times 10^4$ N/C

11 The total momentum of the system before the spring is released must equal the total momentum of the system after the spring is released. Let a positive sign indicate movement to the left and a negative sign indicate movement to the right. Refer to the *Mechanics* section of *Reference Table K*.

Given: $m_A = 7.0$ kg

             $m_B = 3.0$ kg

             $v_{A_i} = 0.0$ m/s

             $v_{B_i} = 0.0$ m/s

             $v_{B_f} = -6.0$ m/s

Find: $v_{A_f} = ?$

Solution:

$$p_{before} = p_{after}$$

$$p = mv$$

$$m_A v_{A_i} + m_B v_{B_i} = m_A v_{A_f} + m_B v_{B_f}$$

$$(7.0 \text{ kg})(0 \text{ m/s}) + (3.0 \text{ kg})(0 \text{ m/s}) = (7.0 \text{ kg})\left(v_{A_f}\right) + (3.0 \text{ kg})(-6.0 \text{ m/s})$$

$$0 \frac{\text{kg} \cdot \text{m}}{\text{s}} = (7.0 \text{ kg})\left(v_{A_f}\right) + \left(-18.0 \frac{\text{kg} \cdot \text{m}}{\text{s}}\right)$$

$$(-7.0 \text{ kg})\left(v_{A_f}\right) = \left(-18.0 \frac{\text{kg} \cdot \text{m}}{\text{s}}\right)$$

$$\left(v_{A_f}\right) = 2.6 \text{ m/s}$$

12 Centripetal force is always directed toward the center of the circular path. When the electron is at point A, the centripetal force is directed toward the left.

13 Refer to the *Electricity* section of *Reference Table K*. The question asks for the energy required to move a charge. So rearrange the equation to solve for W which is the electrical energy:

Given: $q = 8.00 \times 10^{-19}$ C

              $V = 2.00$ V

Find: $W = ?$

Solution: $V = \dfrac{W}{q}$

              $W = Vq$

              $= (2.00 \text{ V})(8.00 \times 10^{-19} \text{ C})$

              $= 1.6 \times 10^{-18}$ J

14 Gravitational force is the universal attraction between two masses. So by definition, gravitational force is always attractive. Electrostatic force may be either attractive or repulsive. Opposite charges attract, but like charges repel.

15 Refer to the *Electricity* section of *Reference Table K:*

$$F_e = \frac{kq_1q_2}{r^2}$$

This equation shows that electrostatic force is inversely proportional to the square of the distance between the centers of the point charges. The question states that the distance between the point charges decreases. Therefore, the electrostatic force between the charges increases. Since both charges are positive, the force is a repulsive force.

WRONG CHOICES EXPLAINED:
(1) Since the charges are both positive, the electrostatic force is that of repulsion, not attraction. Additionally, the distance between the charges decreases. Since electrostatic force is inversely proportional to the square of that distance, the force increases as the distance decreases.

(2) Since the charges are both positive, the electrostatic force is that of repulsion, not attraction.

(3) The distance between the charges decreases. Since electrostatic force is inversely proportional to the square of that distance, the force increases as the distance decreases.

16 Refer to the *Energy Level Diagram for Mercury* on *Reference Table F*. An electron in a mercury atom at energy level *f* has –2.68 eV of energy and at energy level *b* has –5.74 eV of energy. Calculate the difference between energy levels:

Solution: –2.68 eV – (–5.74 eV) = 3.06 eV

17 Frequency is defined as the number of cycles per second (s^{-1} or Hz).

Solution: $f = \dfrac{4}{10.\ \text{seconds}}$
             $= 0.40$ Hz

18 Refer to the *Resistivities at 20°C* chart of the *Reference Table J*. The resistivity of copper at 20°C is $1.72 \times 10^{-8}\ \Omega \bullet$ m. Of the metals listed in the answer choices, only silver has a resistivity less than that of copper. Since the resistivity of silver is $1.59 \times 10^{-8}\ \Omega \bullet$ m at 20°C, silver conducts electricity better than copper.

---

19   The work done on the block is equal to the total change in energy of the block. Refer to the *Mechanics* section of *Reference Table K*. Since we are neglecting the effect of friction, there is no conversion of mechanical energy to heat the block and to absorb energy, therefore, the internal energy of the block does not change. So $Q = 0$:

$$\text{Solution:} \quad W = \Delta E_T$$
$$E_T = PE + KE + Q$$
$$KE = E_T - PE - Q$$
$$= 20 \text{ J} - 15 \text{ J} - 0$$
$$= 5 \text{ J}$$

20   A series circuit, by definition, is a circuit with a single current pathway. Therefore, when only one lightbulb blows out, the circuit is effectively cut. Electricity can no longer flow through the circuit.

WRONG CHOICES EXPLAINED:

(1) By definition, a parallel circuit must have more than one pathway for the current, not one pathway. This is a nonsense distracter.

(2) When one lightbulb goes out in a parallel circuit with multiple current pathways, only the lightbulbs in the portion of the circuit that contains the broken lightbulb go out. Electricity is still able to flow through the other current pathways. The lightbulbs stay on where the electricity is still able to make a complete circuit.

(4) By definition, a series circuit has only one pathway for the current, not multiple pathways. This is a nonsense distracter.

21   Refer to the *Electricity* section of *Reference Table K*:

$$\text{Given:} \quad P = 1200 \text{ W}$$
$$V = 120 \text{ V}$$
$$t = 30 \text{ s}$$

$$\text{Find:} \quad W = ?$$

$$\text{Solution:} \quad W = Pt$$
$$= (1200 \text{ W})(30 \text{ s})$$
$$= 3.6 \times 10^4 \text{ J}$$

22 Refer to *Reference Table A* and *Equations ME11* and *ME23* on *Reference Table K*:

Given: $m = 35 \; kg$

$\bar{v} = 5.0 \; \text{m/s}$

$g = 9.81 \; \text{m/s}^2$

Find: $P = ?$

Solution: $g = \dfrac{F_g}{m}$

$F_g = mg$

$P = F\bar{v}$

$P = mg\bar{v}$

$= (35 \; \text{kg})(9.81 \; \text{m/s}^2)(5.0 \; \text{m/s})$

$= 1700 \; \text{W}$

23 Waves only transfer energy. Waves do not transfer mass.

24 Resonance is the spontaneous vibration of an object at a frequency equal o that of the wave that initiates that resonant vibration. If that spontaneous vibraion is strong enough, it can shatter or break the object.

WRONG CHOICES EXPLAINED:

(1) Refraction is the change in direction of a wave when it passes obliquely ʳrom one medium into another in which the wave moves at a different speed.

(2) Diffraction is the bending of a wave around a barrier.

(4) The Doppler effect is the apparent change in frequency that results when ₐ wave source and an observer are in relative motion with respect to each other.

25 Mechanical waves require a medium for transmission. Sound waves are ᵐechanical waves, so they require a medium through which to travel.

WRONG CHOICES EXPLAINED:

(1), (2), and (4) Light waves, radio waves, and microwaves are all examples of ᵉlectromagnetic waves. Electromagnetic waves do not require a medium ᵗhrough which to travel. They can travel in a vacuum.

26 An electromagnetic wave is a periodic wave consisting of mutually perpendicular electric and magnetic fields that is radiated away from the vicinity of an accelerating charge. Since a moving electron produces both electric and magnetic fields, choice (4) would most likely produce an electromagnetic wave.

WRONG CHOICES EXPLAINED:

(1) A vibrating tuning fork produces sound waves. Sound waves are mechanical waves, not electromagnetic waves.

(2) and (3) Neither a washing machine agitator at work nor a swinging pendulum produces mutually perpendicular electric and magnetic fields. Therefore, neither one produces an electromagnetic wave.

27 When monochromatic light passes from one medium into another medium, the speed of the light wave changes. If the light enters at an oblique angle, it changes direction as it enters the new medium. As a result of the change in speed, the light's wavelength changes. The frequency of the wave does not change.

28 Refer to *Equation W6* on *Reference Table K*:

Given: $n_y = 2n_X$

Find: $v_y = ?$

Solution: $\dfrac{n_2}{n_1} = \dfrac{v_1}{v_2}$

$\dfrac{n_y}{n_x} = \dfrac{v_x}{v_y}$

$v_Y = \dfrac{(v_x)(n_x)}{n_y}$

$= \dfrac{(v_x)(n_x)}{2n_x}$

$= \dfrac{1}{2}v_x$

As a light wave travels from medium X into medium Y, the speed of the light wave is halved.

29 In a transverse wave, the particles of the medium vibrate or exhibit simple harmonic motion about a rest point perpendicular to the direction of the motion of the wave. In the diagram, point P is the disturbance and the wave is moving toward the right. So point P moves up and down. At the instant shown in the diagram, point P is moving toward the top of the page since a wave crest is coming toward point P.

30 Like charges repel, and opposite charges attract. Therefore, the charge of the conductor can be negative. Note that the positive charge can cause a redistribution of charge in a neutral conductor. The electrons in the neutral conductor can be attracted to the side of the conductor closer to the positive charge. That side of the conductor will have a negative charge, and the opposite side will have a positive charge. The entire charge on the conductor will remain neutral, but the negative side of the conductor will be attracted to the positive charge.

WRONG CHOICES EXPLAINED:

(1) and (3) A positive conductor will be repelled by the positive charge.

(2) A negative conductor will be attracted to the positive charge. However, as explained above, a neutral conductor can also be attracted to the positive charge.

31 Refer to the *Classification of Matter* and *Particles of the Standard Model* charts of *Reference Tables G* and *H*. Baryons are composed of combinations of 3 quarks. There are 6 different quarks, but each has a charge of either $+\frac{2}{3}e$ or $-\frac{1}{3}e$. The total charge on a baryon must either be neutral or be a multiple of a whole elementary charge. The total charge cannot be fractional. Only choice (1) is comprised of 3 particles that have accepted quark charge values and add up to a whole number, $+1e$.

WRONG CHOICES EXPLAINED:

(2) The total charge is $+\frac{2}{3}e$, which cannot occur in a baryon.

(3) Quarks have a charge of either $+\frac{2}{3}e$ or $-\frac{1}{3}e$. They never have a charge of 1e or of 0.

(4) Quarks have a charge of either $+\frac{2}{3}e$ or $-\frac{1}{3}e$. They never have a charge of 0.

32 Since the rubber block is originally at rest, use the coefficient of static friction to determine the minimum force needed to start the rubber block moving across dry asphalt. Remember that the normal force, F_N, equals the weight of the rubber block. Refer to the *Mechanics* section of *Reference Table K* and the *Approximate Coefficients of Friction* table of *Reference Table C*:

Given: $\mu_s = 0.85$

$\qquad\qquad F_N = 60 \text{ N}$

Find: $\qquad F_f = ?$

Solution: $\quad F_f = \mu_s F_N$

$\qquad\qquad\quad = (0.85)(60 \text{ N})$

$\qquad\qquad\quad = 51 \text{ N}$

33 Mass is the quantitative measure of inertia. As the mass of an object increases, the object's inertia increases. Since object D has the greatest mass, 8.0 kg, it has the greatest inertia. Speed does not affect inertia. This is extraneous information.

34 Opposite charges attract. Bringing a positively charged rod near the knob of the electroscope causes a redistribution of charge within the electroscope. The negative electrons are attracted to the positive rod. So the electrons move from the leaves and toward the knob. Like charges repel; therefore, the leaves repel each other now that they are both positive from the "missing" electrons. When the rod is removed, the electrons move from the knob back to the leaves so that each portion of the electroscope becomes neutral again. The neutral leaves no longer repel each other and thus return to their original position.

35 An ammeter measures current. It needs to be placed in the current path that is being measured. In choice (1), the ammeter is in the same current path as the resistor on the right side.

WRONG CHOICES EXPLAINED:
(2) A voltmeter measures potential difference across two points in a circuit. This diagram shows a voltmeter, not an ammeter.
(3) Although the diagram shows an ammeter, it is not in the current path that includes the resistor.
(4) An ammeter measures current. This diagram shows a voltmeter, not an ammeter. The voltmeter is correctly connected to measure the potential difference of the resistor, not the current through the resistor.

PART B–1

36 This is an order of magnitude question; $10^0 = 1$. Since 1 meter is approximately equal to 3.3 feet, 10^0 m is a reasonable approximation for the height of a kitchen table.

WRONG CHOICES EXPLAINED:
(1) 10 centimeters is not a reasonable approximation of the height of a kitchen table
(3) 10 meters (approximately 33 feet) is not a reasonable approximation of the height of a kitchen table
(4) 100 meters (approximately 330 feet) is not a reasonable approximation of the height of a kitchen table

37 As the ball rises, the vertical component of its velocity decelerates due to gravity and reaches a velocity of zero at its highest point. At this point, the vertical component of the velocity begins to accelerate due to gravity in the downward directions.

WRONG CHOICES EXPLAINED:
(1) The acceleration of the ball is constant throughout the trajectory. Its acceleration is -9.81 m/s^2 or 98.1 m/s^2 in the downward direction, which is the acceleration due to gravity.
(2) Momentum is equal to mass times velocity. Horizontal velocity is constant until the ball lands on the ground and its motion stops, so the momentum is not zero at the highest point.
(3) Horizontal velocity is constant until the ball lands on the ground and its motion stops. So the horizontal velocity is not zero at the highest point.

38 The shaded area under the line in a velocity versus time graph is equal to displacement. Mathematically, the area under the line can be expressed as $d = vt$.

WRONG CHOICES EXPLAINED:
(3) The slope of the line, $\frac{\Delta v}{t}$, would be equal to the acceleration on a velocity versus time graph.

39 Refer to the *Mechanics* section of *Reference Table K*:

Given: $PE_s = 10$ J

                $x = 0.20$ m

Find: $k = \,?$

Solution: $PE_s = \dfrac{1}{2}Kx^2$

$$k = \dfrac{2PE_s}{x^2}$$

$$= \dfrac{2(10 \text{ J})}{(0.20 \text{ m})^2}$$

$$= 5 \times 10^2 \text{ N/m}$$

40 Refer to the *Mechanics* section of *Reference Table K*:

Given: $m_{ball} = 1.0$ kg

               $m_{cannon} = 500$ kg

               $v_i = 0$ m/s

               $F = 8.0 \times 10^3$ N

               $t = 1.0^{-1}$ s

Find: $\Delta p = \,?$

Solution: $\Delta p = F_{net}t$

$$= \left(8.0 \times 10^3 \text{ N}\right)\left(1 \times 10^{-1} \text{ s}\right)$$

$$= 8.0 \times 10^2 \text{ kg} \bullet \text{m/s}$$

41 Newton's third law states that if object *A* exerts a force on object *B*, then object *B* exerts a force equal in magnitude but opposite in direction to the force on object *A*.

42 When two charged metal spheres of identical size come into contact with one another, the charges redistribute evenly between the spheres. To find the final charge on each, add the 2 original charges and divide by 2:

Solution: $\left(-6 \times 10^{-6} \text{ C}\right) + \left(+2 \times 10^{-6} \text{ C}\right) = -4 \times 10^{-6} \text{ C}$

$$\frac{-4 \times 10^{-6} \text{ C}}{2} = -2 \times 10^{-6} \text{ C}$$

WRONG CHOICES EXPLAINED:
(3) You added up the 2 charges but forgot to divide by 2.

43 The net force is the sum of all the applied forces. The force of friction is always in the opposite direction of motion, so we actually subtract the force of friction from the applied force to get the net force. Refer to the *Mechanics* section of *Reference Table K*.

Given: $F_{applied} = 70 \text{ N}$

$m = 20 \text{ kg}$

$F_f = 10 \text{ N}$

Find: $a = \ ?$

Solution: $F_{net} = F_{applied} - F_f$

$= 70 \text{ N} - 10 \text{ N}$

$= 60 \text{ N}$

$a = \dfrac{F_{net}}{m}$

$= \dfrac{60 \text{ N}}{20 \text{ kg}}$

$= 3.0 \text{ m/s}^2$

44 Refer to the *Modern Physics* section of *Reference Table K*:

$$E_{photon} = \frac{hc}{\lambda}$$

This equation shows that wavelength, *l*, is inversely proportional to the energy of a photon. The graph depicted in choice (4) correctly indicates this type of relationship.

WRONG CHOICES EXPLAINED:

(1) This graph represents a direct proportional relationship.

(2) This graph represents an exponentially proportional relationship.

(3) This graph shows that energy is always the same no matter the wavelength.

45 Point P is a node, the place in a standing wave where there is complete destructive interference. For this to occur, the waves have to be 180° out of phase.

WRONG CHOICES EXPLAINED:

(1) This would result in maximum constructive interference, which is an antinode.

46 Antinodes occur when there is maximum constructive interference between the waves. In this diagram, antinodes occur in 3 places.

47 Refer to the *Electricity* section of *Reference Table K:*

$$R = \frac{\rho L}{A}$$

This equation shows that at constant temperature, resistance is directly proportional to length and inversely proportional to cross-sectional area. Assume that wire 1 has a resistance of R. Then wire 2, which is 2 times the length of wire 1, has 2 times the resistance of wire 1.

WRONG CHOICES EXPLAINED:

(3) Assume that wire 1 has a resistance of R. Then wire 3, which has 2 times the cross-sectional area of wire 1, has half the resistance of wire 1.

(4) Assume that wire 1 has a resistance of R. Then wire 4, which has 2 times the length and 2 times the cross-sectional area of wire 1, has the same resistance as wire 1.

48 Refer to the *List of Physical Constants* on *Reference Table A* to find the value of an elementary charge:

$$e = 1.60 \times 10^{-19} \text{ C}$$

Then convert the charge on sphere A into elementary charges:

Given: $q_A = +4.0 \times 10^{-6}$ C

Solution: excess charges $= \dfrac{\left(+4.0 \times 10^{-6} \text{ C}\right)}{\left(1.6 \times 10^{-19} \text{ C}\right)}$

$= 2.5 \times 10^{13}$

49 Refer to the *List of Physical Constants* (*Reference Table A*) and the *Electricity* section of *Reference Table K*:

Given: $q_A = +4.0 \times 10^{-6}$ C

$q_B = -3.0 \times 10^{-6}$ C

$r = 2.0$ m

Find: $k = 8.99 \times 10^9$ N • m^2/C^2

$F_e = ?$

Solution: $F_e = \dfrac{kq_1 q_2}{r^2}$

$= \dfrac{\left(8.99 \times 10^9 \text{ N • m}^2/\text{C}^2\right)\left(4.0 \times 10^{-6} \text{ C}\right)\left(-3.0 \times 10^{-6} \text{ C}\right)}{\left(2.0 \text{ m}\right)^2}$

$= -2.7 \times 10^{-2}$ N

The negative sign in the answers is ignored because the question asked about the magnitude of the force only.

50 This diagram illustrates the Doppler effect, which is the apparent change in frequency that results when a wave source and an observer are in relative motion to each other. The lines in the diagram represent wave crests. The spaces between the lines represent troughs. At point B, both the wave crests and the troughs are much closer together than at point A. This indicates a shorter wavelength at point B. Refer to the *Waves* section of *Reference Table K*.

$$v = f\lambda$$

This equation shows that when velocity is constant, frequency and wavelength are inversely proportional. If wavelength decreases, frequency increases and vice versa. Therefore, a shorter wavelength results in a higher frequency.

PART B–2

51 One credit is awarded for drawing at least 4 field lines. Magnetic field lines must be drawn from North to South. Each magnetic field line must have an arrowhead to indicate proper direction. Field lines drawn inside the magnet are ignored.

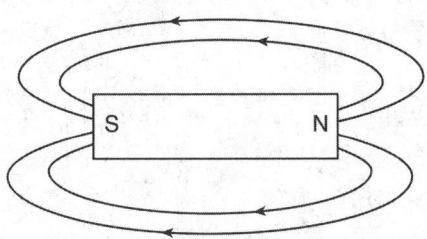

52 Refer to the *Classification of Matter* chart and *Particles of the Standard Model* chart (*Reference Tables G* and *H*). Each proton and each neutron is composed of 3 quarks. Therefore the nucleus of a tritium atom containing 1 proton and 2 neutrons has a total of 9 quarks.

One credit is awarded for an answer of 9.

53 A proton carries an elementary charge of +1. An electron carries an elementary charge of –1. Since a neutrino is neutral, it carries a charge of 0. Add those charges:

$$+1 + (-1) + 0 = 0$$

One credit is awarded for an answer of 0.

54 One credit is awarded for:

- Strong force
- Strong
- Strong nuclear force
- Strong interaction

No credit is awarded for nuclear force since it would be unclear whether you meant strong nuclear force or weak nuclear force.

55 Since both balls are launched from the same height at the same angle, they both fall at the same rate, which is the acceleration due to gravity. Therefore, the second ball travels through the air for the same amount of time as does the first ball. Refer to the *Mechanics* section of *Reference Table K*:

$$v = \frac{d}{t}$$

Since time is constant, the equation indicates that velocity is directly proportional to distance. Since the second ball has twice the velocity of the first ball, the second ball travels twice the distance in the same amount of time.

One credit is awarded for an answer of 2*d*.

56–57   Refer to the *Electricity* section of *Reference Table K*. Solve for time:

Given: $I = 0.71$ A

$V = 120$ V

$W = 3.0 \times 10^5$ J

Find: $t = ?$

Solution: $W = VIt$

$t = \dfrac{W}{VI}$

$= \dfrac{3.0 \times 10^5 \text{ J}}{(120 \text{ V})(0.71 \text{ A})}$

$= 3.5 \times 10^3$ s

56 One credit is awarded for showing the correct equation *and* for substituting the correct values with their units.

57 One credit is awarded for the correct answer, with units, or for an answer that is consistent with the work shown in question 56.

Note: Students will not be penalized more than 1 credit for errors in units in questions 56 and 57.

58–59 Examples of 2 credit responses:

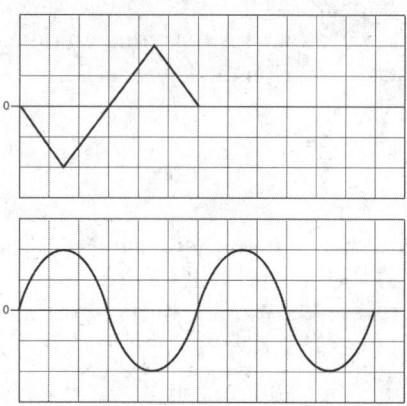

58 One credit is awarded for drawing a transverse wave with an amplitude of 2.0 cm ± 0.2 cm.

59 One credit is awarded for drawing a transverse wave with a wavelength of 6.0 cm ± 0.2 cm.

60 Refer to the *Mechanics* section of *Reference Table K*. Solve for k, which is the spring constant:

Given: $F_s = 35$ N

              $x = 0.10$ m

Find: $k = ?$

Solution: $F_s = kx$

$$k = \frac{F_s}{x}$$

$$= \frac{35 \text{ N}}{0.10 \text{ m}}$$

$$= 350 \text{ N/m}$$

One credit is awarded for an answer of 350 N/m.

61 Refer to the *Prefixes for Powers* chart (*Reference Table B*) and the *Modern Physics* section of *Reference Table K*:

Given: giga $= 10^9$

$E = 1.0 \times 10^9$ J

$c = 3.00 \times 10^8$ m/s

$m = ?$

Solution: $E = mc^2$

$$m = \frac{E}{c^2}$$

$$= \frac{1.0 \times 10^9 \text{ J}}{\left(3.00 \times 10^8 \text{ m/s}\right)^2}$$

$$= 1.1 \times 10^{-8} \text{ kg}$$

One credit is awarded for an answer of 1.1×10^{-8} kg.

62 The speed of light is greater than the speed of sound. Based on the *List of Physical Constants* table (*Reference Table A*), the speed of light is 3.00×10^8 m/s and the speed of sound is 3.31×10^2 m/s. In other words, light travels approximately 1 million times faster than sound!

One credit is awarded for indicating that the speed of light is greater than the speed of sound.

63–64 Refer to the *Electricity* section of *Reference Table K*:

Given: $q = 28$ C

$V = 3.2 \times 10^7$ V

$t = 1.5 \times 10^{-3}$ s

Find: $I = ?$

Solution: $I = \dfrac{\Delta q}{t}$

$$= \frac{28 \text{ C}}{1.5 \times 10^{-3} \text{ s}}$$

$$= 1.9 \times 10^4 \text{ A}$$

63 One credit is awarded for showing the correct equation *and* for substituting the correct values with their units.

64 One credit is awarded for the correct answer, with units, or for an answer that is consistent with the work shown in question 63.
Note: Students will not be penalized more than 1 credit for errors in units in questions 63 and 64.

65 The resultant displacement is the net sum of the individual displacements of each pulse: $+1.00$ m $+ (-0.40$ m$) = +0.60$ m.

One credit is awarded for drawing a square wave between A and B that has a displacement of $+0.6$ m.

PART C

66 Refer to the *Mechanics* section of *Reference Table K*:

Given: $F_{rope} = 158$ N

              $d = 230$ m

Find: $W = ?$

Solution: $W = Fd$

              $= (158 \text{ N}) \times (230 \text{ m})$

              $= 3.63 \times 10^4 \text{ J}$

One credit is awarded for an answer of 3.63×10^4 J or 36,300 J.

67–68   Refer to *Reference Table A* and the *Mechanics* section of *Reference Table K:*

Given: $m = 72.0 \text{ kg}$

$h = 40.0 \text{ m}$

$g = 9.81 \text{ m/s}^2$

Find: $\Delta PE = ?$

Solution: $\Delta PE = mg\Delta h$

$= (72.0 \text{ kg})(9.81 \text{ m/s}^2)(40.0 \text{ m})$

$= 2.83 \times 10^4 \text{ J}$

67   One credit is awarded for showing the correct equation *and* for substituting the correct values with their units.

68   One credit is awarded for the correct answer, with units, or for an answer that is consistent with the work shown in question 67.
Note: Students will not be penalized more than 1 credit for errors in units in questions 67 and 68.

69   Refer to the *Mechanics* section of *Reference Table K:*

$$W = \Delta E_T$$
$$E_T = PE + KE + Q$$

The work, W, done by the rope is greater than the potential energy, PE, gained by the skier. There is no change in kinetic energy for the skier, $KE = 0$. Therefore, the rest of the energy acted against friction. So the internal thermal energy, Q, of the skier-hill system increased.
One credit is awarded for indicating that the internal energy increases.

70 Refer to the *Mechanics* section of *Reference Table K*:

$$W = \Delta E_T$$

The work done by the rope is increasing the mechanical energy of the system. One credit is awarded for indicating that the total mechanical energy increases.

71–72 The resistors are connected in parallel. Refer to the *Electricity* section of *Reference Table K*:

$$
\begin{aligned}
\text{Given:} \quad & R_1 = 15\ \Omega \\
& R_2 = 30\ \Omega
\end{aligned}
$$

$$\text{Find:} \quad R_{eq} = ?$$

$$
\begin{aligned}
\text{Solution} =: \quad \frac{1}{R_{eq}} &= \frac{1}{R_1} + \frac{1}{R_2} \\
&= \frac{1}{15\ \Omega} + \frac{1}{30\ \Omega} \\
&= \frac{2}{30\ \Omega} + \frac{1}{30\ \Omega} \\
&= \frac{3}{30\ \Omega} \\
&= \frac{1}{10\ \Omega} \\
R_{eq} &= 10\ \Omega
\end{aligned}
$$

71 One credit is awarded for showing the correct equation *and* for substituting the correct values with their units.

72 One credit is awarded for the correct answer, with units, or for an answer that is consistent with the work shown in question 71.
Note: Students will not be penalized more than 1 credit for errors in units in questions 71 and 72.

73 Refer to the *Electricity* section of *Reference Table K*. Solve for *I*:

$$\text{Given:} \quad R_{eq} = 10\ \Omega$$
$$V = 60\ V$$

$$\text{Find:} \quad I = \ ?$$

$$\text{Solution:} \quad R = \frac{V}{I}$$
$$I = \frac{V}{R}$$
$$= \frac{60\ V}{10\ \Omega}$$
$$= 6.0\ A$$

One credit is awarded for an answer of 6.0 A or for an answer that is the result of 60 V divided by the answer obtained in question 72.

74–75 Refer to the *Electricity* section of *Reference Table K*. Based on the information originally provided with the diagram as well as the answers found in questions 71 through 73, you can determine the rate (power) in three different ways:

$$\text{Given:} \quad R_{eq} = 10\ \Omega$$
$$V = 60\ V$$
$$I = 6.0\ A$$

$$\text{Find:} \quad P = \ ?$$

$$\text{Solution:} \quad P = \frac{V^2}{R}$$
$$= \frac{(60\ V)^2}{10\ \Omega}$$
$$= 360\ W$$

$$\text{or} \quad P = I^2 R$$
$$= (6.0\ A)^2 (10\ \Omega)$$
$$= 360\ W$$

$$\text{or} \quad P = VI$$
$$= (60\ V)(6.0\ A)$$
$$= 360\ W$$

74 One credit is awarded for showing the correct equation *and* for substituting the correct values with their units.

75 One credit is awarded for the correct answer, with units, or for an answer that is consistent with the work shown in question 74.
Note: Students will not be penalized more than 1 credit for errors in units in questions 74 and 75.

76 Refer to the *Electricity* section of *Reference Table K*:

$$V = V_1 = V_2 = V_3 = \cdots$$
$$R = \frac{V}{I}$$

As shown in the first equation above, all resistors connected in parallel to each other experience the same potential difference provided by the power source. Since the potential difference doesn't change when adding an additional resistor in parallel, neither does the current through the individual resistor. [However, by adding an additional resistor in parallel, the total resistance of the circuit decreases; therefore, the total current on the circuit increases.]
One credit is awarded for:

• The current will remain the same.
• No change.

77–78 Refer to the *Mechanics* section of *Reference Table K*:

Given: $m = 2.50$ kg
                $v = 18.0$ m/s
                $r = 25.0$ m
Find: $KE = ?$
Solution: $KE = \frac{1}{2}mv^2$
               $= \frac{1}{2}(2.50 \text{ kg})(18.0 \text{ m/s})^2$
               $= 405$ J

77 One credit is awarded for showing the correct equation *and* for substituting the correct values with their units.

78 One credit is awarded for the correct answer, with units, or for an answer that is consistent with the work shown in question 77.
Note: Students will not be penalized more than 1 credit for errors in units in questions 77 and 78.

79–80 Refer to the *Mechanics* section of *Reference Table K*:

Given: $m = 2.50$ kg
          $v = 18.0$ m/s
          $r = 25.0$ m

Find: $F_c = ?$

Solution: $F_c = ma_c$

          $a_c = \dfrac{v^2}{r}$

          $F_c = \dfrac{mv^2}{r}$

          $= \dfrac{(2.50 \text{ kg})(18 \text{ m/s})^2}{25.0 \text{ m}}$

          $= 32.4$ N

79 One credit is awarded for showing the correct equation *and* for substituting the the correct values with their units.

80 One credit is awarded for the correct answer, with units, or for an answer that is consistent with the work shown in question 79.
Note: Students will not be penalized more than 1 credit for errors in units in questions 79 and 80.

81–82 Refer to *Reference Table A* and the *Waves* section of *Reference Table K*:

Given: $f = 5.09 \times 10^{14}$ Hz

                $vc = 3.00 \times 10^8$ m/s

Find: $\lambda = ?$

Solution: $v = f\lambda$

                $\lambda = \dfrac{v}{f}$

                $= \dfrac{3.00 \times 10^8 \text{ m/s}}{5.09 \times 10^{14} \text{ Hz}}$

                $= 5.89 \times 10^{-7}$ m

81 One credit is awarded for showing the correct equation *and* for substituting the correct values with their units.

82 One credit is awarded for the correct answer, with units, or for an answer that is consistent with the work shown in question 81.
Note: Students will not be penalized more than 1 credit for errors in units in questions 81 and 82.

83 The angle of incidence is the angle between the incident ray and the normal. Use a protractor to measure the angle of incidence.
One credit is awarded for an answer of $50° \pm 2°$.

84–85   Refer to *Reference Table E* and the *Waves* section of *Reference Table* K. Make sure your calculator is set to degree mode:

Given: $\theta_{air} = 90°$

                  $\theta_x = 50°$

                  $n_{air} = 1.00$

Find: $n_x = ?$

Solution: $n_1 \sin\theta_1 = n_2 \sin\theta_2$

                $n_x \sin\theta_x = n_{air} \sin\theta_{air}$

$$n_x = \frac{n_{air} \sin\theta_{air}}{\sin\theta_x}$$

$$= \frac{(1.00)(\sin 90°)}{\sin 50°}$$

$$= 1.3$$

84   One credit is awarded for showing the correct equation *and* for substituting the correct values with their units.

85   One credit is awarded for the correct answer, with units, or for an answer that is consistent with the work shown in question 84.

Note: Students will not be penalized more than 1 credit for errors in units in questions 84 and 85.

Topic	Question Numbers (total)	Wrong Answers (x)	Grade
Math Skills	2–7, 10–11, 13, 15–17, 19, 21–22, 28, 31–32, 36, 38–40, 42–43, 47–49, 52–53, 55–61, 63–68, 71–75, 77–85: (56)		$\dfrac{100(56-x)}{56}=\%$
Mechanics	1–8, 11, 14, 32–33, 37–38, 40–41, 43, 55, 60, 79–80: (21)		$\dfrac{100(21-x)}{21}=\%$
Energy	19, 22, 39, 66–70, 77–78: (10)		$\dfrac{100(10-x)}{10}=\%$
Electricity/ Magnetism	10, 12–15, 18, 20–21, 26, 30, 34–35, 42, 47–49, 51, 56–57, 63–64, 71–76: (27)		$\dfrac{100(27-x)}{27}=\%$
Waves	9, 17, 23–25, 27–29, 45–46, 50, 58–59, 62, 65, 81–85: (20)		$\dfrac{100(20-x)}{20}=\%$
Modern Physics	16, 31, 44, 52–54, 61: (7)		$\dfrac{100(7-x)}{7}=\%$

Examination
June 2018
Physics: The Physical Setting

PART A
Answer all questions in this part.

Directions (1–35): For *each* statement or question, select the number of the word or expression that, of those given, best completes the statement or answers the question. Some questions may require the use of the *2006 Edition Reference Tables for Physical Setting/Physics.* Record your answers in the spaces provided.

1 Which combination correctly pairs a vector quantity with its corresponding unit?

    (1) weight and kg   (3) speed and m/s

    (2) velocity and m/s   (4) acceleration and m^2/s 1 _____

2 A 12.0-kilogram cart is moving at a speed of 0.25 meter per second. After the speed of the cart is tripled, the inertia of the cart will be

    (1) unchanged   (3) three times greater

    (2) one-third as great   (4) nine times greater   2 _____

3 While taking off from an aircraft carrier, a jet starting from rest accelerates uniformly to a final speed of 40 meters per second on a runway that is 70 meters long. What is the magnitude of the acceleration of the jet?

    (1) $0.29 \ m/s^2$ (3) $1.8 \ m/s^2$

    (2) $0.57 \ m/s^2$ (4) $11 \ m/s^2$ 3 _____

4 A 6.0-kilogram cart initially traveling at 4.0 meters per second east accelerates uniformly at 0.50 meter per second squared east for 3.0 seconds. What is the speed of the cart at the end of this 3.0 second interval?

(1) 1.5 m/s   (3) 3.0 m/s
(2) 5.5 m/s   (4) 7.0 m/s   4 _____

5 A soccer ball is kicked into the air from level ground with an initial speed of 20 meters per second and returns to ground level. At which angle above the horizontal should the ball be kicked in order for the ball to travel the greatest total horizontal distance? [Neglect friction.]

(1) 15°   (3) 45°
(2) 30°   (4) 75°   5 _____

6 Starting from rest, a car travels 18 meters as it accelerates uniformly for 3.0 seconds. What is the magnitude of the car's acceleration?

(1) 6.0 m/s^2 (3) 3.0 m/s^2
(2) 2.0 m/s^2 (4) 4.0 m/s^2 6 _____

7 A ball is rolling horizontally at 3.00 meters per second as it leaves the edge of a tabletop 0.750 meter above the floor. The ball lands on the floor 0.391 second after leaving the tabletop. What is the magnitude of the ball's acceleration 0.200 second after it leaves the tabletop? [Neglect friction.]

(1) 1.96 m/s^2 (3) 9.81 m/s^2
(2) 7.65 m/s^2 (4) 15.3 m/s^2 7 _____

8 A projectile with mass m is fired with initial horizontal velocity v_x from height h above level ground. Which change would have resulted in a greater time of flight for the projectile? [Neglect friction.]

(1) decreasing the mass to $m/2$
(2) decreasing the height to $h/2$
(3) increasing the initial horizontal velocity to $2v_x$
(4) increasing the height to $2h$ 8 _____

9 A golf club hits a stationary 0.050-kilogram golf ball with an average force of 5.0×10^3 newtons, accelerating the ball to a speed of
44 meters per second. What is the magnitude of the impulse
imparted to the ball by the golf club?

(1) 2.2 N • s   (3) 1.1×10^4 N • s

(2) 880 N • s   (4) 2.2×10^5 N • s   9 _____

10 A tennis player's racket applies an average force of 200. newtons to
a tennis ball for 0.025 second. The average force exerted on the
racket by the tennis ball is

(1) 0.025 N   (3) 200. N

(2) 5.0 N   (4) 80.0 N   10 _____

11 The diagram below represents a box sliding down an incline at
constant velocity.

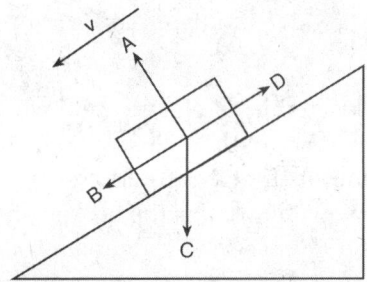

Which arrow represents the direction of the frictional force acting
on the box?

(1) A   (3) C

(2) B   (4) D   11 _____

12 Which diagram represents the directions of the velocity, v, and acceleration, a, of a toy car as it moves in a clockwise, horizontal, circular path at a constant speed?

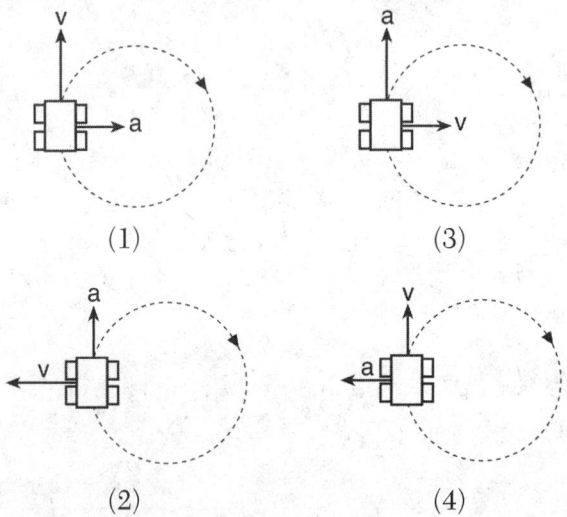

(1)　　　　　　　　　　　(3)

(2)　　　　　　　　　　　(4)

12 _____

13 A charged particle is located in an electric field where the magnitude of the electric field strength is 2.0×10^3 newtons per coulomb. If the magnitude of the electrostatic force exerted on the particle is 3.0×10^{-3} newton, what is the charge of the particle?

(1) 1.6×10^{-19} C　　　　　　(3) 6.0 C

(2) 1.5×10^{-6} C　　　　　　(4) 6.7×10^5 C

13 _____

14 The magnitude of the gravitational field strength near Earth's surface is represented by

(1) $\dfrac{F_g}{m}$　　　　　　　　　(3) mg

(2) G　　　　　　　　　　(4) $\dfrac{Gm_1m_2}{r^2}$

14 _____

15 A car engine supplies 2.0×10^3 joules of energy during the 10 seconds it takes to accelerate the car along a horizontal surface. What is the average power developed by the car engine while it is accelerating?

   (1) 2.0×10^1 W   (3) 2.0×10^3 W

   (2) 2.0×10^2 W   (4) 2.0×10^4 W   15 _____

16 Which forces can be either attractive or repulsive?

   (1) gravitational and magnetic

   (2) electrostatic and gravitational

   (3) magnetic and electrostatic

   (4) gravitational, magnetic, and electrostatic   16 _____

17 Compared to the resistivity of a 0.4-meter length of 1-millimeter-diameter copper wire at 0°C, the resistivity of a 0.8-meter length of 1-millimeter-diameter copper wire at 0°C is

   (1) one-fourth as great   (3) the same

   (2) one-half as great   (4) four times greater   17 _____

18 The work per unit charge required to move a charge between two points in an electric circuit defines electric

   (1) force   (3) field strength

   (2) power   (4) potential difference   18 _____

19 A 2.0-meter length of copper wire is connected across a potential difference of 24 millivolts. The current through the wire is 0.40 ampere. The same copper wire at the same temperature is then connected across a potential difference of 48 millivolts. The current through the wire is

   (1) 0.20 A   (3) 0.80 A

   (2) 0.40 A   (4) 1.6 A   19 _____

20 What is the magnitude of the gravitational force of attraction between two 0.425-kilogram soccer balls when the distance between their centers is 0.500 meter?

(1) 2.41×10^{-11} N   (3) 5.67×10^{-11} N

(2) 4.82×10^{-11} N   (4) 1.13×10^{-10} N   20 _____

21 A sound wave produced by a loudspeaker can travel through water, but not through a vacuum. In comparison, a red light wave produced by a laser can travel through

(1) water, but not through a vacuum
(2) a vacuum, but not through water
(3) both water and a vacuum
(4) neither water nor a vacuum   21 _____

22 As a group of soldiers marches along a road, each soldier steps simultaneously. However, when crossing a bridge, the group does not step simultaneously in order to prevent the bridge from vibrating intensely. The phenomenon responsible for the intense vibrations is

(1) action and reaction   (3) inertia
(2) conservation of momentum   (4) resonance   22 _____

23 Which characteristics of a light wave remain constant when the light wave travels from air into corn oil?

(1) speed and frequency
(2) wavelength and frequency
(3) period and frequency
(4) wavelength and period   23 _____

24 The speed of a light ray ($f = 5.09 \times 10^{14}$ Hz) in corn oil is

(1) 1.47×10^8 m/s   (3) 3.00×10^8 m/s
(2) 2.04×10^8 m/s   (4) 4.41×10^8 m/s   24 _____

25 The spreading out of a wave after passing through an opening in a barrier is an example of

(1) diffraction   (3) reflection
(2) Doppler effect   (4) refraction   25 _____

26 A microwave with a frequency of 5.0×10^{10} hertz has a period of

(1) 2.0×10^{-11} s   (3) 1.7×10^2 s

(2) 6.0×10^{-3} s   (4) 1.5×10^{19} s   26 _____

27 After two light waves have interfered in a vacuum, the two waves will be

(1) changed in frequency

(2) changed in velocity

(3) changed in amplitude

(4) unchanged   27 _____

28 A glass rod is rubbed with silk. During this process, a positive charge is given to the glass rod by

(1) adding electrons to the rod

(2) adding protons to the rod

(3) removing electrons from the rod

(4) removing protons from the rod   28 _____

29 A photon with an energy of 1.33×10^{-21} joule has a frequency of

(1) 5.02×10^{13} Hz   (3) 8.82×10^{14} Hz

(2) 2.01×10^{12} Hz   (4) 5.30×10^{34} Hz   29 _____

30 The speed of a car is increased uniformly from 11 meters per second to 19 meters per second. The average speed of the car during this interval is

(1) 0.0 m/s   (3) 30 m/s

(2) 15 m/s   (4) 4.0 m/s   30 _____

31 The energy equivalent of the rest mass of an electron is

(1) 2.73×10^{-22} J   (3) 1.50×10^{-10} J

(2) 8.20×10^{-14} J   (4) 1.44×10^{-2} J   31 _____

32 A spring has an unstretched length of 0.40 meter. The spring is stretched to a length of 0.60 meter when a 10-newton weight is hung motionless from one end. The spring constant of this spring is

(1) 10 N/m   (3) 25 N/m

(2) 17 N/m   (4) 50 N/m   32 _____

33 An electric circuit contains a battery, three lamps, and an open switch, as represented in the diagram below.

When the switch is open, there is an electric current in

(1) lamp I, only

(2) lamps II and III, only

(3) lamps I, II, and III

(4) none of the lamps   33 _____

34 Which diagram correctly represents an electric field?

      (1)   (2)   (3)   (4)   34 _____

35 Which points on the wave diagram below are 90° out of phase with each other?

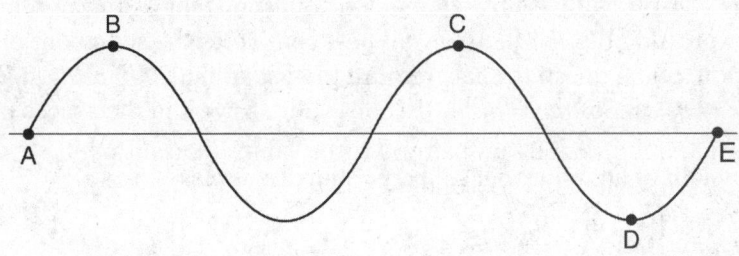

(1) *A* and *E*   (3) *C* and *D*
(2) *B* and *C*   (4) *D* and *E*   35 _____

PART B–1
Answer all questions in this part.

Directions (36–50): For each statement or question, select the number of the word or expression that, of those given, best completes the statement or answers the question. Some questions may require the use of the *2006 Edition Reference Tables for Physical Setting/Physics.* Record your answers in the spaces provided.

36 The height of an individual step on a staircase is closest to

(1) 2.0×10^{-2} m   (3) 2.0×10^{0} m

(2) 2.0×10^{-1} m   (4) 2.0×10^{1} m   36 ____

37 What is the magnitude of the electrostatic force exerted on an electron by another electron when they are 0.10 meter apart?

(1) 2.6×10^{-36} N   (3) 2.3×10^{-26} N

(2) 2.3×10^{-27} N   (4) 1.4×10^{-8} N   37 ____

38 After a 65-newton weight has fallen freely from rest a vertical distance of 5.3 meters, the kinetic energy of the weight is

(1) 12 J   (3) 910 J

(2) 340 J   (4) 1800 J   38 ____

39 A 0.500-kilogram cart traveling to the right on a horizontal, frictionless surface at 2.20 meters per second collides head on with a 0.800-kilogram cart moving to the left at 1.10 meters per second. What is the magnitude of the total momentum of the two-cart system after the collision?

(1) 0.22 kg • m/s   (3) 1.98 kg • m/s

(2) 0.39 kg • m/s   (4) 4.29 kg • m/s   39 ____

40 An object weighing 2.0 newtons is pushed across a horizontal, frictionless surface by a horizontal force of 4.0 newtons. The magnitude of the net force acting on the object is

(1) 0.0 N   (3) 8.0 N

(2) 2.0 N   (4) 4.0 N   40 ____

41 The ratio of the wavelength of AM radio waves traveling in a vacuum to the wavelength of FM radio waves traveling in a vacuum is approximately

(1) 1 to 1
(2) 2 to 1
(3) 10^2 to 1
(4) 10^8 to 1

41 _____

42 A charm quark has a charge of approximately

(1) 5.33×10^{-20} C
(2) 1.07×10^{-19} C
(3) 1.60×10^{-19} C
(4) 2.40×10^{-19} C

42 _____

43 The diagram below represents a 3.0-ohm resistor connected to a 12-volt battery. Meters X and Y are correctly connected in the circuit.

What are the readings on the meters?

(1) $X = 12$ V and $Y = 0.25$ A
(2) $X = 12$ V and $Y = 4.0$ A
(3) $X = 0.25$ A and $Y = 12$ V
(4) $X = 4.0$ A and $Y = 12$ V

43 _____

44 A toy airplane, flying in a horizontal, circular path, completes 10 complete circles in 30 seconds. If the radius of the plane's circular path is 4.0 meters, the average speed of the airplane is

(1) 0.13 m/s
(2) 0.84 m/s
(3) 1.3 m/s
(4) 8.4 m/s

44 _____

45 Which pair of graphs represents the vertical motion of an object falling freely from rest?

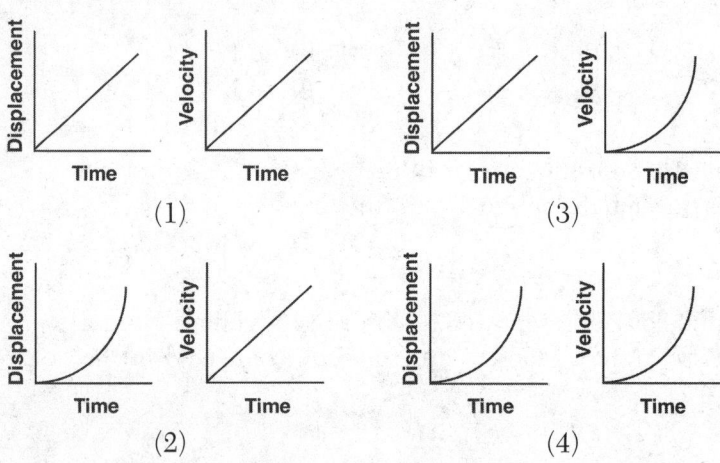

46 An object is thrown straight upward. Which graph best represents the relationship between the object's kinetic energy and the height of the object above its release point? [Neglect friction.]

45 _____

46 _____

7 In the diagram below, X represents a particle in a spring.

Which diagram represents the motion of particle X as a longitudinal wave passes through the spring toward the right?

(1)   (2)   (3)   (4)   47 _____

8 As represented in the diagram below, two wave pulses, X and Y, are traveling toward each other in a rope. Both wave pulses have an amplitude of 0.30 m.

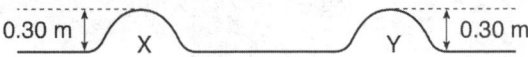

Which diagram shows the pulse produced due to the superposition of pulse X and pulse Y?

(1)   (3)

(2)   (4)   48 _____

49 The horn of a car produces a sound wave of constant frequency. The car, traveling at constant speed, approaches, passes, and then moves away from a stationary observer. Which graph best represents the frequency of this sound wave detected by the observer during the time interval in which the car approaches, passes, and moves away?

        (1)   (2)   (3)   (4)   49 ____

50 A combination of two identical resistors connected in series has an equivalent resistance of 10 ohms. What is the equivalent resistance of the combination of these same two resistors when connected in parallel?

(1) 2.5 Ω   (3) 10. Ω
(2) 5.0 Ω   (4) 20. Ω   50 ____

PART B–2
Answer all questions in this part.

Directions (51–65): Record your answers on the answer sheet provided after the questions. Some questions may require the use of the *2006 Edition Reference Tables for Physical Setting/Physics.*

Base your answers to questions 51 through 53 on the information below and on your knowledge of physics.

The scaled diagram below represents two forces acting concurrently at point *P*. The magnitude of force *A* is 32 newtons and the magnitude of force *B* is 20 newtons. The angle between the directions of force *A* and force *B* is 120°.

51 Determine the linear scale used in the diagram. [1]

52 On the diagram *on the answer sheet*, use a protractor and a ruler to construct a scaled vector to represent the resultant of forces *A* and *B*. Label the vector *R*. [1]

53 Determine the magnitude of the resultant force. [1]

Base your answers to questions 54 through 56 on the information and diagram below and on your knowledge of physics.

A student pushes a box, weighing 50 newtons, 6.0 meters up an incline at a constant speed by applying a force of 25 newtons parallel to the incline. The top of the incline is 2.0 meters higher than the bottom.

54–55 Calculate the total work done on the box by the student while pushing the box from the bottom to the top of the incline. [Show all work, including the equation and substitution with units.] [2]

56 Describe what would happen to the total work done on the box by the student to push the box 6.0 meters up the incline at constant speed if the coefficient of kinetic friction between the box and the incline were increased. [1]

57 In the diagram below, a light ray is incident on an interface between glass and air. When the light strikes the glass-air interface, some of the light is reflected. On the diagram *on the answer sheet*, use a protractor and straightedge to construct the reflected light ray. [1]

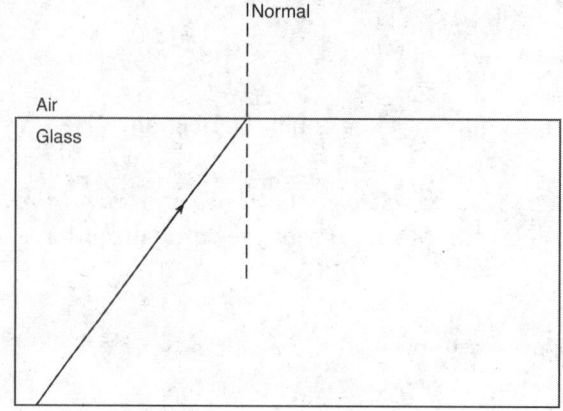

58–59 The current in a wire is 5.0 amperes. Calculate the total amount of charge, in coulombs, that travels through the wire in 36 seconds. [Show all work, including the equation and substitution with units.] [2]

60–61 A spring, with a spring constant of 100 newtons per meter, possesses 2.0 joules of elastic potential energy when compressed. Calculate the spring's change in length from its uncompressed length. [Show all work, including the equation and substitution with units.] [2]

62–63 A monochromatic ray of light ($f = 5.09 \times 10^{14}$ Hz) travels from air into medium X. The angle of incidence of the ray in air is 45.0° and the ray's angle of refraction in medium X is 29.0°. Calculate the absolute index of refraction of medium X. [Show all work, including the equation and substitution with units.] [2]

64–65 An argon-ion laser emits blue-green light having a wavelength of 488 nanometers in a vacuum. Calculate the energy of a photon of this light. [Show all work, including the equation and substitution with units.] [2]

PART C
Answer all questions in this part.

Directions (66–85): Record your answers on the sheet provided in the back. Some questions may require the use of the *2006 Edition Reference Tables for Physical Setting/Physics*.

Base your answers to questions 66 through 70 on the information and diagram below and on your knowledge of physics.

An incandescent lightbulb uses a length of thin tungsten wire as the filament (the part of the operating bulb that produces light).

Incandescent Lightbulb

Filament

One particular lightbulb has a 0.22-meter length of the tungsten wire used as its filament. This tungsten wire filament has a resistance of 19 ohms at a temperature of 20°C. The tungsten wire filament has a resistance of 240 ohms when this bulb is operated at a potential difference of 120 volts.

66–67 Calculate the cross-sectional area of this tungsten wire filament. [Show all work, including the equation and substitution with units.] [2]

68 Explain why the resistance of the tungsten wire filament increases when the bulb is being operated compared to the resistance of the filament at 20°C. [1]

69–70 Calculate the power of this lightbulb when it is being operated at a potential difference of 120 volts. [Show all work, including the equation and substitution with units.] [2]

Base your answers to questions 71 through 75 on the information and diagram below and on your knowledge of physics.

A 150-newton force, applied to a wooden crate at an angle of 30° above the
horizontal, causes the crate to travel at constant velocity across a horizontal
wooden floor, as represented below.

1–72 Calculate the magnitude of the horizontal component of the 150-newton
force. [Show all work, including the equation and substitution with
units.] [2]

73 Determine the magnitude of the frictional force acting on the crate. [1]

4–75 Calculate the magnitude of the normal force exerted by the floor on the
crate. [Show all work, including the equation and substitution with
units.] [2]

Base your answers to question 76 through 80 on the information below and
on your knowledge of physics.

On a flat, level road, a 1500-kilogram car travels around a curve having a constant radius of 45 meters. The centripetal acceleration of the car has a constant
magnitude of 3.2 meters per second squared.

6–77 Calculate the car's speed as it travels around the curve. [Show all work,
including the equation and substitution with units.] [2]

78 Determine the magnitude of the centripetal force acting on the car as it
travels around the curve. [1]

79 What force provides the centripetal force needed for the car to travel
around the curve? [1]

80 Describe what happens to the magnitude of the centripetal force on the car as it travels around the curve if the speed of the car *decreases*. [1]

Base your answers to questions 81 through 85 on the information and diagram below and on your knowledge of physics.

A musician plucks a 0.620-meter-long string on an acoustic guitar, as represented in the diagram below.

The plucked string vibrates, producing a musical note called "G." The waves traveling along the vibrating string produce a standing wave with a frequency of 196 hertz.

81 On the diagram of the standing wave *on the answer sheet*, label *one* node with the letter **N** and one antinode with the letter **A**. [1]

82 Determine the wavelength of the standing wave on the 0.620-meter-long vibrating string. [1]

83–84 Calculate the speed of the wave traveling on the vibrating string. [Show all work, including the equation and substitution with units.] [2]

85 Describe what happens to the frequency when the musician shortens the vibrating portion of the string by pinching the string against the fingerboard while the string continues to vibrate. [1]

Answer Sheet
June 2018
Physics: The Physical Setting

PART B–2

51 1.0 cm = _____ **N**

52

53 _____ **N**

54–55

56 _____

57

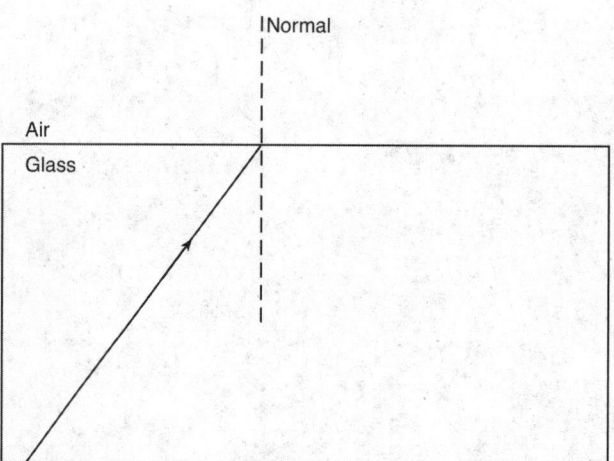

58–59

60–61

62–63

64–65

PART C

66–67

68 _____

69–70

71–72

73 _____ **N**

74–75

76–77

78 _____ **N**

79 _____

80 _____

81

82 _____ **m**

83–84

85 _____

Answers
June 2018
Physics: The Physical Setting

Answer Key

PART A

1. 2	8. 4	15. 2	22. 4	29. 2
2. 1	9. 1	16. 3	23. 3	30. 2
3. 4	10. 3	17. 3	24. 2	31. 2
4. 2	11. 4	18. 4	25. 1	32. 4
5. 3	12. 1	19. 3	26. 1	33. 1
6. 4	13. 2	20. 2	27. 4	34. 1
7. 3	14. 1	21. 3	28. 3	35. 4

PART B–1

36. 2	39. 1	42. 2	45. 2	48. 3
37. 3	40. 4	43. 4	46. 1	49. 3
38. 2	41. 3	44. 4	47. 2	50. 1

PART B–2 and **PART C**. *See* **Answers Explained**.

Answers Explained

PART A

1 The meter/second (m/s) is the unit of measure for both speed and velocity. Velocity is the correct answer because it is a vector quantity, having both magnitude and direction.

WRONG CHOICES EXPLAINED:
(1) Weight is a vector quantity. However, its unit of measure is the newton (N), not the kilogram (kg).
(3) Speed is measured in meters/second (m/s). However, speed is a scalar quantity. It has magnitude but not direction.
(4) Acceleration is a vector quantity. However, its unit of measure is m/s^2 not m^2/s.

2 Mass is the quantitative measure of inertia. In this example, the mass of the cart has not changed. Therefore, the inertia of the cart will be unchanged.

3 Refer to the Mechanics Section of the Physics Reference Tables.

Given: $v_i = 0$ m/s
$v_f = 40$ m/s
$d = 70$ m

Find: $a = ?$

Solution: $v_f^2 = v_i^2 + 2ad$

$$a = \frac{v_f^2 - v_i^2}{2d}$$

$$a = \frac{(40 \text{ m/s})^2 - (0 \text{ m/s})^2}{2(70 \text{ m})}$$

$$a = 11 \text{ m/s}^2$$

4 Refer to the *Mechanics* section of *Reference Table K*.

Given: $m = 6.0$ kg

$v_i = 4.0$ m/s E

$a = 0.50$ m/s^2 E

$t = 3.0$ s

Find: $v_f = ?$

Solution: $v_f = v_i + at$

$v_f = 4.0 \text{ m/s} + \left((0.50 \text{ m/s}^2)(3.0 \text{ s})\right)$

$v_f = 5.5$ m/s

5 The range of a projectile is determined by two parameters—the value of the horizontal component of the initial velocity and the hang time (time in the air).

Refer to the *Mechanics* section of *Reference Table K*. The horizontal component of the velocity is proportional to the velocity and the cosine of the launch angle. The larger the cosine of the angle the greater the horizontal speed.

The hang time (time of flight) is determined by the vertical component of the initial velocity and the acceleration due to gravity. The vertical component of the velocity is proportional to the velocity and the sine of the launch angle. The greater the sine of the angle, the greater the vertical speed, and the longer the time of flight or hang time.

The horizontal distance is proportional to the horizontal speed and time, the greater the horizontal speed and/or the greater the time the greater the distance traveled.

Smaller launch angles have a greater horizontal component of speed but a smaller vertical one. Larger launch angles have a greater vertical component but a smaller horizontal component. An angle of 45° results in neither category larger than the other compared to the other angles but it results it a strong enough showing in both categories to achieve the largest range.

WRONG CHOICES EXPLAINED:

(1) and (2) A smaller angle will result in a larger horizontal velocity but the range will be limited due to the shorter hang time.

(4) A larger angle will result in greater hang time but a smaller horizontal velocity and thus a shorter range.

6 Refer to the *Mechanics* section of *Reference Table K*.

Given: $v_i = 0$ m/s
$d = 18$ m
$t = 3.0$ s

Find: $a = ?$

Solution: $d = v_i t + \frac{1}{2}at^2$

$$a = \frac{2(d - v_i t)}{t^2}$$

$$a = \frac{2\left(18 \text{ m} - \left((0 \text{ m/s})(3.0 \text{ s})\right)\right)}{(3.0 \text{ s})^2}$$

$$a = 4.0 \text{ m/s}^2$$

7 The only acceleration present is the acceleration in the downward direction. This is the acceleration due to gravity, which is approximately equal to 9.81 m/s^2 and is constant.

8 The time of flight, or hang time, is dependent on the motion in the vertical directly (y-direction) only. Increasing the height would increase the amount of time needed for the projectile to hit the ground.

WRONG CHOICES EXPLAINED:
   (1) Changing the mass would not affect the vertical velocity in a frictionless environment.
   (2) Decreasing the height would decrease the hang time.
   (3) Increasing the horizontal velocity would increase the range or horizontal distance traveled but would not affect motion in the vertical direction. Therefore, the hang time would remain the same.

9 Refer to the *Mechanics* section of *Reference Table K.*

Given: $v_i = 0$ m/s

$m = 0.050$ kg

$F = 5.0 \times 10^3$ N

$v_f = 44$ m/s

Find: $J = ?$

Solution: $p = mv$

$J = \Delta p$

$J = m\Delta v$

$J = m\left(v_f - v_i\right)$

$J = (0.050 \text{ kg})(44 \text{ m/s} - 0 \text{ m/s})$

$J = 2.2$ N • s

10 According to Newton's third law of motion, the force that the racket exerts on the tennis ball is equal but opposite in direction to the force that the tennis ball exerts on the racket. Since the racket exerts an average force of 200 N on the tennis ball, the average force exerted on the racket by the tennis ball is also 200 N.

11 The force of friction always acts in the direction opposite to the direction of motion. In the diagram, the box is moving down the incline in direction *B*. Therefore, friction acts in direction *D*.

WRONG CHOICES EXPLAINED:
(1) *A* is the direction of the normal force.
(2) *B* is the direction of motion.
(3) *C* is the direction of the weight of the box.

12 As an object moves in a circular path, its acceleration, *a*, always points toward the center of the circle. The direction of the velocity, *v*, at any point on the circular path is tangent to the circle. Only choice (1) accurately represents both of these conditions.

WRONG CHOICES EXPLAINED:

(2) The arrows are reversed for acceleration, a, and velocity, v.

(3) The arrows for both acceleration, a, and velocity, v, are pointing in incorrect directions.

(4) The arrow for velocity, v, is correct. However, the arrow for acceleration, a, is pointing outward instead of pointing toward the center of the circular path.

13 Refer to the *Electricity* section of *Reference Table K*.

Given:	$E = 2.0 \times 10^3$ N/C
	$F_e = 3.0 \times 10^{-3}$ N
Find:	$q = ?$
Solution:	$E = \dfrac{F_e}{q}$
	$q = \dfrac{F_e}{E}$
	$q = \dfrac{3.0 \times 10^{-3} \text{ N}}{2.0 \times 10^3 \text{ N/C}}$
	$q = 1.5 \times 10^{-6}$ C

WRONG CHOICES EXPLAINED:

(1) This is equivalent to 1 elementary charge.

(3) The magnitudes of the electric field and the electrostatic force were multiplied together instead of being divided appropriately.

14 Refer to the *Mechanics* section of *Reference Table K*.
The gravitational field strength, which is also acceleration due to gravity or g, is equal to $\dfrac{F_g}{m}$.

WRONG CHOICES EXPLAINED:

(2) G is the universal gravitational constant.

(3) mg is equal to weight, which is also the force due to gravity.

(4) $\dfrac{Gm_1 m_2}{r^2}$ is equal to weight, which is also the force due to gravity.

15 Refer to the *Mechanics* section of *Reference Table K.*

Given: $W = 2.0 \times 10^3$ J

$t = 10$ s

Find: $P = ?$

Solution: $P = \dfrac{W}{t}$

$P = \dfrac{2.0 \times 10^3 \text{ J}}{10 \text{ s}}$

$P = 2.0 \times 10^2$ W

16 Both magnetic and electrostatic forces can be either attractive or repulsive

WRONG CHOICES EXPLAINED:
(1), (2), and (4) Gravitational forces are always attractive.

17 Resistivity (ρ) is dependent on a material's internal structure and temperature. In this problem, both wires are made of copper and are at 0°C. Since the internal structure and the temperature of the two wires are the same, the resistivity of the two wires is the same.

The resistance of a wire is proportional to the wire's resistivity and length and is also inversely proportional to the wire's cross-sectional area. Had the question asked about resistance instead of resistivity, doubling the length of the wire would have resulted in double the resistance.

18 Refer to the *Electricity* section of *Reference Table K.*
Work per unit charge, $\dfrac{W}{q}$, is equal to V, which is potential difference.

WRONG CHOICES EXPLAINED:
(1) Electrostatic force can be shown by $F_e = \dfrac{kq_1q_2}{r^2}$, which does not equal $\dfrac{W}{q}$.

(2) Power can be shown by $P = VI = I^2R = \dfrac{V^2}{R}$, none of which equal $\dfrac{W}{q}$.

(3) Field strength can be shown by $E = \dfrac{F_e}{q}$, which does not equal $\dfrac{W}{q}$.

19 Refer to the *Electricity* section of *Reference Table K*. If you have the resistance and voltage, you can find the current. However, the question does not directly give a value for the resistance in either situation.

$$R = \frac{V}{I}$$

$$R = \frac{\rho L}{A}$$

The question states that each situation uses the "same copper wire at the same temperature." This means the resistivity (ρ), length (L), and area (A) of the wire in each situation is the same. Therefore, the resistance in each situation, R_1 and R_2, must be the same, or $R_1 = R_2$.

Given: $V_1 = 24$ mV

            $I_1 = 0.40$ A

            $V_2 = 48$ mV

            $R_1 = R_2$

Find: $I_2 = ?$

Solution: $R = \frac{V}{I}$

            $\frac{V_1}{I_1} = \frac{V_2}{I_2}$

            $I_2 = \frac{V_2 I_2}{V_1}$

            $I_2 = \frac{(48 \text{ mV})(0.40 \text{ A})}{(24 \text{ mv})}$

            $I_2 = 0.80$ A

20 Refer to *Reference Table A* and the *Mechanics* section of *Reference Table K.*

Given: $m_1 = m_2 = 0.425$ kg

$r = 0.500$ m

$G = 6.67 \times 10^{-11}$ N • m^2/kg^2

Find: $F_g = ?$

Solution: $F_g = \dfrac{Gm_1m_2}{r^2}$

$F_g = \dfrac{\left(6.67 \times 10^{-11} \text{ N • m}^2/\text{kg}^2\right)(0.425 \text{ kg})(0.425 \text{ kg})}{(0.500 \text{ m})^2}$

$F_g = 4.82 \times 10^{-11}$ N

21 Light waves are electromagnetic waves and do not need a medium through which to travel. They can travel in a vacuum as well as in a medium, such as water. Sound waves, though, are mechanical waves and require a medium through which to travel. Therefore, sound waves can travel through water but not through a vacuum.

22 Resonance is the spontaneous vibration of an object at a frequency equal to that of the wave that initiates that resonant vibration. If that vibration is strong enough, it can shatter or break the object.

WRONG CHOICES EXPLAINED:
(1) Action and reaction are terms used to describe Newton's third law of motion.
(2) The law of conservation of momentum states that if two objects that are not subjected to external forces interact, the total momentum of the objects before the interaction is equal to the total momentum of the objects after the interaction.
(3) Inertia is the property of matter that resists changes in motion.

23 The frequency of a light wave does not change as the wave travels from one medium to another. Period is the inverse of frequency and, as such, also does not change as a light wave travels from one medium to another.

WRONG CHOICES EXPLAINED:
(1) The speed of a light wave does change as the light travels from one medium to another.
(2) and (4) When the speed of a light wave changes as the wave enters a new medium, the wavelength of that light wave changes proportionally.

24 Refer to the *Reference Table A*, the *Absolute Indices of Refraction (Reference Table E)*, and the *Waves* section of *Reference Table K*. Note that the frequency of the light ray is not needed to solve this problem.

Given: $f = 5.09 \times 10^{14}$ Hz

$n = 1.47$

$c = 3.00 \times 10^8$ m/s

Find: $v = ?$

Solution: $n = \dfrac{c}{v}$

$v = \dfrac{c}{n}$

$v = \dfrac{3.00 \times 10^8 \text{ m/s}}{1.47}$

$v = 2.04 \times 10^8$ m/s

WRONG CHOICES EXPLAINED:
(3) This is the value for the speed of light in a vacuum, not the speed of light in corn oil.
(4) You multiplied the speed of light by the index of refraction instead of dividing the two quantities.

25 Diffraction is the bending of a wave around a barrier.

WRONG CHOICES EXPLAINED:
(2) The Doppler effect is the apparent change in frequency that results when a wave source and an observer are in relative motion with respect to each other.
(3) When light is reflected from a surface, the angle that the incident light ray makes with the normal to the surface is equal to the angle that the reflected ray makes with the normal to the surface.
(4) Refraction is the change in direction of a wave when it passes obliquely from one medium into another, which causes the wave to move at a different speed.

26 Refer to the *Waves* section of *Reference Table K*.

Given: $f = 5.0 \times 10^{10}$ Hz

Find: $T = ?$

Solution: $T = \dfrac{1}{f}$

$T = \dfrac{1}{5.0 \times 10^{10} \text{ Hz}}$

$T = 2.0 \times 10^{-11}$ s

27 Two or more waves passing simultaneously through the same area of a medium affect the medium independently but do not affect each other. Therefore, each wave will remain unchanged.

WRONG CHOICES EXPLAINED:

(1) Apparent changes in frequency, called the Doppler effect, result when a wave source and an observer are in relative motion with respect to each other. This is not the case in this situation.

(2) Changes in velocity occur as a wave moves from one medium into another, which is not the case in this situation.

(3) Changes in amplitude occur at the point of interference. However, each wave returns to its original amplitude when the interference ends and/or after the waves have passed by each other. The question specifically states "after two light waves have interfered." Therefore, the two waves are no longer interfering with each other.

28 When a glass rod is rubbed with silk, electrons leave the rod and transfer to the silk. This leaves the rod with a net positive charge and the silk with a net negative charge.

WRONG CHOICES EXPLAINED:

(1) Adding electrons to the rod would have resulted in a net negative charge on the rod.

(2) and (4) Protons are contained in the nucleus of atoms and therefore are not free to move to or from any object.

29 Refer to *Reference Table* A and the *Modern Physics* section of *Reference Table* K.

Given: $E_{photon} = 1.33 \times 10^{-21}$ J

$h = 6.63 \times 10^{-34}$ J • s

Find: $f = ?$

Solution: $E_{photon} = hf$

$f = E_{photon}$

$f = \dfrac{1.33 \times 10^{-21} \text{ J}}{6.63 \times 10^{-34} \text{ J} \bullet \text{s}}$

$f = 2.01 \times 10^{12}$ Hz

30 This question asks you to find the average of two values. To do so, add the given values and then divide by 2.

Given: $v_i = 11$ m/s

$v_f = 19$ m/s

Find: $\bar{v} = ?$

Solution: $\bar{v} = \dfrac{v_i + v_f}{2}$

$\bar{v} = \dfrac{11 \text{ m/s} + 19 \text{ m/s}}{2}$

$\bar{v} = 15$ m/s

31 Refer to *Reference Table A* and the *Modern Physics* section of *Reference Table K*.

Given: $m_e = 9.11 \times 10^{-31}$ kg

$c = 3.00 \times 10^8$ m/s

Find: $E = ?$

Solution: $E = mc^2$

$E = \left(9.11 \times 10^{-31} \text{ kg}\right)\left(3.00 \times 10^8 \text{ m/s}\right)^2$

$E = 8.20 \times 10^{-14}$ J

32 Refer to the *Mechanics* section of *Reference Table K*.

Given: $x_i = 0.40$ m

$x_f = 0.60$ m

$F = 10$ N

Find: $k = ?$

Solution: $F_s = kx$

$F_s = k\Delta x$

$k = \dfrac{F_s}{\Delta x}$

$k = \dfrac{10 \text{ N}}{(0.60 \text{ m} - 0.40 \text{ m})}$

$k = 50. \text{ N/m}$

33 A closed circuit loop exists between the battery and lamp I, allowing an electric current to flow through only lamp I.

WRONG CHOICES EXPLAINED:
(2) and (3) The open switch does not allow a closed circuit loop that contains lamps II and III. Therefore, no electric current is flowing through those lamps.

(4) Since lamp I is in a closed loop circuit, electric current is flowing through that lamp.

34 A field line represents the path that a small positive charge (known as a test charge) takes while in an electric field. Arrows therefore point toward negative charges and away from positive charges in an electric field. The diagram in choice (1) correctly represents this convention.

WRONG CHOICES EXPLAINED:
(2), (3), and (4) The field lines are pointing in the wrong direction in each of these choices.

35 Points on a periodic wave are in phase if they are at equal displacements from the rest position and are experiencing identical movements. For the points to be experiencing identical movements, they must be moving in the same direction toward or away from the rest position. Points on a periodic wave are 180° degrees out of phase, or completely out of phase, if they are at equal displacements from their rest position but experiencing motion in opposite directions from each other. Points D and E are about halfway to the 180° out-of-phase positions. Therefore, points D and E are $90°$ out of phase with each other.

WRONG CHOICES EXPLAINED:
(1) Points A and E are in phase with each other.
(2) Points B and C are in phase with each other.
(3) Points C and D are $180°$ out of phase with each other.

PART B–1

36 This is an order of magnitude question: 2.0×10^{-1} m ≈ 0.2 m. Since 1 meter is approximately equal to 3.3 feet or 40 inches, 0.2 meters is approximately 8 inches. This is a reasonable approximation for the height of an individual step on a staircase.

WRONG CHOICES EXPLAINED:
(1) 2.0×10^{-1} m = 0.02 m = 2 cm, which is not a reasonable approximation for the height of an individual step.
(3) 2.0×10^{0} m = 2 m ≈ 6.6 ft, which is not a reasonable approximation for the height of an individual step.
(4) 2.0×10^{1} m = 20 m ≈ 66 ft, which is not a reasonable approximation for the height of an individual step.

37 Refer to *Reference Table A* and the *Electricity* section of *Reference Table K*.

Given: $q_1 = q_2 = e = 1.60 \times 10^{-19}$ C

$r = 0.10$ m

$k = 8.99 \times 10^9$ N • m^2/C^2

Find: $F_e = ?$

Solution: $F_e = \dfrac{kq_1q_2}{r^2}$

$F_e = \dfrac{\left(8.99 \times 10^9 \text{ N • m}^2/\text{C}^2\right)\left(1.6 \times 10^{-19} \text{ C}\right)\left(1.6 \times 10^{-19} \text{ C}\right)}{(0.10 \text{ m})^2}$

$F_e = 2.3 \times 10^{-26}$ N

38 Refer to the *Mechanics* section of *Reference Table K*.

Given: $Fg = 65$ N

$h = 5.3$ m

Find: $KE = ?$

Solution: $g = \dfrac{F_g}{m}$ $\quad$ (Law of Conservation of Energy)

$F_g = mg$

$\Delta PE = mg\Delta h$

$KE = \Delta PE$

$KE = F_g\Delta h$

$KE = (65 \text{ N})(5.3 \text{ m})$

$KE = 340$ J

39 Refer to the *Mechanics* section of *Reference Table K*. The carts are moving in opposite directions. Let movement to the right be in the positive direction and movement to the left be in the negative direction.

Given:
$$m_1 = 0.500 \text{ kg}$$
$$v_1 = 2.20 \text{ m/s}$$
$$m_2 = 0.800 \text{ kg}$$
$$v_2 = -1.10 \text{ m/s}$$

Find:
$$p_{after} = ?$$

Solution:
$$p_{before} = p_{after}$$
$$p = mv$$
$$p_{before} = m_1v_1 + m_2v_2$$
$$p_{before} = (0.500 \text{ kg})(2.20 \text{ m/s}) + (0.800 \text{ kg})(-1.10 \text{ m/s})$$
$$p_{before} = 0.22 \text{ kg} \bullet \text{m/s}$$

40 The downward force acting on the object due to the object's weight, or the force due to gravity, is balanced by the normal force acting in the upward direction (Newton's third law of motion). A horizontal force is also acting on the object. Since the object is in a frictionless environment, no force due to friction opposes the horizontal motion. Therefore, the net force acting on the object is the 4.0 N force pushing the object.

41 Refer to the *Electromagnetic Spectrum* in *Reference Table D*. According to the chart, AM radio waves have a wavelength of approximately 10^2 m in a vacuum and FM radio waves have a wavelength of approximately 10^0 m, or 1 m, in a vacuum. Therefore in a vacuum, the ratio of the wavelength of AM radio waves to the wavelength of FM radio waves is 10^2 to 1.

42 Refer to *Reference Table A* and *Reference Table H*. A charm quark carries a charge of $+\frac{2}{3}e$. An elementary charge, e, has a value of 1.6×10^{-19} C. Therefore, a charm quark carries a charge of $(1.6 \times 10^{-19} \text{ C})\left(\frac{2}{3}\right) = 1.07 \times 10^{-19}$ C.

WRONG CHOICES EXPLAINED:

(3) This is the value of one elementary charge, but a charm quark carries a charge of $+\frac{2}{3}e$.

(4) You multiplied the elementary charge by $\frac{3}{2}$ instead of multiplying by $\frac{2}{3}$.

43 Refer to the *Electricity* section of *Reference Table K*. Meter X is properly connected in series in the circuit to measure current.

$$
\begin{aligned}
\text{Given:} \quad & V = 12 \text{ V} \\
& R = 3.0 \ \Omega \\[6pt]
\text{Find:} \quad & I = ? \\[6pt]
\text{Solution:} \quad & R = \frac{V}{I} \\[6pt]
& I = \frac{V}{R} \\[6pt]
& I = \frac{12 \text{ V}}{3.0 \ \Omega} \\[6pt]
& I = 4.0 \text{ A}
\end{aligned}
$$

Meter Y is correctly connected in parallel with the resistor to measure potential difference across the resistor. The resistor is connected in series with the battery and is the only resistor in the circuit. Therefore, the potential difference across the resistor is equal to the potential difference of the battery, which is 12 V.

WRONG CHOICES EXPLAINED:

(1) X is an ammeter, not a voltmeter. You divided 3 by 12 instead of dividing 12 by 3 to obtain the value of the current in the circuit. Additionally, Y is a voltmeter, not an ammeter.

(2) X is an ammeter, not a voltmeter. Y is a voltmeter, not an ammeter.

(3) You divided 3 by 12 instead of dividing 12 by 3 to obtain the value of the current in the circuit.

44 Refer to the *Geometry and Trigonometry* section and the *Mechanics* section of *Reference Table K.*

Given: $d = 10$ complete circles

$t = 30$ s

$r = 4.0$ m

Find: $\bar{v} = ?$

Solution: $\bar{v} = \dfrac{d}{t}$

$C = 2\pi r$

$d = 10C$

$\bar{v} = \dfrac{10(2\pi r)}{t}$

$\bar{v} = \dfrac{10(2\pi \cdot 4.0 \text{ m})}{30 \text{ s}}$

$\bar{v} = 8.4$ m/s

45 An object falling freely experiences a constant acceleration due to gravity. The slope of a displacement versus time graph represents the velocity of the object. If an object experiences acceleration, its instantaneous velocity increases as time passes. Therefore, the curve for displacement versus time becomes steeper as time passes. The slope of a velocity versus time graph is equal to the object's acceleration. Since the acceleration due to gravity is constant, this graph should be a straight line increasing from zero as time goes by. Only the graphs depicted in choice (2) meet these criteria.

WRONG CHOICES EXPLAINED:

(1) The displacement versus time graph indicates constant velocity, while the velocity versus time graph indicates increasing velocity.

(3) The displacement versus time graph indicates constant velocity, while the velocity versus time graph indicates increasing acceleration.

(4) The displacement versus time graph indicates acceleration, while the velocity versus time graph indicates increasing acceleration.

46 As the object rises, it decelerates until it reaches its maximum height. At that point, the object's speed reaches 0 m/s (at which point it begins accelerating downward). As its speed decreases as it rises, the kinetic energy of the object decreases until the kinetic energy reaches zero at the object's maximum height.

WRONG CHOICES EXPLAINED:
(2) This graph indicates that as height increases, kinetic energy increases.
(3) This graph indicates that as height increases, kinetic energy increases exponentially.
(4) This graph indicates that as height increases, kinetic energy remains the same.

47 In a longitudinal wave, a disturbance causes the particles of the material to vibrate in simple harmonic motion in a direction parallel to the direction of motion of the wave. Since the wave is moving to the right, parallel motion would be toward the left and right, which is indicated in the diagram in choice (2).

WRONG CHOICES EXPLAINED:
(1) This would be the type of motion seen for a particle experiencing motion in a surface wave.
(3) These arrows do not indicate any type of simple harmonic motion.
(4) This would be the type of motion observed in a particle experiencing motion in a tranverse wave.

48 The principle of superposition states that the resultant displacement of any point in the medium is the algebraic sum of the displacements of all the individual waves. In this case 0.30 m + 0.30 m = 0.60 m.

WRONG CHOICES EXPLAINED:
(1) This is the displacement due to X or due to Y but not due to both.
(2) This is a displacement equal in magnitude to X or to Y but in the opposite direction.
(4) This is a displacement equal to the sum of the X and Y displacements but in the opposite direction.

49 The Doppler effect is the apparent change in frequency that results when a wave source and an observer are in relative motion with respect to each other. As the distance between the source and the observer decreases, the frequency of the source, as perceived by the observer, is increased. As the distance increases, the observed frequency is decreased.

WRONG CHOICES EXPLAINED:

(1) This graph indicates that as time goes by, the observed frequency increases.

(2) This graph indicates that as time goes by, the observed frequency decreases.

(4) This graph indicates that as time goes by, the observed frequency is initially lower and then the observed frequency is higher.

50 Refer to the *Electricity* section of *Reference Table K*.

Given: $R_{T_{series}} = 10\ \Omega$

$R_1 = R_2$

Find: $R_{T_{parallel}} = ?$

Solution: $R_1 = R_2 = \dfrac{R_{T_{series}}}{2}$

$R_1 = R_2 = \dfrac{10}{2}$

$R_1 = R_2 = 5\ \Omega$

$\dfrac{1}{R_{T_{parallel}}} = \dfrac{1}{R_1} + \dfrac{1}{R_2}$

$\dfrac{1}{R_{T_{parallel}}} = \dfrac{1}{5\ \Omega} + \dfrac{1}{5\ \Omega}$

$\dfrac{1}{R_{T_{parallel}}} = \dfrac{2}{5\ \Omega}$

$R_{T_{parallel}} = \dfrac{5\ \Omega}{2}$

$R_{T_{parallel}} = 2.5\ \Omega$

WRONG CHOICES EXPLAINED:

(2) You forgot to divide by 2 and assumed that both resistors were 10 ohms.

(3) You didn't do any calculations and assumed that both series and parallel circuits with 2 identical resistors have the same resistance.

PART B–2

51 The answer booklet asks for the equivalent to 1.0 cm. Using a centimeter ruler, measure the length of either vector *PB* or vector *PA*. Vector *PB* is approximately 5.0 cm in length and 20 N ÷ 5.0 cm = 4.0 N/cm. Vector *PA* is approximately 8.0 cm in length and 32 N ÷ 8.0 cm = 4.0 N/cm.

One credit is awarded for an answer of 4.0 N ± 0.2 N.

52

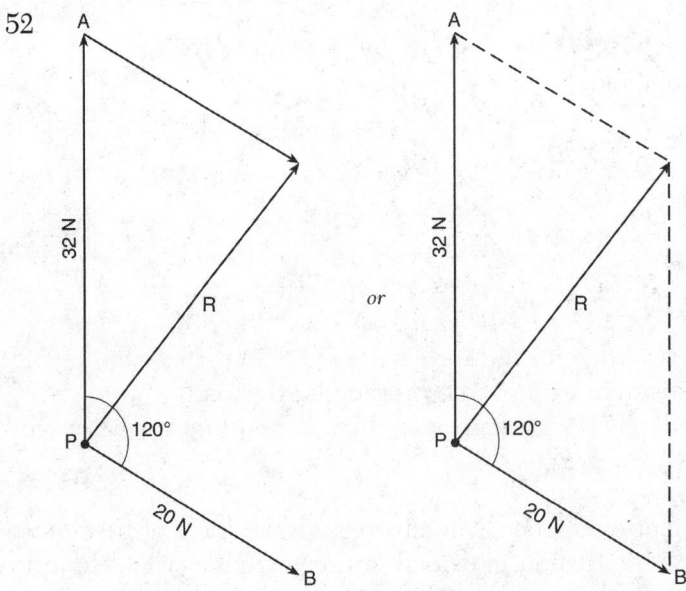

One credit is awarded for constructing the resultant 7.0 cm ± 0.2 cm long at an angle of 38° ± 2° clockwise from force *A*. The resultant vector need not be labeled to receive this credit.

53 Use a centimeter ruler to find the length of resultant vector *R*. Its length is approximately 7.0 cm. Using the scale factor determined in question 51, which is 1.0 cm for every 4.0 N, we can determine that resultant *R* has a value of 7.0 cm × 4.0 N/cm = 28 N.

One credit is awarded for an answer of 28 N ± 2 N or for an answer that is consistent with your response to questions 51 and 52.

54–55   Refer to the *Mechanics* section of *Reference Table K*.

$$\text{Given:} \qquad F_g = 50 \text{ N}$$
$$d = 6.0 \text{ m}$$
$$F_{applied} = 25 \text{ N}$$
$$h = 2.0 \text{ m}$$

Find: $W = ?$

Solution: $W = Fd$
$$W = (25 \text{ N})(6.0 \text{ m})$$
$$W = 150 \text{ N} \bullet \text{m} \quad \text{or} \quad 150 \text{ J}$$

54   One credit is awarded for a correct equation and for substitution of values with units.

55   One credit is awarded for the correct answer with units or for an answer, with units, that is consistent with your answer to question 54.

Note: Students will not be penalized more than 1 credit for errors in units in questions 54 and 55.

56   If the coefficient of kinetic friction increased, the force of friction would increase. If the force of friction increased, a greater force would have to be applied to overcome the force of friction to move the box at constant speed. Hence, the total work done would increase.

One credit is granted for stating that the total work would increase.

57   When you measure the angle that the incident ray makes with the normal, you obtain a value of approximately 37°. Next, draw a reflected ray at an angle of 37° from the normal.

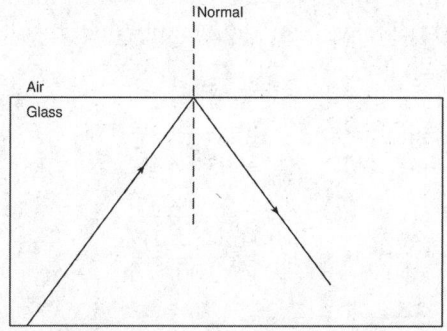

One credit is awarded for a light ray drawn at an angle of $37° \pm 2°$.

Note: No arrowhead is required on the reflected ray. The direction of the ray is implied by the arrowhead on the incident ray. (Rays are not vectors.)

58–59 Refer to the *Electricity* section of *Reference Table K.*

Given: $I = 5.0$ A

               $t = 36$ s

Find: $q = ?$

Solution: $I = \dfrac{\Delta q}{t}$

$$\Delta q = It$$
$$\Delta q = (5.0\ \text{A})(36\ \text{s})$$
$$\Delta q = 180\ \text{A} \bullet \text{s} \quad \text{or} \quad 180\ \text{C}$$

58 One credit is awarded for a correct equation and substitution of values with units.

59 One credit is awarded for the correct answer with units or for an answer, with units, that is consistent with your answer to question 58.

Note: Students will not be penalized more than 1 credit for errors in units in questions 58 and 59.

60–61 Refer to the *Mechanics* section of *Reference Table K.*

Given: $k = 100$ N/m

               $PE_s = 2.0$ J

Find: $\Delta x = ?$

Solution: $PE_s = \dfrac{1}{2}kx^2$

$$x^2 = \dfrac{2PE_s}{k}$$
$$x = \sqrt{\dfrac{2PE_s}{k}}$$
$$x = \sqrt{\dfrac{2(2.0)\ \text{J}}{100\ \text{N/m}}}$$
$$x = 0.20\ \text{m}$$

60 One credit is awarded for a correct equation and substitution of values with units.

61 One credit is awarded for the correct answer with units or for an answer, with units, that is consistent with your answer to question 60.
Note: Students will not be penalized more than 1 credit for errors in units in questions 60 and 61.

62–63 Refer to *Reference Table E* and the *Waves* section of *Reference Table K*.

Given: $\theta_{air} = 45.0°$

$\theta_x = 29.0°$

$n_{air} = 1.00$

Find: $n_x = ?$

Solution: $n_{air} \sin \theta_{air} = n_X \sin \theta_X$

$$n_x = \frac{n_{air} \sin \theta_{air}}{\sin \theta_X}$$

$$n_x = \frac{(1.00) \sin 45.0°}{\sin 29.0°}$$

$$n_x = 1.46$$

62 One credit is awarded for a correct equation and substitution of values with units.

63 One credit is awarded for the correct answer or for an answer, without units, that is consistent with your answer to question 62.
Note: Students will not be penalized more than 1 credit for errors in units in questions 62 and 63.

64–65 Refer to *Reference Table A* and the *Modern Physics* section of *Reference Table K*.

Given:
$$\lambda = 488 \text{ nm} = 4.08 \times 10^{-7} \text{ m}$$
$$h = 6.63 \times 10^{-34} \text{ J} \cdot \text{s}$$
$$c = 3.00 \times 10^{8} \text{ m/s}$$

Find:
$$E_{photon} = ?$$

Solution:
$$E_{photon} = \frac{hc}{\lambda}$$

$$E_{photon} = \frac{\left(6.63 \times 10^{-34} \text{ J} \cdot \text{s}\right)\left(3.00 \times 10^{8} \text{ m/s}\right)}{\left(4.88 \times 10^{-7} \text{ m}\right)}$$

$$E_{photon} = 4.08 \times 10^{-19} \text{ J}$$

64 One credit is awarded for a correct equation and substitution of values with units.

65 One credit is awarded for the correct answer with units or for an answer, with units, that is consistent with your answer to question 64.

Note: Students will not be penalized more than 1 credit for errors in units in questions 64 and 65.

PART C

66–67 Refer to *Reference Table J* and the *Electricity* section of *Reference Table K*.

Given: $L = 0.22$ m

$R_{20°} = 19\ \Omega$

$\rho_{20°} = 5.60 \times 10^{-8}\ \Omega \bullet m$

Find: $A = ?$

Solution: $R = \dfrac{\rho L}{A}$

$A = \dfrac{\rho L}{R}$

$A = \dfrac{(5.60 \times 10 - 8\ \Omega \bullet m)(0.22\ m)}{19}$

$A = 6.5 \times 10^{-10}\ m^2$

66 One credit is awarded for a correct equation and substitution of values with units.

67 One credit is awarded for the correct answer with units or for an answer, with units, that is consistent with your answer to question 66.
Note: Students will not be penalized more than 1 credit for errors in units in questions 66 and 67.

68 Generally, the resistance of a metallic conductor increases with increasing temperature. As the bulb operates, it produces heat and the temperature rises, thus increasing the resistance in the tungsten filament.
One credit is allowed for an explanation that the filament of the operating bulb is at a higher temperature. Acceptable responses include but are not limited to:

- The operating bulb is hotter.
- The filament gets hots when the bulb is operating.
- The resistivity of the tungsten increases.
- The temperature of the filament increases.

69–70 Refer to the *Electricity* section of *Reference Table K.*

Given: $R = 240\ \Omega$
          $V = 120\ V$

Find: $P = ?$

Solution: $P = \dfrac{V^2}{R}$

$P = \dfrac{(120\ V)^2}{240\ \Omega}$

$P = 60\ W$

69 One credit is awarded for a correct equation and substitution of values with units.

70 One credit is awarded for the correct answer with units or for an answer, with units, that is consistent with your answer to question 69.
Note: Students will not be penalized more than 1 credit for errors in units in questions 69 and 70.

71–72 Refer to the *Geometry and Trigonometry* section of *Reference Table K.*

Given: $F = 150\ N$
          $\theta = 30°$

Find: $F_x = ?$

Solution: $A_x = A \cos\theta$ or $\cos\theta = \dfrac{b}{c}$
$F_x = F \cos\theta$ $b = c \cos\theta$
$F_x = (150\ N)(\cos 30.0°)$ $F_x = F \cos\theta$
$F_x = 130\ N$ $F_x = (150\ N)(\cos 30.0°)$
                            $F_x = 130\ N$

71 One credit is awarded for a correct equation and substitution of values with units.

72 One credit is awarded for the correct answer with units or for an answer, with units, that is consistent with your answer to question 72.
Note: Students will not be penalized more than 1 credit for errors in units in questions 71 and 72.

73 The wooden crate is moving at constant velocity in the horizontal direction, and it is not accelerating. Therefore, no net force is acting on the box and $F_f = F_x = 130$ N.
One credit is awarded for 130 N or for an answer that is consistent with your response to question 72.

74–75 Refer to *Reference Table C* and the *Mechanics* section of *Reference Table K*.

$$\text{Given:} \qquad F_f = 130 \text{ N}$$
$$\mu = 0.30$$

$$\text{Find:} \qquad F_N = \text{?}$$

$$\text{Solution:} \qquad F_f = \mu F_N$$

$$F_N = \frac{F_f}{\mu}$$

$$F_N = \frac{130 \text{ N}}{0.30}$$

$$F_N = 430 \text{ N}$$

74 One credit is awarded for a correct equation and substitution of values with units.

75 One credit is awarded for the correct answer with units or for an answer, with units, that is consistent with your answer to question 74.
Note: Students will not be penalized more than 1 credit for errors in units in questions 74 and 75.

76–77 Refer to the *Mechanics* section of *Reference Table K*.

Given: $r = 45$ m

$a_c = 3.2$ m/s^2

Find: $v = ?$

Solution: $a_c = \dfrac{v^2}{r}$

$v^2 = a_c r$

$v = \sqrt{a_c r}$

$v = \sqrt{(3.2 \text{ m/s}^2)(45 \text{ m})}$

$v = 12$ m/s

76 One credit is awarded for a correct equation and substitution of values with units.

77 One credit is awarded for the correct answer with units or for an answer, with units, that is consistent with your answer to question 76.
Note: Students will not be penalized more than 1 credit for errors in units in questions 76 and 77.

78 Refer to the *Mechanics* section of *Reference Table K*.

Given: $m = 1500$ kg

$a_c = 3.2$ m/s^2

Find: $F_c = ?$

Solution: $F_c = ma_c$

$F_c = (1500 \text{ kg})(3.2 \text{ m/s}^2)$

$F_c = 4800$ N

One credit is awarded for an answer of 4800 N.

79 The force of friction between the car and the road provides the centripetal force necessary to keep the car in a circular path.

One credit is awarded for an answer of friction, static friction, or electromagnetic. Credit is not granted for an answer of centripetal force, net force, gravitational force, weight, or normal force.

80 Refer to the *Mechanics* section of *Reference Table K*.

$$F_c = ma_c$$

$$a_c = \frac{v^2}{r}$$

$$F_c = \frac{mv^2}{r}$$

Centripetal force is proportional to the square of the speed. If the speed decreases, the centripetal force decreases.

One credit is awarded for an acceptable response. Acceptable responses include but are not limited to:

• The magnitude of the centripetal force decreases.
• It becomes less.
• It decreases.

81 In a standing wave, adjacent crests and troughs move vertically in opposite directions about points that have no motion. The points that do not move are called nodes, and the crest-trough combinations are called antinodes.

One credit is awarded for labeling one node and one antinode correctly.

82 The diagram indicates the length of a single crest or one-half wavelength. Therefore, the wavelength is 2×0.620 m $= 1.24$ m.

One credit is awarded for an answer of 1.24 m or 1.240 m.

83–84   Refer to the *Waves* section of *Reference Table K*.

Given: $f = 196$ Hz

                $\lambda = 1.24$ m

Find: $v = ?$

Solution: $v = f\lambda$

                $v = (196 \text{ Hz})(1.24 \text{ m})$

                $v = 243$ m/s

83   One credit is awarded for a correct equation and substitution of values with units.

84   One credit is awarded for the correct answer with units or for an answer, with units, that is consistent with your answer to question 83.

Note: Students will not be penalized more than 1 credit for errors in units in questions 83 and 84.

85   Refer to the *Waves* section of *Reference Table K*: $v = f\lambda$. When speed remains constant, frequency and wavelength are inversely proportional to one another. Shortening the string or shortening the wavelength will result in a higher frequency.

One credit is awarded for a statement indicating that the frequency will increase. Acceptable responses include but are not limited to:

• It increases.
• The frequency gets higher.

Topic	Question Numbers (total)	Wrong Answers (x)	Grade
Math Skills	3, 4, 6, 9, 13, 15, 18, 19, 24, 26, 29–32, 36–38, 41–46, 48–55, 57–67, 69–72, 74–78, 80, 82–84: (54)		$\dfrac{100(54 - x)}{54} = \%$
Mechanics	1–12, 14, 20, 30, 32, 40, 44, 45, 52, 53, 71–80: (31)		$\dfrac{100(31 - x)}{31} = \%$
Energy	15, 38, 46, 54, 55, 60, 61: (7)		$\dfrac{100(7 - x)}{7} = \%$
Electricity/ Magnetism	13, 16–19, 28, 33, 34, 37, 43, 50, 56, 58, 59, 66–70: (19)		$\dfrac{100(19 - x)}{19} = \%$
Waves	21–27, 35, 41, 47–49, 57, 62–65, 81–85: (22)		$\dfrac{100(22 - x)}{22} = \%$
Modern Physics	29, 31, 42: (3)		$\dfrac{100(3 - x)}{3} = \%$

Examination June 2019

Physics: The Physical Setting

PART A
Answer all questions in this part.

Directions (1–35): For *each* statement or question, select the *number* of the word or expression that, of those given, best completes the statement or answers the question. Some questions may require the use of the *2006 Edition Reference Tables for Physical Setting/Physics*. Record your answers in the spaces provided.

1 Which pair of quantities represent scalar quantities?

  (1) displacement and velocity   (3) energy and velocity

  (2) displacement and time   (4) energy and time   1 _____

2 A sailboat on a lake sails 40. meters north and then sails 40. meters due east. Compared to its starting position, the new position of the sailboat is

  (1) 40. m due east   (3) 57 m northeast

  (2) 40. m due north   (4) 80. m northeast   2 _____

3 A ball is thrown straight upward from the surface of Earth. Which statement best describes the ball's velocity and acceleration at the top of its flight?

  (1) Both velocity and acceleration are zero.

  (2) Velocity is zero and acceleration is nonzero.

  (3) Velocity is nonzero and acceleration is zero.

  (4) Both velocity and acceleration are not zero.   3 _____

4 As a student runs a plastic comb through her hair, the comb acquires a negative electric charge. This charge results from the transfer of

(1) protons from the comb to her hair
(2) protons from her hair to the comb
(3) electrons from the comb to her hair
(4) electrons from her hair to the comb   4 _____

5 How would the mass and weight of an object on the Moon compare to the mass and weight of the same object on Earth?

(1) Mass and weight would both be less on the Moon.
(2) Mass would be the same but its weight would be less on the Moon.
(3) Mass would be less on the Moon and its weight would be the same.
(4) Mass and weight would both be the same on the Moon.   5 _____

6 An object is moving with constant speed in a circular path. The object's centripetal acceleration remains constant in

(1) magnitude, only
(2) direction, only
(3) both magnitude and direction
(4) neither magnitude nor direction   6 _____

7 As shown in the diagram below, a rope attached to a 500.-kilogram crate is used to exert a force of 45 newtons at an angle of 65 degrees above the horizontal.

The horizontal component of the force acting on the crate is

(1) 19 N   (3) 210 N

(2) 41 N   (4) 450 N   7 _____

8 A spring with a spring constant of 68 newtons per meter hangs from a ceiling. When a 12-newton downward force is applied to the free end of the spring, the spring stretches a total distance of

(1) 0.18 m   (3) 5.7 m

(2) 0.59 m   (4) 820 m   8 _____

9 As a student walks downhill at constant speed, his gravitational potential energy

(1) increases and his kinetic energy increases

(2) increases and his kinetic energy remains the same

(3) decreases and his kinetic energy increases

(4) decreases and his kinetic energy remains the same   9 _____

10 When 150 joules of work is done on a system by an external force of 15 newtons in 20. seconds, the total energy of that system increases by

(1) 1.5×10^2 J   (3) 3.0×10^2 J

(2) 2.0×10^2 J   (4) 2.3×10^3 J   10 _____

11 A person on a ledge throws a ball vertically downward, striking the ground below the ledge with 200 joules of kinetic energy. The person then throws an identical ball vertically upward at the same initial speed from the same point. What is the kinetic energy of the second ball when it hits the ground? [Neglect friction.]

(1) 200 J   (3) less than 200 J

(2) 400 J   (4) more than 400 J   11 _____

12 Two construction cranes are used to lift identical 1200-kilogram loads of bricks the same vertical distance. The first crane lifts the bricks in 20. seconds and the second crane lifts the bricks in 40. seconds. Compared to the power developed by the first crane, the power developed by the second crane is

(1) the same   (3) half as great

(2) twice as great   (4) four times as great   12 _____

13 An ionized calcium atom has a charge of +2 elementary charges. If this ion is accelerated through a potential difference of 2.0×10^3 volts, the ion's change in kinetic energy will be

(1) 1.0×10^3 eV   (3) 3.0×10^3 eV

(2) 2.0×10^3 eV   (4) 4.0×10^3 eV   13 _____

14 A total charge of 100. coulombs flows past a fixed point in a circuit every 500. seconds. What is the current at this point in the circuit?

(1) 0.200 A   (3) 5.00×10^4 A

(2) 5.00 A   (4) 1.25×10^{18} A   14 _____

15 An aluminum wire of length 1.0 meter has a resistance of 9.0×10^{-3} ohm. If the wire were cut into two equal lengths, each length would have a resistance of

(1) $2.8 \times 10^{-8}\ \Omega$ (3) $9.0 \times 10^{-3}\ \Omega$

(2) $4.5 \times 10^{-3}\ \Omega$ (4) $1.8 \times 10^{-2}\ \Omega$ 15 _____

16 In an operating electrical circuit, the source of potential difference could be

(1) a voltmeter   (3) an ammeter

(2) a battery   (4) a resistor   16 _____

17 A lightbulb with a resistance of 2.9 ohms is operated using a 1.5-volt battery. At what rate is electrical energy transformed in the lightbulb?

(1) 0.52 W   (3) 4.4 W

(2) 0.78 W   (4) 6.5 W   17 _____

18 A 40.0-kilogram child exerts a 100.-newton force on a 50.0-kilogram object. The magnitude of the force that the object exerts on the child is

(1) 0.0 N (3) 100. N
(2) 80.0 N (4) 125 N 18 ____

19 Two identical stationary bar magnets are arranged as shown in the diagram below.

What is the direction of the magnetic field at point P?

19 ____

20 A student claps his hands once to produce a sudden loud sound that travels through the air. This sound is classified as a

(1) longitudinal mechanical wave
(2) longitudinal electromagnetic wave
(3) transverse mechanical wave
(4) transverse electromagnetic wave   20 _____

21 A student generates water waves in a pool of water. In order to increase the energy carried by the waves, the student should generate waves with a

(1) greater amplitude   (3) greater wavelength
(2) higher frequency   (4) longer period   21 _____

22 A wave generator produces straight, parallel wave fronts in a shallow tank of uniform-depth water. As the frequency of vibration of the generator increases, which characteristic of the wave will always decrease?

(1) amplitude   (3) wavelength
(2) phase   (4) speed   22 _____

23 A space probe produces a radio signal pulse. If the pulse reaches Earth 12.3 seconds after it is emitted by the probe, what is the distance from the probe to Earth?

(1) 3.71×10^2 m   (3) 4.10×10^8 m
(2) 4.07×10^3 m   (4) 3.69×10^9 m   23 _____

24 The diagram below represents a light ray reflecting from a plane mirror.

The angle of reflection for this light ray is

(1) 20°   (3) 140°
(2) 70°   (4) 160°   24 _____

25 A light wave travels from one medium into a second medium with a greater absolute index of refraction. Which characteristic of the wave can *not* change as the wave enters the second medium?

(1) frequency
(2) speed
(3) direction
(4) wavelength   25 _____

26 The speed of light ($f = 5.09 \times 10^{14}$ Hz) in glycerol is

(1) 1.70×10^6 m/s
(2) 2.04×10^8 m/s
(3) 3.00×10^8 m/s
(4) 4.41×10^8 m/s   26 _____

27 The diagram below represents a standing wave produced in a string by a vibrating wave generator.

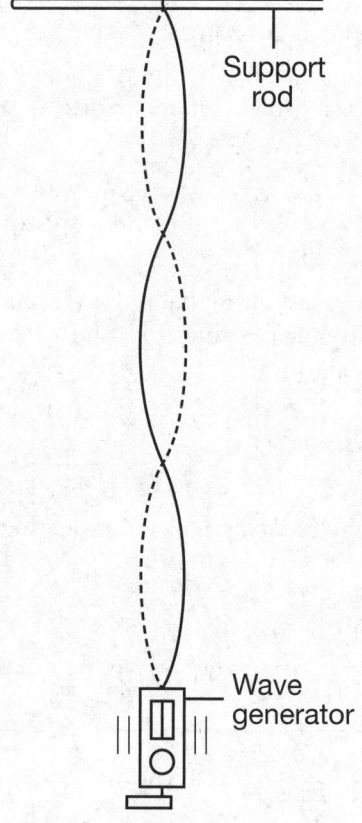

Support
rod

Wave
generator

How many antinodes are shown in this standing wave?

(1) 6
(2) 2
(3) 3
(4) 4   27 _____

28 The Doppler effect is best described as the
  (1) bending of waves as they pass by obstacles or through
       openings
  (2) change in speed of a wave as the wave moves from one
       medium to another
  (3) creation of a standing wave from two waves traveling in
       opposite directions in the same medium
  (4) shift in the observed frequency and wavelength of a wave
       caused by the relative motion between the wave's source and
       an observer   28 _____

29 Which diagram represents diffraction of wave fronts as they
    encounter an obstacle?

(1)   (3)

(2)   (4)   29 _____

30 Which types of forces exist between the two protons in a helium nucleus?

(1) a repulsive electrostatic force and a repulsive gravitational force

(2) a repulsive electrostatic force and an attractive strong nuclear force

(3) an attractive electrostatic force and an attractive gravitational force

(4) an attractive electrostatic force and an attractive strong nuclear force   30 _____

31 A meson could be composed of

(1) a top quark and a bottom quark

(2) an electron and an antielectron

(3) a strange quark and an anticharm quark

(4) an up quark and a muon   31 _____

32 An electron in an excited mercury atom is in energy level g. What is the minimum energy required to ionize this atom?

(1) 0.20 eV   (3) 2.48 eV

(2) 0.91 eV   (4) 7.90 eV   32 _____

33 A student is standing in an elevator that travels from the first floor to the tenth floor of a building. The student exerts the greatest force on the floor of the elevator when the elevator is

(1) accelerating upward as it leaves the first floor

(2) slowing down as it approaches the tenth floor

(3) moving upward at constant speed

(4) at rest on the first floor   33 _____

34 At the bottom of a hill, a car has an initial velocity of +16.0 meters per second. The car is uniformly accelerated at –2.20 meters per second squared for 5.00 seconds as it moves up the hill. How far does the car travel during this 5.00-second interval?

(1) 107 m   (3) 52.5 m

(2) 74.5 m   (4) 25.0 m   34 _____

35 A particle enters the electric field between two oppositely charged parallel plates, as represented in the diagram below.

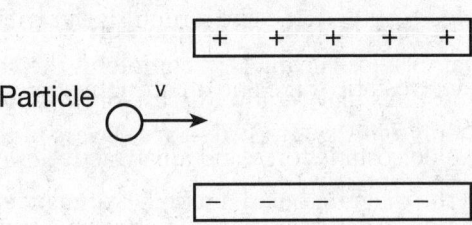

Which particle will be deflected toward the positive plate as it enters the electric field?

(1) photon   (3) electron
(2) proton   (4) neutrino   35 _____

PART B–1
Answer all questions in this part.

Directions (36–50): For *each* statement or question, select the *number* of the word or expression that, of those given, best completes the statement or answers the question. Some questions may require the use of the *2006 Edition Reference Tables for Physical Setting/Physics*. Record your answers in the spaces provided.

36 As represented in the diagram below, an object of mass m, located on the surface of the Moon, is attracted to the Moon with a gravitational force, F.

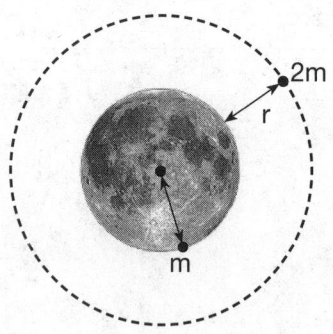

(Not drawn to scale)

An object of mass $2m$, at an altitude equal to the Moon's radius, r, above the surface of the Moon, is attracted to the Moon with a gravitational force of

(1) F (3) $F/2$
(2) $2F$ (4) $F/4$ 36 _____

37 The graph below represents the relationship between velocity and time for an object moving along a straight line.

Velocity vs. Time

What is the magnitude of the object's acceleration?

(1) 5.0 m/s²   (3) 10. m/s²
(2) 8.0 m/s²   (4) 20. m/s²   37 _____

38 Two muons would have a combined charge of

(1) -3.2×10^{-19} C   (3) 0 C
(2) -1.6×10^{-19} C   (4) $+3.2 \times 10^{-19}$ C   38 _____

39 A 1.47-newton baseball is dropped from a height of 10.0 meters and falls through the air to the ground. The kinetic energy of the ball is 12.0 joules the instant before the ball strikes the ground. The maximum amount of mechanical energy converted to internal energy during the fall is

(1) 2.7 J   (3) 14.7 J
(2) 12.0 J   (4) 26.7 J   39 _____

40 A projectile lands at the same height from which it was launched. Which initial velocity will result in the greatest horizontal displacement of the projectile? [Neglect friction.]

(1)

(2)

(3)

(4)

40 _____

41 A 5.0-kilogram box is sliding across a level floor. The box is acted upon by a force of 27 newtons east and a frictional force of 17 newtons west. What is the magnitude of the acceleration of the box?

(1) 0.50 m/s^2

(2) 2.0 m/s^2

(3) 8.8 m/s^2

(4) 10. m/s^2

41 _____

42 The diagram below represents a 2.0-kilogram toy car moving at a constant speed of 3.0 meters per second counterclockwise in a circular path with a radius of 2.0 meters.

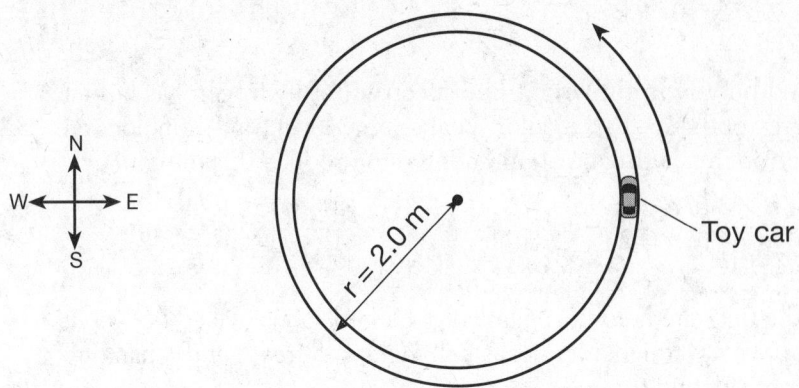

At the instant shown in the diagram, the centripetal force acting on the car is

(1) 4.5 N north   (3) 9.0 N north
(2) 4.5 N west   (4) 9.0 N west   42 _____

43 In which electric circuit would the voltmeter read 10 volts?

(1)   (2)   (3)   (4)

43 _____

44 The lambda baryon has the quark composition *uds*. Which particle has the same electric charge as the lambda baryon?

(1) neutron   (3) proton
(2) electron   (4) antimuon   44 _____

45 How many kilograms of matter would have to be converted into energy to produce 24.0 megajoules of energy?

(1) 2.67×10^{-16} kg   (3) 8.00×10^{-8} kg

(2) 2.67×10^{-10} kg   (4) 8.00×10^{-2} kg   45 _____

46 A red photon in the bright-line spectrum of hydrogen gas has an energy of 3.02×10^{-19} joule. What energy-level transition does an electron in a hydrogen atom undergo to produce this photon?

(1) $n = 3$ to $n = 2$ (3) $n = 5$ to $n = 2$

(2) $n = 4$ to $n = 2$ (4) $n = 6$ to $n = 2$ 46 _____

47 In the diagram below, a negatively charged rod is placed between, but does not touch, identical small metal spheres R and S hanging from insulating threads.

What can be concluded if the rod repels sphere R but attracts sphere S?

(1) Sphere R must be negative and sphere S must be positive.
(2) Sphere R must be negative and sphere S may be positive or neutral.
(3) Sphere R must be positive and sphere S must be negative.
(4) Sphere R must be positive and sphere S may be negative or neutral.

    47 _____

48 The amount of electric energy consumed by a 60.0-watt lightbulb
for 1.00 minute could lift a 10.0 newton object to a maximum vertical height of

(1) 6.00 m   (3) 360. m

(2) 36.7 m   (4) 600. m   48 _____

49 Microwaves can have a wavelength closest to the

(1) radius of Earth   (3) length of a football field

(2) height of Mount Everest   (4) length of a physics student's
                                     thumb
                                                49 _____

50 Two pulses approach each other in a uniform medium, as represented in the diagram below.

Which diagram best represents the superposition of the two pulses
when the pulses overlap?

(1)   (2)   (3)   (4)

                                                            50 _____

PART B–2
Answer all questions in this part.

Directions (51–65): Record your answers on the answer sheet provided after the questions. Some questions may require the use of the *2006 Edition Reference Tables for Physical Setting/Physics*.

Base your answers to questions 51 through 53 on the information below and on your knowledge of physics.

A toy launcher that is used to launch small plastic spheres horizontally contains a spring with a spring constant of 50. newtons per meter. The spring is compressed a distance of 0.10 meter when the launcher is ready to launch a plastic sphere.

51 Determine the elastic potential energy stored in the spring when the launcher is ready to launch a plastic sphere. [1]

52–53 The spring is released and a 0.10-kilogram plastic sphere is fired from the launcher. Calculate the maximum speed with which the plastic sphere will be launched. [Neglect friction.] [Show all work, including the equation and substitution with units.] [2]

54 Two 10.-ohm resistors have an equivalent resistance of 5.0 ohms when connected in an electric circuit with a source of potential difference. Using circuit symbols found in the *Reference Tables for Physical Setting/ Physics*, draw a diagram of this circuit. [1]

55 The graph below shows the relationship between distance, *d*, and time, *t*, for a moving object.

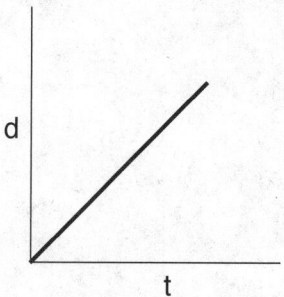

On the axes *on the answer sheet*, sketch the general shape of the graph that shows the relationship between the magnitude of the velocity, v, and time, t, for the moving object. [1]

Base your answers to questions 56 through 58 on the information and diagram below and on your knowledge of physics.

A ray of monochromatic light ($f = 5.09 \times 10^{14}$ Hz) passes from medium X into air. The angle of incidence of the ray in medium X is 25°, as shown.

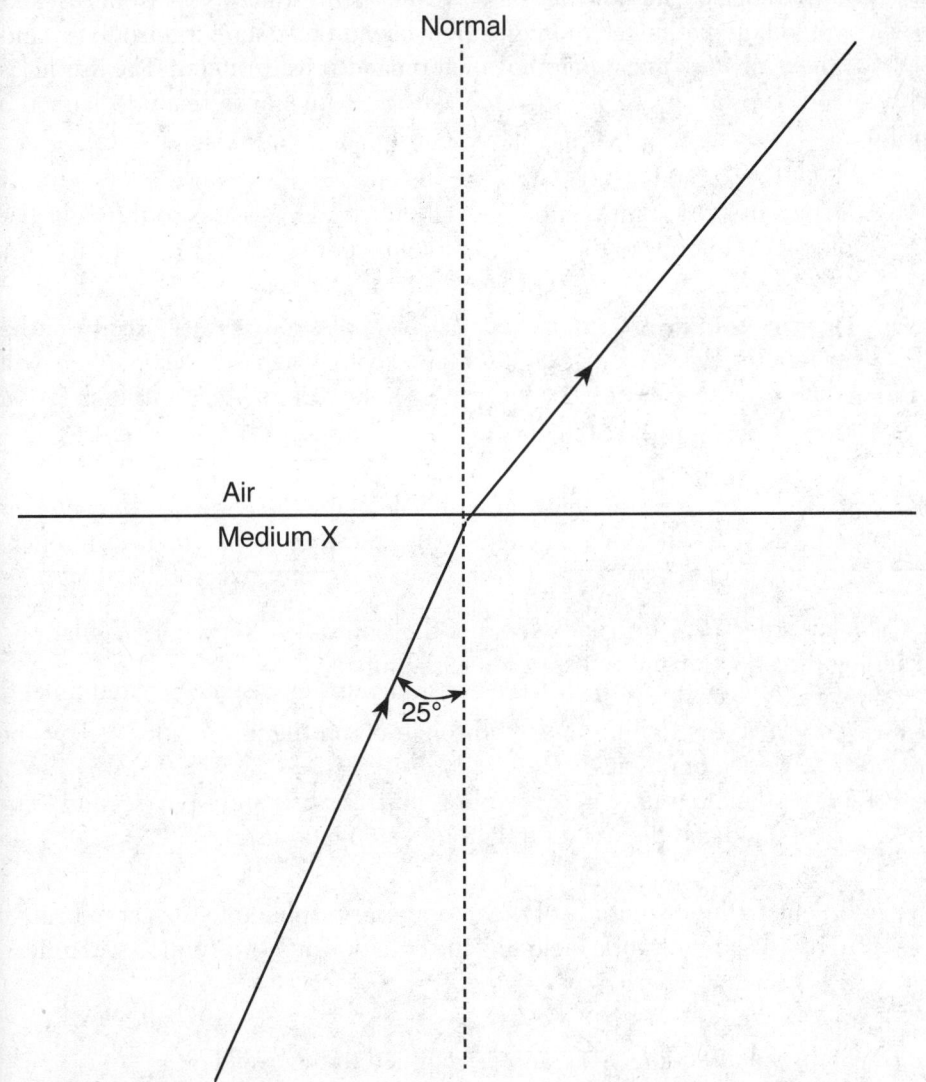

Normal

Air
Medium X

25°

56 Using a protractor, measure and record the angle of refraction in the air to the *nearest degree*. [1]

57–58 Calculate the absolute index of refraction of medium *X*. [Show all work including the equation and substitution with units.] [2]

59–60 A student wishes to record a 7.5-kilogram watermelon colliding with the ground. Calculate how far the watermelon must fall freely from rest so i would be traveling at 29 meters per second the instant it hits the ground [Show all work, including the equation and substitution with units.] [2]

61–62 As represented in the diagram below, block *A* with a mass of 100. gram slides to the right at 4.0 meters per second and hits stationary block *B* with a mass of 150. grams. After the collision, block *B* slides to the right and block *A* rebounds to the left at 1.5 meters per second. [Neglect friction.]

Calculate the speed of block *B* after the collision. [Show all calculations including the equation and substitution with units.] [2]

Base your answers to questions 63 through 65 on the information below and on your knowledge of physics.
A 1.20×10^3-kilogram car is traveling east at 25 meters per second. The brakes are applied and the car is brought to rest in 5.00 seconds.

63–64 Calculate the magnitude of the total impulse applied to the car to bring i to rest. [Show all work, including the equation and substitution with units. [2]

65 State the direction of the impulse applied to the car. [1]

PART C

Answer all questions in this part.

Directions (66–85): Record your answers on the answer sheet provided after the questions. Some questions may require the use of the 2006 *Edition Reference Tables for Physical Setting/Physics*.

Base your answers to questions 66 through 70 on the information and diagram below and on your knowledge of physics.

The diagram shows a negatively charged oil drop that is suspended motionless between two oppositely charged, parallel, horizontal metal plates. The electric field strength between the charged plates is 4.0×10^4 newtons per coulomb. The 1.96×10^{-15}-kilogram oil drop is being acted upon by a gravitational force, F_g, and an electrical force, F_e.

66–67 Calculate the magnitude of the gravitational force, F_g, acting on the oil drop. [Show all work, including the equation and substitution with units.] [2]

68 Determine the magnitude of the upward electrical force, F_e, acting on the oil drop suspended motionless between the charged metal plates. [1]

69–70 Calculate the net electric charge on the oil drop in coulombs. [Show all work, including the equation and substitution with units.] [2]

Base your answers to questions 71 through 75 on the information below and on your knowledge of physics.

In a circuit, a 100.-ohm resistor and a 200.-ohm resistor are connected in parallel to a 10.0-volt battery.

71–72 Calculate the equivalent resistance of the circuit. [Show all work, including the equation and substitution with units.] [2]

73–74 Calculate the current in the 200.-ohm resistor. [Show all work, including the equation and substitution with units.] [2]

 75 Determine the power dissipated by the 100.-ohm resistor. [1]

Base your answers to questions 76 through 80 on the information below and on your knowledge of physics.

A wave traveling through a uniform medium has an amplitude of 0.20 meter, a wavelength of 0.40 meter, and a frequency of 10. hertz.

76–77 On the grid *on the answer sheet*, draw *one* complete cycle of the wave. [2]

78–79 Calculate the speed of the wave. [Show all work, including the equation and substitution with units.] [2]

 80 Determine the period of this wave. [1]

Base your answers to questions 81 through 85 on the information and data table below and on your knowledge of physics.

In an experiment, the potential difference applied across an unmarked resistor was varied while the resistor was held at a constant temperature. The corresponding current through the resistor was measured. The data collected appear in the table below.

Potential Difference (volts)	Current (amperes)
1.5	0.0032
3.0	0.0059
6.0	0.0124
9.0	0.0177
12.0	0.0244

 81 Mark an appropriate scale on the axis labeled "Current (A)." [1]

82 Plot the data points for current versus potential difference. [1]

83 Draw the line or curve of best fit. [1]

84–85 Using your graph, calculate the resistance of the resistor. [Show all work, including the equation and substitution with units.] [2]

Answer Sheet
June 2019

Physics: The Physical Setting

PART B–2

51 _____ J

52–53

54

55

v

t

56 _____ °

57–58

59–60

61–62

63–64

65 _____

PART C

66–67

68 _____ **N**

69–70

71–72

73–74

75 _____ **W**

76–77

78–79

80 _____ s

81–83

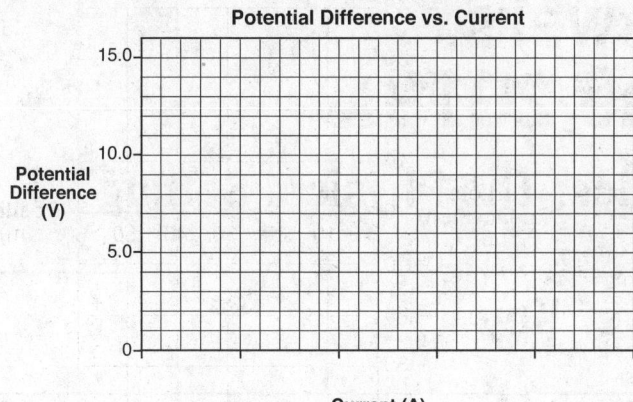

84–85

Answers
June 2019
Physics: The Physical Setting

Answer Key

PART A

1. 4	8. 1	15. 2	22. 3	29. 1
2. 3	9. 4	16. 2	23. 4	30. 2
3. 2	10. 1	17. 2	24. 2	31. 3
4. 4	11. 1	18. 3	25. 1	32. 3
5. 2	12. 3	19. 3	26. 2	33. 1
6. 1	13. 4	20. 1	27. 3	34. 3
7. 1	14. 1	21. 1	28. 4	35. 3

PART B–1

36. 3	39. 1	42. 4	45. 2	48. 3
37. 3	40. 3	43. 4	46. 1	49. 4
38. 1	41. 2	44. 1	47. 2	50. 1

PART B–2 and **PART C**. *See* **Answers Explained**.

Answers Explained

PART A

1 **4** Scalar quantities have magnitude but no direction. Both energy and time are scalar quantities.

WRONG CHOICES EXPLAINED:
(1) Both displacement and velocity are vector quantities.
(2) Displacement is a vector quantity; time is a scalar quantity.
(3) Energy is a scalar quantity; velocity is a vector quantity.

2 **3** When dealing with a word problem requiring a mathematical solution, it is most helpful to translate the words into what quantities/variables are explicitly given and what constants or other variables can be deduced from the situation or words. We also write down the variable that needs to be solved for, or found. Then, it is much easier to determine which formula is required (refer to the appropriate section of the *Reference Tables for Physics*) and then use it to solve the problem. In this case we refer to the *Geometry and Trigonometry* section of *Reference Table K* for the correct equation.

Given: $d_1 = 40.\text{ m }[\text{N}]$

$d_2 = 40.\text{ m }[\text{E}]$

Find: $d_R = ?$

Solution: $c^2 = a^2 + b^2$

$d_R^{\,2} = d_1^{\,2} + d_2^{\,2}$

$d_R = \sqrt{d_1^{\,2} + d_2^{\,2}}$

$d_R = \sqrt{(40.\text{ m})^2 + (40.\text{ m})^2}$

$d_R = 57\text{ m }[\text{NE}]$

3 **2** As the ball travels upward, its velocity decreases; it decelerates at a rate of 9.81 m/s² (acceleration due to gravity) until its velocity is equal to zero at its highest point. Then its velocity increases; it accelerates downward at a rate of 9.81 m/s² (acceleration due to gravity). Acceleration due to gravity is constant throughout the entire path, including at the top or apex of the ball's flight.

WRONG CHOICES EXPLAINED:
(1) This choice is incorrect because acceleration is not zero.
(3) This choice is incorrect because velocity is zero and acceleration is nonzero, which is the exact opposite of what is stated here.
(4) This choice is incorrect because velocity is zero.

4 **4** Electrons carry a negative charge; therefore, electrons transfer from the hair to the comb in order for the comb to acquire a negative charge.

WRONG CHOICES EXPLAINED:
(1), (2) Protons do not transfer between objects.
(3) Electrons carry a negative charge; therefore, if they transferred from the comb to her hair, the hair would acquire a negative charge.

5 **2** The law of conservation of mass states that mass can neither be created nor destroyed, so the mass of the object would remain the same regardless of whether the object is on Earth or on the Moon. However, the acceleration due to gravity is less on the Moon than it is on Earth; so, the weight of the object, which varies depending on gravity ($F_g = mg$), would be less on the Moon than it is on Earth.

WRONG CHOICES EXPLAINED:
(1) Mass would be the same, not less.
(3) Mass would be the same, not less, and weight would be less, not the same, which is the exact opposite of what is stated here.
(4) Weight would be less, not the same.

6 **1** If the object is traveling at constant speed, then the centripetal acceleration would be constant and always pointing toward the center of the circular path. However, depending on the object's position along the circular path, the direction toward the center of the circle would be continually changing.

WRONG CHOICES EXPLAINED:
(2), (3), (4) Magnitude is constant; direction is not constant.

7 1 Refer to the *Mechanics* or the *Geometry and Trigonometry* section of *Reference Table K*.

Given: $F = 45$ N

              $\theta = 65°$

Find: $F_x = ?$

Solution: $A_x = A\cos\theta$ $\cos\theta = \dfrac{b}{c}$

              $F_x = F\cos\theta$ $b = c\cos\theta$

              $F_x = (45\text{ N})\cos 65°$ or $F_x = F\cos\theta$

              $F_x = 19$ N $F_x = (45\text{ N})\cos 65°$

                                            $F_x = 19$ N

8 1 Refer to the *Mechanics* section of *Reference Table K*.

Given: $k = 68$ N/m

              $F_s = 12$ N

Find: $x = ?$

Solution: $F_s = kx$

              $x = \dfrac{F_s}{k}$

              $x = \dfrac{12\text{ N}}{68\text{ N/m}}$

              $x = 0.18$ m

9 4 Refer to the *Mechanics* section of *Reference Table K*.

$$\Delta PE = mg\Delta h$$

Gravitational potential energy is proportional to height; therefore, as the student walks downhill, height decreases.

$$KE = \frac{1}{2}mv^2$$

Kinetic energy is proportional to speed squared. Since the student walks downhill at constant speed, there is no change in the student's kinetic energy. Choice (4) correctly states that gravitational potential energy decreases and kinetic energy remains the same.

10 1 If 150 joules of work is done on a system, then the total energy of the system increases by 150 joules. Choice (1) correctly represents the amount in scientific notation of 150 joules.

11 1 As the ball leaves the person's hand and travels upward, its speed decreases due to deceleration due to gravity until it reaches its highest point where its velocity will be equal to zero. The ball then begins to accelerate downward due to acceleration due to gravity and its speed increases until it reaches its starting point, where the magnitude of the velocity will be equal to the magnitude of the velocity at the start, only in the opposite direction (downward instead of upward). Given that the first ball was thrown downward with an equal velocity, and that resulted in the ball striking the ground below with 200 joules of kinetic energy, the second ball will also strike the ground with 200 joules of kinetic energy.

12 3 Refer to the *Mechanics* section of *Reference Table K*.

Given: $m_1 = m_2 = 1200.\ \text{kg}$

$d_1 = d_2$

$t = 20.\ \text{s}$

$t = 40.\ \text{s}$

Find: $P_2 : P_1$?

Solution: $g = \dfrac{F_g}{m}$

Since $m_1 = m_2$ then $F_{g1} = F_{g2}$

$W = Fd$

Since $d_1 = d_2$ and $F_{g1} = F_{g2}$ then $W_1 = W_2$

$P = \dfrac{W}{t}$

$W = Pt$

$P_2 t_2 = P_1 t_1$

$P_2 = \dfrac{P_1 t_1}{t_2}$

$P_2 = \dfrac{P_1(20\ \text{s})}{(40\ \text{s})}$

$P_2 = \dfrac{1}{2}P_1$

WRONG CHOICE EXPLAINED:
(2) You compared the power of the first crane to the second crane instead of the second crane to the first crane.

13 **4** Refer to *Reference Table* A and the *Electricity* section of *Reference Table K*.

Given: $\qquad q = +2e$

$$e = 1.60 \times 10^{-19} \text{ C}$$

$$q = 3.20 \times 10^{-19} \text{ C}$$

$$V = 2.0 \times 10^{3} \text{ V}$$

$$1 \text{ eV} = 1.60 \times 10^{-19} \text{ J}$$

Find: $\qquad \Delta KE = ?$

Solution:
$$V = \frac{W}{q}$$

$$W = Vq$$

$$W = (2.0 \times 10^{3} \text{ V})(3.20 \times 10^{-19} \text{ C})$$

$$W = 6.4 \times 10^{-16}$$

$$W = \left(6.4 \times 10^{-16} \text{ J}\right)\left(\frac{1 \text{ eV}}{1.6 \times 10^{-19} \text{ J}}\right)$$

$$W = 4.0 \times 10^{3} \text{ eV}$$

WRONG CHOICE EXPLAINED:
(1) This is equivalent to having completed the calculation using a +1 elementary charge instead of +2 elementary charges.

14 **1** Refer to the *Electricity* section of *Reference Table K.*

Given: $q = 100$ C

                $t = 500$ s

Find: $I = ?$

Solution:

$$I = \frac{\Delta q}{t}$$

$$I = \frac{100 \text{ C}}{500 \text{ s}}$$

$$I = 0.200 \text{ A}$$

WRONG CHOICES EXPLAINED:
(2) You divided time by charge instead of charge by time.
(3) You multiplied the charge by time instead of dividing.

15 **2** Refer to the *Electricity* section of *Reference Table K.*

Given: $L_1 = 1.0$ m

              $R_1 = 9.0 \times 10^{-3}$ Ω

              $L_2 = \frac{1}{2}L_1$

Find: $R_2 = ?$

Solution:

$$R = \frac{\rho L}{A}$$

Resistance is proportional to length. Halving the length would result in the resistance being halved as well, since both the resistivity and cross-sectional area have not changed.

WRONG CHOICE EXPLAINED:
(4) You doubled the resistance instead of halving it.

16 **2** A battery is a combination of two or more electric cells and is the source of potential difference in a circuit.

WRONG CHOICES EXPLAINED:
(1) A voltmeter is a device used to measure potential difference.
(3) An ammeter is a device used to measure electric current.
(4) A resistor is a device that supplies resistance to a circuit.

17 **2** Refer to the *Electricity* section of *Reference Table K.*

Given: $R = 2.9\ \Omega$

                $V = 1.5\ \text{V}$

Find: $P = ?$

Solution: $P = \dfrac{V^2}{R}$

            $P = \dfrac{(1.5\ \text{V})^2}{2.9\ \Omega}$

            $P = 0.78\ \text{W}$

WRONG CHOICE EXPLAINED:
(1) You forgot to square the voltage.

18 **3** Newton's Third Law states that if object A exerts a force on object B, then object B exerts an equal in magnitude but opposite in direction force on object A. Since the child exerts a 100. N force on the object, the object in return exerts a 100. N force on the child.

19 **3** Magnetic field lines can be used to visualize magnetic fields. By convention, field lines are drawn pointing away from north magnetic poles and toward south poles. At point *P* we could draw a field line horizontally from *P*, left toward the south pole of the magnet at the top of the diagram, and a vertical field line from *P* downward toward the south pole of the magnet below *P*. The net resultant magnetic force would be diagonally left and downward.

WRONG CHOICE EXPLAINED:
(1) You drew or assumed field lines away from south and toward north.

20 **1** Sound waves travel as longitudinal mechanical waves.

WRONG CHOICES EXPLAINED:
(3) Waves on a string is an example of transverse mechanical waves.
(4) Light travels as transverse electromagnetic waves.

21 **1** Amplitude is related to the energy carried by the wave. The greater the energy, the greater the amplitude of the wave.

WRONG CHOICES EXPLAINED:
(2) Frequency is the number of waves per unit time.
(3) Wavelength is the length of one complete wave cycle.
(4) Period is the time for one complete repetition of a cycle.

22 **3** Refer to the *Waves* section of *Reference Table K*.

$$v = f\lambda$$

The speed of a wave in a given medium is constant; therefore, frequency and wavelength are inversely proportional to each other. If frequency increases, wavelength will decrease and vice versa.

WRONG CHOICES EXPLAINED:
(1) Amplitude is related to the energy of the wave. Increasing the frequency that the wave is generated will not affect the amplitude.
(2) Points on a periodic wave that are at equal displacements from their rest positions and that are experiencing identical movements are said to be in phase.
(4) The speed of a mechanical wave is determined by the medium it is traveling in. The medium remains the same; therefore, the speed of the wave remains the same.

23 **4** Refer to the *Mechanics* section of *Reference Table K* and *Reference Table A*.

Given: $t = 12.3$ s

$\bar{v} = c = 3.00 \times 10^8$ m/s

Find: $d = ?$

Solution:

$$\bar{v} = \frac{d}{t}$$

$$d = \bar{v}t$$

$$d = \left(3.0 \times 10^8 \text{ m/s}\right)(12.3 \text{ s})$$

$$d = 3.69 \times 10^9 \text{ m}$$

WRONG CHOICE EXPLAINED:
(2) You calculated your answer using the speed of sound in air instead of the speed of light in a vacuum.

24 **2** Angles of incidence and reflection are calculated in reference to the normal. In this case, the angle that the reflected ray makes with the normal is 70°.

WRONG CHOICES EXPLAINED:
(1) This is the angle that the ray makes with the plane mirror.
(3) You added the angles of incidence and refraction to get the measure of the angle between the rays.
(4) You added all the angles except the angle that the reflected ray makes with the surface.

25 **1** The frequency of the light wave is determined by the source that produces it. As the light travels from one medium into another, the frequency does not change.

WRONG CHOICES EXPLAINED:
(2) As light travels from one medium to another with different absolute indices of refraction, its speed changes as it enters the new medium.
(3) Refraction, a change in direction of a wave when it passes obliquely from one medium into another, will occur due to the change in speed resulting from moving into a medium with a different absolute index of refraction.

(4) The speed of a wave is proportional to its frequency and wavelength. If frequency is constant, then, as the speed of a wave changes as it enters a new medium with a different absolute index of refraction, the wavelength will also proportionately change.

26 **2** Refer to the *Waves* section of *Reference Table K*, *Reference Table A*, and *Reference Table E*.

Given: $f = 5.09 \times 10^{10}$ Hz

$c = 3.00 \times 10^8$ m/s

$n = 1.47$

Find: $v = ?$

Solution: $n = \dfrac{c}{v}$

$v = \dfrac{c}{n}$

$v = \dfrac{3.0 \times 10^8 \text{ m/s}}{1.47}$

$v = 2.04 \times 10^8$ m/s

WRONG CHOICES EXPLAINED:

(3) You just used the speed of light in a vacuum without adjusting for change in speed in the new medium.

(4) You multiplied the speed of light by the absolute index of refraction instead of dividing.

27 **3** An antinode is the point or locus of points on an interference pattern that results in maximum constructive interference. In this diagram of a standing wave, there are 3 antinodes.

WRONG CHOICE EXPLAINED:

(4) These are the number of nodes, or points of maximum destructive interference.

28 **4** The Doppler effect is the apparent change in frequency that results when a wave source and an observer are in relative motion with respect to each other. As the distance between the source and the observer decreases, the frequency of the source, as perceived by the observer, is increased; as the distance increases, the apparent frequency is decreased.

WRONG CHOICE EXPLAINED:
(1) Diffraction is the bending of a wave around a barrier.

29 **1** Diffraction is the bending of a wave around a barrier.

WRONG CHOICES EXPLAINED:
(2) This diagram does not include the bending of the wave around the obstacle.
(3) This diagram depicts reflection of a wave on a surface.
(4) This diagram is similar to the reflection diagram; however, the reflected waves have changed wavelength. This would be what occurs if the wave were refracted as it traveled from one medium into another, but not from simple reflection.

30 **2** Protons carry a positive charge. Like charges repel; therefore, the protons experience a repulsive electrostatic force. The nuclear force is the attractive, short-range force responsible for binding protons and neutrons in the nucleus; therefore, the protons experience an attractive strong nuclear force. Choice (2) correctly indicates these two forces.

31 **3** Refer to *Reference Table G* and *Reference Table H*. A meson is composed of a quark and an antiquark. Choice (3) correctly lists a quark particle and an antiquark particle.

WRONG CHOICES EXPLAINED:
(1) These two particles are both quarks.
(2) These two particles are both leptons.
(4) One particle is a quark and the other is a lepton.

32 **3** Refer to *Reference Table F*. Energy level g is at -2.48 eV. In order to ionize the atom, 2.48 eV of energy needs to be provided.

WRONG CHOICES EXPLAINED:
(1) This is the difference between levels f and g.
(2) This is the difference between levels g and h.
(4) This is the difference between the ground state and g.

33 **1** Newton's Third Law states that if object A exerts a force on object B, then object B exerts an equal in magnitude but opposite in direction force on object A. Normally, the student's weight or force due to gravity is balanced by the upward normal force. In order for the elevator to accelerate upward, there has to be a

net upward force applied to the elevator and, thus, to the student inside the elevator. Therefore, at the time of acceleration, the student applies an equal downward force to the floor of the elevator.

WRONG CHOICES EXPLAINED:

(2) When the elevator decelerates, there would be a net downward force being applied to the elevator and, thus, to the student inside. Therefore, the net downward force that the student exerts on the elevator would be less than what it would be if the elevator were at rest.

(3), (4) Moving at constant speed would be the same as being at rest on the ground floor (no net force), so the student would only be exerting a downward force due to gravity or his or her weight.

34 **3** Refer to the *Mechanics* section of *Reference Table K*.

Given: $v_i = 16.0$ m/s

$a = -2.20$ m/s^2

$t = 5.00$ s

Find: $d = ?$

Solution: $d = v_i t + \dfrac{1}{2} at^2$

$d = (16.0 \text{ m/s})(5.00 \text{ s}) + \dfrac{1}{2}(-2.20 \text{ m/s}^2)(5.00 \text{ s})^2$

$d = (80.0 \text{ m}) + (-27.5 \text{ m})$

$d = 52.5$ m

The negative acceleration or deceleration indicates that it is opposite to the direction of motion.

WRONG CHOICES EXPLAINED:

(1) You used positive 2.20 for the acceleration instead of −2.20, or you added the second half of the equation instead of subtracting it properly.

(2) You forgot to square the time in the second half of the equation.

(4) You forgot to multiply the second half of the equation by $\dfrac{1}{2}$.

35 **3** Electrons carry a negative charge. Like charges attract and opposites repel, so the electron would be deflected toward the positive plate and away from the negative one.

WRONG CHOICES EXPLAINED:
(1), (4) Photons and neutrinos do not carry charge and would not be deflected toward either plate.
(2) Protons carry a positive charge and a proton would be deflected toward the negative plate and away from the positive one.

PART B–1

36 **3** Refer to the *Mechanics* section of *Reference Table K*.

Given:
$$m_1 = m$$
$$r_1 = r$$
$$\text{mass of moon} = m_{moon}$$
$$F_{g1} = F$$
$$m_2 = 2m$$
$$r_2 = 2r$$

Find:
$$F_{g2} = ?$$

Solution:
$$F_{g1} = \frac{Gmm_{moon}}{r^2} = F$$
$$F_{g2} = \frac{G(2m)m_{moon}}{(2r)^2}$$
$$F_{g2} = \frac{2Gmm_{moon}}{4r^2}$$
$$F_{g2} = \frac{2}{4}\left(\frac{Gmm_{moon}}{r^2}\right)$$
$$F_{g2} = \frac{1}{2}F_{g1} \text{ or } \frac{1}{2}F$$

WRONG CHOICES EXPLAINED:
(1) You forgot to square the 2 in $2r$ in the denominator.
(2) You used r instead of $2r$ in the denominator.
(4) You forgot to use $2m$ instead of m in the numerator.

37 **3** The slope of the velocity versus time graph is given by the relationship $\Delta v/\Delta t$, the rate at which velocity has changed over the time interval—in other

words, the acceleration. We can pick any two points on the line in the graph and substitute them into the formula to calculate the acceleration.

Given: $v_1 = 0.0$ m/s

              $v_2 = 10.$ m/s

              $t_1 = 1.0$ s

              $t_1 = 2.0$ s

Find: $a = ?$

Solution: $a = \dfrac{\Delta v}{\Delta t}$

              $a = \dfrac{\left(v_2 - v_1\right)}{\left(t_2 - t_1\right)}$

              $a = \dfrac{(10.0 \text{ m/s} - 0 \text{ m/s})}{(2.0 \text{ s} - 1.0 \text{ s})}$

              $a = \dfrac{10.0 \text{ m/s}}{1.0 \text{ s}}$

              $a = 10.0$ m/s^2

38 **1** Refer to *Reference Table H* and *Reference Table A*. A muon carries a -1 elementary charge. An elementary charge is equivalent to 1.6×10^{-19} C; therefore, 2 muons would carry a charge of $-2 \times 1.6 \times 10^{-19}$ C or -3.2×10^{-19} C.

WRONG CHOICES EXPLAINED:
(2) You forgot to multiply by 2.
(3) You most likely used the charge for a muon neutrino and not a muon.
(4) You multiplied by +2 instead of −2.

---

39 **1** Refer to the *Mechanics* section of *Reference Table K*.

Given:
$$F_g = 1.47 \text{ N}$$
$$\Delta h = 10.0 \text{ m}$$
$$KE_f = 12 \text{ J}$$

Find:
$$\Delta Q = ?$$

Solution:
$$E_{T_f} = E_{T_i}$$
$$E_T = PE + KE + Q$$
$$PE_f + KE_f + Q_f = PE_i + KE_i + Q_i$$

The initial kinetic energy is zero since the ball is not moving at the start.
The final potential energy is zero since the ball is about to hit the ground (zero height).

$$PE_f + Q_f = KE_i + Q_i$$
$$Q_f - Q_i = PE_i - KE_f$$
$$\Delta Q = mg\Delta h - KE_f$$
$$g = \frac{F_g}{m}$$
$$F_g = mg$$
$$\Delta Q = F_g\Delta h - KE_f$$
$$\Delta Q = \left((1.47 \text{ N})(10.0 \text{ m})\right) - 12 \text{ J}$$
$$\Delta Q = 14.7 \text{ J} - 12 \text{ J}$$
$$\Delta Q = 2.7 \text{ J}$$

40 **3** Horizontal displacement of the projectile depends on the horizontal component of the velocity and the amount of time the projectile travels.
Refer to the *Mechanics* section of *Reference Table K* for the equation to calculate the horizontal component of the velocity and the equation for calculating the horizontal distance traveled.

$$A_x = A\cos\theta$$
$$v_x = v_i\cos\theta$$
$$\overline{v} = \frac{d}{t}$$
$$d = \overline{v}t$$
$$d_x = \overline{v}_x t$$

The amount of time that the projectile travels in the horizontal direction is equal to the time it takes for the projectile to reach its highest point in the air, vertically, and come back down to the ground; in this case, it is the starting point, also known as the hang time. This time depends on the starting vertical component of the velocity and acceleration due to gravity.

Refer to the *Mechanics* section of *Reference Table K* for the equation to calculate the vertical component of the velocity and the hang time.

$$A_y = A\sin\theta$$
$$v_{iy} = v_i\sin\theta$$
$$v_f = v_i + at$$
$$v_{fy} = v_{iy} + a_y t$$

Once the total hang time is calculated, it can be substituted into the distance equation for the horizontal direction.

A smaller angle will result in a greater x component and smaller y component; i.e., it will be able to travel faster in the x direction but for less time. A larger angle will result in a smaller x component and a larger y component of velocity; i.e., it will be able to travel for longer in the x direction but at a slower speed. A 45 degree angle mathematically turns out to be the optimal angle for the greatest range—i.e., travel in the x direction for a given initial velocity.

Choice (1) and choice (2) both have a starting velocity of 10. m/s, but choice (1), which is launched at a 45 degree angle, will yield a longer range, so we can eliminate choice (2).

Choice (1) and choice (3) both have a launch angle of 45 degrees, but choice (3) will yield a longer range because it has a larger initial velocity, so we can eliminate choice (1).

Choice (3) and choice (4) both have the same initial velocity, but, since choice (3) is launched at a 45 degree angle and choice (4) is launched at a 60 degree angle, choice (3) will yield a larger range, so we can eliminate choice (4). Therefore, choice (3) will therefore result in the largest horizontal displacement.

41 2 Refer to the *Mechanics* section of *Reference Table K.*

Given: $m = 5.0$ kg

$F = 27$ N east

$F_f = 17$ N west

Find: $a = ?$

Solution: $F_{net} = F_{applied} - F_f$

$F_{net} = 27$ N $- 17$ N

$F_{net} = 10$ N

$$a = \frac{F_{net}}{m}$$

$$a = \frac{10 \text{ N}}{5.0 \text{ kg}}$$

$a = 2.0$ N/kg $= 2.0$ m/s^2

WRONG CHOICES EXPLAINED:

(1) You divided the mass by the net force instead of the other way around.

(3) You added the applied force and the frictional force instead of subtracting.

(4) You calculated the net force and didn't continue to solve for acceleration using mass.

42 4 Refer to the *Mechanics* section of *Reference Table K.*

Given: $m = 2.0$ kg

$v = 3.0$ m/s

$r = 2.0$ m

Find: $F_c = ?$

Solution: $F_c = ma_c$

$$a_c = \frac{v^2}{r}$$

$$F_c = \frac{mv^2}{r}$$

$$F_c = \frac{(2.0 \text{ kg})(3.0 \text{ m/s})^2}{(2.0 \text{ m})}$$

$F_c = 9.0$ N

Centripetal force and acceleration are always directed toward the center of the circular path, which in this diagram is west.

WRONG CHOICES EXPLAINED:
(1) You solved for centripetal acceleration instead of centripetal force and chose the direction of the instantaneous velocity instead of the centripetal force.
(2) You chose the correct direction for centripetal force but solved for centripetal acceleration instead of centripetal force.
(3) You calculated the correct magnitude but chose the direction of the instantaneous velocity instead of the centripetal force.

43 **4** Refer to the *Electricity* section of *Reference Table K*.

series circuit
$$V_T = V_1 + V_2$$
parallel circuit
$$V_T = V_1 = V_2$$

Choice (4) shows a series circuit with a 10 V battery and a voltmeter measuring the total potential difference across both resistors which, in a series circuit, is equal to the potential difference supplied by the power source, in this case 10 volts.

WRONG CHOICES EXPLAINED:
(1) This circuit has the voltmeter connected to one of the resistors, which is connected in parallel to the second resistor. In a parallel circuit, the potential difference in each resistor is equal to each other and to that of the power source, in this case 20 volts.
(2) This is a series circuit connected similarly to the correct choice (4). However, the power source is supplying 20 volts; therefore, the voltmeter connected across both resistors will also read 20 volts.
(3) This is a series circuit with the voltmeter connected to only one of the resistors. In a series circuit, the sum of the potential differences across each resistor will equal the total provided by the battery, which in this diagram is 10 V. Given that there are two equal resistors, each voltmeter will read 5.0 volts.

44 **1** Refer to *Reference Table H*.

Given: lambda baryon—made up of *uds* quarks

Find: particle with same electric charge as a lambda baryon

Solution:

$$\text{up quark} = +\frac{2}{3}e$$

$$\text{down quark} = -\frac{1}{3}e$$

$$\text{strange quark} = -\frac{1}{3}e$$

$$\frac{2}{3} + \left(-\frac{1}{3}\right) + \left(-\frac{1}{3}\right) = 0$$

The correct choice is (1); neutrons also carry zero electric charge.

WRONG CHOICES EXPLAINED:
(2) Electrons carry −1e charge.
(3) Protons carry +1e charge.
(4) Antimuons carry +1e charge.

45 **2** Refer to the *Modern Physics* section of *Reference Table K*, *Reference Table A*, and *Reference Table B*.

Given: $E = 24.0 \text{ MJ} = 24.0 \times 10^6 \text{ J} = 2.40 \times 10^7 \text{ J}$

Find: $m = ?$

Solution: $E = mc^2$

$$m = \frac{E}{c^2}$$

$$m = \frac{2.40 \times 10^7 \text{ J}}{\left(3.00 \times 10^8 \text{ m/s}\right)^2}$$

$$m = 2.67 \times 10^{-10} \text{ kg}$$

WRONG CHOICES EXPLAINED:
(1) You didn't convert the 24.0 megajoules into joules before substituting into the equation.
(3) You didn't convert the 24.0 megajoules into joules before substituting into the equation, and you forgot to square the speed of light in the denominator.
(4) You forgot to square the speed of light in the denominator.

46 **1** Refer to the *Modern Physics* section of *Reference Table K*, *Reference Table F*, and *Reference Table A*.

Given: $E_{photon} = 3.02 \times 10^{-19}$ J

Find: equivalent energy level transition—
$n = ?$ to $n = ?$

Solution:
$$E_{photon} = 3.02 \times 10^{-19} J \left(\frac{1 \text{ eV}}{1.60 \times 10^{-19} J} \right)$$

$$E_{photon} = 1.89 \text{ eV}$$

$$n_3 - n_2 = -1.51 \text{ eV} - (-3.41 \text{ eV}) = 1.89 \text{ eV}$$

WRONG CHOICES EXPLAINED:
(2) This produces a photon with 2.55 eV of energy.
(3) This produces a photon with 2.86 eV of energy.
(4) This produces a photon with 3.02 eV of energy.

47 **2** A negatively charged rod brought near a neutral object causes a redistribution of charge within the object. It is still neutral but some of the charges have been separated; the negative electrons farther away from the rod on the far side of the object leave the side closest to the rod more positive. Given that like charges repel and opposite charges attract, if both objects were neutral, they would both be attracted to the rod. If an object were negative, it would be repelled by the rod. If an object were positive, it would be attracted to the rod. Since object R is repelled by the rod, we must conclude that it carries a negative charge. Since object S is attracted, it can either be positive or neutral.

48 **3** Refer to the *Electricity* and *Mechanics* sections of *Reference Table K*.

Given: $P = 60.0$ W

$t = 1.0$ min $= 60.0$ s

$F_g = 10.0$ N

Find: $d = ?$

Solution: $W = Pt$

$W = (60.0 \text{ W})(60.0 \text{ s})$

$W = 3600.$ J

$W = Fd$

$d = \dfrac{W}{F}$

$d = \dfrac{3600. \text{ J}}{10.0 \text{ N}}$

$d = 360.$ m

WRONG CHOICES EXPLAINED:

(1) You forgot to convert 1 minute into 60 seconds before substituting into the equation.

(4) You just multiplied 60 watts by 10 newtons.

49 **4** Refer to the *Electromagnetic Spectrum* section of *Reference Table D*.
Microwaves have wavelengths between 10^{-4} and 10^0 meters in length. This is an order of magnitude question. Ten to the -4 power is equivalent to 0.0001 meter or 0.004 inches. Ten to the zero power or one meter is approximately equal to 3.3 feet or 40 in. Answer choice (4), a student's thumb, falls within this range.

WRONG CHOICES EXPLAINED:

(1) The radius of Earth is on a 10^6 order of magnitude.

(2) Mount Everest is on a 10^4 order of magnitude at almost 10,000 meters high.

(3) A football field is on a 10^2 order of magnitude at almost 100 meters wide.

50 **1** The effect produced when two or more wave pulses pass through the same point simultaneously is called interference. We add the amplitudes of each pulse together to get the combined amplitude pattern. Choice (1) correctly represents what the pulse would look like at the instant that they are superimposed at the same location in the medium.

WRONG CHOICES EXPLAINED:
(2) This is the mirror image of the correct answer choice.
(3) You flipped the bottom pulse to the top and combined the two into a rectangular shape.
(4) You flipped the top pulse to the bottom and combined the two into a rectangular shape.

PART B–2

51   Refer to the *Mechanics* section of *Reference Table K.*

Given: $k = 50.$ N/m

                    $x = 0.10$ m

Find: $PE_s = ?$

Solution:

$$PE_s = \frac{1}{2}kx^2$$

$$PE_s = \frac{1}{2}(50.\ \text{N/m})(0.10\ \text{m})^2$$

$$PE_s = 0.25\ \text{J}$$

Note: One credit is awarded for an answer of 0.25 J.

52–53 Refer to the *Mechanics* section of *Reference Table K.*

$$\text{Given:} \qquad m = 0.10 \text{ kg}$$

$$\text{Find:} \qquad v_{max} = ?$$

$$\text{Solution:} \qquad \Delta PE_s = \Delta KE$$

$$\Delta PE_s = \frac{1}{2}mv^2$$

$$v^2 = \frac{2(\Delta PE_s)}{m}$$

$$v = \sqrt{\frac{2(\Delta PE_s)}{m}}$$

$$v = \sqrt{\frac{2(0.25 \text{ J})}{0.10 \text{ kg}}}$$

$$v = 2.2 \text{ m/s}$$

52 One credit is awarded for a correct equation and substitution of values with units *or* for an answer with units that is consistent with your answer to question 51.

53 One credit is awarded for the correct answer with units or for an answer, with units, that is consistent with your answer to question 52.
Note: Students will not be penalized more than 1 credit for errors in units in questions 52 and 53.

54

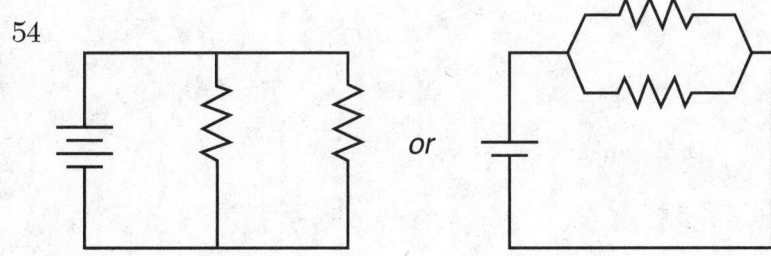

or

One credit is awarded for a circuit diagram showing two resistors connected in parallel with a cell or a battery.
Note: Credit is allowed for lines not touching the battery if the space between the lines and the battery is less than or equal to the distance between the battery symbol lines.

55 You are given a distance versus time graph for a moving object. The slope on a distance versus time graph is equal to the velocity of the object. This graph has a constant slope; hence, the velocity of this object is constant over time. You are asked to draw a graph for velocity versus time for this object. A horizontal line would represent a constant velocity over time.

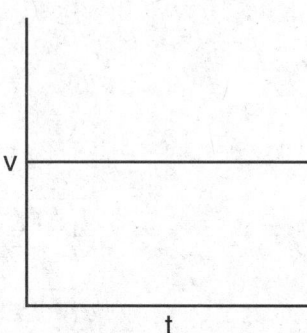

One credit is awarded for a line that approximates a horizontal line representing constant velocity.

56 You are asked to measure the angle of refraction with a protractor. Angles of refraction are measured with respect to the normal.
One credit is awarded for an answer of $40° \pm 2°$.

57–58 Refer to the *Waves* section of *Reference Table K*.

Given: $\theta_{air} = 40.0°$

$\theta_X = 25.0°$

$n_{air} = 1.00$

Find: $n_x = ?$

Solution: $n_{air} \sin\theta_{air} = n_x \sin\theta_x$

$$n_x = \frac{n_{air} \sin\theta_{air}}{\sin\theta_x}$$

$$n_x = \frac{(1.00)\sin 40°}{\sin 25°}$$

$$n_x = 1.5$$

57 One credit is awarded for a correct equation and substitution of values with units *or* for an answer with units that is consistent with your answer to question 56.

58 One credit is awarded for the correct answer *or* for an answer, without units, that is consistent with your answer to question 57.
Note: Students will not be penalized more than 1 credit for errors in units in questions 57 and 58.

59–60 Refer to the *Mechanics* section of *Reference Table K* and *Reference Table A*.

Given: $m = 7.5$ kg

$v_i = 0.0$ m/s

$v_f = 29$ m/s

$a = 9.81$ m/s^2

Find: $d = ?$ or $\Delta h = ?$

Solution: $\Delta PE = \Delta KE$

$$mg\Delta h = \frac{1}{2}mv^2$$

$$v_f^2 = v_i^2 + 2ad$$

$$\Delta h = \frac{1}{2}\frac{mv^2}{mg}$$

$$v_f^2 - v_i^2 = 2ad$$

$$d = \frac{v_f^2 - v_i^2}{2a}$$

$$\Delta h = \frac{1}{2}\frac{v^2}{g}$$

$$d = \frac{(29 \text{ m/s})^2 - (0 \text{ m/s})^2}{2(9.81 \text{ m/s}^2)}$$

$$\Delta h = \frac{(29 \text{ m/s})^2}{2(9.81 \text{ m/s}^2)}$$

$$d = 43 \text{ m}$$

$$\Delta h = 43 \text{ m}$$

59 One credit is awarded for a correct equation and substitution of values with units.

60 One credit is awarded for the correct answer with units *or* for an answer, with units, that is consistent with your answer to question 59.
Note: Students will not be penalized more than 1 credit for errors in units in questions 59 and 60.

61–62 Refer to the *Mechanics* section of *Reference Table K.*

Given: $m_A = 100.\ \mathrm{g}$

$v_{Ai} = 4.0\ \mathrm{m/s}$

$m_B = 150\ \mathrm{g}$

$v_{Bi} = 0\ \mathrm{m/s}$

$v_{Af} = -1.5\ \mathrm{m/s}$

Find: $v_{Bf} = ?$

Solution: $p_{before} = p_{after}$

$p = mv$

$m_1 v_{1i} + m_2 v_{2i} = m_1 v_{1f} + m_2 v_{2f}$

$m_1 v_{1i} = m_1 v_{1f} + m_2 v_{2f}$

$$v_{2f} = \frac{m_1 v_{1i} - m_1 v_{1f}}{m_2}$$

$$v_{2f} = \frac{(100.\ \mathrm{g})(4.0\ \mathrm{m/s}) - (100.\ \mathrm{g})(-1.5\ \mathrm{m/s})}{150\ \mathrm{g}}$$

$$v_{2f} = \frac{400\ \mathrm{gm/s} - (-150.\ \mathrm{gm/s})}{150\ \mathrm{g}}$$

$$v_{2f} = 3.7\ \mathrm{m/s}$$

61 One credit is awarded for a correct equation and substitution of values with units.

62 One credit is awarded for the correct answer with units or for an answer, with units, that is consistent with your answer to question 61.

Note: Students will not be penalized more than 1 credit for errors in units in questions 61 and 62.

63–65 Refer to the *Mechanics* section of *Reference Table K*.

Given:

$$m = 1.20 \times 10^3 \text{ kg}$$
$$v_i = 25 \text{ m/s}$$
$$t = 5.00 \text{ s}$$
$$v_f = 0 \text{ m/s}$$

Find:

$$J = ?$$

Solution:

$$p = mv$$
$$J = \Delta p$$
$$J = m\Delta v$$
$$J = m(v_f - v_i)$$
$$J = 1.20 \times 10^3 \text{ kg}(0 \text{ m/s} - 25 \text{ m/s})$$
$$J = -3.0 \times 10^4 \text{ kg} \cdot \text{m/s or } 3.0 \times 10^4 \text{ N} \cdot \text{s}$$

63 One credit is awarded for a correct equation and substitution of values with units.

64 One credit is awarded for the correct answer or for an answer, with units, that is consistent with your answer to question 63.
Note: Students will not be penalized more than 1 credit for errors in units in questions 63 and 64. Credit is allowed for a correct answer with units that is positive or negative.

65 The car is traveling east. In order to stop the car's motion, the impulse, or force, over time would need to be applied opposite to the car's direction of motion, or in this case west.
Acceptable responses for credit include but are not limited to
   – West
   – Opposite to the direction of the car's motion
   – Backward

PART C

66–67 Refer to the *Mechanics* section of *Reference Table K* and *Reference Table A*.

Given: $m = 1.56 \times 10^{-15}$ kg

                  $g = 9.81$ m/s^2

Find: $F_g = ?$

Solution: $g = \dfrac{F_g}{m}$

                  $F_g = mg$

                  $F_g = \left(1.96 \times 10^{-15} \text{ kg}\right)\left(9.81 \text{ m/s}^2\right)$

                  $F_g = 1.92 \times 10^{-14}$ N

66 One credit is awarded for a correct equation and substitution of values with units.

67 One credit is awarded for the correct answer with units or for an answer, with units, that is consistent with your answer to question 66.

Note: Students will not be penalized more than 1 credit for errors in units in questions 66 and 67.

68 Given that the oil drop is motionless, the forces acting on the oil drop must be in equilibrium. We calculated the force due to gravity in question 67; therefore, the upward electrical force must be equal to 1.92×10^{-14} N.

One credit is awarded for an answer of 1.92×10^{-14} N *or* for an answer that is consistent with your answer to question 67.

69–70 Refer to the *Electricity* section of *Reference Table K.*

Given: $E = 4.0 \times 10^4$ N/C

                   $F_e = 1.92 \times 10^{-14}$ N

Find: $q = ?$

Solution: $E = \dfrac{F_e}{q}$

$$q = \dfrac{F_e}{E}$$

$$q = \dfrac{1.92 \times 10^{-14} \text{ N}}{4.0 \times 10^4 \text{ N/C}}$$

$$q = 4.8 \times 10^{-19} \text{ C}$$

69 One credit is awarded for a correct equation and substitution of values with units or for an answer, with units, that is consistent with your responses to questions 67 and 68.

70 One credit is awarded for the correct answer with units *or* for an answer, with units, that is consistent with your answer to question 69.

Note: Students will not be penalized more than 1 credit for errors in units in questions 69 and 70.

71–72 Refer to the *Electricity* section of *Reference Table K*.

Given: $R_1 = 100.\ \Omega$

$R_2 = 200.\ \Omega$

$V_T = 10.0\ \text{V}$

Find: $R_{T_{parallel}} = ?$

Solution:

$$\frac{1}{R_{T_{parallel}}} = \frac{1}{R_1} + \frac{1}{R_2}$$

$$\frac{1}{R_{T_{parallel}}} = \frac{1}{100.\ \Omega} + \frac{1}{200.\ \Omega}$$

$$\frac{1}{R_{T_{parallel}}} = \frac{3}{200.\ \Omega}$$

$$R_{T_{parallel}} = \frac{200.\ \Omega}{3}$$

$$R_{T_{parallel}} = 66.7\ \Omega$$

71 One credit is awarded for a correct equation and substitution of values with units.

72 One credit is awarded for the correct answer with units or for an answer with units, that is consistent with your answer to question 72.
Note: Students will not be penalized more than 1 credit for errors in units in questions 71 and 72.

73–74 Refer to the *Electricity* section of *Reference Table K*.

Given: $R_2 = 200.\ \Omega$

$V_T = 10.0\ \text{V} = V_2$

Find: $I = ?$

Solution:

$$R = \frac{V}{I}$$

$$I = \frac{V}{R}$$

$$I = \frac{10.0\ \text{V}}{200.\ \Omega}$$

$$I = 0.0500\ \text{A or } 5.00 \times 10^{-2}\ \text{A}$$

73 One credit is awarded for a correct equation and substitution of values with units.

74 One credit is awarded for the correct answer with units or for an answer, with units, that is consistent with your answer to question 73.
Note: Students will not be penalized more than 1 credit for errors in units in questions 73 and 74.

75 Refer to the *Electricity* section of *Reference Table K.*

$$\text{Given:} \quad R = 100. \ \Omega$$
$$V = 10.0 \text{ V}$$

$$\text{Find:} \quad P = ?$$

$$\text{Solution:} \quad P = \frac{V^2}{R}$$

$$P = \frac{(10.0 \text{ V})^2}{100. \ \Omega}$$

$$P = 1.00 \text{ W}$$

One credit is awarded for an answer of 1.00 W.

76 One credit is awarded for at least one complete wave with a wavelength of 0.40 meter regardless of phase or shape.

77 One credit is awarded for at least one complete wave with an amplitude of 0.20 meter regardless of phase or shape.
Example of a 2-credit response for questions 76 and 77:

78–79 Refer to the *Waves* section of *Reference Table K.*

Given: $\lambda = 0.40$ m
$f = 10.$ Hz

Find: $v = ?$

Solution: $v = f\lambda$
$v = (10.\text{ Hz})(0.40\text{ m})$
$v = 4.0$ m/s

78 One credit is awarded for a correct equation and substitution of values with units.

79 One credit is awarded for the correct answer with units or for an answer, with units, that is consistent with your answer to question 78.
Note: Students will not be penalized more than 1 credit for errors in units in questions 78 and 79.

80 Refer to the *Waves* section of *Reference Table K.*

Given: $f = 10.$ Hz

Find: $T = ?$

Solution: $T = \dfrac{1}{f}$

$T = \dfrac{1}{10.\text{ Hz}}$
$T = 0.10$ s

One credit is awarded for an answer of 0.10 s.

81 One credit is awarded for using an appropriate linear scale on the axis marked "Current (A)" on the x-axis. An appropriate scale would use the length of the axis to represent the range of current found in the data table, and each box would represent equivalent amounts.

82 One credit is awarded for plotting all of the points accurately ±0.3 grid space.

83 One credit is awarded for drawing the best fit line or curve that is consistent with your responses to questions 81 and 82. A best fit line or curve is not a "connect the dots" but, rather, the **line** or curve that **best** represents the data on a scatter plot. This **line** may pass through some of the points, none of the points, or all of the points (ideally as many as possible that the line passes through), and there will be a similar number of points above or below at similar distances away from the line if the line is not passing through the points.

Example of a 3-credit response for questions 81 through 83:

84–85 The slope of a potential difference versus current graph is given by the relationship $\Delta V/\Delta I$ or, in other words, the resistance. We can pick any two points on the line in the graph and substitute them into the formula to calculate the resistance.

Given: $V_1 = 0$ V
            $V_2 = 3.0$ V
            $I_1 = 0$ A
            $I_2 = 0.006$ A

Find: $R = ?$

Solution: $R = \dfrac{\Delta V}{\Delta I}$

            $R = \dfrac{(3.0\text{ V} - 0\text{ V})}{(0.006\text{ A} - 0\text{ A})}$

            $R = 5.0 \times 10^2 \ \Omega$

84 One credit is awarded for the correct answer with units or for an answer, with units, that is consistent with your answers to questions 81–83.

85 One credit is awarded for the correct answer with units or for an answer, with units, that is consistent with your answer to question 84.

Note: Students will not be penalized more than 1 credit for errors in units in questions 84 and 85.

Topic	Question Numbers (total)	Wrong Answers (x)	Grade
Math Skills	2, 7, 8, 10, 12–18, 23, 26, 32, 34, 36–42, 45, 46, 48, 51–53, 56–64, 66–75, 78–85: (55)		$\dfrac{100(55 - x)}{55} = \%$
Mechanics	1–3, 5–7, 11, 12, 18, 33, 34, 36, 37, 40–42, 55, 59–67: (26)		$\dfrac{100(26 - x)}{26} = \%$
Energy	1, 8–11, 39, 48, 51–53: (10)		$\dfrac{100(10 - x)}{10} = \%$
Electricity/ Magnetism	4, 13–17, 19, 35, 43, 47, 48, 54, 68–75, 84, 85: (22)		$\dfrac{100(22 - x)}{22} = \%$
Waves	20–29, 50, 56–58, 76–80: (19)		$\dfrac{100(19 - x)}{19} = \%$
Modern Physics	30–32, 38, 44–46, 49: (8)		$\dfrac{100(8 - x)}{8} = \%$

Let's Review Regents:

Physics—Physical Setting 2020

Miriam A. Lazar

Principal
Archimedes Academy
for
Math, Science & Technology Applications
Bronx, New York

formerly
Chemistry/Physics Teacher
Stuyvesant High School
New York, New York

Illustration Acknowledgment

page 289   From Physics, by Halliday and Resnick. Copyright
                           © 1962 by John Wiley & Sons, Inc. Reprinted by
                           permission of John Wiley & Sons, Inc.

Published by Kaplan, Inc., d/b/a Barron's Educational Series
750 Third Avenue
New York, NY 10017
www.barronseduc.com

ISBN: 978-1-5062-5410-4

Printed in Canada

10 9 8 7 6 5 4 3 2 1

Kaplan, Inc. d/b/a Barron's Educational Series print books are available at
special quantity discounts to use for sales promotions, employee premiums, or
educational purposes. For more information or to purchase books, please call
the Simon & Schuster special sales department at 866-506-1949.

TABLE OF CONTENTS

PREFACE

To the Student:

This book has been written to help you understand and review high school physics. Physics is not a particularly easy subject, and no book—no matter how well written—can give you *instant* insight into the subject. Nevertheless, if you read this book carefully and do all of the problems and review questions, you will have a pretty good understanding of what physics is all about.

I have designed this book to be your "physics companion." It is somewhat more detailed than some review books but is probably much less detailed than your hardcover text. This book is divided into 14 chapters. Each chapter begins with *Key Ideas* and ends with multiple-choice questions (part A and B-1 Regents style questions) and some (where available) short-constructed response and free-response questions (part B-2 and C Regents style questions) drawn from past NYS Regents Physics examinations. The answers to these questions appear after the glossary. The appendices contain various items of information that are useful in the study of physics and in answering short-constructed response and free-response questions.

This book covers all of the concepts and skills listed in the NYS Physical Setting: Physics core curriculum. In this edition, a significant amount of material that I had included in previous editions as review for a general physics course that went beyond the core (the core being only the minimum to be taught and what is testable by NYS) has at the request of reviewers and my publisher been deleted from this edition. There is only a small amount of material included that is not part of the core, but which I felt helped to clarify concepts that were a part of the core and so remain. A star icon (✪) appears next to chapter subheadings that are not part of the core curriculum. For those students taking the NYS Physical Setting: Physics Regents Examination at the end of your course, I have included general information about this examination and a summary of concepts to be mastered and skills to be demonstrated. These items can be found in Appendix 4. The correlation of chapter material with the NYS Physical Setting: Physics core curriculum Regents is clearly indicated in the Appendix.

If you have any comments about this book, I would appreciate hearing from you. Please write to me in care of the publisher, whose name and address are given on the copyright page.

To the Teacher:

Another review book in Physics? Every author feels that he or she has a unique contribution to make, and I'm certainly no exception to this rule. It is my belief that a physics review book should be more than an embellished outline of a particular syllabus with questions and problems added.

Introductory physics is a difficult subject that requires diligent effort on the part of the student, and any book worth its salt must provide careful and detailed explanations of the material. I wrote the book with these thoughts in mind and I hope that it will be successful for both you and your students.

This book covers all of the concepts and skills listed in the NYS Physical Setting: Physics core curriculum. I have divided the book into 14 chapters because I believe that the material is best presented in this fashion. In this edition, a significant amount of material, that I had included in previous editions as review for a general physics course that went beyond the core (the core being only the minimum to be taught and what is testable by NYS) has at the request of reviewers and my publisher been deleted from this edition. There is only a small amount of material included which is not part of the core but which I felt helped to clarify concepts that were a part of the core and so remain. A star icon (✪) appears next to chapter subheadings that are not part of the core curriculum. You may not feel that my division of the material is appropriate for your teaching style or course, but there is no one correct way to order the material in a physics text. If you are more comfortable with another approach, then use it by all means.

I spent considerable time in working out short-constructed response and free-response problems throughout the text because I believe that is the area where students experience their greatest difficulties with physics. I chose *not* to pad each section with multiple-choice questions but reserved them for the end of each chapter. In this edition I have included a section of short-constructed response and free-response questions at the end of each chapter. These questions were drawn from past NYS Regents examinations spanning more than 30 years.

I hope that this book will give you some new insights and enable you to plan and execute your physics lessons with a spirit of inquiry. If you have any comments or corrections, please write to me in care of the publisher, whose name and address are given on the copyright page or you can e-mail me at archimedesacademy@gmail.com.

I wish to thank my editors at Barron's for their patience and understanding, Mr. Joseph Lazar (my father) for his excellent art work and all my former students at Stuyvesant, who contributed much to the first edition that continues to be included in this edition. I wish to thank Albert S. Tarendash, my teacher, my mentor, and my friend, for his invaluable hard work and effort that went into the creation of the first edition that lives on in this edition.

I continue to dedicate this book to the memories of Dr. Murray Kahn and Dr. Matthew Litwin. Both men were friends and colleagues and outstanding chemistry teachers whose first and last thoughts were always of their students.

Miriam Lazar

INTRODUCTION TO PHYSICS

KEY IDEAS

Science depends on our ability to measure quantities. The SI metric system is used internationally as the standard for scientific measurement. It consists of seven basic quantities (e.g., length and time) and many derived quantities (e.g., speed and density).

Every measurement has a degree of uncertainty that is related to the limits of the measuring instrument. The concept of significant digits helps us to evaluate the degree of uncertainty in a particular measurement.

A graph of data points can indicate whether there is regularity within a set of data and can be used to predict how variables will behave under given conditions.

KEY OBJECTIVES

At the conclusion of this chapter you will be able to:

• State the fundamental quantities of measurement in the Système International (SI) and the metric units associated with them.
• Perform calculations using scientific notation.
• Determine the number of significant digits in a measurement.
• Incorporate significant digits within calculations.
• Determine the order of magnitude of a measurement.
• Plot a graph from a series of data points.
• Determine proportional relationships within data.
• Calculate the slope of a straight-line graph.
• State the common mathematical relationships in a right triangle.

1.1 PHYSICS IS . . .?

It's not easy to come up with a precise definition of physics because this subject is so broad. The best one we've heard so far is this: "Physics is what physicists do." But *what* do they do? Physicists study the universe, from the smallest parts of matter (the particles that make up atoms) to the largest (the galaxies and beyond). They study the interactions of matter and energy, using both experimental (laboratory) and theoretical (mathematical) techniques. In this chapter, we will introduce some of these techniques in order to prepare us for the material that is covered in later chapters.

1.2 MEASUREMENT AND THE METRIC SYSTEM

The basis of all science lies in the ability to measure quantities. For example, we can easily measure the length of a table. In principle, we can even design an experiment to measure the distance from Earth to the nearest star. Therefore, these lengths have scientific meaning to us. However, modern physics theorizes that an electron has no measurable diameter, and consequently no experiment or apparatus can ever measure this length. As a result, we say that the diameter of an electron has no scientific meaning.

To be able to measure quantities we must have a *system* of measurement. We use the Système International (SI) because scientists all over the world express measurements in these metric units. The SI recognizes seven fundamental quantities upon which all measurement is based: (1) length, (2) mass, (3) time, (4) temperature, (5) electric current, (6) luminous intensity, and (7) number of particles. Each of these quantities has a unit of measure based on a standard that can be duplicated easily and does not vary appreciably.

Length

The unit of length is the *meter* (m), which is approximately 39 inches. The standard is based on the speed of light, which is absolutely constant (in a vacuum) and has an assigned value of 2.99792458×10^8 meters per second.

Mass

The unit of mass is the *kilogram* (kg), which has an approximate weight (on Earth) of 2.2 pounds. The standard is a platinum-iridium cylinder that is kept at constant temperature and humidity in a dustless vault in Sèvres, France. (Note that mass and weight are *not* the same quantities. Mass is the measure of the matter an object contains, while weight is the force with which gravity attracts matter.)

Time

The unit of time is the *second* (s). The standard is based on the frequency of vibrations of cesium-133 atoms under certain defined conditions.

✪ Temperature

The unit of temperature is the *kelvin* (K). The standard is based on the point at which solid, liquid, and gaseous water coexist simultaneously (the "triple" point, which has an assigned value of 273.16 K).

Electric Current

The unit of electric current is the *ampere* (A). The standard is based on the mutual forces experienced by parallel current-carrying wires.

✪ Luminous Intensity

The unit of luminous intensity is the *candela* (cd). The standard is based on the amount of radiation emitted by a certain object, known as a black-body radiator, at the freezing temperature of platinum (2046 K).

✪ Number of Particles

The unit of number of particles is the *mole* (mol). The standard is based on the number of atoms contained in 0.012 kilogram of carbon-12 (6.02×10^{23} atoms).

1.3 SCIENTIFIC NOTATION

Once we have established a system of measurement, we need to be able to express small and large numbers easily. Scientific notation accomplishes this purpose. In scientific notation, a number is expressed as a power of 10 and takes the form

$$M \times 10^n$$

M is the *mantissa*; it is greater than or equal to 1 and is less than 10 ($1 \le M < 10$). The *exponent*, *n*, is an integer. For example, the number 2300 is written in scientific notation as 2.3×10^3 (not as 23×10^2 or 0.23×10^4). The number 0.0000578 is written as 5.78×10^{-5}.

To write a number in scientific notation, we move the decimal place until the mantissa is a number between 1 and 10. If we move the decimal place to the left, the exponent is a positive number; if we move it to the right, the exponent is a negative number.

✪ indicates material that is not part of the core curriculum.

If we wish to *multiply* two numbers expressed in scientific notation, we *multiply* the mantissas and *add* the exponents. The final result must always be expressed in proper scientific notation. Here are two examples:

$$(2.0 \times 10^5)(3.0 \times 10^{-2}) = \underline{6.0 \times 10^3}$$
$$(4.0 \times 10^4)(5.0 \times 10^3) = 20. \times 10^7 = \underline{2.0 \times 10^8}$$

To *add* two numbers expressed in scientific notation, both numbers must have the same exponent. The mantissas are then *added* together. The following example illustrates the application of this rule:

$$(2.0 \times 10^3) + (3.0 \times 10^2) = (2.0 \times 10^3) + (0.30 \times 10^3) = \underline{2.3 \times 10^3}$$

To *divide* two numbers expressed in scientific notation, we *divide* the mantissas and *subtract* the exponents. The following example illustrates the application of this rule:

$$\frac{(3.0 \times 10^3)}{(2.0 \times 10^{-2})} = 1.5 \times 10^5$$

1.4 ACCURACY, PRECISION, AND SIGNIFICANT DIGITS

As stated in Section 1.2, every scientific discipline, including physics, is concerned with making measurements. Since no instrument is perfect, however, every measurement has a degree of uncertainty associated with it. A well-designed experiment will reduce the uncertainty of each measurement to the smallest possible value.

Accuracy refers to how well a measurement agrees with an accepted value. For example, if the accepted density of a material is 1220 kilograms per meter^3 and a student's measurement is 1235 kilograms per meter^3, the difference (15 kg/m^3) is an indication of the accuracy of the measurement. The smaller the difference, the more accurate is the measurement.

Precision describes how well a measuring device can produce a measurement. The limit of precision depends on the design and construction of the measuring device. No matter how carefully we measure, we can *never* obtain a result more precise than the limit of our measuring device. A good general rule is that the limit of precision of a measuring device is equal to plus or minus one-half of its smallest division. For example, in the diagram below:

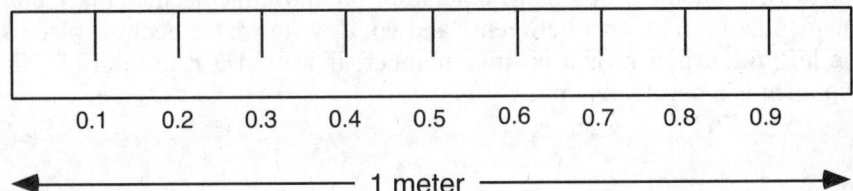

$$0.1 \quad 0.2 \quad 0.3 \quad 0.4 \quad 0.5 \quad 0.6 \quad 0.7 \quad 0.8 \quad 0.9$$

1 meter

4

the smallest division of the meter stick is 0.1 meter, and the limit of its precision is ±0.05 meter. When we read any measurement using this meter stick, we must attach this limit to the measurement, for example, 0.27 ± 0.05 meter.

Significant digits are the digits that are part of any valid measurement. The number of significant digits is a direct result of the number of divisions the measuring device contains. The following diagram represents a meter stick with no divisions on it. How should the measurement, indicated by the arrow, be reported?

Since there are no divisions, all we know is that the measurement is somewhere between 0 and 1 meter. The best we can do is to make an *educated guess* based on the position of the arrow, and we report our measurement as 0.3 meter. This meter stick allows us to measure length to *one* significant digit.

Suppose we now use a meter stick that has been divided into tenths and repeat the measurement:

Now we can report the measurement with less uncertainty because we *know* that the indicated length lies between 0.3 and 0.4 meter. If we allow ourselves one guess, we could report the length as 0.33 meter. This measurement has *two* significant digits. The more significant digits a measurement has, the more confidence we have in our ability to reproduce the measurement because only the *last* digit is in doubt.

Measurements that contain zeros can be particularly troublesome. For example, we say that the average distance between Earth and the Moon is 238,000 miles. Do we *really* know this number to *six* significant digits? If so, we would have measured the distance to the nearest *mile*. Actually, this measurement contains only *three* significant figures. The distance is being reported to the nearest *thousand miles*. The zeros simply tell us how *large* the measurement is.

To avoid confusion, a number of rules have been established for determining how many significant digits a measurement has.

Rules for Determining the Number of Significant Digits in a Measurement

1. All nonzero numbers are significant. The measurement 2.735 meters has four significant digits.
2. Zeros located *between* nonzero numbers are also significant. The measurements 1.0285 kilograms and 202.03 seconds each have five significant digits.
3. For numbers greater than or equal to 1, zeros located at the *end* of the measurement are significant only if a decimal point is present. The measurement 60 amperes has one significant digit. In this case, the zero indicates the *size* of the number, not its significance. The measurements 60. amperes and 60.000 amperes, however, have two and five significant digits, respectively.
4. For numbers less than 1, the leading zeros are *not* significant; they indicate the size of the number. Thus, the measurements 0.00**2** kilogram, 0.0**20** kilogram, and 0.000**200** kilogram have one, two, and three significant digits, respectively. (The significant digits are indicated in **bold** type.)

If we use scientific notation, we need not become involved with the preceding rules because the mantissas always contains the proper number of significant digits; the size of the number is absorbed into the exponent. For example, the measurement 3.10×10^{-4} meter has three significant digits.

Using Significant Digits in Calculations

Significant digits are particularly important in calculations involving measured quantities and it is crucial that the *result* of a calculation does not imply a greater precision than any of the individual measurements. Calculators routinely give us answers with ten digits. It is incorrect to believe that the results of most of our calculations have this many significant digits.

When two measurements are multiplied (or divided), the answer should contain as many significant digits as the *least* precise measurement. For example, if the measurement 2.3 meters (two significant digits) is multiplied by 7.45 meters (three significant digits), the answer will contain two significant digits:

$$(2.3 \text{ m})(7.45 \text{ m}) = 17.135 \text{ m}^2 = \underline{17 \text{ m}^2}$$

(Note that the units are also multiplied together.)

When two measurements are added (or subtracted), the answer should contain as many *decimal places* as the measurement with the *smallest* number of decimal places. For example, when 8.11 kilograms and 2.476 kilograms are added, the answer will be taken to the second decimal place:

$$8.11 \text{ kg} + 2.476 \text{ kg} = 10.586 \text{ kg} = \underline{10.59 \text{ kg}}$$

(Note that the answer has been *rounded* to two decimal places.)

If *counted* numbers (such as 6 atoms) or defined numbers (such as 273.16 K) are used in calculations, they are treated as though they had an infinite number of significant digits or decimal places.

1.5   ORDER OF MAGNITUDE

There are times when we are interested in the *size* of a measurement rather than its actual value. The *order of magnitude* of a measurement is the power of 10 closest to its value. For example, the order of magnitude of 1284 kilograms (1.284×10^3) is 10^3, while the order of magnitude of 8756 kilograms (8.756×10^3) is 10^4. Orders of magnitude are very useful for comparing quantities, such as mass or distance, and for estimating the answers to problems involving complex calculations.

For a real world example, the size of an oak leaf is on an order of magnitude of 10^{-1} m while the size of a single proton is on an order of magnitude of 10^{-15} m. By subtracting the exponents we can determine how many orders of magnitude difference there are between two quantities. In the above example there is a difference in order of magnitude on the scale of 10^{14}. That is, an oak leaf is on order of magnitude 10^{14} times larger than a single proton.

1.6   GRAPHING DATA

We have all heard the expression "A picture is worth a thousand words." This adage is particularly true when we wish to present experimental data. The following table of data gives the stretched lengths of a spring (in meters) when different weights (in a unit called newtons) are placed on it.

Weight (N)	Length (m)
0.0	0.0
5.0	0.04
10.	0.11
15	0.13
20.	0.18
25	0.27

Now plot these points and draw a graph.

Note how the graph is constructed. First, the axes are drawn so that the data can be displayed over the entire graph. Next, the axes are labeled with the names of the quantities and their units of measure. It is traditional to place the quantity that is varied by the experimenter (the *independent variable*) along the x-axis, and the result of the experiment (the *dependent variable*) along the y-axis. Each item of data is then entered in the correct place on the graph.

Note that we do *not* play "connect the dots." Rather, we look for some regular relationship between the data points. We then draw a smooth line (or curve) that will fit this relationship as closely as possible. The "scatter" of the data in this example implies that we are dealing with a straight-line relationship. Although in most cases the graphed line will pass through a maximum number of data points, it is entirely possible that the graph may not pass through *any* of the data points. We require only that the data points above and below the graph be evenly distributed, as shown in the example. This graph is known as a *best-fit straight line*.

8

If a graph yields a straight line, we can conclude that the variables change uniformly. In the example given above, the length of the spring increases regularly as heavier weights are applied to the spring, and the steadily rising line of the graph reflects this direct relationship. If a graph is a curve, however, the variables (here, weight and length) change at a rate that is not uniform.

1.7 DIRECT AND INVERSE PROPORTIONS

Two quantities are *directly proportional* to one another if a change in one quantity is accompanied by an identical change in the other. For example, the table given below represents the *ideal* relationship between the mass of a substance and its corresponding volume. If the mass changes by a factor of 2 or 3, the volume changes by the same factor. This means that the *ratio* of the two quantities is a constant.

Mass (kg)	Volume (m³)
0.0	0.0
2.0	0.040
4.0	0.080
6.0	0.12
8.0	0.16
10.	0.20

When we plot this relationship, we obtain a straight-line graph that passes through the origin.

Frequently, the *slope* of a straight line provides us with additional information. Recall that we calculate the slope of a line by selecting two data points (x_1, y_1 and x_2, y_2) and finding the ratio $\Delta y/\Delta x$:

PHYSICS CONCEPTS

$$\text{Slope} = m = \frac{\Delta y}{\Delta x} = \frac{y_2 - y_1}{x_2 - x_1}$$

*Note that this slope equation does not appear on the Reference Tables. However, it is essential when solving problems. Memorize this equation.

The slope of the line representing a direct proportion is known as the *constant of proportionality,* and it provides the ratio between the variables. (In the mass-volume graph, the constant is known as the *density* of the substance.)

Two quantities are *inversely proportional* to one another if a change in one quantity is accompanied by a reciprocal change in the other. For example, the table given below represents the ideal relationship between the pressure of a gas and its corresponding volume at constant temperature. If the pressure changes by a factor of 2 or 3, the volume changes by a factor of ½ or ⅓. This means that the *product* of the two quantities is a constant.

Pressure (atm)	Volume (m³)
0.5	0.24
1.0	0.12
2.0	0.060
3.0	0.040
4.0	0.030
5.0	0.024

If we plot this relationship we will obtain a curve known as a *hyperbola.* As shown in the diagram, this curve will approach both the *x*- and the *y*-axis but will not intersect the axes.

10

Other types of relationships include direct-squared proportions and inverse-squared proportions. Two quantities exhibit a direct-squared proportion if an increase in one causes a squared increase in the other.

$y = x^2$

Two quantities exhibit an inverse-squared proportion if an increase in one causes a squared decrease in the other.

$y = \dfrac{1}{x^2}$

1.8 MATHEMATICAL RELATIONSHIPS WITHIN RIGHT TRIANGLES

Right triangles play an important part in the solution of physics problems. In this section, we summarize some of the more important relationships common to these triangles.

Consider the right triangle drawn below, where h indicates the hypotenuse and x and y are the two shorter sides.

Three well-known trigonometric relationships relate the value of acute angle θ to the lengths of the sides of the triangle:

PHYSICS CONCEPTS

$$\sin \theta = \frac{a}{c} \quad \cos \theta = \frac{b}{c} \quad \tan \theta = \frac{a}{b}$$

In addition, the Pythagorean theorem relates the lengths of the sides of the triangle:

PHYSICS CONCEPTS

$$a^2 + b^2 = c^2$$

PART A AND B–1 QUESTIONS

1. The unit of mass in the SI metric system is the
(1) gram   (2) kilogram (3) newton   (4) meter

2. A meter stick has millimeter divisions marked on it. The limit of precision of this instrument is
(1) 0.005 m   (2) 0.01 m   (3) 0.0005 m   (4) 0.001 m

3. What is the order of magnitude for the measurement 72 meters per second?
(1) 10^{-2} (2) 10^1 (3) 10^2 (4) 10^4

4. How many significant digits are contained in the measurement 500,000 kilometers?
(1) 1   (2) 2   (3) 3   (4) 6

5. How many significant digits are contained in the measurement 406.200 seconds?
(1) 6   (2) 5   (3) 3   (4) 4

6. How many significant digits are contained in the measurement 0.000300 volt?
(1) 1   (2) 2   (3) 3   (4) 6

7. When the measurements 33.972 kilograms and 0.21 kilogram are added, the answer, to the correct number of significant digits, is
(1) 34 kg   (2) 34.2 kg (3) 34.18 kg   (4) 34.182 kg

8. When the measurements 8.14 meters and 2.1 meters are multiplied, the answer, to the correct number of significant digits, is
(1) 17 m^2 (2) 17.0 m^2 (3) 17.09 m^2 (4) 17.094 m^2

9. How can the measurement 0.00567 liter be expressed to one significant digit?
(1) 0.005 L   (2) 0.0050 L   (3) 0.006 L   (4) 0.0060 L

10. The approximate height of a high school physics student is
(1) 10^1 m (2) 10^2 m (3) 10^0 m (4) 10^{-2} m

11. The mass of a paper clip is approximately
(1) $1 \times 10^6 \text{ kg}$ (2) $1 \times 10^3 \text{ kg}$ (3) $1 \times 10^{-3} \text{ kg}$ (4) $1 \times 10^{-6} \text{ kg}$

12. What is the approximate length of a baseball bat?
(1) 10^{-1} m (2) 10^0 m (3) 10^1 m (4) 10^2 m

13. The length of a dollar bill is approximately
(1) $1.5 \times 10^{-2} \text{ m}$ (3) $1.5 \times 10^1 \text{ m}$
(2) $1.5 \times 10^{-1} \text{ m}$ (4) $1.5 \times 10^2 \text{ m}$

14. What is the approximate mass of an automobile?
(1) 10^1 kg (2) 10^2 kg (3) 10^3 kg (4) 10^6 kg

15. The reading of the ammeter in the diagram below should be recorded as

(1) 1 A   (2) 0.76 A   (3) 0.55 A   (4) 0.5 A

16. Which is the most likely mass of a high school student?
(1) 10 kg   (2) 50 kg   (3) 600 kg   (4) 2500 kg

17. What is the approximate thickness of this piece of paper?
(1) 10^1 m (2) 10^0 m (3) 10^{-2} m (4) 10^{-4} m

18. What is the approximate mass of a pencil?
(1) $5.0 \times 10^{-3} \text{ kg}$ (3) $5.0 \times 10^0 \text{ kg}$
(2) $5.0 \times 10^{-1} \text{ kg}$ (4) $5.0 \times 10^1 \text{ kg}$

Answers to Part A and B–1 questions can be found on page 454.

MOTION IN ONE DIMENSION

Chapter Two

KEY IDEAS

Motion is the change of position in time. *Displacement* is the directed change of an object's position. *Velocity* is the time rate of change of displacement, and *acceleration* is the time rate of change of velocity.

Under conditions of constant acceleration (also known as uniform acceleration), the motion of an object is governed by a set of interrelated equations. Objects that fall freely near the surface of the Earth are uniformly accelerated by gravity.

The motion of an object can be described by a series of motion graphs. The object's position, velocity, and acceleration can be plotted as functions of time. These graphs can then be used to illustrate various aspects of the object's motion at every point in time.

KEY OBJECTIVES

At the conclusion of this chapter you will be able to:

- Define the terms *motion, distance, displacement, average velocity, speed, instantaneous velocity,* and *acceleration,* and state their SI units.
- Solve problems involving average velocity and constant velocity.
- Distinguish between average velocity and instantaneous velocity, and relate these terms to a position-time graph.
- Solve problems involving the equations of uniformly accelerated motion.
- Solve problems involving freely falling objects.
- Interpret the data provided by motion graphs and solve problems related to them.

2.1   MOTION DEFINED

How do we know when an object is in motion? If we look at the hour hand of a watch, it does not appear to be moving, yet over a period of time we see a change in its position. Therefore, a reasonable definition of **motion** is the change of an object's position in time.

2.2 GRAPHING AN OBJECT'S MOTION

Graphs are especially useful for analyzing an object's motion. The position of the object is plotted along the y-axis, and the elapsed time along the x-axis. Here is a graph of very general motion in one dimension:

The origin of the graph $(0, 0)$ marks the reference point for both position and time. When we say "zero time" we mean the time when we begin the event—the time when we "start the clock," so to speak. Similarly, "zero position" means the specific place where we begin measuring. It may be the ground or a table top or a spot on the wall. Generally, we use the letters d to represent position and t to represent time. If we know that we are measuring *horizontal* position, we may use the letter x, or d_x in place of d. If we are measuring a vertical position we can substitute the letter y or d_y in place of d.

The dotted lines on the graph tell us where the object is at a given time. At time t_1, the object is at position d_1; at time t_2, the object is at position d_2.

2.3 DISPLACEMENT

The **displacement** of an object is the change in its position and is measured in units of length (such as meters or inches). In the graph in Section 2.2, the displacement of the object between times t_1 and t_2 is given by the relationship $\mathbf{d} = \mathbf{d}_2 - \mathbf{d}_1$. Since we are subtracting two coordinates ($\mathbf{d}_1$ from $\mathbf{d}_2$), we are not primarily concerned about the exact path taken by the object between these points. We assume that the magnitude of the displacement is given by the length of a straight line between $\mathbf{d}_1$ and $\mathbf{d}_2$ along the axis representing position.

Displacement is known as a *vector* quantity because, in addition to magnitude, it has *direction*. (Vector quantities, which are discussed in detail in Chapter 4, are set in boldface type.) The magnitude of displacement is known as *distance*.

15

In the first graph above, the displacement (**d**) from t_1 to t_2 is positive because its magnitude (indicated by the arrow) is measured in the positive direction; in the second graph, the displacement is negative. Whenever we are working with one-dimensional motion, we can use positive and negative signs to represent opposite directions.

In each of the graphs, the *elapsed time* is given by the relationship $t = t_2 - t_1$ and is measured in units of time, such as seconds or hours. We always read the time axis from left to right since normally we do not travel backward in time.

2.4   VELOCITY

In the graph below, the slope of the straight line connecting the two points on the graph is given by the relationship **d**/t, and it indicates *how rapidly* the position of the object (**d**) has changed over the time interval (t). This quantity is known as the **average velocity** ($\bar{v}$) of the object and is measured in units such as meters per second (m/s) or miles per hour (mph). Mathematically, the average velocity of the object is defined by the equation

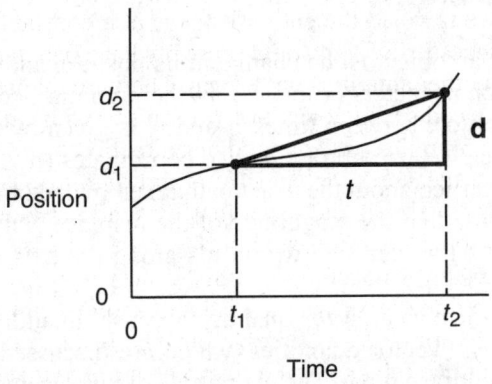

$$\bar{v} = \frac{d}{t} = \frac{d_2 - d_1}{t}$$

Velocity, like displacement, is a vector quantity because it has both magnitude and direction. As with displacement, we can use positive and negative signs to represent motion in opposite directions. The magnitude of the velocity is known as **speed**.

PROBLEM

The position of an object is +35 meters at 2.0 seconds and is +87 meters at 15 seconds. Calculate the average velocity of the object.

SOLUTION

$$\bar{v} = \frac{d}{t} = \frac{d_2 - d_1}{t_2 - t_1}$$

$$= \frac{+87 \text{ m} - (+35 \text{ m})}{15 \text{ s} - 2.0 \text{ s}} = \frac{+52 \text{ m}}{13 \text{ s}} = +4.0 \text{ m/s}$$

This object is traveling in the positive direction with an average speed of 4.0 m/s.

We use the term *average velocity* because we do not know exactly what is happening *between* the two points in question. For example, suppose we traveled by automobile due west 1000 miles and the trip took 20 hours. When we calculate our average velocity for the trip, we obtain

$$\bar{v} = \frac{1000 \text{ mi [W]}}{20 \text{ h}}$$

$$= 50 \text{ mph [W]}$$

Does this mean that we traveled the entire distance at a constant speed of 50 miles per hour? Not necessarily! We would probably have had to add fuel, eat, pay tolls, or engage in other activities on the trip. There might have been construction delays or reduced speed zones. All we can say with certainty is that our average speed was 50 miles per hour and our direction of travel was west.

How then could we measure our velocity at any *point* on our trip—our **instantaneous velocity**? (This is the value that we read on the speedometer of our car.) One way would be to measure our average velocity over *smaller and smaller* time intervals. To accomplish this, however, we would need to use mathematical techniques that are beyond the scope of this book. The other way is to use a position versus time graph. The instantaneous velocity at any point on the graph is the slope of a line drawn *tangent* to the graph at that point, as shown in the diagram.

17

Slope of tangent line
equals instantaneous
velocity at this point.

2.5 ACCELERATION

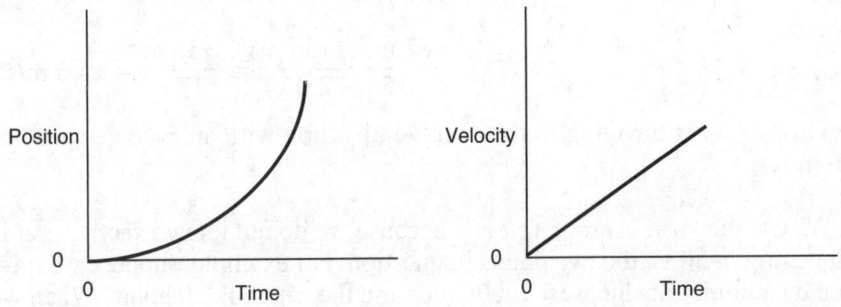

Consider the two graphs shown above. The first graph represents the position of an automobile as a function of time. Note that the graph becomes steeper (curves upward) as time passes. This occurs because the automobile's instantaneous velocity is *increasing.*

The second graph represents the instantaneous velocity of the same automobile as a function of time. Note that the graph is a straight line directed upward. This graph also shows that the instantaneous velocity of the automobile is increasing with time.

Actually, the graphs represent the motion of the automobile from two different viewpoints: that of position (as measured by the automobile's *odometer*) and that of velocity (as measured by the automobile's *speedometer*).

The *slope* of the velocity–time graph is given by the relationship $\Delta v/t$ and indicates the rate at which the velocity of the object (Δv) has changed over the time interval (t). This quantity is known as the **acceleration (a)** of the object and is measured in units such as meters per second^2 (m/s^2). Since the graph is a straight line, the acceleration in this case is constant or *uniform.* Mathematically, the uniform acceleration of the object is defined by the equation

=== **PHYSICS CONCEPTS** ===

$$a = \frac{\Delta v}{t} = \frac{v_f - v_i}{t}$$

Acceleration is also a vector quantity because it has both magnitude and direction. A positive acceleration means that the velocity of an object is becoming more positive with time; a negative acceleration, that the velocity of the object is becoming more negative with time.

PROBLEM

The velocity of an object is +47 meters per second at 3.0 seconds and is +65 meters per second at 12.0 seconds. Calculate the acceleration of the object.

SOLUTION

$$a = \frac{\Delta v}{t} = \frac{v_f - v_i}{t_2 - t_1}$$

$$= \frac{+65 \text{ m/s} - (+47 \text{ m/s})}{12.0 \text{ s} - 3.0 \text{ s}} = \frac{+18 \text{ m/s}}{9.0 \text{ s}}$$

$$= +2.0 \text{ m/s}^2$$

This object's velocity is becoming more positive by 2.0 m/s each second. Since the magnitude of the velocity is increasing, the object is speeding up. The table below shows how this occurs.

t (s)	0.0	1.0	2.0	3.0	4.0	5.0	6.0	7.0	8.0	9.0
v (m/s)	+47	+49	+51	+53	+55	+57	+59	+61	+63	+65

2.6 THE EQUATIONS OF UNIFORMLY ACCELERATED MOTION

In the problem we solved in Section 2.5, an object accelerates uniformly, at 2.0 meters per second^2, from 47 meters per second to 65 meters per second in 9.0 seconds. There is a great deal of additional information about the object we might wish to learn. For example:

1. What is the *average velocity* of the object over 9.0 seconds?
2. What is the *displacement* of the object at the end of 9.0 seconds?
3. What is the *instantaneous velocity* of the object at any given time (at 6.0 s, for example)?

To solve these problems, we use a set of five equations that describe the motion of an object undergoing uniform acceleration. In each of these equations, we use the subscripts i (for *initial* value), f (for *final* value), and $\overline{v}$ to indicate average velocity. Your textbook or teacher may use different subscripts or notation, but they all yield the same results. The equations for uniformly accelerated motion are as follows:

PHYSICS CONCEPTS

1. $\overline{v} = \dfrac{d}{t}$

2. $\overline{v} = \dfrac{v_i + v_f}{2}$

3. $v_f = v_i + a \cdot t$

4. $d = v_i \cdot t + \dfrac{1}{2} a \cdot t^2$

5. $v_f^2 = v_i^2 + 2 \cdot a \cdot d$

*Note that equation 2 does not appear on the Reference Tables.

Equation 1 is the definition of average velocity. Equation 2 tells us that, under uniform acceleration, the average velocity lies midway between the initial and final velocities. It should be noted that although this equation is not included on the Reference Table, it is essential to solving problems on the exam. Equation 3 is just the definition of acceleration ($a = \Delta v / t$) rearranged in a more convenient form for solving problems. Equations 4 and 5 are relationships that have been derived from the first three equations.

Which equation should you use to solve a particular problem? The answer depends on the data you are given. In the problem we have been considering, an object with an initial velocity of 47.0 meters per second accelerates uniformly at 2.0 meters per second^2 for 9.0 seconds. Suppose we wish to calculate the *displacement* of this object at the end of 9.0 seconds. We list the variables that are part of the problem, along with their values:

$$v_i = 47.0 \text{ m/s}$$
$$a = 2.0 \text{ m/s}^2$$
$$t = 9.0 \text{ s}$$
$$d = \text{???}$$

If we examine the list of equations given above, we see that equation 4 contains the four variables that form the basis of our problem. We solve the problem by substituting the values and calculating the answer:

$$d = v_i \cdot t + \frac{1}{2} a \cdot (t)^2$$

$$= (47.0 \text{ m/s})(9.0 \text{ s}) + \frac{1}{2}(2.0 \text{ m/s}^2)(9.0 \text{ s})^2$$

$$= 504 \text{ m} = 500 \text{ m}$$

2.7 FREELY FALLING OBJECTS

The following table represents the motion of an object falling from rest near the surface of the Earth when air resistance is ignored.

t (s)	0.00	1.00	2.00	3.00	4.00	5.00	6.00
v (m/s)	0.00	9.80	19.6	29.4	39.2	49.0	58.8
d (m)	0.00	4.90	19.6	44.1	78.4	122	176

If we analyze this motion, we see that the speed of the object increases uniformly by 9.8 meters per second for each second of travel. This suggests that the object is subject to a constant acceleration of 9.8 meters per second^2. The distance traveled by the object over time verifies that the object's acceleration is constant.

PROBLEM
How does the distance traveled by the object described above over time verify that the object's acceleration is constant?

SOLUTION
The distance traveled by an object under uniform acceleration is given by the equation:

$$\mathbf{d} = \mathbf{v}_i \cdot t + \frac{1}{2}\,\mathbf{a} \cdot (t)^2$$

Since the object starts from rest, this equation reduces to:

$$\mathbf{d} = \frac{1}{2}\,\mathbf{a} \cdot (t)^2$$

If we substitute each of the corresponding values of **d** and t given in the table, we find that the acceleration in each case is 9.8 m/s^2.

If we were to investigate further, we would find that *all* objects falling near the surface of the Earth experience a constant acceleration of 9.8 meters per second^2 if air resistance is ignored. This phenomenon is due to the presence of gravity, which affects each and every object. If we were to travel to the Moon, we would find that all objects also fall to its surface with a uniform acceleration. However, this acceleration is only 1.6 meters per second^2 because of the Moon's weaker gravitational forces.

Since free fall involves uniform acceleration, the five equations we developed in Section 2.6 can be used to solve all free-fall problems. We need only remember that objects can move up as well as down in the presence of gravity. Therefore, we assign the up direction as positive and the down direction as negative. Since gravity *always* points downward (i.e., toward the Earth), its value is taken to be –9.8 meters per second^2. Gravitational acceleration is denoted by the lowercase letter g.

PROBLEM

An object is dropped from rest from a height of 49 meters.
(a) How long does the object take to hit the ground?
(b) What is its speed as it hits the ground?

SOLUTION

(a) Since the initial velocity is zero, we can use the equation

$$\mathbf{d} = \frac{1}{2}\,\mathbf{a} \cdot (t)^2$$

The *displacement* is −49 m (since we measure the distance in a downward direction), and the acceleration due to gravity is −9.8 m/s² (since gravity points downward). Therefore:

$$\mathbf{d} = \frac{1}{2}\,\mathbf{a} \cdot (t)^2$$

$$-49 \text{ m} = \frac{1}{2}\left(-9.8\ \frac{\text{m}}{\text{s}^2}\right)(t)^2$$

$$(t)^2 = 10.\ \text{s}^2$$

$$t = 3.2 \text{ s}$$

(b) We can use the relationship $\mathbf{v}_f = \mathbf{v}_i + \mathbf{a} \cdot t$, which reduces to $\mathbf{v}_f = \mathbf{a} \cdot t$ since the initial velocity is zero. Then

$$\mathbf{v}_f = (-9.8 \text{ m/s}^2)\,(3.2 \text{ s}) = -31 \text{ m/s}$$

2.8   MOTION GRAPHS REVISITED

Throughout this chapter we have used motion graphs as aids to understanding the concept of motion. In this section, we take a more detailed look at these graphs and the information they can provide. We shall examine three types of graphs: position–time, velocity–time, and acceleration–time.

Position–Time Graphs

The following graph illustrates the position of an object as a function of time.

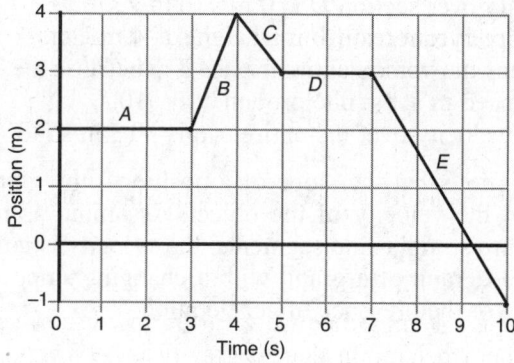

We recall from Section 2.2 that "zero time" represents the start of an event and that "zero position" represents some arbitrary reference point. We have divided the position–time graph into five sections: *A, B, C, D,* and *E*. Since each section is a straight-line segment, the velocity within each section is *constant* and the acceleration over each section is *zero*. We will learn how to interpret this graph by considering the following problem.

PROBLEM

1. What is the displacement over each section of the graph?
2. What is the velocity over each section of the graph?
3. What is the displacement over the entire trip (0–10 seconds)?
4. What is the average velocity over the entire trip (0–10 seconds)?

SOLUTION

1. To calculate the displacement (**d**) we *subtract* the initial position from the final position.
   - The displacement over section *A* is 0 m because the object has not changed its position.
   - The displacement over section *B* is +2 m because the object has changed its position from +2 m to +4 m.
   - The displacement over section *C* is –1 m because the object has changed its position from +4 m to +3 m.
   - The displacement over section *D* is 0 m because the object has not changed its position.
   - The displacement over section *E* is –4 m because the object has changed its position from +3 m to –1 m.
2. The velocity over each section is found by *dividing* the displacement by the elapsed time. $\left(\dfrac{\mathbf{d}}{t}\right)$
   - The velocity over section *A* is 0 m/s (0 m/3 s).
   - The velocity over section *B* is +2 m/s (+2 m/1 s).
   - The velocity over section *C* is –1 m/s (–1 m/1 s).

- The velocity over section D is 0 m/s (0 m/2 s).
- The velocity over section E is –1.3 m/s (–4 m/3 s).

3. The displacement over the entire trip is –3 m because the object changed position from +2 m at $t = 0$ s to –1 m at $t = 10$ s.
4. The average velocity over the entire trip is –0.3 m/s (–3 m/10 s).

It is important to note that the slope of a position–time graph or a **d** vs. t graph is equal to the velocity of the object. Assuming as in the example above that there are straight-line segments, the velocity is constant for each segment. A curved graph or a graph with a changing slope would indicate changing velocity or the presence of acceleration.

Velocity–Time and Acceleration–Time Graphs

The graph below illustrates the velocity of an object as a function of time.

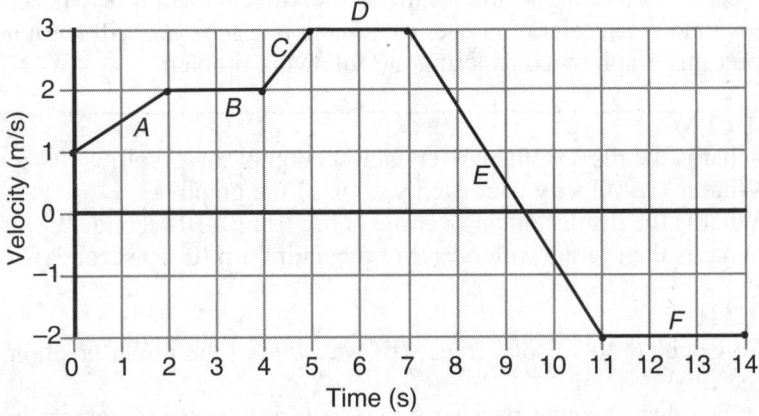

The values on the y-axis represent the *instantaneous* velocities of the object at the times marked on the x-axis. It is as though we were looking at a car's speedometer at various times. We have divided the graph into six sections: A, B, C, D, E, and F. Since each section is a straight-line segment, the object's acceleration within each section is *constant*. We will learn how to interpret this graph by considering the following problem.

PROBLEM

1. What is the average velocity within each section of the graph?
2. What is the acceleration within each section of the graph?
3. When does the object come to rest?
4. When does the object reverse the direction of its motion?
5. What is the displacement within each section of the graph?
6. What is the displacement over the entire trip (0–14 seconds)?
7. What is the average velocity over the entire trip (0–14 seconds)?
8. What is the shape of the corresponding *acceleration versus time* graph?

SOLUTION

1. The average velocity for each section is calculated by finding the *midpoint* of each line, that is, by adding the initial and final velocities within each section and dividing this sum by 2:

$$\left(\overline{\mathbf{v}} = \frac{\mathbf{v}_i + \mathbf{v}_f}{2} \right)$$

We must remember to take both positive and negative signs into account when we add the velocities. For example, in section *E*, the initial velocity is +3.0 m/s and the final velocity is –2.0 m/s. Therefore, the average velocity is

$$\frac{+3.0 \text{ m/s} + (-2.0 \text{ m/s})}{2} = +0.5 \text{ m/s}$$

The table summarizes the results of the calculations for the six sections:

section	A	B	C	D	E	F
$\overline{\mathbf{v}}$ (m/s)	+1.5	+2.0	+2.5	+3.0	+0.5	–2.0

2. The acceleration within each section is found by calculating the *slope* of each of the lines:

$$\left(\frac{\Delta \mathbf{v}}{t} = \frac{\mathbf{v}_f - \mathbf{v}_i}{t} \right)$$

For example, in section *A*, $\mathbf{v}_i$ = 1.0 m/s, $\mathbf{v}_f$ = 2.0 m/s, and Δt = 2.0 s. The acceleration is calculated to be

$$\left(\frac{+2.0 \text{ m/s} - 1.0 \text{ m/s}}{2.0 \text{ s}} \right) = +0.50 \text{ m/s}^2$$

The table summarizes the results of the calculations for the six sections:

section	A	B	C	D	E	F
$\mathbf{a}$ (m/s²)	+0.50	0	+1.0	0	–1.25	0

3. The object comes to rest when its velocity is *zero*. Referring to the graph, we estimate that zero velocity corresponds to an approximate time of 9.5 s. (Actually, the time is 9.4 s; we could have *calculated* this value by using the graph and the equation $\mathbf{v}_f = \mathbf{v}_i + \mathbf{a} \cdot t$.)
4. Before 9.4 s, the velocity of the object is always positive; after 9.4 s, its velocity is negative. Therefore the object reverses the direction of its motion at 9.4 s.
5. There are two ways to calculate the displacement of the object.

First, we could multiply the average velocity of each section by the time elapsed in that section. For example, in section E the average velocity is +0.5 m/s and the elapsed time is 4.0 s. Therefore, the displacement of the object is +2.0 m (+0.5 m/s · 2.0 s). A *positive* displacement means that the object traveled the distance in the positive direction.

Second, we could calculate the displacement by measuring the *area* between the section line and the *x*-axis. (In mathematics, this is known as calculating the area "under the curve."). We will use this method to calculate the displacement for section F of the graph, as follows.

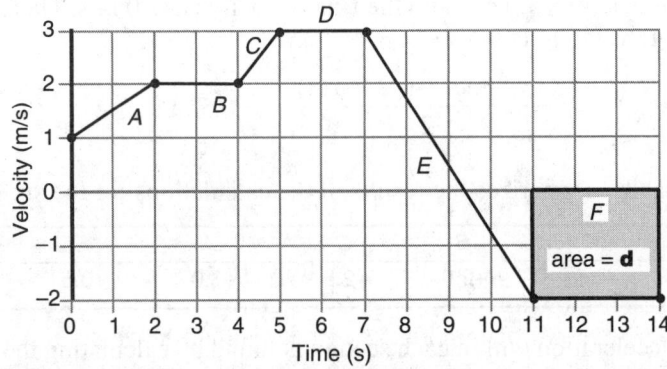

The shaded area is defined by a rectangle whose length is 3.0 s and whose height is –2.0 m/s (this value is negative because section F lies *under* the *x*-axis). The area is the *product* of these two values (3.0 s · –2.0 m/s), that is, –6.0 m.

The table summarizes the results of the calculations for the six sections:

section	A	B	C	D	E	F
ds (m)	+3.0	+4.0	+2.5	+6.0	+2.0	–6.0

6. The displacement over the entire trip is found by *adding* the displacements for all the sections:

$$d_{total} = +11.5 \text{ m}$$

(Refer to the table above.)

7. The average velocity for the entire trip is found by *dividing* d_{total} by the total time (14 s):

$$\overline{v}_{(total)} = +0.82 \text{ m/s}$$

8. The acceleration versus time graph for this object is constructed by referring to the accelerations over all of the sections. (See the table on page 25 for part 2 of this problem.)

26

Note that the accelerations are drawn as straight-line segments *within* each section, but they are not connected *between* sections. The reason is that the change in velocity between sections is so abrupt that the acceleration cannot be calculated accurately. (Contrast the velocity versus time graph shown in part 5 of this problem.)

The first point of importance to note is that the slope of a velocity–time graph or a **v** vs. *t* graph is equal to the acceleration of the object. Assuming as in the graph on page 26 that there are straight-line segments, the acceleration is constant for each segment. A curved graph or a graph with a changing slope would indicate changing acceleration.

The second point of importance to note is that the area under the curve of a velocity–time graph is equal to the displacement of the object.

Lastly, in an acceleration–time graph or an **a** vs. *t* graph, it is important to note that the area under the curve is equal to the object's change in velocity.

PART A AND B–1 QUESTIONS

✪ 1. A flashing light of constant 0.20-second period is situated on a lab cart. The diagram below represents a photograph of the light as the cart moves across a tabletop.

A   B

• • •   •   •   •   •

How much time elapsed as the cart moved from position *A* to position *B*?
(1) 1.0 s   (2) 5.0 s   (3) 0.80 s   (4) 4.0 s

✪ indicates material that is not part of the core curriculum.

2. Approximately how much time does it take light to travel from the Sun to Earth?
   (1) 2.00×10^{-3} s   (3) 5.00×10^2 s
   (2) 1.28×10^0 s   (4) 4.50×10^{19} s

3. An object travels for 8.00 seconds with an average speed of 160. meters per second. The distance traveled by the object is
   (1) 20.0 m   (2) 200. m   (3) 1280 m   (4) 2560 m

4. A blinking light of constant period is situated on a lab cart. Which diagram best represents a photograph of the light as the cart moves with constant velocity?
   (1) • • •   •   •   •
   (2) •   •   •   • • •
   (3) •   •   •   •   •   •   •
   (4) • •   •   •   •   • •

5. A car is accelerated at 4.0 meters per second² from rest. The car will reach a speed of 28 meters per second at the end of
   (1) 3.5 s   (2) 7.0 s   (3) 14 s   (4) 24 s

Base your answers to questions 6 through 8 on the information and diagram below. The diagram represents a block sliding along a frictionless surface between points A and G.

6. As the block moves from point A to point B, the speed of the block will be
   (1) decreasing
   (2) increasing
   (3) constant, but not zero
   (4) zero

7. Which expression represents the magnitude of the block's acceleration as it moves from point C to point D?
   (1) $\dfrac{m}{F}$ (2) $\dfrac{v}{t}$ (3) $m \, \Delta v$ (4) $\dfrac{2d}{t}$

8. Which formula represents the velocity of the block as it moves along the horizontal surface from point E to point F?
   (1) $\bar{v} = \dfrac{d}{t}$ (2) $\bar{v} = \dfrac{\Delta v}{2}$ (3) $v_f^2 = 2a \, \Delta d$ (4) $\Delta v = \dfrac{1}{2} a \, (\Delta t)^2$

28

9. An object that is originally moving at a speed of 20. meters per second accelerates uniformly for 5.0 seconds to a final speed of 50. meters per second. What is the acceleration of the object?
   (1) 14 m/s²   (2) 10. m/s²   (3) 6.0 m/s²   (4) 4.0 m/s²

10. A block starting from rest slides down the length of an 18-meter plank with a uniform acceleration of 4.0 meters per second². How long does the block take to reach the bottom?
   (1) 4.5 s   (2) 2.0 s   (3) 3.0 s   (4) 9.0 s

Base your answers to questions 11 through 15 on the information below.

A toy projectile is fired from the ground vertically upward with an initial velocity of +29 meters per second. The projectile arrives at its maximum altitude in 3.0 seconds. [Neglect air resistance.]

11. The greatest height the projectile reaches is approximately
   (1) 23 m   (2) 44 m   (3) 87 m   (4) 260 m

12. What is the velocity of the projectile when it hits the ground?
   (1) 0. m/s   (2) –9.8 m/s   (3) –29 m/s   (4) +29 m/s

13. What is the displacement of the projectile from the time it left the ground until it returned to the ground?
   (1) 0. m   (2) 9.8 m   (3) 44 m   (4) 88 m

14. Which graph best represents the relationship between velocity (v) and time (t) for the projectile?

       (1)   (2)   (3)   (4)

15. As the projectile rises and then falls back to the ground, its acceleration
   (1) decreases, then increases   (3) increases, only
   (2) increases, then decreases   (4) remains the same

16. A freely falling object near the Earth's surface has a constant
   (1) velocity of –1.00 m/s   (3) acceleration of –1.00 m/s²
   (2) velocity of –9.81 m/s   (4) acceleration of –9.81 m/s²

17. The speed of an object undergoing constant acceleration increases from 8.0 meters per second to 16.0 meters per second in 10. seconds. How far does the object travel during the 10. seconds?
(1) 3.6×10^2 m   (3) 1.2×10^2 m
(2) 1.6×10^2 m   (4) 8.0×10^1 m

18. A 2.0-kilogram stone that is dropped from the roof of a building takes 4.0 seconds to reach the ground. Neglecting air resistance, the maximum speed of the stone will be approximately
(1) 8.0 m/s   (2) 9.8 m/s   (3) 29 m/s   (4) 39 m/s

Base your answers to questions 19 and 20 on the diagram below, which shows a 1-kilogram aluminum sphere and a 3-kilogram brass sphere, both having the same diameter and both at the same height above the ground. Both spheres are allowed to fall freely. [Neglect air resistance.]

19. Both spheres are released at the same instant. They will reach the ground at
(1) the same time but with different speeds
(2) the same time with the same speeds
(3) different times but with the same speeds
(4) different times and with different speeds

20. If the spheres are 19.6 meters above the ground, the time required for the aluminum sphere to reach the ground is
(1) 1 s   (2) 2 s   (3) 8 s   (4) 4 s

21. A softball is thrown straight up, reaching a maximum height of 20 meters. Neglecting air resistance, what is the ball's approximate vertical velocity when it hits the ground?
(1) –10 m/s   (2) –20 m/s   (3) –15 m/s   (4) –40 m/s

22. An object is allowed to fall freely near the surface of a planet. The object has an acceleration due to gravity of 24 meters per second^2. How far will the object fall during the first second?
(1) 24 m   (2) 12 m   (3) 9.8 m   (4) 4.9 m

23. Starting from rest, an object rolls freely down an incline that is 10 meters long in 2 seconds. The acceleration of the object is approximately
(1) 5 m/s (2) 5 m/s^2 (3) 10 m/s (4) 10 m/s^2

Base your answers to questions 24 and 25 on the information below.

An object starting from rest moves down an incline with an acceleration of 2 meters per second^2 for 2 seconds.

24. How far does the object move during the 2 seconds?
(1) 1 m (2) 2 m (3) 8 m (4) 4 m

25. The final speed of the object after 2 seconds is
(1) 1 m/s (2) 2 m/s (3) 8 m/s (4) 4 m/s

26. An object initially traveling in a straight line with a speed of 5.0 meters per second is accelerated at 2.0 meters per second^2 for 4.0 seconds. The total distance traveled by the object in the 4.0 seconds is
(1) 36 m (2) 24 m (3) 16 m (4) 4.0 m

27. Starting from rest, object A falls freely for 2.0 seconds, and object B falls freely for 4.0 seconds. Compared with object A, object B falls
(1) one-half as far (3) 3 times as far
(2) twice as far (4) 4 times as far

28. An object starts from rest and falls freely. What is the speed of the object at the end of 3.00 seconds?
(1) 9.81 m/s (2) 19.6 m/s (3) 29.4 m/s (4) 88.2 m/s

29. An object is allowed to fall freely near the surface of a planet. The object falls 54 meters in the first 3.0 seconds after it is released. The acceleration due to gravity on that planet is
(1) –6.0 m/s^2 (2) –12 m/s^2 (3) –27 m/s^2 (4) –108 m/s^2

30. An object, initially at rest, falls freely near the Earth's surface. How long does it take the object to attain a speed of 98 meters per second?
(1) 0.1 s (2) 10 s (3) 98 s (4) 960 s

31. A car accelerates uniformly from rest at 3.2 meters per second^2. When the car has traveled a distance of 40. meters, its speed will be
(1) 8.0 m/s (2) 12.5 m/s (3) 16 m/s (4) 128 m/s

32. An object initially at rest accelerates at 5 meters per second^2 until it attains a speed of 30 meters per second. What distance does the object move while accelerating?

(1) 30 m   (2) 90 m   (3) 3 m   (4) 600 m

Base your answers to questions 33 through 36 on the graph below, which is a velocity versus time graph for an object moving in a straight line.

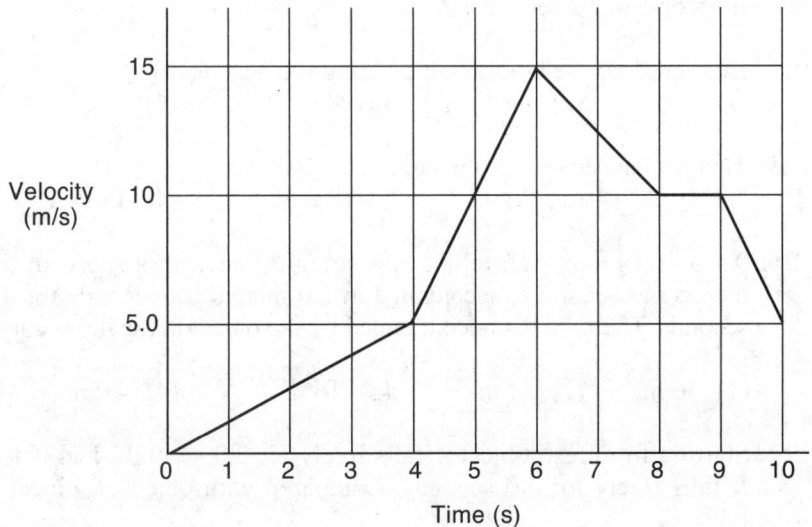

33. During which time interval did the object travel the greatest distance?

(1) 0 s to 4 s   (2) 4 s to 6 s   (3) 6 s to 8 s   (4) 9 s to 10 s

34. During which time interval did the object have the greatest positive acceleration?

(1) 0 s to 4 s   (2) 4 s to 6 s   (3) 6 s to 8 s   (4) 9 s to 10 s

35. What was the acceleration of the object during the time interval between 4 seconds and 6 seconds?

(1) 20. m/s^2 (2) 10. m/s^2 (3) 5.0 m/s^2 (4) 0.20 m/s^2

36. What distance did the object travel during the time interval between 9 seconds and 10 seconds?

(1) 5.0 m   (2) 7.5 m   (3) 10. m   (4) 20. m

Base your answers to questions 37 through 41 on the graph below, which represents the relationship between speed and time for an object in motion along a straight line.

37. What is the acceleration of the object during the time interval $t = 3$ seconds to $t = 5$ seconds?
    (1) 5.0 m/s² (2) 7.5 m/s²   (3) 12.5 m/s²   (4) 17.5 m/s²

38. What is the average speed of the object during the time interval $t =$ 6 seconds to $t = 8$ seconds?
    (1) 7.5 m/s (2) 10 m/s   (3) 15 m/s   (4) 17.5 m/s

39. What is the total distance traveled by the object during the first 3 seconds?
    (1) 15 m   (2) 20 m   (3) 25 m   (4) 30 m

40. During the interval $t = 8$ seconds to $t = 10$ seconds, the speed of the object is
    (1) zero (2) increasing (3) decreasing (4) constant, but not zero

41. What is the maximum speed reached by the object during the 10 seconds of travel?
    (1) 10 m/ s (2) 25 m/s   (3) 150 m/s   (4) 250 m/s

42. An object is thrown vertically upward from the surface of the Earth. Which graph best represents the relationship between velocity and time for the object as it rises and then returns to the Earth?

33

Base your answers to questions 43 and 44 on the graph and information below.

Cars *A* and *B* both start from rest at the same location at the same time.

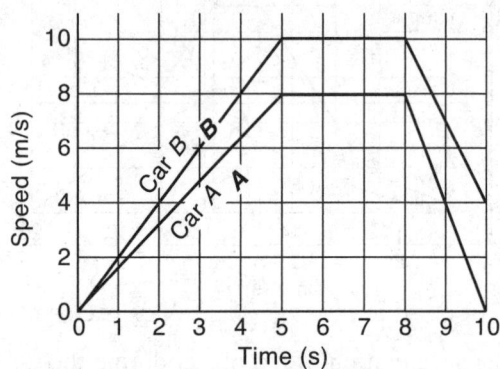

43. What is the magnitude of the acceleration of car *A* during the period between *t* = 8 seconds and *t* = 10 seconds?
(1) 20 m/s^2 (2) 16 m/s^2 (3) 18 m/s^2 (4) 4 m/s^2

44. Compared to the speed of car *B* at 6 seconds, the speed of car *A* at 6 seconds is
(1) less   (2) greater   (3) the same

45. The graph represents the acceleration acting on an object as a function of time. During which time interval is the velocity of the object constant?

(1) 0 to 2   (2) 2 to 3   (3) 3 to 4   (4) 4 to 5

34

Base your answers to questions 46 through 50 on the accompanying graph, which represents the motions of four cars on a straight road.

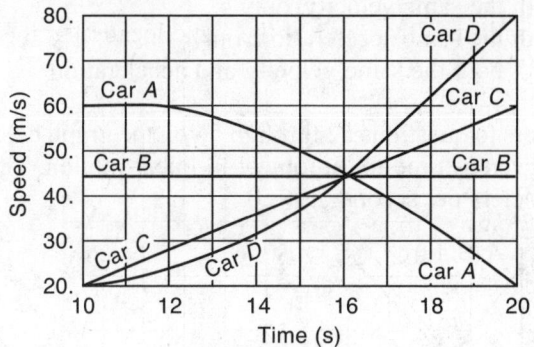

46. The speed of car C at time $t = 20$ seconds is closest to
    (1) 60 m/s   (2) 45 m/s   (3) 3.0 m/s   (4) 4.0 m/s

47. Which car has zero acceleration?
    (1) A (2) B (3) C (4) D

48. Which car has a negative acceleration?
    (1) A (2) B (3) C (4) D

49. Which car moves the greatest distance in the time interval $t = 10$ seconds to $t = 16$ seconds?
    (1) A (2) B (3) C (4) D

50. Which graph best represents the relationship between distance and time for car C?

    (1)   (2)   (3)   (4)

51. The motions of cars A, B, and C in a straight path are represented by the graph below.

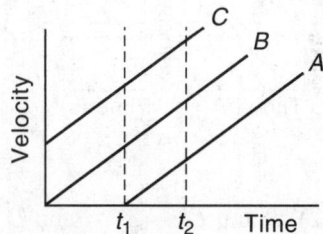

During the time interval from t_1 to t_2, the three cars travel
(1) the same distance
(2) with the same velocity, only
(3) with the same acceleration, only
(4) with both the same velocity and acceleration

Base your answers to questions 52 through 56 on the graph below, which represents velocity versus time for an object in linear motion. The object has a velocity of 20 meters per second at $t = 0$.

52. During which interval is the magnitude of the acceleration greatest?
(1) *EF*   (2) *FG*   (3) *GH*   (4) *IJ*

53. The acceleration of the object at point *D* on the curve is
(1) 0 m/s²   (2) 5 m/s²   (3) –10 m/s²   (4) –20 m/s²

54. During what interval does the object have zero acceleration?
(1) *BC*   (2) *EF*   (3) *GH*   (4) *HI*

55. At what point is the distance from start zero?
(1) *C*   (2) *E*   (3) *F*   (4) *J*

56. At what point is the distance from start a maximum?
(1) *C*   (2) *E*   (3) *G*   (4) *J*

57. Which combination of graphs best describes free-fall motion? [Neglect air resistance.]

       A   B   C   D

(1) *A* and *C*   (2) *B* and *D*   (3) *A* and *D*   (4) *B* and *C*

58. The graph below represents the relationship between velocity and time for an object moving in a straight line. What is the acceleration of the object?

(1) 0 m/s²   (2) 5 m/s²   (3) 3 m/s²   (4) 15 m/s²

59. The graph below represents the velocity versus time relationship for a ball thrown vertically upward. Time zero represents the time of release.

During the time interval between 1.0 second and 2.0 seconds, the displacement of the ball from the point where it was released
(1) decreases   (2) increases   (3) remains the same

60. The area under a speed versus time curve is a measure of
(1) acceleration   (2) distance   (3) momentum   (4) velocity

61. The graph below represents the displacement of an object moving in a straight line as a function of time.

37

What was the total distance traveled by the object during the 10.0-second time interval?
(1) 0 m   (2) 8 m   (3) 16 m   (4) 24 m

62. A race car starting from rest accelerates uniformly at a rate of 4.90 meters per second^2. What is the car's speed after it has traveled 200. meters?
(1) 1960 m/s   (2) 62.6 m/s   (3) 44.3 m/s   (4) 31.3 m/s

63. A ball is thrown straight downward with a speed of 0.50 meter per second from a height of 4.0 meters. What is the speed of the ball 0.70 second after it is released? [Neglect friction.]
(1) 0.50 m/s   (2) 7.4 m/s   (3) 9.8 m/s   (4) 15 m/s

64. An astronaut standing on a platform on the Moon drops a hammer. If the hammer falls 6.0 meters vertically in 2.7 seconds, what is its acceleration?
(1) 1.6 m/s^2 (2) 2.2 m/s^2 (3) 4.4 m/s^2 (4) 9.8 m/s^2

65. An observer recorded the following data for the motion of a car undergoing constant acceleration.

Time (s)	Speed (m/s)
3.0	4.0
5.0	7.0
6.0	8.5

What was the magnitude of the acceleration of the car?
(1) 1.3 m/s^2 (2) 2.0 m/s^2 (3) 1.5 m/s^2 (4) 4.5 m/s^2

66. Which graph best represents the relationship between the velocity of an object thrown straight upward from Earth's surface and the time that elapses while it is in the air? [Neglect friction.]

(1)   (2)   (3)   (4)

67. A car increases its speed from 9.6 meters per second to 11.2 meters per second in 4.0 seconds. The average acceleration of the car during this 4.0-second interval is
(1) 0.40 m/s²   (2) 2.4 m/s²   (3) 2.8 m/s²   (4) 5.2 m/s²

68. What is the speed of a 2.5-kilogram mass after it has fallen freely from rest through a distance of 12 meters?
(1) 4.8 m/s   (2) 15 m/s   (3) 30. m/s   (4) 43 m/s

69. A cart travels with a constant nonzero acceleration along a straight line. Which graph best represents the relationship between the distance the cart travels and time of travel?

   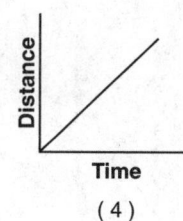

     (1)   (2)   (3)   (4)

70. Which graph best represents the relationship between the acceleration of an object falling freely near the surface of Earth and the time that it falls?

  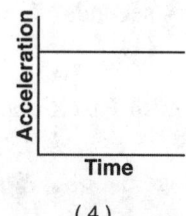

     (1)   (2)   (3)   (4)

71. A rock falls from rest off a high cliff. How far has the rock fallen when its speed is 39.2 meters per second? [Neglect friction.]
(1) 16.9 m   (2) 44.1 m   (3) 78.3 m   (4) 123 m

72. A rocket initially at rest on the grond lifts off vertically with a constant acceleration of 2.0×10^1 meters per second². How long will it take the rocket to reach an altitude of 9.0×10^3 meters?
(1) 3.0×10^1 s   (3) 4.5×10^2 s
(2) 4.3×10^1 s   (4) 9.0×10^2 s

73. The speed of a wagon increases from 2.5 meters per second to 9.0 meters per second in 3.0 seconds as it accelerates uniformly down

a hill. What is the magnitude of the acceleration of the wagon during his 3.0-second interval?
(1) 0.83 m/s^2 (2) 2.2 m/s^2 (3) 3.0 m/s^2 (4) 3.8 m/s^2

74. A 1.0-kilogram ball is dropped from the roof of a building 40. meters tall. What is the approximate time of fall? [Neglect air resistance.]
(1) 2.9 s   (2) 2.0 s   (3) 4.1 s   (4) 8.2 s

75. The graph below represents the relationship between speed and time for an object moving along a straight line.

Speed vs. Time

What is the total distance traveled by the object during the first 4 seconds?
(1) 5 m   (2) 20 m   (3) 40 m   (4) 80 m

Answers to Part A and B–1 questions can be found on page 454.

PART B–2 AND C QUESTIONS

Base your answers to questions 1 through 3 on the speed-time graph below, which represents the linear motion of a cart.

1. Determine the magnitude of the acceleration of the cart during interval *AB*. [Show all calculations, including the equation and substitution with units.]

40

2. Calculate the distance traveled by the cart during interval *BC*. [Show all calculations, including the equation and substitution with units.]

3. What is the average speed of the cart during interval *CD*?

Base your answers to questions 4 through 6 on the information and data table below.

A car is traveling due north at 24.0 meters per second when the driver sees an obstruction on the highway. The data table below shows the velocity of the car at 1.0-second intervals as it is brought to rest on the straight, level highway.

Time (s)	Velocity (m/s)
0.0	24.0
1.0	19.0
2.0	14.0
3.0	10.0
4.0	4.0

Using the information in the data table, construct a graph on the grid provided following the directions below.

4. Plot the data points for velocity versus time.

5. Draw the best-fit line.

41

6. Using your graph, determine the acceleration of the car. [Show all calculations, including the equation and substitution with units.]

Base your answers to questions 7 through 10 on the data table below, which describes the motion of an object moving in a straight line.

Data Table

Time (s)	Speed (m/s)
0.0	0.0
1.0	1.2
2.0	2.7
3.0	3.3
4.0	5.0
5.0	5.6

Directions (7–10): Using the information in the data table, construct a graph on the grid provided below, following the directions below.

7. Plot the data points.

8. Draw the line of best-fit.

9. On the same grid, sketch a line representing an object decelerating uniformly in a straight line.

Speed vs. Time

10. Based on your line of best-fit, what is the acceleration of the object?

42

Base your answers to questions 11 through 14 on the information and data table below.

A 1.00-kilogram mass was dropped from rest from a height of 25.0 meters above Earth's surface. The speed of the mass was determined at 5.0-meter intervals and recorded in the data table below.

Data Table

Height Above Earth's Surface (m)	Speed (m/s)
25.0	0.0
20.0	9.9
15.0	14.0
10.0	17.1
5.0	19.8
0	22.1

Speed vs. Height Above Earth's Surface

Height Above Earth's Surface (m)

Directions (11–14): Using the information in the data table, construct a graph on the grid above following the directions below.

11. Mark an appropriate scale on the axis labeled "Height Above Earth's Surface (m)."

12. Plot the data points for speed versus height above Earth's surface.

13. Draw the line or curve of best fit.

14. Using your graph, determine the speed of the mass after it has fallen a vertical distance of 12.5 meters.

Base your answers to questions 15 through 18 on the information and diagram below.

A spark timer is used to record the position of a lab cart accelerating uniformly from rest. Each 0.10 second, the timer marks a dot on a recording tape to indicate the position of the cart at that instant, as shown.

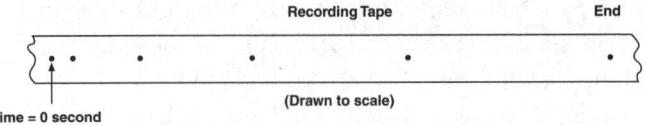

Recording Tape   End

(Drawn to scale)

time = 0 second

15. Using a metric ruler, measure the distance the cart traveled during the interval $t = 0$ second to $t = 0.30$ second. Record your answer to the *nearest tenth of a centimeter*.

16. Calculate the magnitude of the acceleration of the cart during the time interval $t = 0$ second to $t = 0.30$ second. [Show all work, including the equation and substitution with units.]

17. Calculate the average speed of the cart during the time interval $t = 0$ second to $t = 0.30$ second. [Show all work, including the equation and substitution with units.]

18. On the diagram below, mark *at least four* dots to indicate the position of a cart traveling at a constant velocity.

Recording Tape

Answers to Part B–2 and C questions can be found on pages 460–466.

| Chapter Three | # FORCES AND NEWTON'S LAWS |

KEY IDEAS

A *force* is a push or a pull in a given direction. If the force is unbalanced, an acceleration will result. The relationship between force and motion is given by Newton's first two laws of motion.

Special categories of forces include weight and friction. *Weight* is the force on an object as a result of gravitational attraction by a massive body such as a planet. Its direction is always toward the center of the body. *Friction* is a force resulting from the contact between two surfaces. Frictional forces are always directed so that they oppose any relative motion.

Forces always occur in pairs; if an object exerts a force on another object, the second object exerts an equal and opposite force on the first object. This relationship is known as Newton's third law of motion.

KEY OBJECTIVES

At the conclusion of this chapter you will be able to:

- Define the term *force* and state its SI unit.
- State Hooke's law, and use relevant data to measure a force.
- State Newton's first law of motion.
- State Newton's second law of motion, and use it to solve problems.
- Define the term *weight*, and relate it to Newton's second law of motion.
- Define the term *normal force*.
- Define the term *frictional force*, and solve simple problems involving kinetic friction.
- Define the term *coefficient of friction*, and use it in the solution of problems.
- State Newton's third law of motion, and apply it to common situations.

3.1   INTRODUCTION

In Chapter 2 we examined the one-dimensional motion of an object in great detail, a study known as *kinematics*. Now we focus our attention on the relationship between forces and motion. This is the study of *dynamics*.

3.2   WHAT IS A FORCE?

A **force** is a push or a pull in a given direction. Since a force has both magnitude (the "strength" of the force) and direction, it is a vector quantity. We will use the letter **F** to represent force.

3.3   MEASURING FORCES USING HOOKE'S LAW

We can measure the magnitude of a force very simply by recognizing that an applied force will stretch or compress a spring. The English scientist Robert Hooke was able to show that the magnitude of a force (**F**) is directly proportional to the elongation (stretch) or compression of a spring (x) within certain limits. The relationship, known as *Hooke's law*, is given by the equation

===== **PHYSICS CONCEPTS** =====

$$\mathbf{F}_s = kx$$

In the SI system of measurement, if x is measured in meters, $\mathbf{F}_s$ is measured in units called *newtons* (N). One newton is equivalent to approximately ¼ pound of force. The constant of proportionality, k, is known as the *spring constant*. Its unit is the newton per meter (N/m), and it is related to the stiffness of the spring; the greater the constant, the stiffer the spring. The following problem illustrates how Hooke's law can be used.

PROBLEM
The following set of data was obtained by a student while investigating Hooke's law in the laboratory:

Force (N)	Elongation of Spring (m)
0.0	0.0
3.0	0.9
6.0	2.2
9.0	2.8
12	4.1

(a) Graph this set of data.
(b) Using the graph, determine the force applied to the spring if an elongation of 2.3 meters is measured.
(c) Calculate the spring constant *from the graph*.

SOLUTION

(a)

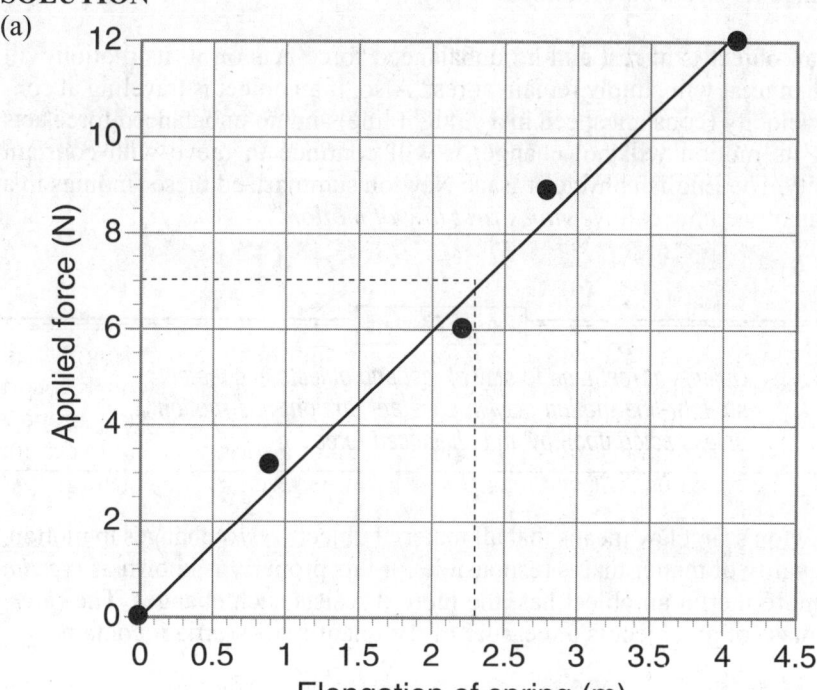

Note that the graph does not need to pass through the data points; it is a "best-fit" straight line.

(b) Reading directly from the graph, we see that an applied force of 7.0 N corresponds to an elongation of 2.3 m.

(c) The spring constant k can be found by calculating the *slope* of the best-fit straight line $\left(\dfrac{\Delta F}{\Delta x}\right)$. A careful calculation of the slope yields a value of 3.0 N/m for the spring constant.

3.4 NEWTON'S FIRST LAW OF MOTION

Originally, it was believed that forces were necessary to keep all objects in motion. The Italian astronomer Galileo Galilei, however, reasoned that a force *changed* the motion of an object as long as the force was not "balanced" by other forces. For example, if you push (lightly!) on a wall, the wall will not move appreciably because your force is balanced by a number of other forces present. However, if you apply an unbalanced force to a chair on a smooth floor, you may see a number of possible effects:

- If the chair is at rest, it will begin to move.
- If the chair is in motion, it may speed up, slow down, come to rest, or change its *direction* of motion.

If an object is at rest and no unbalanced force acts on it, its motion will not change; it will simply remain at rest. Also, if an object is traveling at constant velocity (constant speed in a straight line) and no unbalanced force acts on it, its motion will not change; it will continue to move with constant velocity. The English physicist Isaac Newton summarized these findings in a statement we now call *Newton's first law of motion*:

PHYSICS CONCEPTS

Objects at rest tend to stay at rest and objects in constant straight-line motion stay in constant straight-line motion unless acted upon by an unbalanced force.

Newton's first law means that all material objects *resist* changes in motion. The quality of matter that is responsible for this property is known as *inertia*. The more inertia an object has, the more it resists such changes. The (*inertial*) mass of an object is a measure of the quantity of inertia it contains.

PROBLEM
In which of the following situations do unbalanced forces act?
(a) A car is traveling at 20 meters per second, and the driver steps on the brake.
(b) A chair is dragged across a floor at constant velocity.
(c) A plane makes a turn at constant speed.
(d) An object is dropped vertically toward the ground.

SOLUTION
(a) An unbalanced force acts since the speed of the car is lowered.
(b) No unbalanced force acts on the chair since its velocity is constant.
(c) An unbalanced force acts since the plane's direction is changing.
(d) An unbalanced force acts since the object's speed is increasing.

In every case, an unbalanced force gives rise to an *acceleration* (i.e., a change in velocity over time).

3.5 NEWTON'S SECOND LAW OF MOTION

Although an unbalanced force will give an object an acceleration, the magnitude of the acceleration is determined by two quantities: the magnitude of the force and the *mass* of the object. It has been verified countless times that the acceleration is *directly proportional* to the magnitude of the force and *inversely proportional* to the mass of the object. If we choose our units carefully, we can write this relationship, which we call *Newton's second law of motion,* as follows:

=== **PHYSICS CONCEPTS** ===

$$a = \frac{F_{net}}{m} \text{ or } F_{net} = m \cdot a$$

We measure the mass (m) in kilograms, and the acceleration (a) in meters per second^2. The unbalanced force (F_{net}) is measured in newtons.

PROBLEM
What are the equivalent basic SI units for the newton?

SOLUTION
We can answer this question by using the second equation for Newton's second law with units instead of numbers:

$$F_{net} = m \cdot a$$

$$N = kg \cdot \frac{m}{s^2} = \frac{kg \cdot m}{s^2}$$

PROBLEM
Complete the following table (fill in the blank spaces) by applying Newton's second law of motion:

F_{net} (N)	m (kg)	a (m/s^2)
3.0	6.0	?
?	4.5	2.0
12	?	1.2

SOLUTION
We substitute the values for each row into the equation $F_{net} = m \cdot a$.
Row 1: $3.0 \text{ N} = 6.0 \text{ kg} \cdot a$; $a = 0.50 \text{ m/s}^2$
Row 2: $F_{net} = 4.5 \text{ kg} \cdot 2.0 \text{ m/s}^2$; $F_{net} = 9.0 \text{ N}$
Row 3: $12 \text{ N} = m \cdot 1.2 \text{ m/s}^2$; $m = 10. \text{ kg}$

3.6 WEIGHT

When we hold an object, we feel its heaviness as it pushes into our hand. As we learned in Chapter 2, when the object drops to the ground, it falls with an acceleration of 9.8 meters per second^2 (if we ignore air resistance). These two observations lead us to conclude that a force is present on the object. We call this force **weight**, and its origin is the gravitational attraction between the object and the Earth.

How can we calculate the weight of an object? When an object falls freely, the force on it (its weight!) is unbalanced. We can apply Newton's second law to calculate this force:

PHYSICS CONCEPTS

$$\mathbf{g} = \frac{\mathbf{F}_g}{m} \text{ or } \mathbf{F}_g = m \cdot \mathbf{g}$$

We use $\mathbf{F}_g$ to represent the force due to gravity (the weight of the object) and $\mathbf{g}$ to represent the gravitational acceleration the object experiences in free fall.

PROBLEM
What is the weight of an object with a mass of 15 kilograms?

SOLUTION

$$\mathbf{F}_g = m \cdot \mathbf{g}$$
$$= (15 \text{ kg})(9.8 \text{ m/s}^2) = 149 \text{ N}$$

Helpful hint: 9.8 is very close to 10. In many cases, for most multiple-choice problems involving calculations of weight from mass or mass from weight, it is much easier to multiply and/or divide by 10 instead of 9.8 to estimate an answer quickly.

3.7 NORMAL FORCES

When an object is at rest on a horizontal surface, such as a table, it has weight but is not accelerating. In this situation, the weight of the object is not an unbalanced force. In fact, the net force on the object is zero! Therefore, another force, which serves to balance the effects of gravity, must be present on the object. The origin of this supporting force is the surface itself, and this force is responsible for the contact between the object and the surface.

Since this second force is perpendicular to both the object and the surface, it is called a **normal force**. (In mathematics, the word *normal* is used to describe lines that are perpendicular.) We represent the normal force by the symbol $\mathbf{F}_N$. The diagram below represents the relationship between the weight of an object and the normal force on it.

F_N (From the surface)

F_g (From Earth)

PROBLEM
How does the magnitude of the normal force compare with the weight of the object in the diagram shown above?

SOLUTION
Since the object is at rest, the net force on it is zero. Therefore, the magnitude of the normal force must be *equal* to the weight of the object.
(*Note*: This is not always the case, as we will see in Chapter 4.)

3.8 FRICTIONAL FORCES

Consider the following situation: A 5.0-kilogram object is pulled across a floor with an applied force of 20. newtons. The acceleration of the object is measured to be 3.0 meters per second². This situation is illustrated in the diagram below:

3.0 m/s²

5.0 kg ⟶ 20. N

Something seems to be wrong here: $F \uparrow ma$! But we know from Newton's second law that *ma* must equal the *unbalanced* force ($\mathbf{F}_{net}$). In this situation, $\mathbf{F}_{net}$ is equal to 15 newtons (5.0 kg · 3.0 m/s²), *not* to the applied force of 20. newtons. Where have 5.0 newtons gone?

There is another force present that we have not considered: friction. **Frictional forces** are always present when two surfaces are in contact. The direction of a frictional force on an object is always *opposite* to the direction of the object's motion. We will represent a frictional force by the symbol $\mathbf{F}_f$. We now complete the diagram above by adding in the 5.0-newton frictional force:

PROBLEM

A force of 50. newtons is used to drag a 10.-kilogram object across a horizontal table. If a frictional force of 15. newtons is present on the object, calculate (a) the unbalanced force on the object and (b) the acceleration of the object.

SOLUTION

We begin the solution by drawing a diagram of the situation presented in the problem:

$$\longrightarrow \quad \mathbf{F}_{net}$$

$$\longrightarrow \quad a$$

$$15\ N\ (\mathbf{F}_f) \longleftarrow \boxed{10.\ kg\ (m)} \longrightarrow 50.\ N\ (\mathbf{F}_{applied})$$

(a) We find the unbalanced force ($\mathbf{F}_{net}$) by *subtracting* the frictional force ($\mathbf{F}_f$) from the applied force ($\mathbf{F}_{applied}$). We subtract because the forces are in *opposite* directions:

$$\mathbf{F}_{net} = \mathbf{F}_{applied} - \mathbf{F}_f = 50.\ N - 15\ N = 35\ N$$

(b) We calculate the acceleration of the object by applying Newton's second law:

$$\mathbf{a} = \frac{\mathbf{F}_{net}}{m} = \frac{35\ N}{10.\ kg} = 3.5\ m/s^2$$

PROBLEM

A student drags an object across a laboratory table at constant velocity using an applied force of 12 newtons. Calculate the frictional force present on the object.

SOLUTION

Since the object is moving with constant velocity, its acceleration is zero, as is the unbalanced force on it (why?). We can now apply this relationship:

$$0\ N = \mathbf{F}_{net} = \mathbf{F}_{applied} - \mathbf{F}_f = 12\ N - \mathbf{F}_f$$
$$\mathbf{F}_f = 12\ N$$

There are two ways in which we can view frictional forces. First, when an object is at rest on a surface, a certain minimum force is needed to get the object started. This frictional force is known as *static friction*. Second, there is the frictional force present on an object already in motion; this is known as *kinetic friction*. The force of static friction is usually larger than the force of kinetic friction because surfaces at rest tend to form stronger intermolecular attractions than surfaces in relative motion.

Friction is a contact force whose magnitude depends on two factors: the nature of the surfaces in contact and the magnitude of the contact force (i.e., the *normal force*) present on the object. To reduce friction, we generally try to reduce the amount of contact between the surfaces by using lubricants, such as oil or Teflon, or by employing rolling rather than sliding. Very smooth surfaces, such as glass on glass, may have extremely large frictional forces because of the higher degree of contact.

Friction is always with us. When we solve physics problems, however, we usually try to ignore frictional forces in order to make the problems simpler. For example, we assume that an object falling near the Earth's surface will have a constant acceleration of 9.8 meters per second per second, but this is not so. The friction between the air and the falling object will cause the object to reach a constant *terminal speed.*

3.9   THE COEFFICIENT OF FRICTION

The **coefficient of kinetic friction**, represented by the symbol μ_k, is one way of predicting how much friction will be produced on an object because of its contact with another surface while in motion relative to each other. Experimentally, it is found that the frictional force of an object in motion is *directly proportional* to the normal force present on the object. We can write this relationship as follows:

PHYSICS CONCEPTS

$$F_f = \mu \cdot F_N$$

We can calcuate μ by measuring the frictional force on an object (see the preceding problem) and then dividing this value by the normal force present on the object. The next problem demonstrates this procedure.

The **coefficient of static friction,** represented by the symbol μ_s, is one way of predicting how much friction will be produced on an object because of its contact with another surface while at rest.

PROBLEM
As illustrated in the diagram below, a 5.0-kilogram object slides across a horizontal surface. If a 5.0-newton frictional force is present, calculate the coefficient of kinetic friction between the two surfaces.

$$\mathbf{F}_N = \mathbf{F}_g = 49 \text{ N}$$

5.0 N ($\mathbf{F}_f$) ← | 5.0 kg (m) | ⟶ (direction of motion)

$$\mathbf{F}_g = mg = (5.0 \text{ kg})(9.8 \text{ m/s}^2) = 49 \text{ N}$$

SOLUTION

The first step is to calculate the weight of the object, as shown above, which is 49 N. Since this surface is horizontal and the object is not falling downward, the weight is balanced by the normal force, which is also 49 N.

We calculate μ_k from the relationship $\mu_k = \dfrac{\mathbf{F}_f}{\mathbf{F}_N} = \dfrac{5.0 \text{ N}}{49 \text{ N}} = 0.10$

Note that μ_k does not have units.

The larger the value of μ_k, the greater the frictional force on the object. A table of possible values of μ_k is given below for reference:

Surface – Surface	μ_k
Steel – steel (nonlubricated)	0.57
Steel – steel (lubricated)	0.06
Rubber – concrete (dry)	0.68
Wood – wood	0.30
Waxed wood – dry snow	0.05

3.10 NEWTON'S THIRD LAW OF MOTION— ACTION AND REACTION

Consider the diagram below: An angry woman kicks a wall with her foot and is taken to the hospital to have her fractured toe set in a cast.

Force of wall on toe ← → Force of toe on wall

Why was the woman's toe broken? After all, it was she who exerted a force *on the wall*! As a consequence of her kick, however, the wall returned the favor and exerted a force back *on her toe*. Unfortunately, her toe was not as strong as the wall, and it broke under the stress.

This situation describes a very general principle in physics, one that has been verified countless times and is known as *Newton's third law of motion*:

PHYSICS CONCEPTS

If object A exerts a force on object B, then
object B exerts an equal and opposite force back on object A.

The third law has also been stated equivalently in two other ways: (1) "Forces occur in pairs" and (2) "To every action, there is an equal and opposite reaction" (where the terms *action* and *reaction* refer to the two forces).

According to Newton's third law, *two objects* must always be present and only one force of the pair is present on each object. The two forces are always equal in magnitude and opposite in direction.

We can apply the third law to many situations:

- The recoil of all objects when they collide (e.g., a ball bouncing off a wall) is a consequence of this law.
- When we step on a scale, we place a downward force on the scale. The scale, in turn, exerts an upward force which supports us (the normal force). It is this force that registers our weight.

• We can walk because of Newton's third law! When we take a step, we push *backward* on the Earth. The Earth, in turn, pushes *forward* on us. (If you don't believe this, think of yourself stepping out of a boat onto the shore: Which way does the boat move as you move forward?)

PART A AND B–1 QUESTIONS

1. The diagram below represents a block suspended from a spring. The spring is stretched 0.200 meter. If the spring constant is 200 newtons per meter, what is the weight of the block?

   (1) 40.0 N   (2) 20.0 N   (3) 8.00 N   (4) 4.00 N

2. Compared to the inertia of a 1-kilogram mass, the inertia of a 4-kilogram mass is
   (1) one-fourth as great   (3) 16 times as great
   (2) one-sixteenth as great   (4) 4 times as great

3. An 80-kilogram skier slides on waxed skis along a horizontal surface of snow at constant velocity while pushing with his poles. What is the horizontal component of the force pushing him forward?
   (1) 0.05 N   (2) 0.4 N   (3) 40 N   (4) 4 N

4. If the mass of a moving object could be doubled, the inertia of the object would be
   (1) halved (2) doubled (3) unchanged (4) quadrupled

5. The fundamental units for a force of 1 newton are
   (1) meters per second^2
   (2) kilograms
   (3) meters per second^2 per kilogram
   (4) kilogram · meters per second^2

6. An object with a mass of 2 kilograms is accelerated at 5 meters per second^2. The net force acting on the mass is
   (1) 5 N   (2) 2 N   (3) 10 N   (4) 20 N

7. A cart is uniformly accelerating from rest. The net force acting on the cart is
   (1) decreasing   (2) zero   (3) constant   (4) increasing

Base your answers to questions 8 and 9 on the information and diagram below.

A force is applied to a 5.0-kilogram object. The force is always applied in the same direction, but its magnitude varies with time according to the graph below. [Neglect friction.]

8. What is the acceleration of the object at time $t = 2.5$ seconds?
   (1) 1.0 m/s^2 (2) 0.2 m/s^2 (3) 5.0 m/s^2 (4) 9.8 m/s^2

9. During which time interval did the object have a constant velocity?
   (1) 0 s to 1 s   (2) 1 s to 2 s   (3) 2 s to 3 s   (4) 3 s to 4 s

Note that question 10 has only three choices.

10. A carpenter hits a nail with a hammer. Compared to the magnitude of the force the hammer exerts on the nail, the magnitude of the force the nail exerts on the hammer during contact is
   (1) less   (2) greater   (3) the same

11. What force is needed to give an electron an acceleration of 1.00×10^{10} meters per second^2?
    (1) 9.11×10^{-41} N   (3) 9.11×10^{-21} N
    (2) 9.11×10^{-31} N   (4) 1.10×10^{43} N

12. When an unbalanced force of 10. newtons is applied to an object whose mass is 4.0 kilograms, the acceleration of the object will be
    (1) 40. m/s^2 (2) 2.5 m/s^2 (3) 9.8 m/s^2 (4) 0.40 m/s^2

13. In the graph below the acceleration of an object is plotted against the unbalanced force on the object. The mass of the body is

    (1) 1.0 kg   (2) 2.0 kg   (3) 0.5 kg   (4) 8.0 kg

14. Assume that an object has no unbalanced force acting on it. Which statement about the object is true?
    (1) The object may be in motion.
    (2) The object must be slowing down.
    (3) The object must be at rest.
    (4) The object may be speeding up.

15. A cart is uniformly accelerating from rest. The net force acting on the cart is
    (1) decreasing (2) zero   (3) constant   (4) increasing

16. If the mass of an object were doubled, its acceleration due to gravity would be
    (1) halved   (2) doubled   (3) unchanged   (4) quadrupled

⭐ 17. A car's performance is tested on various horizontal road surfaces. The brakes are applied, causing the rubber tires of the car to slide along the road without rolling. The tires encounter the greatest force of friction to stop the car on
    (1) dry concrete   (3) wet concrete
    (2) dry asphalt   (4) wet asphalt

⭐ indicates material that is not part of the core curriculum.

✪ 18. If a 65-kilogram astronaut exerts a force with a magnitude of 50. newtons on a satellite that she is repairing, the magnitude of the force that the satellite exerts on her is
(1) 0 N
(3) 50. N more than her weight
(2) 50. N less than her weight
(4) 50. N

19. What is the weight of a 5.0-kilogram object at the surface of the Earth?
(1) 0.5 N
(2) 5.0 N
(3) 49 N
(4) 490 N

20. For a freely falling object, the ratio of the force of gravity to its acceleration is
(1) weight
(2) momentum
(3) kinetic energy
(4) mass

21. The weight of a chicken egg is most nearly equal to
(1) 10^{-3} N
(2) 10^{-2} N
(3) 10^0 N
(4) 10^2 N

22. A force of 10. newtons applied to mass M accelerates the mass at 2.0 meters per second^2. The same force applied to a mass of $2M$ would produce an acceleration of
(1) 1.0 m/s^2
(2) 2.0 m/s^2
(3) 0.5 m/s^2
(4) 4.0 m/s^2

23. An object with a mass of 0.5 kilogram starts from rest and achieves a maximum speed of 20 meters per second in 0.01 second. What average unbalanced force accelerates this object?
(1) 1000 N
(2) 10 N
(3) 0.1 N
(4) 0.001 N

24. An object is acted upon by a constant unbalanced force. Which graph best represents the motion of this object?

Time (1)   Time (2)   Time (3)   Time (4)

25. At a given location on the Earth's surface, which graph best represents the relationship between an object's mass (M) and weight (W)?

(1)   (2)   (3)   (4)

59

26. Two forces of 5 newtons and 10 newtons act on a 10-kilogram body as shown. The magnitude of the body's acceleration will be

(1) 50 m/s²   (2) 2.0 m/s²   (3) 1.5 m/s²   (4) 0.5 m/s²

27. If an object is moving north, the direction of the frictional force is
   (1) north   (2) south   (3) east   (4) west

28. In order to keep an object weighing 20 newtons moving at constant speed along a horizontal surface, a force of 10 newtons is required. The force of friction between the surface and the object is
   (1) 0 N   (2) 10 N   (3) 20 N   (4) 30 N

29. A box is sliding down an inclined plane as shown. The force of friction is directed toward point

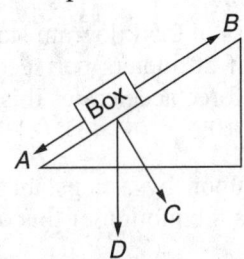

(1) A   (2) B   (3) C   (4) D

30. A horizontal force of 15 newtons pulls a 5-kilogram block along a horizontal surface. If the force produces an acceleration of 2 meters per second², the frictional force acting on the block is
   (1) 1 N   (2) 2 N   (3) 5 N   (4) 15 N

31. A 2-kilogram mass moving along a horizontal surface at 10 meters per second is acted upon by a 5-newton force of friction. The time required to bring the mass to rest is
   (1) 10 s   (2) 2 s   (3) 5 s   (4) 4 s

32. A force of 40 newtons applied horizontally is required to push a 20-kilogram box at constant velocity across the floor. The coefficient of friction between the box and the floor is
   (1) 0.2   (2) 2   (3) 0.5   (4) 5

33. A 100-newton box is moving on a horizontal surface. A force of 10 newtons applied parallel to the surface is required to keep the box moving at constant velocity. What is the maximum coefficient of kinetic friction between the box and the surface?
(1) 0.1   (2) 10   (3) 0.5   (4) 1000

Base your answers to questions 34 and 35 on the diagram and information below.

The diagram shows a 2.0 kilogram mass suspended from a string that is attached to a 3.0 kilogram lab cart. (The system is kept at rest by means of hook *H*.) Assume that the system is frictionless, that the pulley has no mass, and that the table is level.

34. The force exerted on the cart by the hook is closest to
(1) 6.0 N   (2) 2.0 N   (3) 20. N   (4) 10. N

35. After the cart is released, the total mass being accelerated is
(1) 6.0 kg   (2) 2.0 kg   (3) 5.0 kg   (4) 4.0 kg

36. What is the net force on the block shown?

(1) 0 N   (2) 9.8 N   (3) 10 N   (4) 20 N

37. In the diagram below, surface A of the wooden block has twice the area of surface B. If a force of F newtons is required to keep the block moving at constant speed across the table when it slides on surface A, what force is needed to keep the block moving at constant speed when it slides on surface B?

    (1) F (2) $2F$ (3) $\dfrac{1}{2}F$ (4) $4F$

38. A l-kilogram bird stands on a limb. The force that the limb exerts on the bird is
(1) 1 N   (2) 2 N   (3) 0 N   (4) 9.8 N

39. A test booklet is sitting at rest on a desk. Compared to the magnitude of the force of the booklet on the desk, the magnitude of the force of the desk on the booklet is
(1) less   (2) greater   (3) the same

40. A 2.00-kilogram mass is at rest on a horizontal surface. The force exerted by the surface on the mass is approximately
(1) 0 N   (2) 2.00 N   (3) 9.80 N   (4) 19.6 N

41. As masses are added to a box resting on a table, the force exerted by the table on the box
(1) decreases   (2) increases   (3) remains the same

42. A baseball bat moving at high velocity strikes a feather. If air resistance is neglected, compared to the force exerted by the bat on the feather, the force exerted by the feather on the bat will be
(1) smaller   (2) larger   (3) the same

43. Which graph best represents the motion of an object on which the net force is zero?

      (1)   (2)   (3)   (4)

44. The diagram below represents a spring hanging vertically that stretches 0.075 meter when a 5.0-newton block is attached. The spring-block system is at rest in the position shown.

The value of the spring constant is
(1) 38 N/m   (2) 67 N/m   (3) 130 N/m   (4) 650 N/m

45. A force of 1 newton is equivalent to 1

(1) $\dfrac{\text{kg} \cdot \text{m}}{\text{s}^2}$

(2) $\dfrac{\text{kg} \cdot \text{m}}{\text{s}}$

(3) $\dfrac{\text{kg} \cdot \text{m}^2}{\text{s}^2}$

(4) $\dfrac{\text{kg}^2 \cdot \text{m}^2}{\text{s}^2}$

46. The graph below represents the relationship between the force applied to a spring and spring elongation for four different springs.

Force vs. Elongation

Which spring has the greatest spring constant?
(1) *A*   (2) *B*   (3) *C*   (4) *D*

Base your answers to questions 47 and 48 on the diagram below, which shows a 1.0-newton metal disk resting on an index card that is balanced on top of a glass.

47. What is the net force acting on the disk?
(1) 1.0 N   (2) 2.0 N   (3) 0 N   (4) 9.8 N

48. When the index card is quickly pulled away from the glass in a horizontal direction, the disk falls straight down into the glass. This action is a result of the disk's
(1) inertia   (2) charge   (3) shape   (4) temperature

49. A vertical spring 0.100 meter long is elongated to a length of 0.119 meter when a 1.00-kilogram mass is attached to the bottom of the spring. The spring constant of this spring is
(1) 9.8 N/m   (2) 82 N/m   (3) 98 N/m   (4) 520 N/m

Note that question 50 has only three choices.

50. Compared to the force needed to start sliding a crate across a rough level floor, the force needed to keep it sliding once it is moving is
(1) less   (2) greater   (3) the same

51. A 400-newton girl standing on a dock exerts a force of 100 newtons on a 10 000-newton sailboat as she pushes it away from the dock. How much force does the sailboat exert on the girl?
(1) 25 N   (2) 100 N   (3) 400 N   (4) 10 000 N

52. What is the magnitude of the force needed to keep a 60.-newton rubber block moving across level, dry asphalt in a straight line at a constant speed of 2.0 meters per second?
(1) 40. N   (2) 51 N   (3) 60. N   (4) 120 N

Answers to Part A and B–1 questions can be found on page 454.

PART B-2 AND C QUESTIONS

1. The diagram below shows a spring compressed by a force 6.0 newtons from its rest position to its compressed position.

Calculate the spring constant for this spring. [Show all calculations, including equations and substitutions with units.]

Base your answers to questions 2 through 5 on the information in the data below. The data were obtained by varying the force applied to a spring and measuring the corresponding elongation of the spring.

Applied Force (N)	Elongation of Spring (m)
0.0	0.00
4.0	0.16
8.0	0.27
12.0	0.42
16.0	0.54
20.0	0.71

Directions (2–4): Using the information in the table, construct a graph on the grid provided following the directions below.

2. Mark an appropriate scale on the axis labeled "Elongation (m)."

3. Plot the data points for force versus elongation.

4. Draw the best-fit line.

Applied Force vs. Elongation

5. *Using the best-fit line*, determine the spring constant of the spring. [Show all calculations, including the equation and substitution with units.]

Base your answers to questions 6 and 7 on the information given below.

Friction provides the centripetal force that allows a car to round a circular curve.

6. Find the minimum coefficient of friction needed between the tires and the road to allow a 1,600-kilogram car to round a curve of radius 80. meters at a speed of 20. meters per second. [Show all work, including formulas and substitutions with units.]

7. If the mass of the car were increased, how would that affect the maximum speed at which it could round the curve?

8. A 1500-kilogram car accelerates at 5.0 meters per second^2 on a level, dry, asphalt road. Determine the magnitude of the net horizontal force acting on the car.

_____ N

Base your answers to questions 9 through 11 on the information and data table below.

A student performed an experiment in which the weight attached to a suspended spring was varied and the resulting total length of the spring measured. The data for the experiment are in the table below.

Attached Weight vs. Total Spring Length

Attached Weight (N)	Total Spring Length (m)
0.98	0.37
1.96	0.42
2.94	0.51
3.92	0.59
4.91	0.64

Directions (9–11): Using the information in the data table, construct a graph following the directions below.

9. Plot the data points for the attached weight versus total spring length.

10. Draw the line or curve of best fit.

11. Using your graph, determine the length of the spring before any weight was attached.

12. Objects in free fall near the surface of the Earth accelerate downward at 9.81 meters per second^2. Explain why a feather does *not* accelerate at this rate when dropped near the surface of Earth.

13. A skier on waxed skis is pulled at constant speed across level snow by a horizontal force of 39 newtons. Calculate the normal force exerted on the skier. [Show all work, including the equation and substitution with units.]

Answers to Part B–2 and C questions can be found on pages 466–470.

Chapter Four	VECTOR QUANTITIES AND THEIR APPLICATIONS

KEY IDEAS

Vectors are quantities that have both magnitude and direction. Forces and displacements are examples of vector quantities. A vector is best represented by an arrow: it points in the vector's direction, and its length is proportional to the vector's magnitude.

Vector quantities can be combined with one another. The process is known as vector addition, and the sum is equal to the combined result of the interaction of the individual vectors. The vector sum is also known as the resultant vector. Vectors can be added geometrically by using a measured scale or mathematically by using the laws of algebra and trigonometry.

The inverse process of vector addition is known as resolution. In resolution, a single vector is separated into a number of components. Resolution usually simplifies the solution of vector-related problems such as two-dimensional motion, static equilibrium, and motion on an inclined plane.

KEY OBJECTIVES

At the conclusion of this chapter you will be able to:

- Define the terms *scalar* and *vector* and list scalar and vector quantities.
- Represent a vector quantity by an arrow drawn to scale.
- Relate the direction of a vector to compass directions.
- Define the term *resultant vector*.
- Add vector quantities (1) graphically and (2) algebraically.
- Relate vector subtraction to vector addition.
- Define the term *vector resolution*, and resolve a vector into its *x*- and *y*-components. (Treatment is limited to two-dimensional analysis.)
- Add vector quantities by adding their *x*- and *y*-components.
- Define the term *static equilibrium*.
- Solve static equilibrium problems.
- Identify and calculate the parallel and perpendicular components of an object's weight when the object is on an inclined plane.
- Solve problems involving motion on an inclined plane.
- Solve problems involving the motion of an object in two dimensions.

4.1 INTRODUCTION

If we want to measure the mass of an object, it makes no difference whether we face in any particular direction while taking the measurement. Quantities such as mass, time, and temperature are called **scalar** quantities because they can be described solely in terms of their magnitudes (sizes).

If we are traveling in a car at 60 miles per hour, however, it makes a great deal of difference which way the car is facing: A car traveling east from Los Angeles might end up in New York, but if the car faced west it could end up in the Pacific! Quantities such as velocity, acceleration, and force are called vector quantities because they must be described in terms of their magnitudes *and* directions.

4.2 DISPLACEMENT, AND REPRESENTATION OF VECTOR QUANTITIES

In our study of motion in Chapter 2, we defined the simplest vector, called *displacement*, as a directed change in the position of an object. In other words, displacement is a distance (magnitude) in a given direction. For example, the quantity 30 meters [east] represents a displacement. (In this book, we will indicate the direction in brackets following the magnitude of the vector.)

We can easily represent a vector quantity by using an arrow. The length of the arrow represents the magnitude of the vector, and the direction of the vector points from the tail of the arrow toward its tip. The diagrams illustrate two displacement vectors.

It should be obvious that the arrows shown above are not really 30 meters and 50 meters long. They are drawn *to scale*, as a map maker does when creating a map. In this instance, 1 centimeter of arrow length represents 20 meters of distance, and the arrows are 1.5 and 2.5 centimeters long, respectively.

PROBLEM

If 0.20 meter represents a displacement of 30. kilometers, what displacement is represented by a vector 0.80 meter long?

SOLUTION

$$0.80 \; \cancel{m} \left(\frac{30. \; km}{0.20 \; \cancel{m}} \right) = 120 \; km$$

There are many ways to represent the direction of a vector. We will use the method that we believe to be simplest in the long run. The direction is indicated by the angle that the vector makes with the x-axis (horizontal). We use the compass points to describe how many degrees the vector is north (N) or south (S) of the east-west (E-W) baseline. The diagram illustrates how we do this.

On occasion, it may be necessary to measure the angle that the vector makes with the y-axis (vertical). The conversion is an easy matter since the angles are complementary to one another. (In the diagram above, the angles would be 60° east of north, 20° west of north, and 45° west of south.)

4.3   VECTOR ADDITION

An airplane traveling at 300 meters per second [east] enters the jet stream, whose velocity is 100 meters per second [north]. What is the velocity of the airplane when measured by a person on the ground?

This problem, as well as a host of related problems, can be solved by a process called *vector addition*. When applied to vector quantities, the term *addition* means calculating the net effect of two or more vectors acting on the same object. It does not matter whether the vector quantities are velocities, displacements, forces, or others. The techniques of vector addition are the same in each case.

Suppose a person walks 3.0 meters [east] and then walks 4.0 meters [east]. What is the net displacement of the person? This is a simple problem in vector addition.

The addition is performed by placing the two vectors (drawn to scale, 1 cm = 1 m) in a line, head to tail, as shown in the diagram. The sum of the vectors (indicated by the bold arrow) is called the **resultant vector (R)** and is determined by drawing an arrow beginning at the tail of the first vector and extending to the head of the second vector. The vector equation for this sum is:

$$\begin{array}{ccccc} \textbf{A} & + & \textbf{B} & = & \textbf{R} \\ 3.0 \text{ m [E]} & & 4.0 \text{ m [E]} & & 7.0 \text{ m [E]} \end{array}$$

Note that symbols representing vector quantities are set in boldface type (**A**). Another way of representing a vector quantity is to place an arrow above the symbol ($\vec{A}$). In this book we will use boldface type for vectors.

When vectors are oriented in opposite directions, their magnitudes are *subtracted*, as shown in the following problem.

PROBLEM
A bird flies north 3.0 kilometers and then south 4.0 kilometers. What is the resultant displacement of the bird?

SOLUTION

In this case, as the diagram (drawn to scale, 1 cm = 1 km) shows, the addition yields a resultant of 1.0 kilometer south:

$$\mathbf{A} \qquad + \qquad \mathbf{B} \qquad = \qquad \mathbf{R}$$
$$\text{3.0 m [N]} \qquad \text{4.0 km [S]} \qquad \text{1.0 km [S]}$$

Now we will consider how vectors at right angles are added.

PROBLEM
A plane flies with a velocity of 300. meters per second [east] and enters the jet stream, whose velocity is 100. meters per second [north]. What is the resultant velocity of the plane?

SOLUTION
The solution to this problem also involves a diagram drawn to scale (1 cm = 100 m/s):

If we used a ruler to measure the magnitude of the resultant velocity $\mathbf{R}$, we would find that the length of the resultant arrow translates to 320 m/s. Using a protractor, we see that the angle made by the resultant with the x-axis is approximately 18° N of E.

We could have solved this problem algebraically by recognizing that the resultant is the hypotenuse of a right triangle. Using the Pythagorean theorem, we have

$$R^2 = P^2 + J^2$$
$$R^2 = (300. \text{ m/s})^2 + (100. \text{ m/s})^2 = 100,000 \text{ m/s}^2$$
$$R = 316.2 \text{ m/s} = 316 \text{ m/s}$$

To calculate angle θ, we note that

$$\tan\theta = \frac{\text{opposite side}}{\text{adjacent side}} = \frac{100. \text{ m/s [N]}}{300. \text{ m/s [E]}} = 0.333$$

$$\theta = \tan^{-1}(0.33) = 18° \text{ [N of E]}$$

Finally, let's consider the effect of two forces that act *concurrently* (i.e., at the same point) at an angle other than 90°.

PROBLEM
*Note: Solving by this method is not required in Physical Setting Physics Core.

As shown in the diagram, an object is subjected to two concurrent forces: **A** = 50. newtons [east] and **B** = 30. newtons [30.° north of east]. What is the resultant force on the object?

30. N [30.° N of E]
B
30.°
A
50. N [E]

SOLUTION
We place the vectors head to tail by displacing vector **B** to the right, as shown in the diagram:

R
θ
A
B
30.°

Then we draw **R** from the tail of **A** to the head of **B**. Using a ruler, a protractor, and a suitable scale factor (1 cm = 10 N), we find that the magnitude of **R** is 77 N and the angle θ is 11°.

This problem may also be solved algebraically by using the relationships known as the law of cosines and the law of sines. First we relabel the triangle as shown:

c
A
b
B
C
a

Then we have:

Law of cosines: $c^2 = a^2 + b^2 - 2ab \cos C$

Law of sines: $\dfrac{a}{\sin A} = \dfrac{b}{\sin B} = \dfrac{c}{\sin C}$

Using both diagrams and applying these two laws, we get:

$$R^2 = (50.\ N)^2 + (30.\ N)^2 - 2(50.\ N)(30.\ N)\ (\cos 150°)$$

$$R = 77\ N$$

$$\frac{77\ N}{\sin 150°} = \frac{30\ N}{\sin \theta} \Rightarrow \theta = 11°$$

This is a complicated procedure indeed! We will soon see that there is an easier—and more powerful—way to add vectors algebraically.

4.4 VECTOR SUBTRACTION

Subtraction is really a special case of addition. In mathematics, the expression $a - b$ is equivalent to the expression $a + (-b)$, where $-b$ is called the negative of b. Similarly, we subtract two vectors by adding one of them to the negative of the other:

$$\mathbf{A} - \mathbf{B} = \mathbf{A} + (-\mathbf{B})$$

Since vectors have both magnitude and direction, the negative of a vector **V** is equal to **V** in magnitude but opposite in direction, as illustrated below.

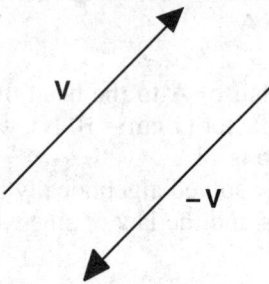

The following diagram compares vector addition and subtraction:

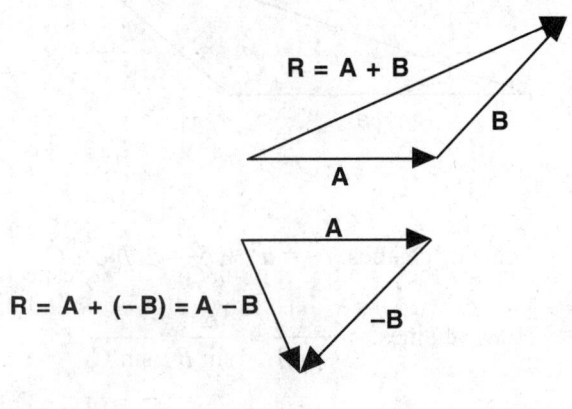

4.5 RESOLUTION OF VECTORS

If an automobile is traveling at 20 meters per second at 45° north of east, we know that the vehicle is traveling north and east at the same time. But *how fast* is it going in each direction? To solve this problem we use a technique known as **vector resolution**.

Resolving a vector means breaking it down into a number of components that will add to produce the original vector, as shown in the diagram.

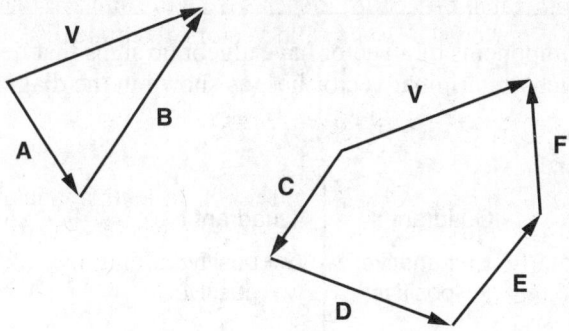

Here **A** and **B** are components of **V**, as are **C**, **D**, **E**, and **F**. As the diagram implies, a vector can be resolved into any number of components in virtually any direction. However, it is usually most helpful to resolve a vector into just *two* components that lie along the *x*- and the *y*-axis (i.e., the components are *perpendicular* to each other).

This resolution is illustrated in the following diagram:

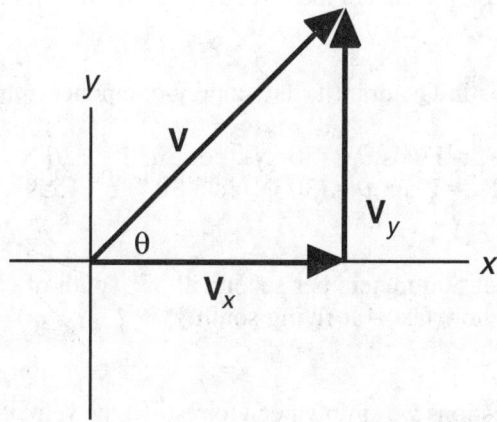

V_x and V_y are called the *x*- and *y*-components of **V**, respectively, and θ is the angle that **V** makes with the *x*-axis. Since the vector and its components form a right triangle, it follows that:

$$\frac{V_x}{V} = \cos \theta \implies V_x = V \cos \theta$$

$$\frac{V_y}{V} = \sin \theta \implies V_y = V \sin \theta$$

The x- and y-components of a vector have algebraic signs that depend on the quadrant in which the original vector lies, as shown in the diagram below:

Quadrant II
x = negative
y = positive

Quadrant I
x = positive
y = positive

Quadrant III
x = negative
y = negative

Quadrant IV
x = positive
y = negative

PROBLEM
Find the x- and y-components of the force $\mathbf{F}$ = 30. newtons [30.° north of east].

SOLUTION
Since the force lies in Quadrant I, its x- and y-components are positive:

$$\mathbf{F}_x = \mathbf{F} \cos \theta = (30.\ \text{N})\ (\cos 30.°) = +26\ \text{N}$$
$$\mathbf{F}_y = \mathbf{F} \sin \theta = (30.\ \text{N})\ (\sin 30.°) = +15\ \text{N}$$

PROBLEM
A plane is flying at 500 meters per second at 37° south of east. How fast is it flying east, and how fast is it flying south?

SOLUTION
To answer the questions we simply need to resolve the velocity vector into its x- and y-components. Since the velocity lies in Quadrant IV, its x-component is positive while its y-component is negative:

$$\mathbf{v}_x = \mathbf{v} \cos \theta = +(500\ \text{m/s}) \cdot (\cos 37°) = +400\ \text{m/s (i.e., 400 m/s [E])}$$
$$\mathbf{v}_y = \mathbf{v} \sin \theta = -(500\ \text{m/s}) \cdot (\sin 37°) = -300\ \text{m/s (i.e., 300 m/s [S])}$$

4.6   USING RESOLUTION TO ADD VECTORS

To solve a vector problem graphically (i.e., using a ruler and protractor) is usually fairly easy. However, we cannot achieve the precision of a mathematical solution. Unfortunately, mathematical solutions using the laws of cosines and sines are tedious and difficult.

   We now develop a technique for vector addition that is relatively simple and yet gives us precision. The rules for this method are as follows:

1. Resolve each of the vectors to be added into its x- and y-components. Remember to include the proper sign (positive or negative), depending on the quadrant of each vector.
2. Be aware that, if a vector lies on the x-axis, its y-component is zero; if a vector lies on the y-axis, its x-component is zero.
3. Add the x-components together to produce the x-component of the resultant vector ($\mathbf{R}_x$).
4. Add the y-components together to produce the y-component of the resultant vector ($\mathbf{R}_y$).
5. Calculate the magnitude of the resultant vector by means of the Pythagorean theorem: $R = \sqrt{\mathbf{R}_x^2 + \mathbf{R}_y^2}$
6. Find the angle that the resultant vector makes with the x-axis from this relationship: $\tan \theta = \dfrac{\mathbf{R}_y}{\mathbf{R}_x}$.

   Let's add three vectors using this method.

PROBLEM
Find the resultant of the force vectors illustrated below:

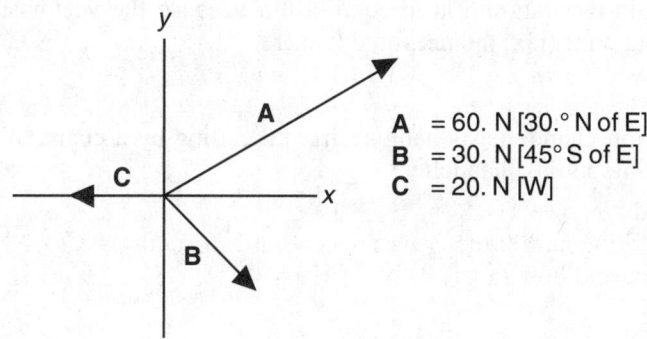

$\mathbf{A}$ = 60. N [30.° N of E]
$\mathbf{B}$ = 30. N [45° S of E]
$\mathbf{C}$ = 20. N [W]

SOLUTION
We begin by calculating the x- and y-components of each force vector:

$$\mathbf{A}_x = \mathbf{A} \cos \theta = + (60.\ \text{N})(\cos 30.°) = + 52\ \text{N}$$
$$\mathbf{A}_y = \mathbf{A} \sin \theta = + (60.\ \text{N})(\sin 30.°) = + 30.\ \text{N}$$
$$\mathbf{B}_x = \mathbf{B} \cos \theta = + (30.\ \text{N})(\cos 45°) = + 21\ \text{N}$$

$$B_y = B \sin \theta = -(30 \text{ N})(\sin 45°) = -21 \text{ N}$$
$$C_x = -20 \text{ N}$$
$$C_y = 0 \text{ N}$$

Next we add the x-components: $(+52 \text{ N}) + (+21 \text{ N}) + (-20. \text{ N}) = +53 \text{ N} = R_x$
and the y-components: $(+30. \text{ N}) + (-21 \text{ N}) + (0 \text{ N}) = +9.0 \text{ N} = R_y$

R_x and R_y are the respective x- and y-components of the resultant force. Since both values are positive, the resultant force lies in Quadrant I (N of E). The magnitude of R is calculated from the equation

$$R = \sqrt{R_x{}^2 + R_y{}^2} = \sqrt{(+53 \text{ N})^2 + (+9.0 \text{ N})^2} = \underline{54 \text{ N}}$$

and the angle θ is calculated from

$$\tan \theta = \frac{R_y}{R_x} = \frac{9.0 \text{ N [N]}}{53 \text{ N [E]}} = 0.17 \Rightarrow \theta = 9.6°[\text{N of E}]$$

4.7 STATIC EQUILIBRIUM

An object is at equilibrium if the vector sum of all the forces acting on it equals zero. If the object is at rest, we say that it is in **static equilibrium**.

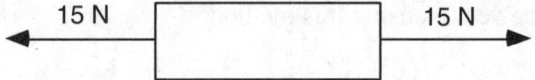

The object in the diagram is in equilibrium because the vector sum of all forces acting on it (i.e., the net force) is zero.

PROBLEM
A 500-newton chandelier is hanging from a ceiling by a chain. What other force is acting on the chandelier?

SOLUTION

The term *tension* is used to describe the force that causes the chain to be taut. The direction of the tension (**T**) in this problem is upward because the tension supports the chandelier.

Since the chandelier is in equilibrium, the vector sum of the forces on it must equal zero. We can write the vector equation for the situation as follows:

$$\mathbf{T} + \mathbf{F}_g = 0$$
$$\mathbf{T} + (-500 \text{ N}) = 0$$
$$\mathbf{T} = +500 \text{ N}$$

(Note that the direction of $\mathbf{F}_g$, which is downward, is represented as a negative quantity. Since tension is a positive quantity, its direction is upward, as stated above.)

PROBLEM

Jesse, whose weight ($\mathbf{F}_g$) is 650 newtons, hangs by his arms from a horizontal bar in the gym. What is the tension in each of his arms?

SOLUTION

It is reasonable to assume that there is equal tension in each of Jesse's arms, and we use **T** to represent this tension. We can write the vector quantity as

$$\mathbf{T} + \mathbf{T} + \mathbf{F}_g = 0$$
$$\mathbf{T} + \mathbf{T} + (-650 \text{ N}) = 0$$
$$2\mathbf{T} = +650 \text{ N}$$
$$\mathbf{T} = +325 \text{ N}$$

The tension in each of Jesse's arms is equal to 325 N.

PROBLEM

Angela, whose weight ($\mathbf{F}_g$) is 400 newtons, sits on a tire swing in her backyard. Linda pulls Angela horizontally to the right so that the rope of the swing makes a 30° angle with the horizontal. If Angela is at rest in this position, what is the tension (**T**) in the rope and what force (**P**) does Linda exert on Angela?

SOLUTION

The vector equation for the forces on Angela is

$$\mathbf{T} + \mathbf{P} + \mathbf{F}_g = 0$$

The easiest way to solve this problem is to resolve the vectors into their x- and y-components. We will use the techniques developed earlier in this chapter.

Since the resultant is zero, both the x- and y-component of the resultant must also be zero. We can write these equations:

$$\mathbf{F}_{g_x} + \mathbf{P}_x + \mathbf{T}_x = \mathbf{R}_x = 0$$
$$\mathbf{F}_{g_y} + \mathbf{P}_y + \mathbf{T}_y = \mathbf{R}_y = 0$$

Force	x-component (N)	y-component (N)
$\mathbf{F}_g$	0	−400
$\mathbf{P}$	+P	0
$\mathbf{T}$	−T cos 30°	+T sin 30°
R	0	0

Substituting in the y-direction, we obtain

$$-400 \text{ N} + 0 + T \sin 30° = 0$$
$$T \sin 30° = 400 \text{ N}$$
$$T = \frac{400 \text{ N}}{\sin 30°} = 800 \text{ N}$$

80

Substituting in the x-direction, we obtain

$$0 + P + (-T \cos 30°) = 0$$
$$P = +T \cos 30° = (800)(\cos 30°)$$
$$= 693 \text{ N} = 700 \text{ N}$$

4.8   INCLINED PLANES

Let us now consider an object whose weight is $\mathbf{F}_g$, sliding down a friction-less inclined plane that makes an angle of θ with the ground. The diagram illustrates the situation.

If we examined the object closely, we would find that it has an acceleration directly parallel to the surface of the plane. To analyze the forces acting on the object, we choose a set of axes that are parallel and perpendicular to the surface of the plane as shown in the diagram below.

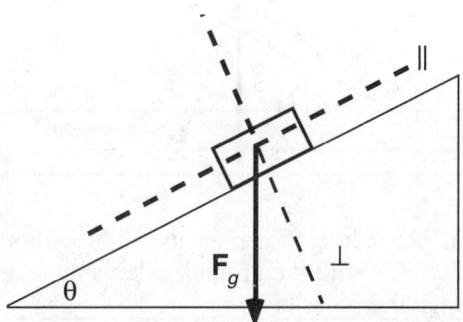

It is most convenient to resolve the weight of the object into components that are parallel and perpendicular to the inclined plane.

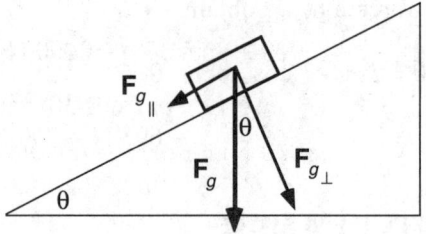

Simple trigonometry yields these relationships

$$\mathbf{F}_{g_\parallel} = \mathbf{F}_g \sin \theta \quad \text{and} \quad \mathbf{F}_{g_\perp} = \mathbf{F}_g \cos \theta$$

Hint: The above equations as they appear are not printed on the Reference Table and should therefore be memorized.

Since the object's acceleration is parallel to the plane, force $\mathbf{F}_{g_\parallel}$ is unbalanced. The object does not accelerate in the perpendicular direction, therefore, $\mathbf{F}_{g_\perp}$ is balanced by the normal force $\mathbf{F}_N$, which also acts perpendicularly to the surface of the plane and is equal to $\mathbf{F}_g \cos \theta$.

The diagram below illustrates all of the forces acting on the object as it travels down the plane.

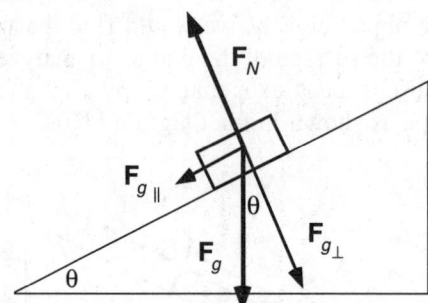

PROBLEM

A 5000-newton polar bear slides down an ice slide without friction. If the slide makes an angle of 37° with the ground, calculate the accelerating force and the normal force on the polar bear.

SOLUTION

$$\mathbf{F}_{g\parallel} = \mathbf{F}_g \sin \theta$$
$$= (5000 \text{ N})(\sin 37°) = 3000 \text{ N}$$

$$\mathbf{F}_N = \mathbf{F}_{g\perp} = \mathbf{F}_g \cos \theta$$
$$= (5000 \text{ N})(\cos 37°) = 4000 \text{ N}$$

Suppose an object is sliding down an inclined plane at constant speed. In this situation, the forces in the parallel direction, as well as the forces in the perpendicular direction, are balanced. It is the force due to friction ($\mathbf{F}_f$) that balances $\mathbf{F}_g$, as shown in the diagram.

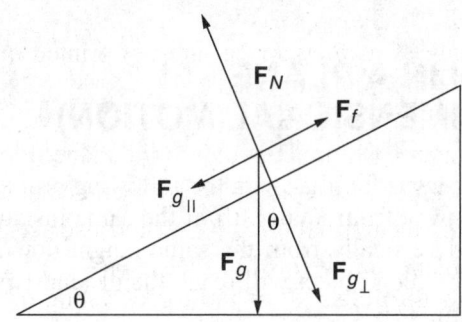

PROBLEM
A wooden crate whose weight ($\mathbf{F}_g$) is 1200 newtons slides down a metal ramp at constant speed as it is unloaded from an airplane. The ramp makes an angle of 30° with the horizontal. Calculate the force of friction ($\mathbf{F}_f$), the normal force ($\mathbf{F}_N$), and the coefficient of kinetic friction (μ_k) between the crate and the ramp.

SOLUTION

$$\mathbf{F}_f = \mathbf{F}_{g\parallel} = \mathbf{F}_g \sin \theta$$
$$= (1200 \text{ N})(\sin 30°) = 600 \text{ N}$$

$$\mathbf{F}_N = \mathbf{F}_{g\perp} = \mathbf{F}_g \cos\theta$$
$$= (1200\ \text{N})(\cos 30°) = 1040\ \text{N}$$

$$\mathbf{F}_f = \mu_k \mathbf{F}_N$$

$$\mu_k = \frac{F_f}{F_N}$$
$$= \frac{600\ \text{N}}{1040\ \text{N}} = 0.57$$

4.9 MOTION IN A PLANE (TWO-DIMENSIONAL MOTION)

Every student of physics is familiar with the following problem: An object is thrown horizontally away from a tall cliff at the same instant that an identical object is dropped vertically from the same height down the cliff. If air resistance is ignored, which object will reach the ground first? The diagram illustrates the problem.

As unreasonable as it may seem, both objects hit the ground at exactly the same instant. Why does this happen? The explanation is based on the fact that the force due to gravity is the only factor that governs the downward motion of both objects. Since both objects fall at the same rate, they reach the ground at the same time. The difference is that the object thrown outward travels horizontally as well as vertically.

The following diagram represents a time lapse picture of this situation.

Notice that at any instant the vertical displacement of each object is the same. Also notice that in each time interval the horizontal displacement traveled by the object thrown outward is constant.

In the y-direction, both objects can be considered as falling freely from rest (with an acceleration of -9.8 meter per second per second) and are governed by the equations of motion developed in Chapter 2:

$$\mathbf{d}_y = \mathbf{v}_{iy}\, t + \frac{1}{2}\mathbf{a}_y\, (t)^2$$

$$\mathbf{v}_{fy}{}^2 = \mathbf{v}_{iy}{}^2 + 2\mathbf{a}_y\, \mathbf{d}_y$$

$$\mathbf{a}_y = \frac{\Delta \mathbf{v}_y}{t} = \frac{\mathbf{v}_{fy} - \mathbf{v}_{iy}}{t}$$

In the x-direction, the thrown object travels at constant speed. Since $\mathbf{v}_x$ does not change, it is equal to the average speed, and we can use a variation of the equation $\mathbf{v} = \mathbf{d}/t$ to describe its motion:

$$\mathbf{v}_x = \frac{\mathbf{d}_x}{t}$$

It is important to note that the x- and y-motions of the two objects are *completely independent* of one another while occurring in the same amount of time.

PROBLEM
An object is thrown outward from a cliff with a horizontal velocity of 20. meters per second. With air resistance ignored, the object takes 15 seconds to reach the bottom of the cliff. Calculate (a) the height of the cliff and (b) the

horizontal distance that was traveled by the object by the time it reaches the ground.

SOLUTION

(a) We solve this part by considering the vertical motion of the object: its initial velocity is zero, and its acceleration is –9.8 m/s². Then:

$$\mathbf{d}_y = \mathbf{v}_{iy}\, t + \frac{1}{2}\, \mathbf{a}_y\, (t)^2$$

$$= 0 + \frac{1}{2}\, \mathbf{a}_y\, (t)^2$$

$$= \frac{1}{2} \left(-9.8\, \frac{m}{s^2} \right) (15\ s)^2$$

$$= -1102.5\ \text{m (i.e., 1100 m } downward)$$

(b) Since the horizontal velocity is constant, we can apply this equation:

$$\mathbf{v}_x = \frac{\mathbf{d}_x}{t}$$

$$\mathbf{d}_x = \mathbf{v}_x\, t$$

$$= \left(20.\frac{m}{s} \right) (15\ s) = 300\ \text{m}$$

Now let's consider an object such as a cannonball that is fired at an angle to the ground, as shown in the diagram.

The shape of the path (known as the *trajectory*) is parabolic. If the initial velocity of the object is $\mathbf{v}_i$ and the angle is θ, we must resolve the velocity into its x- and y-components before we can solve any problem undergoing this type of motion:

$$\mathbf{v}_{i_x} = \mathbf{v}_i \cos \theta$$
$$\mathbf{v}_{i_y} = \mathbf{v}_i \sin \theta$$

In the x-direction, the object travels at constant speed. In the y-direction, we treat the object as if it were tossed directly upward and then returned to the ground: Its speed decreases as it rises and increases as it returns to the ground.

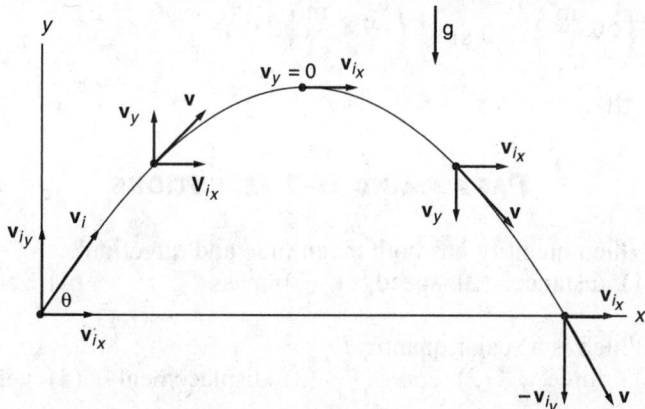

The motion equations developed in Chapter 2 can be used to solve situations involving this type of motion.

It is important to note that the up portion of the motion graph is symmetrical to the down half; therefore, $t_{up} = t_{down}$. In addition, the velocity in the vertical direction at any point on the up side is equal in magnitude and opposite in direction to the symmetrical point on the down side of the graph. This information can be used to quickly calculate the total time of the object's flight using the equation $\mathbf{a} = \mathbf{v}_y/t$. The initial velocity can be divided by 9.81 m/s^2 to solve for the time it takes for the up portion. The total time is calculated simply by multiplying the time obtained by 2.

PROBLEM

An object is fired from the ground at 100. meters per second at an angle of 30.° with the horizontal.

(a) Calculate the horizontal and vertical components of the initial velocity.

(b) After 2.0 seconds, how far has the object traveled in the horizontal direction?

(c) How high is the object at this point?

SOLUTION

(a) $\mathbf{v}_{i_x} = \mathbf{v}_i \cos \theta = (100.\ \text{m/s})(\cos 30.°) = 87\ \text{m/s}$

   $\mathbf{v}_{i_y} = \mathbf{v}_i \sin \theta = (100.\ \text{m/s})(\sin 30.°) = 50.\ \text{m/s}$

(b) $\mathbf{v}_x = \dfrac{\mathbf{d}_x}{t}$

   $\mathbf{d}_x = \mathbf{v}_x \cdot t$

   $= \left(87\ \dfrac{\text{m}}{\text{s}}\right)(2.0\ \text{s}) = 174\ \text{m}$

(c) $\mathbf{d}_y = \mathbf{v}_{i_y} t + \dfrac{1}{2}\,\mathbf{a}_y\,(t)^2$

   $\mathbf{d}_y = \left(50.\ \dfrac{\text{m}}{\text{s}}\right)(2.0\ \text{s}) + \dfrac{1}{2}\left(-9.8\ \dfrac{\text{m}}{\text{s}^2}\right)(2.0\ \text{s})^2$

   $= 80.\ \text{m}$

PART A AND B-1 QUESTIONS

1. Which quantity has both magnitude and direction?
   (1) distance   (2) speed   (3) mass   (4) velocity

2. Which is a scalar quantity?
   (1) force   (2) energy   (3) displacement   (4) velocity

3. Which is a vector quantity?
   (1) time   (2) work   (3) displacement   (4) distance

4. Two concurrent forces of 6 newtons and 12 newtons could produce the same effect as a single force of
   (1) 5.0 N   (2) 15 N   (3) 20 N   (4) 72 N

5. A girl attempts to swim directly across a stream 15 meters wide. When she reaches the other side, she is 15 meters downstream. The magnitude of her displacement is closest to
   (1) 30 m   (2) 21 m   (3) 17 m   (4) 15 m

6. If a woman runs 100 meters north and then 70 meters south, her total displacement will be
   (1) 30 m [N]   (2) 30 m [S]   (3) 170 m [N]   (4) 170 m [S]

7. A 5-newton force directed north and a 5-newton force directed west both act on the same point. The resultant of these two forces is approximately
   (1) 5 N [NW]   (2) 7 N [NW]   (3) 5 N [SW]   (4) 7 N [SW]

8. Two forces of 5 newtons and 15 newtons acting concurrently could have a resultant with a magnitude of
   (1) 5 N   (2) 10 N   (3) 25 N   (4) 75 N

9. Two 10.0-newton forces act concurrently on a point at an angle of 180° to each other. The magnitude of the resultant of the two forces is
   (1) 0 N   (2) 10.0 N   (3) 18.0 N   (4) 20.0 N

10. The diagram below represents two forces acting concurrently on an object. The magnitude of the resultant force is closest to

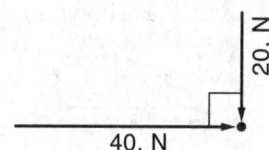

   (1) 20 N   (2) 40 N   (3) 45 N   (4) 60 N

11. Forces **A** and **B** have a resultant **R**. Force **A** and resultant **R** are shown in the diagram below.

   Which vector below best represents force **B**?

12. Which pair of concurrent forces could have a resultant of 5?
    (1) 2 N and 2 N   (3) 2 N and 8 N
    (2) 2 N and 4 N   (4) 2 N and 16 N

13. Two concurrent forces of 40 newtons and **X** newtons have a resultant of 100 newtons. Force **X** could be
(1) 20 N   (2) 40 N   (3) 80 N   (4) 150 N

14. Which vector best represents the resultant of the two vectors shown below?

15. Which vector best represents the resultant of forces F_1 and F_2 acting concurrently on point P as shown in the diagram below?

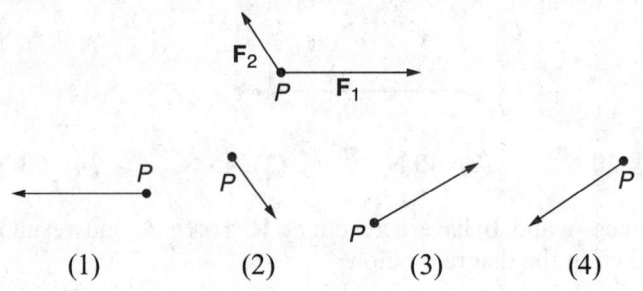

16. Four forces act on a point as shown.

The resultant of the four forces is
(1) 0 N   (2) 5 N   (3) 14 N   (4) 20 N

17. The resultant of a 12-newton force and a 7-newton force is 5 newtons. The angle between the forces is
(1) 0°   (2) 45°   (3) 90°   (4) 180°

18. Three forces act concurrently on an object in equilibrium. These forces are 10 newtons, 8 newtons, and 6 newtons. The resultant of the 6-newton and 8-newton forces is
(1) 0   (3) 10 N
(2) between 0 and 10 N   (4) greater than 10 N

19. The resultant of two concurrent forces is minimum when the angle between them is
(1) 0°   (2) 45°   (3) 90°   (4) 180°

20. As the angle between two concurrent forces decreases from 180°, their resultant
(1) decreases   (2) increases   (3) remains the same

21. As the angle between two concurrent forces of 5.0 newtons and 7.0 newtons increases from 0° to 180°, the magnitude of their resultant changes from
(1) 0 N to 35 N   (3) 12 N to 2.0 N
(2) 2.0 N to 12 N   (4) 12 N to 0 N

22. The maximum number of components that a single force may be resolved into is
(1) one   (2) two   (3) three   (4) unlimited

23. A constant force is exerted on a box as shown in the diagram.

As angle θ decreases to 0°, the magnitude of the horizontal component of the force
(1) decreases   (2) increases   (3) remains the same

24. Which diagram represents the vector with the largest downward component? [Assume each vector has the same magnitude.]

25. What is the magnitude of the vertical component of the velocity vector shown below?

Horizontal

(1) 10. m/s   (2) 69 m/s   (3) 30. m/s   (4) 40. m/s

26. In the diagram below, the numbers 1, 2, 3, and 4 represent possible directions in which a force could be applied to a cart. If the force applied in each direction has the same magnitude, in which direction will the vertical component of the force be the least?

(1) 1   (2) 2   (3) 3   (4) 4

27. A resultant force of 10. newtons is made up of two component forces acting at right angles to each other. If the magnitude of one of the components is 6.0 newtons, the magnitude of the other component must be

(1) 16 N   (2) 8.0 N   (3) 6.0 N   (4) 4 N

28. A ball is fired vertically upward at 5.0 meters per second from a cart moving horizontally to the right at 2.0 meters per second. Which vector best represents the resultant velocity of the ball when fired?

(1)   (3)

(2)   (4)

29. An object weighing 24 newtons is placed on a 30° slope as shown. The component of the weight acting parallel to the slope is closest to

   (1) 12 N   (2) 17 N   (3) 22 N   (4) 24 N

Base your answer to question 30 on the diagram below, which shows a cart held motionless by an external force **F** on a frictionless incline.

30. What is the magnitude of the external force **F** necessary to hold the cart motionless at point *C*?
   (1) 4.9 N   (2) 2.0 N   (3) 9.8 N   (4) 19.6 N

31. Weight **W** is supported by a rod and a cable, as shown.

If angle *A* is made smaller, the tension in the cable will
   (1) decrease   (2) increase   (3) remain the same

32. An object weighing 600 newtons is pulled up a frictionless incline at a constant speed. If the incline makes an angle of 30° with the horizontal, the force on the object parallel to the incline is
(1) 200 N   (2) 300 N   (3) 520 N   (4) 600 N

33. An object rests on an incline. As the angle between the incline and the horizontal increases, the force needed to prevent the object from sliding down the incline
(1) decreases   (2) increases   (3) remains the same

34. A projectile is fired at an angle of 53° to the horizontal with a speed of 80. meters per second. What is the vertical component of the projectile's initial velocity?
(1) 130 m/s   (2) 100 m/s   (3) 64 m/s   (4) 48 m/s

35. A ball is fired with a velocity of 12 meters per second from a cannon pointing north, while the cannon is moving eastward at a velocity of 24 meters per second. Which vector best represents the resultant velocity of the ball as it leaves the cannon?

36. A batted softball leaves the bat with an initial velocity of 44 meters per second at an angle of 37° above the horizontal. What is the magnitude of the initial vertical component of the softball's velocity?
(1) 0 m/s   (2) 26 m/s   (3) 35 m/s   (4) 44 m/s

37. A ball rolls down a curved ramp as shown in the diagram below. Which dotted line best represents the path of the ball after leaving the ramp?

   (1) *A*   (2) *B*   (3) *C*   (4) *D*

Base your answers to questions 38 and 39 on the diagram below, which shows a ball projected horizontally with an initial velocity of 20. meters per second east, off a cliff 100. meters high. [Neglect air resistance.]

38. How many seconds does the ball take to reach the ground?
   (1) 4.5   (2) 20.   (3) 9.8   (4) 2.0

39. During the flight of the ball, what is the direction of its acceleration?
   (1) downward (2) upward   (3) westward   (4) eastward

Base your answers to questions 40 and 41 on the diagram below, which shows a ball thrown toward the east and upward at an angle of 30° to the horizontal. Point *X* represents the ball's highest point.

40. What is the direction of the ball's velocity at point *X*? [Neglect friction.]
   (1) down   (2) up   (3) west   (4) east

41. What is the direction of the ball's acceleration at point *X*? [Neglect friction.]
   (1) down   (2) up   (3) west   (4) east

42. A bullet is fired horizontally from the roof of a building 100. meters tall with a speed of 850. meters per second. Neglecting air resistance, how far will the bullet drop in 3.00 seconds?
(1) 29.4 m   (2) 44.1 m   (3) 100. m   (4) 2,550 m

Base your answers to questions 43 and 44 on the diagram below, which represents a ball being kicked by a foot and rising at an angle of 30.° from the horizontal. The ball has an initial velocity of 5.0 meters per second. [Neglect friction.]

43. What is the magnitude of the horizontal component of the ball's initial velocity?
(1) 2.5 m/s   (2) 4.3 m/s   (3) 5.0 m/s   (4) 8.7 m/s

44. As the ball rises, the vertical component of its velocity
(1) decreases   (2) increases   (3) remains the same

Base your answers to questions 45 and 46 on the information below.

A rocket is launched at an angle of 60° to the horizontal. The initial velocity of the rocket is 500 meters per second. [Neglect friction.]

45. The vertical component of the initial velocity is
(1) 250 m/s   (2) 433 m/s   (3) 500 m/s   (4) 1000 m/s

46. Compared to the horizontal component of the rocket's initial velocity, the horizontal component after 10 seconds would be
(1) less   (2) greater   (3) the same

Base your answers to questions 47 through 49 on the information below.

An object is thrown horizontally off a cliff with an initial velocity of 5.0 meters per second. The object strikes the ground 3.0 seconds later.

47. What is the vertical speed of the object as it reaches the ground? [Neglect friction.]
(1) 130 m/s   (2) 29 m/s   (3) 15 m/s   (4) 5.0 m/s

48. How far from the base of the cliff will the object strike the ground? [Neglect friction].
(1) 2.9 m   (2) 9.8 m   (3) 15 m   (4) 44 m

49. What is the horizontal speed of the object 1.0 second after it is released? [Neglect friction.]
(1) 5.0 m/s   (2) 10. m/s   (3) 15 m/s   (4) 30. m/s

50. Which graph best represents the motion of an object sliding down a frictionless inclined plane?

   (1)   (2)   (3)   (4)

51. Four forces are acting on an object as shown in the diagram below.

If the object is moving with a constant velocity, the magnitude of force **F** must be
(1) 0 N   (2) 20 N   (3) 100 N   (4) 40 N

52. An elevator containing a man weighing 800 newtons is rising at a constant speed. The force exerted by the man on the floor of the elevator is
(1) less than 80 N   (3) 800 N
(2) between 80 and 800 N   (4) more than 800 N

Base your answers to questions 53 through 55 on the information and diagram below.

The reading of the spring balance is *t* newtons. [Neglect friction and the weight of the spring balance.]

53. What is the direction of the net force on object *B*?

(1) ←

(3) →

(2) ↑

(4) ↓

54. The net force on object *B* is equal to
(1) *t* N   (2) 40 N   (3) $(40 + t)$ N   (4) $(40 - t)$ N

55. The net force on object *A* is equal to
(1) *t* N   (2) 120 N   (3) $(120 + t)$ N   (4) $(120 - t)$ N

56. The speedometer in a car does *not* measure the car's velocity because velocity is a
(1) vector quantity and has a direction associated with it
(2) vector quantity and does not have a direction associated with it
(3) scalar quantity and has a direction associated with it
(4) scalar quantity and does not have a direction associated with it

57. A projectile launched at an angle of 45° above the horizontal travels through the air. Compared to the projectile's theoretical path with no air friction, the actual trajectory of the projectile with air friction is
(1) lower and shorter   (3) higher and shorter
(2) lower and longer   (4) higher and longer

58. Two stones, *A* and *B*, are thrown horizontally from the top of a cliff. Stone *A* has an initial speed of 15 meters per second and stone *B* has an initial speed of 30. meters per second. Compared to the time it takes stone *A* to reach the ground, the time it takes stone *B* to reach the ground is
(1) the same   (3) half as great
(2) twice as great   (4) four times as great

59. Which diagram represents a box in equilibrium?

60. An airplane flies with a velocity of 750. kilometers per hour, 30.0° south of east. What is the magnitude of the eastward component of the plane's velocity?
(1) 866 km/h　(2) 650. km/h　(3) 433 km/h　(4) 375 km/h

61. A block weighing 10.0 newtons is on a ramp inclined at 30.0° to the horizontal. A 3.0-newton force of friction, F_f, acts on the block as it is pulled up the ramp at constant velocity with force F, which is parallel to the ramp, as shown in the diagram below.

What is the magnitude of force F?
(1) 7.0 N　　(2) 8.0 N　　(3) 10. N　　(4) 13 N

62. A 25-newton horizontal force northward and a 35-newton horizontal force southward act concurrently on a 15-kilogram object on a frictionless surface. What is the magnitude of the object's acceleration?
(1) 0.67 m/s²   (2) 1.7 m/s²   (3) 2.3 m/s²   (4) 4.0 m/s²

63. The diagram below represents two concurrent forces.

Which vector represents the force that will produce equilibrium with these two forces?

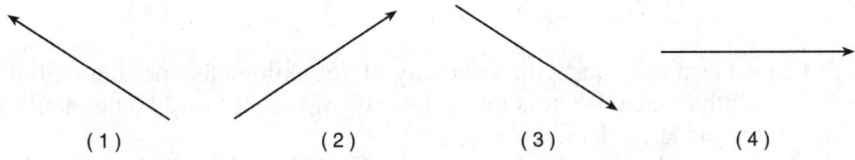

    (1)   (2)   (3)   (4)

64. Which is a vector quantity?
(1) speed   (2) work   (3) mass   (4) displacement

65. A soccer player kicks a ball with an initial velocity of 10. meters per second at an angle of 30.° above the horizontal. The magnitude of the horizontal component of the ball's initial velocity is
(1) 5.0 m/s   (2) 8.7 m/s   (3) 9.8 m/s   (4) 10. m/s

66. Two spheres, *A* and *B*, are simultaneously projected horizontally from the top of a tower. Sphere *A* has a horizontal speed of 40. meters per second and sphere *B* has a horizontal speed of 20. meters per second. Which statement best describes the time required for the spheres to reach the ground and the horizontal distance they travel? [Neglect friction and assume the ground is level.]
(1) Both spheres hit the ground at the same time and at the same distance from the base of the tower.
(2) Both spheres hit the ground at the same time, but sphere *A* lands twice as far as sphere *B* from the base of the tower.
(3) Both spheres hit the ground at the same time, but sphere *B* lands twice as far as sphere *A* from the base of the tower.
(4) Sphere *A* hits the ground before sphere *B*, and sphere *A* lands twice as far as sphere *B* from the base of the tower.

67. In the diagram below, a 20.-newton force due north and a 20.-newton force due east act concurrently on an object, as shown in the diagram below.

North

20. N

East

20. N

The additional force necessary to bring the object into a state of equilibrium is
(1) 20. N, northeast   (3) 28 N, northeast
(2) 20. N, southwest   (4) 28 N, southwest

68. Two forces act concurrently on an object. Their resultant force has the largest magnitude when the angle between the forces is
(1) 0°   (2) 30°   (3) 90°   (4) 180°

69. Which is *not* a vector quantity?
(1) electric charge   (3) velocity
(2) magnetic field strength   (4) displacement

Note that question 70 has only three choices.

70. As the angle between two concurrent forces decreases, the magnitude of the force required to produce equilibrium
(1) decreases   (2) increases   (3) remains the same

71. A child walks 5.0 meters north, then 4.0 meters east, and finally 2.0 meters south. What is the magnitude of the resultant displacement of the child after the entire walk?
(1) 1.0 m   (2) 5.0 m   (3) 3.0 m   (4) 11.0 m

Base your answers to questions 72 and 73 on the information below.

A stream is 30. meters wide and its current flows southward at 1.5 meters per second. A toy boat is launched with a velocity of 2.0 meters per second eastward from the west bank of the stream.

72. What is the magnitude of the boat's resultant velocity as it crosses the stream?
(1) 0.5 m/s   (2) 2.5 m/s   (3) 3.0 m/s   (4) 3.5 m/s

101

73. How much time is required for the boat to reach the opposite bank of the stream?
(1) 8.6 s   (2) 12 s   (3) 15 s   (4) 60. s

74. Which is a vector quantity?
(1) electric charge   (3) electric potential difference
(2) electric field strength   (4) electric resistance

Note that question 75 has only three choices.

75. A 6.0-newton force and an 8.0-newton force act concurrently on a point. As the angle between these forces increases from 0° to 90°, the magnitude of their resultant
(1) decreases   (2) increases   (3) remains the same

76. A machine launches a tennis ball at an angle of 25° above the horizontal at a speed of 14 meters per second. The ball returns to level ground. Which combination of changes *must* produce an increase in time of flight of a second launch?
(1) decrease the launch angle and decrease the ball's initial speed
(2) decrease the launch angle and increase the ball's initial speed
(3) increase the launch angle and decrease the ball's initial speed
(4) increase the launch angle and increase the ball's initial speed

77. A plane flying horizontally above Earth's surface at 100. meters per second drops a crate. The crate strikes the ground 30.0 seconds later. What is the magnitude of the horizontal component of the crate's velocity just before it strikes the ground? [Neglect friction.]
(1) 0 m/s   (2) 100. m/s   (3) 294 m/s   (4) 394 m/s

78. A woman with horizontal velocity v_1 jumps off a dock into a stationary boat. After landing in the boat, the woman and the boat move wth velocity v_2. Compared to velocity v_1, velocity v_2 has
(1) the same magnitude and the same direction
(2) the same magnitude and opposite direction
(3) smaller magnitude and the same direction
(4) larger magnitude and the same direction

79. The diagram below shows a 4.0-kilogram object accelerating at 10. meters per second^2 on a rough horizontal surface.

Acceleration = 10. m/s^2 ⟶

Frictional force = F_f

m = 4.0 kg

Applied force = 50. N

(Not drawn to scale)

What is the magnitude of the frictional force F_f acting on the object?
(1) 5.0 N   (2) 10. N   (3) 20. N   (4) 40. N

80. The diagram below represents a force vector, A, and a resultant vector, R.

A

R

Which force vector B below could be added to force vector A to produce resultant vector R?

B   B   B   B
(1)   (2)   (3)   (4)

81. A golf ball is propelled with an initial velocity of 60. meters per second at 37° above the horizontal. The horizontal component of the golf ball's initial velocity is
(1) 30. m/s   (2) 36 m/s   (3) 40. m/s   (4) 48 m/s

82. A 3-newton force and a 4-newton force are acting concurrently on a point. Which force could *not* produce equilibrium with these two forces?
(1) 1 N   (2) 7 N   (3) 9 N   (4) 4 N

83. A volleyball hit into the air has an initial speed of 10. meters per second. Which vector best represents the angle above the horizontal that the ball should be hit to remain in the air for the greatest amount of time?

| (1) | (2) | (3) | (4) |

84. Which is a scalar quantity?
(1) acceleration   (3) speed
(2) momentum   (4) displacement

85. A projectile is fired with an initial velocity of 120. meters per second at an angle, θ, above the horizontal. If the projectile's initial horizontal speed is 55 meters per second, then angle θ measures approximately
(1) 13°   (2) 27°   (3) 63°   (4) 75°

86. A 2.0-kilogram laboratory cart is sliding across a horizontal frictionless surface at a constant velocity of 4.0 meters per second east. What will be the cart's velocity after a 6.0-newton westward force acts on it for 2.0 seconds?
(1) 2.0 m/s east   (3) 10. m/s east
(2) 2.0 m/s west   (4) 10. m/s west

Note that question 87 has only three choices.

87. A student on her way to school walks four blocks east, three blocks north, and another four blocks east, as shown in the diagram.

Compared to the distance she walks, the magnitude of her displacement from home to school is
(1) less   (2) greater   (3) the same

88. Which vector diagram best represents a cart slowing down as it travels to the right on a horizontal surface?

(1)   (2)   (3)   (4)

89. Two 30.-newton forces act concurrently on an object. In which diagram would the forces produce a resultant with a magnitude of 30. newtons?

(1)   (2)   (3)   (4)

Answers to Part A and B–1 questions can be found on page 455.

PART B–2 & C QUESTIONS

1. Base your answers to parts *a* through *c* on the information below.

   A student pulls a cart across a horizontal floor by exerting a force of 50. newtons at an angle of 35° to the horizontal.

   a On the diagram provided, *using a protractor and a straightedge,* construct a scaled vector showing the 50.-newton force acting on the cart at the appropriate angle. The force *must* be drawn to a scale of 1.0 centimeter = 10. N.

   Label the 50.-newton force and the 35° angle on your diagram. *Be sure your final answer appears with the correct labels (numbers and units).*

   b Construct the horizontal component of the force vector to scale on your diagram, and label it *H*.

   c What is the magnitude of the horizontal component of the force?

2. Explain how to find the coefficient of kinetic friction between a wooden block of unknown mass and a tabletop in the laboratory. Include the following in your explanation:

   • Measurements required

   • Equipment needed

   • Procedure

   • Equation(s) needed to calculate the coefficient of friction

3. Base your answers to parts *a* through *c* on the information below.

   A newspaper carrier on her delivery route travels 200. meters due north and then turns and walks 300. meters due east.

   a *On a separate paper*, draw a vector diagram following the directions below.
     (1) Using a ruler and protractor and starting at point *P*, construct the sequence of two displacement vectors for the newspaper carrier's route. Use a scale of 1.0 centimeters = 100. meters. Label the vectors.

     (2) Construct and label the vector that represents the carrier's resultant displacement from point *P*.

   b What is the magnitude of the carrier's resultant displacement?

   c What is the angle (in degrees) between north and the carrier's resultant displacement?

Base your answers to questions 4 through 5 on the diagram and data table on page 107. The diagram shows a worker moving a 50.0-kilogram safe up a ramp by applying a constant force of 300. newtons parallel to the ramp. The data table shows the position of the safe as a function of time.

m = 50.0 kg

= 300. N

Ramp

Horizontal

Time (s)	Distance Moved up the Ramp (m)
0.0	0.0
1.0	2.2
2.0	4.6
3.0	6.6
4.0	8.6
5.0	11.0

4. Using the information in the data table, construct a line graph on the grid provided. Plot the data points *and* draw the best-fit line.

Distance vs. Time

Distance Moved Up the Ramp (m)

Time (s)

5. Using one or more complete sentences, explain the physical significance of the slope of the graph.

Base your answers to questions 6 through 10 on the information and vector diagram below.

A 20.-newton force due north and a 40.-newton force due east act concurrently on a 10.-kilogram object, located at point P.

6. Using a ruler, determine the scale used in the vector diagram by finding the number of newtons represented by each centimeter.

7. On the vector diagram provided, use a ruler and protractor to construct the vector that represents the resultant force.

8. What is the magnitude of the resultant force?

9. What is the measure of the angle (in degrees) between east and the resultant force?

10. Calculate the magnitude of the acceleration of the object. [Show all calculations, including the equation and substitution with units.]

Base your answers to questions 11 and 12 on the information below.

A 5.0-kilogram block weighing 49 newtons sits on a frictionless, horizontal surface. A horizontal force of 20. newtons toward the right is applied to the block. [Neglect air resistance.]

11. On the diagram below, draw a vector to represent each of the three forces acting on the block. Use a ruler and a scale of 1.0 centimeter = 20. newtons. Begin each vector at point *C* and label its magnitude in newtons.

Frictionless surface

12. Calculate the magnitude of the acceleration of the block. [Show all calculations, including the equation and substitution with units.]

13. A 160.-newton box sits on a 10.-meter-long frictionless plane inclined at an angle of 30.° to the horizontal as shown. Force (*F*) applied to a rope attached to the box causes the box to move with a constant speed up the incline.

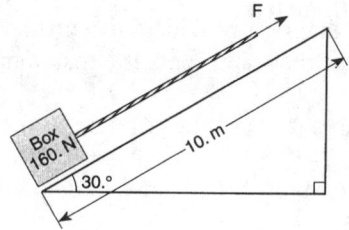

On the diagram below construct a vector to represent the weight of the box. Use a metric ruler and a scale of 1.0 centimeter = 80. newtons. Begin the vector at point *B* and label its magnitude in newtons.

14. Using dimensional analysis, show that the expression v^2/d has the same units as acceleration. [Show all the steps used to arrive at your answer.]

Base your answers to questions 15 through 17 on the information and diagram below.

A child is flying a kite, *K*. A student at point *B*, located 100. meters away from point *A* (directly underneath the kite), measures the angle of elevation of the kite from the ground as 30.°.

15. In the space below, use a metric ruler and protractor to draw a triangle representing the positions of the kite, *K*, and point *A* relative to point *B* that is given. Label points *A* and *K*. Use a scale of 1.0 centimeter = 20. meters.

Scale
1.0 cm = 20. m

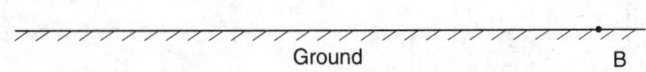

16. Use a metric ruler and your scale diagram to determine the height, *AK*, of the kite.

17. A small lead sphere is dropped from the kite. Calculate the amount of time required for the sphere to fall to the ground. [Show all calculations, including the equation and substitution with units. Neglect air resistance.]

18. The diagram below shows a 5.0-kilogram block accelerating at 6.0 meters per second² along a rough horizontal surface by the application of a horizontal force, F, of 50. newtons.

What is the magnitude in newtons of the force of friction, F_f, acting on the block?

19. A box of mass m is held motionless on a frictionless inclined plane by a rope that is parallel to the surface of the plane. On the diagram provided, draw and label all of the force vectors acting on the box.

Base your answers to questions 20 through 22 on the information and vector diagram below.

A dog walks 8.0 meters due north and then 6.0 meters due east.

20. Using a metric ruler and the vector diagram, determine the scale used in the diagram.

$$1.0 \text{ cm} = \underline{\hspace{3cm}} \text{ m}$$

21. On the diagram above, construct the resultant vector that represents the dog's total displacement.

22. Determine the magnitude of the dog's total displacement.

Base your answers to questions 23 through 25 on the information below.

A kicked soccer ball has an initial velocity of 25 meters per second at an angle of 40.° above the horizontal, level ground. [Neglect friction.]

23. Calculate the magnitude of the vertical component of the ball's initial velocity. [Show all work, including the equations and substitution with units.]

24. Calculate the maximum height the ball reaches above its initial position. [Show all work, including the equation and substitution with units.]

25. On the diagram below, sketch the path of the ball's flight from its initial position at point P until it returns to level ground.

P   Level ground

Base your answers to questions 26 and 27 on the information and diagram below.

A soccer ball is kicked from point P_i at an angle above a horizontal field. The ball follows an ideal path before landing on the field at point P_f.

P_i Horizontal field P_f

26. On the diagram above, draw an arrow to represent the direction of the net force on the ball when it is at position X. Label the arrow F_{net}. [Neglect friction.]

112

27. On the diagram *in your answer booklet*, draw an arrow to represent the direction of the acceleration of the ball at position *Y*. Label the arrow *a*. [Neglect friction.]

Base your answers to questions 28 and 29 on the information below.

A 747 jet, traveling at a velocity of 70. meters per second north, touches down on a runway. The jet slows to rest at the rate of 2.0 meters per second^2.

28. Calculate the total distance the jet travels on the runway as it is brought to rest. [Show all work, including the equation and substitution with units.]

29. On the diagram below, point *P* represents the position of the jet on the runway. Beginning at point *P*, draw a vector to represent the magnitude and direction of the acceleration of the jet as it comes to rest. Use a scale of 1.0 centimeter = 0.50 meter/second^2.

Base your answers to questions 30 and 31 on the information and diagram below.

A force of 60. newtons is applied to a rope to pull a sled across a horizontal surface at a constant velocity. The rope is at an angle of 30. degrees above the horizontal.

30. Calculate the magnitude of the component of the 60.-newton force that is parallel to the horizontal surface. [Show all work, including the equation and substitution with units.]

31. Determine the magnitude of the frictional force acting on the sled.

Base your answers to questions 32 through 34 on the information and diagram below.

A projectile is launched into the air with an initial speed of v_i at a launch angle of 30.° above the horizontal. The projectile lands on the ground 2.0 seconds later.

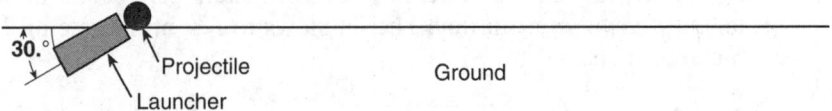

32. On the diagram above, sketch the ideal path of the projectile.

33. How does the maximum altitude of the projectile change as the launch angle is increased from 30.° to 45° above the horizontal? [Assume the same initial speed, v_i.]

34. How does the total horizontal distance traveled by the projectile change as the launch angle is increased from 30.° to 45° above the horizontal? [Assume the same initial speed, v_i.]

Base your answers to questions 35 through 39 on the information and diagram below.

A horizontal force of 8.0 newtons is used to pull a 20.-newton wooden box moving toward the right along a horizontal, wood surface, as shown.

35. Starting at point P on the diagram below, use a metric ruler and a scale of 1.0 cm = 4.0 N to draw a vector representing the normal force acting on the box. Label the vector F_N.

36. Calculate the magnitude of the frictional force acting on the box. [Show all work, including the equation and substitution with units.]

37. Determine the magnitude of the net force acting on the box.

_____ N

38. Determine the mass of the box.

_____ N

39. Calculate the magnitude of the acceleration of the box. [Show all work, including the equation and substitution with units.]

Base your answers to questions 40 through 42 on the information and diagram below.

Force *A* with a magnitude of 5.6 newtons and force *B* with a magnitude of 9.4 newtons act concurrently on point *P*.

40. Determine the scale used in the diagram.

41. Use a ruler and protractor to construct a vector representing the resultant of forces *A* and *B*.

42. Determine the magnitude of the resultant force.

43. Explain the difference between a scalar and a vector quantity.

44. A 10.-kilogram rubber block is pulled horizontally at constant velocity across a sheet of ice. Calculate the magnitude of the force of friction acting on the block. [Show all work, including the equation and substitution with units.]

Base your answers to questions 45 through 47 on the information below.

A projectile is fired from the ground with an initial velocity of 250. meters per second at an angle of 60.° above the horizontal.

45. On the diagram below, use a protractor and ruler to draw a vector to represent the initial velocity of the projectile. Begin the vector at point *P*, and use a scale of 1.0 centimeter = 100. meters per second.

P   Horizontal

46. Determine the horizontal component of the initial velocity.

47. Explain why the projectile has *no* acceleration in the horizontal direction. [Neglect air friction.]

Answers to Part B–2 and C questions can be found on pages 470–485.

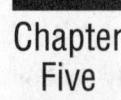
Chapter
Five

CIRCULAR MOTION AND GRAVITATION

An object that travels in a circular path experiences an acceleration directed toward the center of the circle and known as a centripetal acceleration. The unbalanced force responsible for this acceleration is called a centripetal force and is also directed toward the center of the circle.

The study of planetary motion was greatly advanced by Kepler, who deduced three laws that provided a mathematical basis for this motion. Kepler's three laws were later proved by Newton when he developed his law of universal gravitation: two masses attract each other with a force proportional to their masses and inversely proportional to the square of the distance between them.

KEY OBJECTIVES
At the conclusion of this chapter you will be able to:
- Identify the direction of an object's velocity when it is undergoing uniform circular motion.
- Define the terms *centripetal acceleration* and *centripetal force*, and identify the directions of these quantities when an object undergoes uniform circular motion.
- State the equations for calculating centripetal force and centripetal acceleration.
- Solve problems involving uniform circular motion.
- Define the term *period of revolution* and relate it to the equations of uniform circular motion.
- State Newton's law of universal gravitation and solve problems related to it.
- Solve simple problems involving satellites in (a circular) orbit.
- Define the term *geosynchronous orbit*.
- Relate weight to gravitational force.
- Describe the field concept of gravitation.
- Relate the strength of a gravitational field with the acceleration due to gravity.

5.1 INTRODUCTION

In this chapter we study another aspect of motion in a plane, namely, motion in a circle. The subject of circular motion leads, in turn, to a study of gravitation because both natural and artificial satellites travel in nearly circular paths.

5.2 CIRCULAR MOTION

An object moves in a circular path at constant speed as indicated in the diagram below.

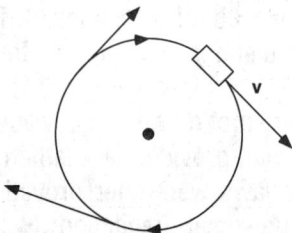

At various points in the path, the direction of the velocity, **v**, of the object is tangent to the circle, as shown in the diagram. Since the *direction* of the object's motion is changing, the object must be subjected to an unbalanced force and is, therefore, accelerating. Here is an example of an object that accelerates even though its speed does not change. This is due to the fact that acceleration is a change in the *velocity* of an object, and this change can be in the magnitude and/or direction of the velocity.

As the object moves in its circular path, its acceleration always points toward the center of the circle. For this reason we call the acceleration a **centripetal** ("center-seeking") acceleration (**a**$_c$).

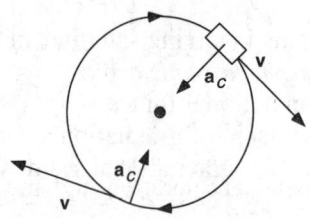

The centripetal acceleration of an object is calculated by means of this equation:

$$a_c = \frac{v^2}{r}$$

where v is the tangential speed of the object and r is the radius of the circular path.

The unbalanced force that causes the centripetal acceleration is called the **centripetal force** (F_c). It also points toward the center of the circular path and is given by this relationship:

$$F_c = ma_c = \frac{mv^2}{r}$$

Note that the format of the last equation above results from the substitution of the a_c formula into the F_c formula. The formula in this format does not appear on the Reference Tables; however, it is the most common form of the formula necessary for solving problems involving circular motion.

We can think of the centripetal force as the force needed to keep the object in its circular path. If the mass of the object were increased, or if the speed of the object were increased, more force would be needed to keep the object in its circular path. If, however, the radius of the path were increased, less force would be required for this purpose.

PROBLEM

A 5.0-kilogram object travels clockwise in a horizontal circle with a speed of 20. meters per second, as shown in the diagram. The radius of the circular path is 25 meters.

(a) Calculate the centripetal acceleration and the centripetal force on the object.
(b) In the position shown, indicate the *direction* of the velocity and of the centripetal force of the object.

SOLUTION

(a) $\quad a_c = \dfrac{v^2}{r}$

$$= \dfrac{\left(20.\dfrac{m}{s}\right)^2}{25 \text{ m}} = 16 \text{ m/s}^2$$

$$\mathbf{F}_c = m\mathbf{a}_c = (5.0 \text{ kg})\left(16 \dfrac{m}{s^2}\right) = 80. \text{ N}$$

(b)

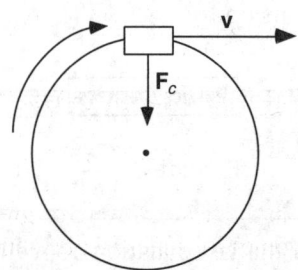

Measuring the speed of an object in circular motion is not always easy, but we can measure the speed of the object indirectly by making use of a quantity known as the period (T). The period of revolution is the time an object takes to complete one revolution in a circular path. In one revolution, the distance the object travels equals the circumference of the circle ($2\pi r$). Using the equation $v = \dfrac{d}{t}$, we substitute and obtain

═══════════ PHYSICS CONCEPTS ═══════════

$$v = \dfrac{2\pi r}{T}$$

───────────────────────────────

*Note that this formula does not appear on the Reference Tables but is frequently needed to solve problems involving circular motion.

───────────────────────────────

PROBLEM
An object traveling in a circular path makes 1200 revolutions in 1.0 hour. If the radius of the path is 10. meters, calculate the speed of the object.

SOLUTION
If the object makes 1200 revolutions in 1.0 h (3600 s), the time (T) for one revolution is

$$T = \dfrac{3600 \text{ s}}{1200 \text{ rev}} = 3.0 \text{ s}$$

The speed of the object is

$$v = \frac{2\pi r}{T} = \frac{(2)\,(\pi)\,(10.\text{ m})}{3.0\text{ s}} = 21\text{ m/s}$$

The equation for the centripetal acceleration can be rewritten in terms of the period of revolution rather than the speed of the object:

PHYSICS CONCEPTS

$$a_c = \frac{v^2}{r} = \frac{4\pi^2 r}{T^2}$$

5.3 NEWTON'S LAW OF UNIVERSAL GRAVITATION

The motions of the planets in the heavens were of great interest in the sixteenth century. The German astronomer Johannes Kepler summarized the laws of planetary motion. More than half a century after Kepler proposed his laws, Isaac Newton was able to prove them mathematically by means of his law of universal gravitation. Newton recognized that the force responsible for pulling an apple toward the Earth has the same origin as the force that keeps the Moon in its orbit about the Earth. This force, which we call *gravitation*, is present between all bodies of mass in the universe, from the largest galaxies to the smallest atoms.

Newton's law of universal gravitation can be explained as follows. Consider two point masses, m_1 and m_2, separated by a distance r as shown in the diagram (the distance between object centers).

The force of gravitation is an *attractive* force. The force on either mass ($\mathbf{F}_g$) is given by this relationship:

PHYSICS CONCEPTS

$$\mathbf{F}_g = \frac{Gm_1 m_2}{r^2}$$

According to *Newton's universal law*:

The gravitational force between two masses is directly proportional to each mass and inversely proportional to the square of the distance between the masses.

This is known as an *inverse square* law. The constant G is called the *universal gravitational constant*. Its measured value is 6.67×10^{-11} newton·meter^2 per kilogram^2. The table below illustrates how the relative gravitational force between two objects depends on their relative masses and the relative distance between them.

Change	Relative Mass$_1$	Relative Mass$_2$	Relative Distance	Relative Force
Original condition	m_1	m_2	r	F
Double m_1	$2m_1$	m_2	r	2F
Triple m_2	m_1	$3m_2$	r	3F
Double m_1, Triple m_2	$2m_1$	$3m_2$	r	6F
Double r	m_1	m_2	$2r$	¼F
Triple r	m_1	m_2	$3r$	⅑F
Quarter r	m_1	m_2	¼r	16F
⅕ of r	m_1	m_2	⅕r	25F

PROBLEM

Calculate the gravitational force of attraction between the Earth and the Moon, given that the mass of the Earth is 6.0×10^{24} kilograms, the mass of the Moon is 7.4×10^{22} kilograms, and the average Earth–Moon distance is 3.8×10^8 meters.

SOLUTION

$$\mathbf{F}_g = \frac{Gm_{\text{Earth}}m_{\text{Moon}}}{(r_{\text{Earth–Moon}})}$$

$$= \frac{\left(6.67 \times 10^{-11} \frac{\text{N} \cdot \text{m}^2}{\text{kg}^2}\right)(6.0 \times 10^{24} \text{ kg})(7.4 \times 10^{22} \text{ kg})}{(3.8 \times 10^8 \text{ m})^2}$$

$$= 2.1 \times 10^{20} \text{ N}$$

This force is present on both the Earth and the Moon (in keeping with Newton's third law). It may seem like an outrageously large force, but consider the masses of the two objects involved! For "normal, Earth-sized" objects, on the other hand, the force of gravitational attraction is quite small.

✪ 5.4 SATELLITE MOTION

Suppose a satellite is traveling around the Earth in a circular orbit at a distance r from the center of the planet. The satellite is kept in orbit by a centripetal force whose magnitude is given by the relationship:

$$F_c = \frac{m_s v^2}{r}$$

where m_s is the mass of the satellite, **v** is its speed in orbit, and r is the distance from the center of the Earth. We know that the origin of this centripetal force is gravitational force; therefore, we can equate the centripetal force relationship with Newton's law of universal gravitation:

$$F_c = \frac{m_s v^2}{r} = \frac{Gm_{\text{Earth}}m_s}{r^2} = F_g$$

$$\frac{v^2}{r} = \frac{Gm_{\text{Earth}}}{r^2}$$

$$v = \sqrt{\frac{Gm_{\text{Earth}}}{r}}$$

We can use the relationship to find the speed necessary for the satellite to be in a given orbit. If the satellite were to move faster, its orbit would be closer to the Earth. If the satellite were to move slower, its orbit would be further from the Earth.

There is a specific distance at which the period of a satellite matches the period of the Earth's rotation (1 day). This type of orbit is known as a **geosynchronous orbit**. Satellites in geosynchronous orbits are especially useful for communication purposes because their positions with respect to the ground do not vary.

5.5 WEIGHT AND GRAVITATIONAL FORCE

We have measured an object's weight by the relationship $\mathbf{F}_g = m\mathbf{g}$, and we can measure an object's gravitational attraction to the Earth by the equation

$$\mathbf{F}_g = \frac{Gm_1 m_2}{r^2}$$

✪ indicates material that is not part of the core curriculum.

Weight and gravitational attraction, however, are one and the same force since an object's weight is gravitational in origin. Accordingly, we can set the two terms equal to each other:

$$\mathbf{F}_g = mg = \frac{Gmm_{\text{Earth}}}{r^2} = \mathbf{F}_g$$

$$\mathbf{g} = \frac{Gm_{\text{Earth}}}{r^2}$$

We can now appreciate the fact that the gravitational acceleration of an object is independent of its mass; it depends only on the mass of the Earth and the distance of the object from the Earth's center.

PROBLEM
Calculate the gravitational acceleration of a satellite that is in orbit at a distance of 1.0×10^8 meters from the center of the Earth.

SOLUTION

$$\mathbf{g} = \frac{Gm_{\text{Earth}}}{r^2}$$

$$= \frac{\left(6.67 \times 10^{-11} \, \dfrac{\text{N} \cdot \text{m}^2}{\text{kg}^2}\right)(6.0 \times 10^{24} \, \text{kg})}{(1.0 \times 10^8 \, \text{m})^2}$$

$$= 0.040 \text{ m/s}^2 \text{ (toward the center of the Earth)}$$

5.6 THE GRAVITATIONAL FIELD

A gravitational field is a region of space that attracts masses with a gravitational force. The gravitational-field concept assumes that one mass somehow changes the space around it and a second mass then interacts with the field. The result is that an attractive gravitational force is exerted on the second mass. One type of interaction is represented in the diagram below.

Gravitational Field Due to M

The gravitational field is a vector quantity: its direction is the direction of the force on the mass m_0, which we call the *test mass*. The test mass is assumed to be much smaller than the mass M, whose field is being determined. The strength of a gravitational field (g) is defined to be the ratio of the gravitational force on the test mass (F_g) to the test mass (m_0):

PHYSICS CONCEPTS

$$g = \frac{F_g}{m_0}$$

The unit of gravitational field strength is the *newton per kilogram* (N/kg). A little "fiddling" with units will show that this unit is equivalent to the unit *meter per second²*—the unit of (gravitational) acceleration. The gravitational field is simply another way of viewing the action of gravity. If we drop an object near the surface of the Earth, it will accelerate at 9.8 meters per second²; since it is in the Earth's gravitational field, it experiences a gravitational force of 9.8 newtons per kilogram of mass. Numerically, the gravitational acceleration and the gravitational field strength are always equal.

PART A AND B-1 QUESTIONS

1. An object travels in a circular path of radius 5.0 meters at a uniform speed of 10. meters per second. What is the magnitude of the object's centripetal acceleration?
   (1) 10.0 m/s² (2) 2.0 m/s² (3) 5.0 m/s² (4) 20. m/s²

Base your answers to questions 2 and 3 on the information and diagram below.

A roller coaster cart starts from rest and accelerates, due to gravity, down a track. The cart starts at a height that enables it to complete a loop in the track. [Neglect friction.]

2. The magnitude of the centripetal force keeping the cart in circular motion would be greatest at point
(1) *A*   (2) *B*   (3) *C*   (4) *D*

3. Which diagram best represents the path followed by an object that falls off the cart when the cart is at point *D*?

(1)   (2)   (3)   (4)

4. A motorcycle of mass 100 kilograms travels around a flat, circular track of radius 10 meters with a constant speed of 20 meters per second. What force is required to keep the motorcycle moving in a circular path at this speed?
(1) 200 N   (2) 400 N   (3) 2000 N   (4) 4000 N

5. Two masses, *A* and *B*, move in circular paths as shown in the diagram below.

The centripetal acceleration of mass *A*, compared to that of mass *B*, is
(1) the same   (3) one-half as great
(2) twice as great   (4) 4 times as great

Base your answers to questions 6 through 9 on the diagram below, which represents a ball of mass *M* attached to a string. The ball moves at a constant speed around a flat horizontal circle of radius *R*.

6. When the ball is in the position shown, the direction of the centripetal force is toward point
    (1) *A*   (2) *B*   (3) *C*   (4) *D*

7. The centripetal acceleration of the ball is
    (1) zero
    (2) constant in direction, but changing in magnitude
    (3) constant in magnitude, but changing in direction
    (4) changing in both magnitude and direction

8. If the string is shortened while the speed of the ball remains the same, the centripetal acceleration will
    (1) decrease   (2) increase   (3) remain the same

9. If the mass of the ball is decreased, the centripetal force required to keep it moving in the same circle at the same speed will
    (1) decrease   (2) increase   (3) remain the same

Base your answers to questions 10 through 12 on the diagram below which represents a 2.0-kilogram mass moving in a circular path on the end of a string 0.50 meter long. The mass moves in a horizontal plane at a constant speed of 4.0 meters per second.

10. The force exerted on the mass by the string is
    (1) 8 N   (2) 16 N   (3) 32 N   (4) 64 N

11. In the position shown in the diagram, the velocity of the mass is directed toward point
    (1) *A*   (2) *B*   (3) *C*   (4) *D*

12. The speed of the mass is changed to 2.0 meters per second. Compared to the centripetal acceleration of the mass when moving at 4.0 meters per second, its centripetal acceleration when moving at 2.0 meters per second would be
    (1) half as great   (3) one-fourth as great
    (2) twice as great   (4) 4 times as great

Base your answers to questions 13 and 14 on the information and diagram below.

A pendulum with a 10-kilogram bob is released at point A and allowed to swing without friction, as shown in the diagram.

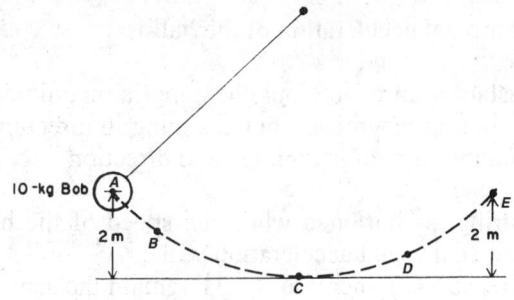

13. What is the weight of the bob?
    (1) 0.1 N   (2) 0.98 N   (3) 10 N   (4) 98 N

14. The centripetal acceleration of the bob is greatest at point
    (1) E (2) B (3) C (4) D

Base your answers to questions 15 through 17 on the diagram below, which shows a section of a level road viewed from above. The road has one curve (AB) with a radius of $2R$ and a second curve (CD) with a radius of R. A car of mass M moves, with constant speed, v, along the road.

15. What is the magnitude of the net force acting on the car as it moves from B to C?
    (1) Mv (3) $\dfrac{Mv^2}{2}$

    (2) Mv^2 (4) zero

16. What is the magnitude of the net force acting on the car when it is moving from C to D?
    (1) $\dfrac{Mv^2}{R}$ (3) MvR

    (2) Mv^2R (4) zero

17. Compared to the centripetal acceleration of the car when it is moving from C to D, the centripetal acceleration when it is moving from A to B is
    (1) less   (2) greater   (3) the same

Base your answers to questions 18 through 21 on the diagram and information below.

At an amusement park, a passenger whose mass is 50 kilograms rides in a cage. The cage has a constant speed of 10 meters per second in a vertical circular path of radius R equal to 10 meters.

18. What is the magnitude of the centripetal acceleration of the passenger?
    (1) 1.0 m/s^2 (3) 5.0 $\times10^2$ m/s^2
    (2) 2.0 $\times$ 10^3 m/s^2 (4) 10. m/s^2

19. What is the direction of the centripetal acceleration of the passenger at the instant the cage reaches the highest point in the circle?
    (1) to the left (2) to the right (3) up   (4) down

20. What does the 50.-kilogram passenger weigh at rest?
    (1) 1600 N   (2) 490 N   (3) 50 N   (4) 0 N

21. What is the magnitude of the centripetal force acting on the passenger?
    (1) 0 N   (2) 50 N   (3) 4.9 $\times$ 10^2 N (4) 5.0 $\times$ 10^2 N

Base your answers to questions 22 through 24 on the diagram below, which shows a 10-newton centripetal force causing a l-kilogram mass to move in a circular path with a speed of 10 meters per second.

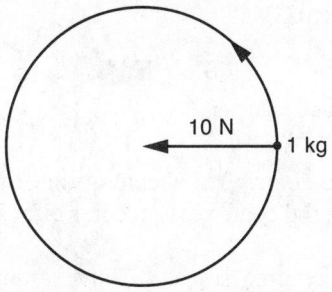

22. What is the radius of the circular path?
(1) 1 m   (2) $\sqrt{10}$ m   (3) 10 m   (4) 100 m

23. The centripetal acceleration of the object is equal to
(1) 100 m/s² (2) 1000 m/s² (3) $\dfrac{100}{r}$ m/s² (4) $\dfrac{1000}{r}$ m/s²

24. If the speed of the mass increases and the radius of its path remains constant, the centripetal force will
(1) decrease   (2) increase   (3) remain the same

Base your answers to questions 25 through 27 on the diagram below, which shows the path of an object moving counterclockwise in a circle of radius 2.0 meters. The speed of the object is 6.0 meters per second, and the mass of the object is 0.2 kilogram.

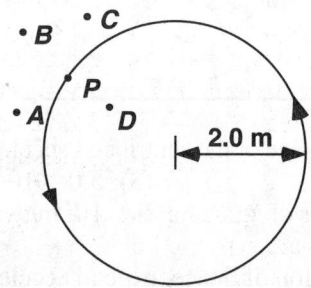

25. What is the magnitude of the acceleration of the object
(1) 60 m/s²   (2) 12 m/s²   (3) 3.0 m/s²   (4) 18 m/s²

26. What is the magnitude of the the the centripetal on the object?
(1) 2.4 N   (2) 3.6 N   (3) 6.0 N   (4) 12 N

27. If the speed of the object increases, the magnitude of the centripetal force needed to keep the object moving in the same circle will
(1) decrease   (2) increase   (3) remain the same

Base your answers to questions 28 through 30 on the diagram below, which represents a car of mass 1000 kilograms traveling around a horizontal circular track of radius 200 meters at a constant speed of 20 meters per second.

28. If the speed of the car were doubled, the centripetal acceleration of the car would be
    (1) the same   (3) one-half as great
    (2) doubled   (4) 4 times as great

29. The magnitude of the centripetal force acting on the car is closest to
    (1) 100 N   (2) 1000 N   (3) 2000 N   (4) 4000 N

30. If additional passengers were riding in the car, at the original speed, the car's centripetal acceleration would be
    (1) less   (2) greater   (3) the same

31. If the mass of one of two particles is doubled and the distance between them is doubled, the force of attraction between the two particles will
    (1) decrease   (2) increase   (3) remain the same

32. If the mass of an object near the surface of the Earth is increased from M to $3M$, the acceleration of the object due to gravity will be
    (1) one-third as great   (3) 3 times as great
    (2) 9 times as great   (4) unchanged

33. Gravitational force of attraction **F** exists between two point masses, A and B, when they are separated by a fixed distance. After mass A is tripled and mass B is halved, the gravitational attraction between the two masses is
    (1) 1/6 **F**   (2) 2/3 **F**   (3) 3/2 **F**   (4) 6 **F**

34. A rocket weighs 10,000 newtons at the Earth's surface. If the rocket rises to a height equal to the Earth's radius, its weight will be
    (1) 2500 N   (2) 5000 N   (3) 10,000 N   (4) 40,000 N

35. Compared to the mass of a 10-newton object on the Earth, the mass of the same object on the Moon is
    (1) less   (2) greater   (3) the same

36. As a space ship from Earth goes toward the Moon, the force it exerts on the Earth
    (1) decreases   (2) increases   (3) remains the same

37. Two masses of 10.0 kilograms and 1.0 kilogram, respectively, are located 1.0 meter apart. The gravitational force that each mass exerts on the other is
    (1) 6.7×10^{-9} N   (3) 6.7×10^{-11} N
    (2) 6.7×10^{-10} N   (4) 6.7×10^{-12} N

38. The diagram below shows spheres A and B with masses of M and $3M$, respectively.

If the gravitational force of attraction of sphere A on sphere B is 2 newtons, then the gravitational force of attraction of sphere B on sphere A is
(1) 9 N   (2) 2 N   (3) 3 N   (4) 4 N

39. Which graph best represents the gravitational force between two point masses as a function of the distance between the masses?

40. Which graph best represents the relationship between the mass of an object and its distance from the center of the Earth?

41. Two objects of fixed mass are moved apart so that they are separated by 3 times their original distance. Compared to the original gravitational force between them, the new gravitational force is
(1) one-third as great   (3) 3 times greater
(2) one-ninth as great   (4) 9 times greater

42. Three equal masses, A, B, and C, are arranged as shown in the diagram.

If the gravitational force between A and B is 3 newtons, then the gravitational force between A and C is
(1) 1 N   (2) 9 N   (3) 3 N   (4) 27 N

43. An object has a weight W at the surface of the Earth. At a distance of 3 Earth radii from the center of the Earth, the weight of the object will be
(1) $W/9$ (2) $W/3$ (3) $3W$ (4) $9W$

44. The acceleration due to gravity at a point near the surface of the Moon is ⅙ that near the surface of the Earth. The weight of a 2.0-kilogram mass at that same point near the surface of the Moon is approximately
(1) 1.6 N   (2) 2.0 N   (3) 3.3 N   (4) 0.33 N

45. The gravitational field at a given location is 9.73 newtons per kilogram. The gravitational acceleration at this location is
(1) 4.90 m/s² (2) 9.73 m/s² (3) 9.81 m/s² (4) 19.6 m/s²

46. A satellite is moving at constant speed in a circular orbit about the Earth, as shown in the diagram below.

The net force acting on the satellite is directed toward point
(1) A (2) B (3) C (4) D

47. What is the gravitational acceleration on a planet where a 2-kilogram mass has a weight of 16 newtons on the planet's surface?
(1) ⅛ m/s²   (2) 8 m/s²   (3) 10 m/s²   (4) 32 m/s²

48. On planet Gamma, a 4.0-kilogram mass experiences a gravitational force of 24 newtons. What is the acceleration due to gravity on planet Gamma?
(1) 0.17 m/s² (2) 6.0 m/s² (3) 9.8 m/s² (4) 96 m/s²

49. A rock falls from rest a vertical distance of 0.72 meter to the surface of a planet in 0.63 second. The magnitude of the acceleration due to gravity on the planet is
(1) 1.1 m/s² (2) 2.3 m/s² (3) 3.6 m/s² (4) 9.8 m/s²

50. A 1200-kilogram space vehicle travels at 4.8 meters per second along the level surface of Mars. If the magnitude of the gravitational field strength on the surface of Mars is 3.7 newtons per kilogram, the magnitude of the normal force acting on the vehicle is
(1) 320 N   (2) 930 N   (3) 4400 N   (4) 5800 N

Base your answers to questions 51 through 54 on the diagram below, which represents a satellite orbiting the Earth. The satellite's distance from the center of the Earth equals 4 Earth radii.

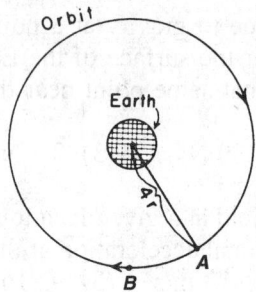

51. Which vector best represents the velocity of the satellite at point *B*?
   (1) ↑   (3) ←
   (2) ↓   (4) →

52. The original satellite is replaced by one with twice the mass, and the orbit speed and radius are unchanged. Compared to the magnitude of the acceleration of the original satellite, the magnitude of the acceleration of the new satellite is
   (1) one-half as great   (3) twice as great
   (2) the same   (4) 4 times as great

53. Which vector best represents the acceleration of the satellite at point *A* in its orbit?
   (1) ↗   (3) ↘
   (2) ↙   (4) ↖

54. If the satellite's distance from the center of the Earth were increased to 5 Earth radii, the centripetal force on the satellite would
   (1) decrease   (2) increase   (3) remain the same

134

55. The diagram below shows an object moving counterclockwise around a horizontal, circular track.

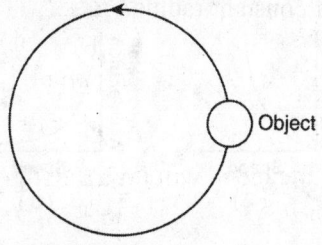

Horizontal track

Which diagram represents the direction of both the object's velocity and the centripetal force acting on the object when it is in the position shown?

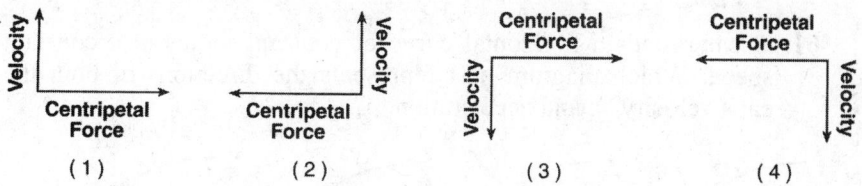

56. A 1750-kilogram car travels at a constant speed of 15.0 meters per second around a horizontal, circular track with a radius of 45.0 meters. The magnitude of the centripetal force acting on the car is
(1) 5.00 N   (2) 583 N   (3) 8750 N   (4) 3.94×10^5 N

57. As a meteor moves from a distance of 16 Earth radii to a distance of 2 Earth radii from the center of Earth, the magnitude of the gravitational force between the meteor and Earth becomes
(1) ⅛ as great   (3) 64 times as great
(2) 8 times as great   (4) 4 times as great

58. Which diagram best represents the gravitational forces, F_g, between a satellite, S, and Earth?

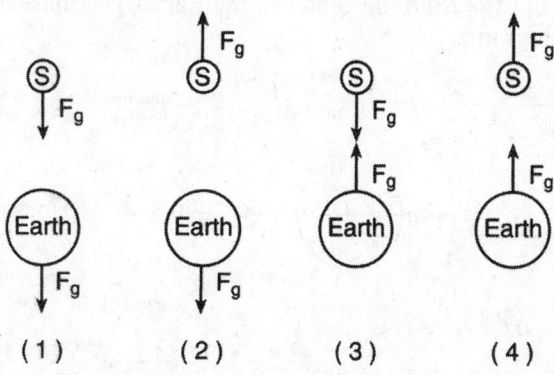

59. Which graph best represents the relationship between the magnitude of the centripetal acceleration and the speed of an object moving in a circle of constant radius?

60. A 60.-kilogram physics student would weigh 1560 newtons on the surface of planet X. What is the magnitude of the acceleration due to gravity on the surface of planet X?
(1) 0.038 m/s^2 (2) 6.1 m/s^2 (3) 9.8 m/s^2 (4) 26 m/s^2

61. A car rounds a horizontal curve of constant radius at a constant speed. Which diagram best represents the directions of both the car's velocity, v, and acceleration, a?

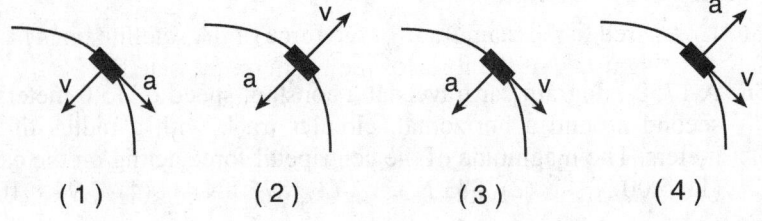

62. A 2.00-kilogram object weighs 19.6 newtons on Earth. If the acceleration due to gravity on Mars is 3.71 meters per second^2, what is the object's mass on Mars?
(1) 2.64 kg (2) 2.00 kg (3) 19.6 N (4) 7.42 N

Base your answers to questions 63 through 66 on the diagram below which represents a satellite in an elliptical orbit about the Earth. The highest point, A, is four Earth radii ($4R$) from the center of the Earth. The lowest point, B. is two Earth radii ($2R$) from the center of the Earth. The mass of the satellite is 3.0×10^6 kilograms.

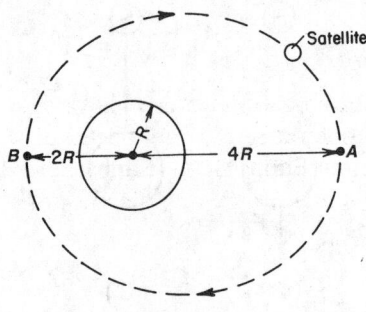

63. Which vector represents the direction of the satellite's velocity at point *A*?

(1) ⟶   (3) ↓

(2) ⟵   (4) ↑

64. Which vector represents the direction of the centripetal force on the satellite at point *B*?

(1) ⟶   (3) ↑

(2) ⟵   (4) ↓

65. As the satellite moves from point *A* toward point *B*, the velocity of the satellite
(1) decreases   (2) increases   (3) remains the same

66. Compared to the magnitude of the force of the satellite on the Earth, the magnitude of the force of the Earth on the satellite is
(1) less   (2) greater   (3) the same

67. A car moves with a constant speed in a clockwise direction around a circular path of radius *r*, as represented in the diagram below.

When the car is in the position shown, its acceleration is directed toward the
(1) north   (2) west   (3) south   (4) east

68. A 0.50-kilogram object moves in a horizontal circular path with a radius of 0.25 meter at a constant speed of 4.0 meters per second. What is the magnitude of the object's acceleration?
(1) 8.0 m/s^2 (2) 16 m/s^2 (3) 32 m/s^2 (4) 64 m/s^2

69. Earth's mass is approximately 81 times the mass of the Moon. If Earth exerts a gravitational force of magnitude F on the Moon, the magnitude of the gravitational force of the Moon on Earth is

(1) F (2) $\frac{F}{81}$ (3) $9F$ (4) $81F$

70. The diagram shows two bowling balls, A and B, each having a mass of 7.00 kilograms, placed 2.00 meters apart.

What is the magnitude of the gravitational force exerted by ball A on ball B?

(1) 8.17×10^{-9} N   (3) 8.17×10^{-10} N
(2) 1.63×10^{-9} N   (4) 1.17×10^{-10} N

71. A ball attached to a string is moved at constant speed in a horizontal circular path. A target is located near the path of the ball as shown in the diagram.

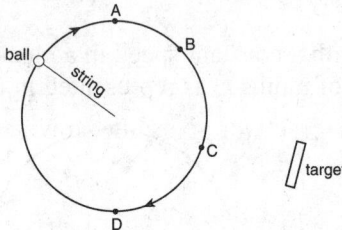

At which point along the ball's path should the string be released, if the ball is to hit the target?

(1) A (2) B (3) C (4) D

72. As an astronaut travels from the surface of Earth to a position that is four times as far away from the center of Earth, the astronaut's

(1) mass decreases   (3) weight increases
(2) mass remains the same   (4) weight remains the same

Base your answers to questions 73 and 74 on the information and diagram below.

The diagram shows the top view of a 65-kilogram student at point A on an amusement park ride. The ride spins the student in a horizontal circle of radius 2.5 meters, at a constant speed of 8.6 meters per second. The floor is lowered and the student remains against the wall without falling to the floor.

73. Which vector best represents the direction of the centripetal acceleration of the student at point *A*?

(1)   (2)   (3)   (4)

74. The magnitude of the centripetal force acting on the student at point *A* is approximately
(1) 1.2×10^4 N   (3) 2.2×10^2 N
(2) 1.9×10^3 N   (4) 3.0×10^1 N

75. Which diagram best represents the gravitational field lines surrounding Earth?

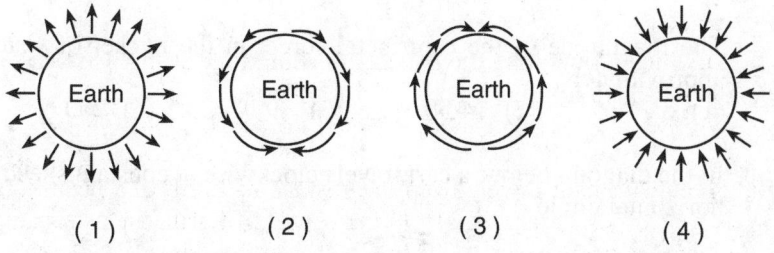

(1)   (2)   (3)   (4)

76. A 2.0-kilogram object is falling freely near Earth's surface. What is the magnitude of the gravitational force that Earth exerts on the object?
(1) 20. N   (2) 2.0 N   (3) 0.20 N   (4) 0.0 N

77. A 25.0-kilogram space probe fell freely with an acceleration of 2.00 meters per second² just before it landed on a distant planet. What is the weight of the space probe on that planet?
(1) 12.5 N   (2) 25.0 N   (3) 50.0 N   (4) 250. N

78. The diagram below represents two satellites of equal mass, A and B, in circular orbits around a planet.

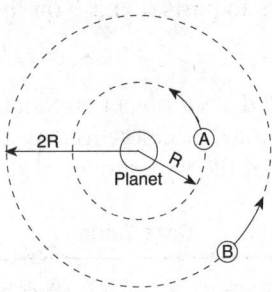

Compared to the magnitude of the gravitational force of attraction between satellite A and the planet, the magnitude of the gravitational force of attraction between satellite B and the planet is
(1) half as great
(2) twice as great
(3) one-fourth as great
(4) four times as great

79. The diagram below shows a 5.0-kilogram bucket of water being swung in a horizontal circle of 0.70-meter radius at a constant speed of 2.0 meters per second.

The magnitude of the centripetal force on the bucket of water is approximately
(1) 5.7 N   (2) 14 N   (3) 29 N   (4) 200 N

80. In the diagram below, a cart travels clockwise at constant speed in a horizontal circle.

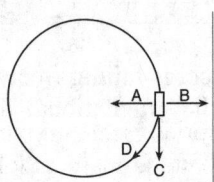

At the position shown in the diagram, which arrow indicates the direction of the centripetal acceleration of the cart?
(1) A (2) B (3) C (4) D

Answers to Part A and B–1 questions can be found on page 455.

PART B-2 AND C QUESTIONS

1. Base your answers to parts *a* and *b* on the information and the data table below.

   An astronaut on a distant planet conducted an experiment to determine the gravitational acceleration on that planet. The data table shows the results of the experiment.

Data Table

Mass (kilograms)	Weight (newtons)
15	106
20.	141
25	179
30.	216
35	249

a Using the information in the data table, construct a graph on the grid provided, following the directions.

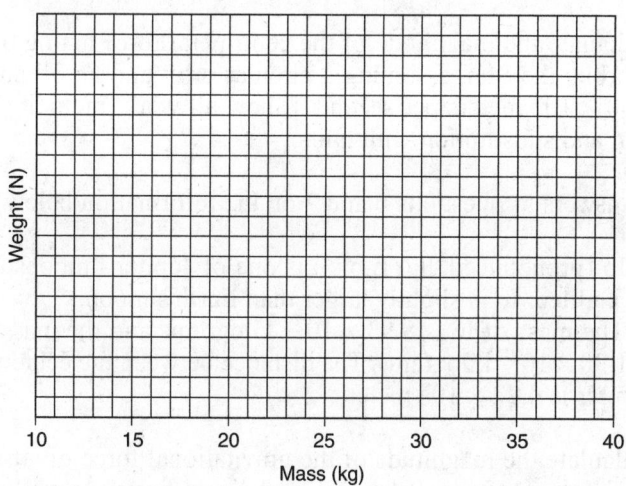

   (1) Mark an appropriate scale on the axis labeled "Weight (N)."
   (2) Plot a weight versus mass graph for the astronaut's data and draw the best-fit line.

   b *Using your graph,* determine the planet's gravitational acceleration. [Show all calculations, including equations and substitutions with units.]

141

Base your answers to questions 2 through 4 on the information and diagram below.

A 1.50-kilogram cart travels in a horizontal circle of radius 2.40 meters at a constant speed of 4.00 meters per second.

2. Calculate the time required for the cart to make one complete revolution. [Show all work, including the equation and substitution with units.]

3. Describe a change that would quadruple the magnitude of the centripetal force.

4. On the diagram, draw an arrow to represent the direction of the acceleration of the cart in the position shown. Label the arrow *a*.

5. Calculate the magnitude of the centripetal force acting on Earth as it orbits the Sun, assuming a circular orbit and an orbital speed of 3.00×10^4 meters per second. [Show all work, including the equation and substitution with units.]

Base your answers to questions 6 and 7 on the information below.

Io (pronounced "EYE oh") is one of Jupiter's moons discovered by Galileo. Io is slightly larger than Earth's moon.
The mass of Io is 8.93×10^{22} kilograms and the mass of Jupiter is 1.90×10^{27} kilograms. The distance between the centers of Io and Jupiter is 4.22×10^8 meters.

6. Calculate the magnitude of the gravitational force of attraction that Jupiter exerts on Io. [Show all work, including the equation and substitution with units.]

7. Calculate the magnitude of the acceleration of Io due to the gravitational force exerted by Jupiter. [Show all work, including the equation and substitution with units.]

Base your answers to questions 8 through 10 on the passage and data table below.

The net force on a planet is due primarily to the other planets and the Sun. By taking into account all the forces acting on a planet, investigators calculated the orbit of each planet.

A small discrepancy between the calculated orbit and the observed orbit of the planet Uranus was noted. It appeared that the sum of the forces on Uranus did not equal its mass times its acceleration, unless there was another force on the planet that was not included in the calculation. Assuming that this force was exerted by an unobserved planet, two scientists working independently calculated where this unknown planet must be in order to account for the discrepancy. Astronomers pointed their telescopes in the predicted direction and found the planet we now call Neptune.

Data Table

Mass of the Sun	1.99×10^{30} kg
Mass of Uranus	8.73×10^{25} kg
Mass of Neptune	1.03×10^{26} kg
Mean distance of Uranus to the Sun	2.87×10^{12} m
Mean distance of Neptune to the Sun	4.50×10^{12} m

8. What fundamental force is the author referring to in this passage as a force between planets?

9. The diagram below represents Neptune, Uranus, and the Sun in a straight line. Neptune is 1.63×10^{12} meters from Uranus.

(Not drawn to scale)

Calculate the magnitude of the interplanetary force of attraction between Uranus and Neptune at this point. [Show all work, including the equation and substitution with units.]

10. The magnitude of the force the Sun exerts on Uranus is 1.41×10^{21} newtons. Explain how it is possible for the Sun to exert a greater force on Uranus than Neptune exerts on Uranus.

Answers to Part B–2 and C questions can be found on pages 486–490.

MOMENTUM AND ITS CONSERVATION

KEY IDEAS

Momentum is the product of mass and velocity. It is a vector quantity that measures the tendency of an object to remain in motion. An *impulse* is the product of force and time and is responsible for changing the momentum of an object. It is also a vector quantity.

If two objects interact (and are not subject to external forces) the momentum of the system (i.e., the two objects) is conserved. This basic law of physics can be used to describe collision and explosion phenomena.

KEY OBJECTIVES

At the conclusion of this chapter you will be able to:
- Define the term *momentum*, and state its SI unit.
- Solve problems involving mass, velocity, and momentum.
- Define the term *impulse*, and state its SI unit.
- Relate impulse to change in momentum.
- Solve impulse-momentum problems.
- State the law of conservation of momentum, and solve problems based on this law.
- Relate the law of conservation of momentum to Newton's third law of motion.

6.1 MOMENTUM

Suppose we wanted to measure the "tendency" of an object to remain in motion. We would find that two factors are necessary to describe this tendency: the mass and the velocity of the object. The more mass an object has, the more force is required to bring it to rest. Similarly, the greater the velocity of the object, the more force is necessary to bring it to rest. These two factors, mass and velocity, are combined into a single quantity that we call **momentum**. We define momentum, symbolized as **p**, as the product of mass and velocity:

===== **PHYSICS CONCEPTS** =====

$$\mathbf{p} = m\mathbf{v}$$

Momentum is a vector quantity and its direction is the direction of the velocity of the object. Its unit is the *kilogram · meter per second* (kg · m/s).

PROBLEM

An object whose mass is 3.5 kilograms is traveling at 20. meters per second [east]. Calculate the momentum of the object.

SOLUTION

$$\mathbf{p} = m\mathbf{v}$$
$$= (3.5 \text{ kg})(20. \text{ m/s [E]})$$
$$= 70. \text{ kg} \cdot \text{m/s [E]}$$

6.2 NEWTON'S SECOND LAW AND MOMENTUM

When Newton developed his second law of motion, he recognized that the unbalanced force on an object caused a change in the object's momentum:

$$\mathbf{F}_{net} = \frac{\Delta \mathbf{p}}{t} = \frac{\Delta(m\mathbf{v})}{t}$$

Since mass is usually constant:

$$\mathbf{F}_{net} = \frac{m\Delta \mathbf{v}}{t} = m\mathbf{a}$$

We can rewrite Newton's second law in the following form:

$$\mathbf{F}_{net}\, t = \Delta \mathbf{p}$$

We call the quantity F *t*, the **impulse** delivered to the object. Impulse is symbolized by the letter **J**, and its unit is the *newton · second* (N · s). Impulse is a vector quantity, and its direction is the direction of the net force.

From Newton's second law it follows that the impulse delivered to the object changes its momentum, and the unit newton · second is equivalent to the unit kilogram · meter per second.

PROBLEM

A 5.0-kilogram object traveling at 3.0 meters per second [east] is subjected to a force that increases its velocity to 7.0 meters per second [east]. Calculate: (a)

145

the initial momentum of the object, (b) the final momentum of the object, (c) the change in momentum of the object, and (d) the impulse delivered to the object. (e) If the force acts for 0.20 second, what are its magnitude and its direction?

SOLUTION

(a) $\mathbf{p}_i = m\mathbf{v}_i$
$\quad = (5.0 \text{ kg})(3.0 \text{ m/s [E]})$
$\quad = 15 \text{ kg} \cdot \text{m/s [E]}$

(b) $\mathbf{p}_f = m\mathbf{v}_f$
$\quad = (5.0 \text{ kg}) (7.0 \text{ m/s [E]})$
$\quad = 35 \text{ kg} \cdot \text{m/s [E]}$

(c) $\Delta \mathbf{p} = \mathbf{p}_f - \mathbf{p}_i$
$\quad = 35 \text{ kg} \cdot \text{m/s [E]} - 15 \text{ kg} \cdot \text{m/s [E]}$
$\quad = 20. \text{ kg} \cdot \text{m/s [E]}$

(d) $\mathbf{J} = \mathbf{F} t = \Delta \mathbf{p}$
$\quad = 20. \text{ N} \cdot \text{s [E]}$

(e) $\mathbf{J} = \mathbf{F} t$

$\mathbf{F} = \dfrac{\mathbf{J}}{t}$

$\quad = \dfrac{20. \text{ N} \cdot \text{s [E]}}{0.20 \text{ s}}$

$\quad = 100 \text{ N [E]}$

6.3 CONSERVATION OF MOMENTUM

If two objects that are not subjected to any external forces interact (e.g., they collide), the total momentum of the objects before the interaction is equal to their total momentum after the interaction:

===== **PHYSICS CONCEPTS** =====

$$\mathbf{p}_{before} = \mathbf{p}_{after}$$
$$\mathbf{p}_{1i} + \mathbf{p}_{2i} = \mathbf{p}_{1f} + \mathbf{p}_{2f}$$

Note that this equation in this form does not appear on the Reference Tables but is used to solve problems.

This relationship, known as *conservation of momentum*, is a fundamental law of physics. The diagram below illustrates the conservation of momentum.

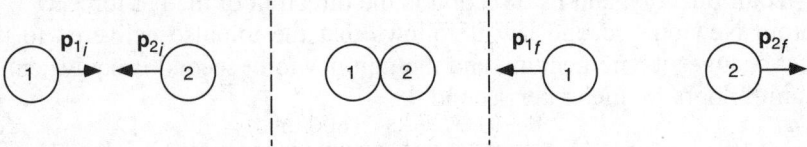

PROBLEM

A 0.50-kilogram object traveling at 2.0 meters per second [east] collides with a 0.30-kilogram object traveling at 4.0 meters per second [west]. After the collision, the 0.30-kilogram object is traveling at 2.0 meters per second [east]. What are the magnitude and the direction of the velocity of the first object?

SOLUTION

$$\mathbf{p}_{1i} + \mathbf{p}_{2i} = \mathbf{p}_{1f} + \mathbf{p}_{2f}$$

$$m_1\mathbf{v}_{1i} + m_2\mathbf{v}_{2i} = m_1\mathbf{v}_{1f} + m_2\mathbf{v}_{2f}$$

$$\mathbf{v}_{1f} = \frac{m_1\mathbf{v}_{1i} + m_2\mathbf{v}_{2i} - m_2\mathbf{v}_{2f}}{m_1}$$

$$= \frac{(0.50 \text{ kg})(2.0 \text{ m/s}) + (0.30 \text{ kg})(-4.0 \text{ m/s}) - (0.30 \text{ kg})(2.0 \text{ m/s})}{0.50 \text{ kg}}$$

$$= -1.6 \text{ m/s} = 1.6 \text{ m/s [W]}$$

Consider a gun about to fire a bullet. The initial momentum of the gun-bullet system is zero. Since the bullet moves in one direction, the gun recoils in the opposite direction, so that the momentum after the "explosion" is also zero.

PROBLEM

A 5.0-kilogram gun fires a 0.0020-kilogram bullet. If the bullet exits the gun at 800. meters per second [east], calculate the recoil velocity of the gun.

SOLUTION

$$\mathbf{p}_{1i} + \mathbf{p}_{2i} = \mathbf{p}_{1f} + \mathbf{p}_{2f}$$

$$0 = \mathbf{p}_{1f} + \mathbf{p}_{2f}$$

$$0 = m_1\mathbf{v}_{1f} + m_2\mathbf{v}_{2f}$$

$$\mathbf{v}_{1f} = \frac{0 - m_2\mathbf{v}_{2f}}{m_1}$$

$$= \frac{0 - (0.0020 \text{ kg})(+800. \text{ m/s})}{5.0 \text{ kg}}$$

$$= -0.32 \text{ m/s} = 0.32 \text{ m/s [W]}$$

PROBLEM

A bullet with a mass of 0.0020 kilogram is fired at a speed of 600. meters per second [east] at a block with a mass of 0.500 kilograms at rest on a table. Calculate the velocity of the bullet-block system after the bullet embeds itself in the block.

SOLUTION

$$\mathbf{p}_{1_i} + \mathbf{p}_{2_i} = \mathbf{p}_{1_f} + \mathbf{p}_{2_f}$$

$$m_1\mathbf{v}_{1_i} + 0 = (m_1 + m_2)\mathbf{v}_{2_f}$$

$$\mathbf{v}_{2_f} = \frac{m_1\mathbf{v}_{1_i}}{(m_1 + m_2)}$$

$$= \frac{(0.0020 \text{ kg})(600. \text{ m/s})}{(0.0020 \text{ kg} + 0.500 \text{ kg})}$$

$$= 1.99 \text{ m/s[E]}$$

6.4 CONSERVATION OF MOMENTUM AND NEWTON'S THIRD LAW

We can rewrite the law of conservation of momentum as follows:

$$\mathbf{p}_{1_i} + \mathbf{p}_{2_i} = \mathbf{p}_{1_f} + \mathbf{p}_{2_f}$$

$$-(\mathbf{p}_{2_f} - \mathbf{p}_{2_i}) = (\mathbf{p}_{1_f} - \mathbf{p}_{1_i})$$

$$-\Delta\mathbf{p}_2 = \Delta\mathbf{p}_1$$

The new equation means that the objects undergo equal but opposite changes in momentum. Because momentum change is equal to impulse, the two objects deliver equal but opposite impulses to one another in the same time interval. *Therefore, for any arbitrarily small interval of time, they must exert equal and opposite forces on one another.*

Does this seem familiar? We have just shown that Newton's third law is a direct consequence of the law of conservation of momentum!

PART A AND B-1 QUESTIONS

1. If the direction of the momentum of an object is west, the direction of the velocity of the object is
   (1) north   (2) south   (3) east   (4) west

2. A 2-newton force acts on a mass. If the momentum of the mass changes by 120 kilogram · meters per second, the force acts for a time of
   (1) 8 s   (2) 30 s   (3) 60 s   (4) 120 s

3. As an object falls freely toward the Earth, its momentum
(1) decreases (2) increases (3) remains the same

4. A 50-kilogram mass has a momentum of 100 kilogram · meters per second. What is the velocity of the mass?
(1) 0.5 m/s (2) 2 m/s (3) 2500 m/s (4) 5000 m/s

5. An impulse **J** is applied to an object. The change in the momentum of the object is

(1) **J** (2) 2**J** (3) $\dfrac{\textbf{J}}{2}$ (4) 4**J**

6. The momentum of an object is the product of its
(1) mass and acceleration
(2) mass and velocity
(3) force and displacement
(4) force and distance

7. Momentum may be expressed in
(1) joules (3) kilogram · meters per second^2
(2) watts (4) Newton · seconds

8. If a 3.0-kilogram object moves 10. meters in 2.0 seconds, its average momentum is
(1) 60. kg · m/s (3) 15 kg · m/s
(2) 30. kg · m/s (4) 10. kg · m/s

9. The direction of an object's momentum is always the same as the direction of the object's
(1) inertia (3) velocity
(2) potential energy (4) weight

10. A mass is moving with a uniform velocity. If the time required to stop the mass increases, the force required to stop it must
(1) decrease (2) increase (3) remain the same

11. A force of 3.0 newtons applied to an object produces a change in velocity of 12 meters per second in 0.40 second. The mass of the object is
(1) 1.0 kg (2) 0.10 kg (3) 10. kg (4) 0.31 kg

12. An object traveling at 4.0 meters per second has a momentum of 16 kilogram · meters per second. What is the mass of the object?
(1) 64 kg (2) 20 kg (3) 12 kg (4) 4.0 kg

13. What is the magnitude of the velocity of a 25-kilogram mass that is moving with a momentum of 100. kilogram · meters per second?
(1) 0.25 m/s   (2) 2500 m/s   (3) 40. m/s   (4) 4.0 m/s

14. A 15-newton force acts on an object in a direction due east for 3.0 seconds. What will be the change in momentum of the object?
(1) 45 kg · m/s [E]   (3) 5.0 kg · m/s [E]
(2) 45 kg · m/s [W]   (4) 0.20 kg · m/s [W]

15. An unbalanced 6.0-newton force acts eastward on an object for 3.0 seconds. The impulse produced by the force is
(1) 18 N · s [E]   (3) 18 N · s [W]
(2) 2.0 N · s [E]   (4) 2.0 N · s [W]

16. As the unbalanced force applied to an object increases, the time rate of change of the object's momentum
(1) decreases (2) increases   (3) remains the same

17. A rocket with a mass of 1000 kilograms is moving at a speed of 20 meters per second. The magnitude of the momentum is
(1) 50 kg · m/s   (3) 20,000 kg · m/s
(2) 200 kg · m/s   (4) 400,000 kg · m/s

18. An impulse of 30.0 newton · seconds is applied to a 5.00-kilogram mass. If the mass had a speed of 100. meters per second before the impulse, its speed after the impulse could be
(1) 250. m/s   (2) 106 m/s   (3) 6.00 m/s   (4) 0 m/s

19. What is the magnitude of the change in momentum produced when a force of 5.0 newtons acts on a 10-kilogram object for 3.0 seconds?
(1) 1.5 kg · m/s   (3) 10 kg · m/s
(2) 5.0 kg · m/s   (4) 15 kg · m/s

20. An object is brought to rest by a constant force. Which factor other than the mass and velocity of the object must be known in order to determine the magnitude of the force required to stop the object?
(1) the time that the force acts on the object
(2) the gravitational potential energy of the object
(3) the density of the object
(4) the weight of the object

21. A constant unbalanced force acts on an object for 3.0 seconds, producing an impulse of 6.0 newton · seconds. What is the magnitude of the force?
(1) 6.0 N   (2) 2.0 N   (3) 3.0 N   (4) 18 N

22. A car with a mass of 1.0×10^3 kilograms is moving with a speed of 1.4×10^2 meters per second. The impulse required to bring the car to rest is
(1) 1.4×10^2 N · s   (3) 7.0×10^4 N · s
(2) 1.4×10^4 N · s   (4) 1.4×10^5 N · s

Base your answers to questions 23 through 26 on the information and graph below.

A force acts on a 3.0-kg cart that is on a frictionless horizontal surface. The resulting motion is shown by the graph.

23. How long does the force act on the cart?
(1) 1.0 s   (2) 2.0 s   (3) 3.0 s   (4) 7.0 s

24. What is the acceleration of the cart during the first 2 seconds?
(1) 1.0 m/s^2 (2) 0 m/s^2 (3) 4.5 m/s^2 (4) $10.$ m/s^2

25. What is the momentum of the cart at the end of 6 seconds?
(1) $10.$ N · s   (2) $90.$ N · s   (3) 210 N · s   (4) 630 N · s

26. What is the total distance traveled by the cart during 7 seconds?
(1) 45 m   (2) 105 m   (3) 165 m   (4) 210 m

27. When two objects collide, there will be no net change in the
(1) velocity of each object
(2) displacement of each object
(3) kinetic energy of each object
(4) total momentum of the objects

28. An 80.-kilogram skater and a 60.-kilogram skater stand at rest in the center of a skating rink. The two skaters push each other apart. The 60.-kilogram skater moves with a velocity of 10. meters per second east. What is the velocity of the 80.-kilogram skater? [Neglect any frictional effects.]
   (1) 0.13 m/s [W]   (3) 10. m/s [E]
   (2) 7.5 m/s [W]   (4) 13. m/s [E]

29. A 1.0-kilogram object moving east with a velocity of 10 meters per second collides with a 0.50-kilogram object that is at rest. Neglecting friction, what is the momentum of the system after the collision?
   (1) 15 kg · m/s [E]   (3) 10 kg · m/s [E]
   (2) 15 kg · m/s [W]   (4) 10 kg · m/s [W]

30. Two carts having masses of 5.0 kilograms and 1.0 kilogram, respectively, are pushed apart by a compressed spring. If the 5.0-kilogram cart moves westward at 2.0 meters per second, the magnitude of the velocity of the 1.0-kilogram cart will be
   (1) 2.0 kg · m/s   (3) 10. kg · m/s
   (2) 2.0 m/s   (4) 10. m/s

31. A 20.-kilogram cart traveling east with a speed of 6.0 meters per second collides with a 30.-kilogram cart traveling west. If both carts come to rest after the collision, what was the speed of the westbound cart before the collision?
   (1) 0 m/s   (2) 9.0 m/s   (3) 3.0 m/s   (4) 4.0 m/s

32. As a ball falls freely toward the Earth, the momentum of the Earth-ball system
   (1) decreases (2) increases   (3) remains the same

33. A cart has a momentum of 10. kilogram · meters per second east before a collision and a momentum of 5.0 kilogram · meters per second west after the collision. The net change in momentum of this cart is
   (1) 5.0 kg · m/s [E]   (3) 15. kg · m/s [E]
   (2) 5.0 kg · m/s [W]   (4) 15. kg · m/s [W]

34. A projectile with a mass of 0.01 kilogram has a muzzle velocity of 1000 meters per second when fired from a rifle weighing 5 kilograms. The recoil velocity of the rifle is
   (1) 0.1 m/s   (2) 2 m/s   (3) 5 m/s   (4) 50 m/s

Base your answers to questions 35 through 39 on the diagram below which represents carts A and B being pushed apart by a spring that exerts an average force of 50. newtons for a period of 0.20 second. [Assume frictionless conditions.]

35. What is the magnitude of the impulse applied by the spring on cart A?
    (1) 5.0 N · s   (2) 10. N · s   (3) 50. N · s   (4) 100 N · s

36. Compared to the magnitude of the impulse acting on cart A, the magnitude of the impulse acting on cart B is
    (1) one-half as great   (3) the same
    (2) twice as great   (4) 4 times as great

37. Compared to the velocity of cart B at the end of the 0.20-second interaction, the velocity of cart A is
    (1) one-half as great   (3) the same
    (2) twice as great   (4) 4 times as great

38. What is the average acceleration of cart B during the 0.20-second interaction?
    (1) 0 m/s²   (2) 10. m/s²   (3) 25 m/s²   (4) 50. m/s²

39. Compared to the total momentum of the carts before the spring is released, the total momentum of the carts after the spring is released is
    (1) one-half as great   (3) the same
    (2) twice as great   (4) 4 times as great

Base your answers to questions 40 and 41 on the information and diagram below.

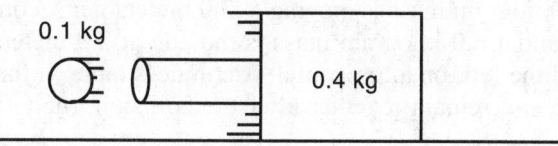

A 0.4-kilogram toy cannon is at rest on a horizontal, frictionless surface. When a 0.1-kilogram projectile is fired horizontally from the barrel of the cannon, the cannon recoils with a speed of 2.5 meters per second.

40. The speed of the projectile as it leaves the barrel is
    (1) 0.125 m/s (2) 2.5 m/s   (3) 10.0 m/s   (4) 4.0 m/s

41. The total change in the momentum of the cannon as it is fired is
    (1) 1.0 N · s (2) 0.25 N · s (3) 0.6 N · s   (4) 2.9 N · s

42. A 2-kilogram car and a 3-kilogram car are originally at rest on a horizontal, frictionless surface as shown in the diagram below. A compressed spring is released, causing the cars to separate. The 3-kilogram car reaches a maximum speed of 2 meters per second. What is the maximum speed of the 2-kilogram car?

    (1) 1 m/s   (2) 2 m/s   (3) 3 m/s   (4) 6 m/s

43. A 2.0-kilogram cart traveling east with a speed of 6 meters per second collides with a 3.0-kilogram cart traveling west. If both carts come to rest immediately after the collision, what was the speed of the westbound cart before the collision?
    (1) 6 m/s   (2) 2 m/s   (3) 3 m/s   (4) 4 m/s

44. Two carts resting on a frictionless surface are forced apart by a spring. One cart has a mass of 6 kilograms and moves to the left at a speed of 3 meters per second. If the second cart has a mass of 9 kilograms, it will move to the right at a speed of
    (1) 1 m/s   (2) 2 m/s   (3) 3 m/s   (4) 6 m/s

45. A 2.0-kilogram rifle initially at rest fires a 0.002-kilogram bullet. As the bullet leaves the rifle with a velocity of 500 meters per second, what is the momentum of the rifle-bullet system?
    (1) 2.5 kg · m/s   (3) 0.5 kg · m/s
    (2) 2.0 kg · m/s   (4) 0 kg · m/s

46. A 4.0-kilogram mass is moving at 3.0 meters per second toward the right, and a 6.0-kilogram mass is moving at 2.0 meters per second toward the left, on a horizontal, frictionless table. If the two masses collide and remain together after the collision, their final momentum is
    (1) 1.0 kg · m/s   (3) 12 kg · m/s
    (2) 24 kg · m/s   (4) 0 kg · m/s

47. Cart *A* has a mass of 2 kilograms and a speed of 3 meters per second. Cart *B* has a mass of 3 kilograms and a speed of 2 meters per second. Compared to the inertia and magnitude of momentum of cart *A*, cart *B* has
    (1) the same inertia and a smaller magnitude of momentum
    (2) the same inertia and the same magnitude of momentum
    (3) greater inertia and a smaller magnitude of momentum
    (4) greater inertia and the same magnitude of momentum

48. A 0.45-kilogram football traveling at a speed of 22 meters per second is caught by an 84-kilogram stationary receiver. If the football comes to rest in the receiver's arms, the magnitude of the impulse imparted to the receiver by the ball is
    (1) 1800 N•s   (2) 9.9 N•s   (3) 4.4 N•s   (4) 3.8 N•s

49. The diagram below represents two masses before and after they collide. Before the collision, mass m_A is moving to the right with speed **v**, and mass m_B is at rest. Upon collision, the two masses stick together.

Before Collision   **After Collision**

Which expression represents the speed, **v'**, of the masses after the collision? [Assume no outside forces are acting on m_A or m_B.]

(1) $\dfrac{m_A + m_B \mathbf{v}}{m_A}$ (2) $\dfrac{m_A + m_B}{m_A \mathbf{v}}$ (3) $\dfrac{m_B \mathbf{v}}{m_A + m_B}$ (4) $\dfrac{m_A \mathbf{v}}{m_A + m_B}$

50. Which object has the greatest inertia?
    (1) a 5.00-kg mass moving at 10.0 m/s
    (2) a 10.0-kg mass moving at 1.00 m/s
    (3) a 15.0-kg mass moving at 10.0 m/s
    (4) a 20.0-kg mass moving at 1.00 m/s

51. A 6.0-kilogram block, sliding to the east across a horizontal, frictionless surface with a momentum of 30. kilogram•meters per second, strikes an obstacle. The obstacle exerts an impulse of 10. newton•seconds to the west on the block. The speed of the block after the collision is
    (1) 1.7 m/s   (2) 3.3 m/s   (3) 5.0 m/s   (4) 20. m/s

52. A 1.0-kilogram laboratory cart moving with a velocity of 0.50 meter per second due east collides with and sticks to a similar cart initially at rest. After the collision, the two carts move off together with a velocity of 0.25 meter per second due east. The total momentum of this frictionless system is
    (1) zero before the collision
    (2) zero after the collision
    (3) the same before and after the collision
    (4) greater before the collision than after the collision

53. A bicycle and its rider have a combined mass of 80. kilograms and a speed of 6.0 meters per second. What is the magnitude of the average force needed to bring the bicycle and its rider to a stop in 4.0 seconds?
    (1) 1.2×10^2 N   (3) 4.8×10^2 N
    (2) 3.2×10^2 N   (4) 1.9×10^3 N

54. Which situation will produce the greatest change of momentum for a 1.0-kilogram cart?
    (1) accelerating it from rest to 3.0 m/s
    (2) accelerating it from 2.0 m/s to 4.0 m/s
    (3) applying a net force of 5.0 N for 2.0 s
    (4) applying a net force of 10.0 N for 0.5 s

55. In the diagram below, scaled vectors represent the momentum of each of two masses, A and B, sliding toward each other on a frictionless, horizontal surface.

Mass A   Frictionless surface   Mass B

Which scaled vector best represents the momentum of the system after the masses collide?

(1)   (3)

(2)   (4)

56. A 60-kilogram student jumps down from a laboratory counter. At the instant he lands on the floor his speed is 3 meters per second. If the student stops in 0.2 second, what is the average force of the floor on the student?
    (1) 1×10^{-2} N   (3) 9×10^2 N
    (2) 1×10^2 N   (4) 4 N

57. A force of 6.0 newtons changes the momentum of a moving object by 3.0 kilogram•meters per second. How long did the force act on the mass?

   (1) 1.0 s   (2) 2.0 s   (3) 0.25 s   (4) 0.50 s

58. A 3.0-kilogram steel block is at rest on a frictionless horizontal surface. A 1.0-kilogram lump of clay is propelled horizontally at 6.0 meters per second toward the block as shown in the diagram below.

Frictionless surface

Upon collision, the clay and steel block stick together and move to the right with a speed of

   (1) 1.5 m/s   (2) 2.0 m/s   (3) 3.0 m/s   (4) 6.0 m/s

59. In the diagram below, a block of mass M initially at rest on a frictionless horizontal surface is struck by a bullet of mass m moving with horizontal velocity $\mathbf{v}$.

What is the velocity of the bullet-block system after the bullet embeds itself in the block?

   (1) $\left(\dfrac{M+\mathbf{v}}{M}\right)m$ (2) $\left(\dfrac{m+M}{m}\right)\mathbf{v}$ (3) $\left(\dfrac{m+\mathbf{v}}{M}\right)m$ (4) $\left(\dfrac{m}{m+M}\right)\mathbf{v}$

Answers to Part A and B–1 questions can be found on page 456.

PART B–2 AND C QUESTIONS

1. Base your answers to parts *a* through *c* on the diagram and information below.

   Two railroad carts, *A* and *B*, are on a frictionless, level track. Cart *A* has a mass of 2.0×10^3 kilograms and a velocity of 3.0 meters per second toward the right. Cart *B* has a velocity of 1.5 meters per second toward the left. The magnitude of the momentum of cart *B* is 6.0×10^3 kilogram-meters per second. When the two carts collide, they lock together.

   a What is the magnitude of the momentum of cart *A* before the collision? [Show all calculations, including equations and subtitutions with units.]

   b On the diagram, construct a scaled vector that represents the momentum of cart *A* before the collision. The momentum vector *must* be drawn to a scale of 1.0 centimeter = 1,000 kilogram-meters per second. *Be sure your final answer appears with correct labels (numbers and units).*

   c In one or more *complete sentences*, describe the momentum of the two carts after the collision and justify your answer based on the initial momenta of both carts.

2. A 1000-kilogram car traveling due east at 15 meters per second is hit from behind and receives a forward impulse of 6000 newton-seconds. Determine the magnitude of the car's change in momentum due to this impulse.

Answers to Part B–2 and C questions can be found on page 490.

WORK AND ENERGY

**Chapter
Seven ·**

Λ
KEY IDEAS
╱___Λ

Work is the product of force and the component of displacement in the direction of the force; work is a scalar quantity. Without motion there can be no work.

Power is the rate at which work is done, and it is also a scalar quantity. If work is done on an object, the work may be used to change the object's kinetic energy (the energy associated with its motion), its potential energy (the energy associated with its position), or its internal energy (the energy associated with its atoms and molecules).

An elastic collision is one in which momentum and kinetic energy are conserved. When gas molecules collide with the walls of a container, these collisions are very nearly elastic.

Heat energy is the energy associated with changes in internal energy.

KEY OBJECTIVES

At the conclusion of this chapter you will be able to:

- Define the following terms: *kinetic energy; gravitational potential energy; elastic potential energy; internal energy; partially inelastic collision; totally inelastic collision; elastic collision.*
- Define the term *work*, and state its SI unit.
- Solve problems involving force, displacement, and work.
- Define the term *power*, and state its SI unit.
- Solve problems involving power and work.
- State the equation for calculating kinetic energy, and solve problems using this equation.
- State the equation for calculating gravitational potential energy, and solve problems using this equation.
- Solve problems that relate changes in kinetic energy to changes in gravitational potential energy.
- State the equation for calculating the elastic potential energy of a spring, and solve problems using this equation.

7.1 WORK

Would you pay a person $10 an hour to do this? The typical answer is "No!" because the person isn't doing any work. What exactly do we mean by the term *work*? In physics, **work** is defined as the product of force and displacement, assuming that both are in the *same direction*. As in the case of the man above, if the displacement of the object in the direction of the force is equal to zero, then the force does no work.

PHYSICS CONCEPTS

$$W = \mathbf{F} \cdot \mathbf{d}$$

Even though force and displacement are vector quantities, work is a scalar quantity. The unit of work is the *newton · meter*, which is called a joule (**J**) in honor of English scientist James Prescott Joule.

PROBLEM

How much work is done on an object if a force of 30 newtons [south] displaces the object 200 meters [south]?

SOLUTION

$$W = \mathbf{F} \cdot \mathbf{d}$$
$$= (30 \text{ N[S]})(200 \text{ m [S]})$$
$$= 6000 \text{ J}$$

Now we will suppose that the force and the displacement are not in the same direction. We define work to be the product of the component of the force in the direction of the displacement and the displacement.

The diagram illustrates this relationship.

$$W = (\mathbf{F} \cos\theta) \cdot \mathbf{d}$$

PROBLEM

As Alex pulls his red wagon down the sidewalk, the handle of the wagon makes an angle of 60° with the pavement. If Alex exerts a force of 100 newtons along the direction of the handle, how much work is done when the displacement of the wagon is 20 meters along the ground?

SOLUTION

$$W = (\mathbf{F}\cos\theta) \cdot \mathbf{d}$$
$$= (100 \text{ N})(\cos 60°)(20 \text{ m})$$
$$= (50 \text{ N})(20 \text{ m})$$
$$= 1000 \text{ J}$$

PROBLEM

A constant force of 50 newtons is applied over a distance of 10 meters.
(a) Prepare a graph of force versus displacement.
(b) Calculate the area underneath the curve.
(c) Calculate the work done.

SOLUTION
(a)

(b) *Area = base · height*
      = (10 m)(50 N)
      = 500 J

(c) $W = \mathbf{F} \cdot \mathbf{d}$
      = (50 N)(10 m)
      = 500 J

It is evident that the area under the force-versus-displacement curve, shown below, is equal to the work done on the object.

In general, the area under *any* force-versus-displacement curve is equal to the work done.

7.2   POWER

Power is a term that is frequently misused. We define **power** as the *rate* at which work is done:

PHYSICS CONCEPTS

$$P = \frac{W}{t} = \frac{Fd}{t} = F\bar{v}$$

Power is also a scalar quantity, and its unit is joules per second (J/s), also known as the watt (W).

PROBLEM
If 3000 joules of work is performed on an object in 1.0 minute, what is the power expended on the object?

SOLUTION

$$P = \frac{W}{t}$$

$$= \frac{3000 \text{ J}}{60. \text{ s}}$$

$$= 50 \text{ W}$$

PROBLEM
A 200-newton force is applied to an object that moves in the direction of the force. If the object travels with a constant velocity of 10 meters per second, calculate the power expended on the object.

SOLUTION

$$P = \frac{W}{t} = \frac{\mathbf{F} \cdot \mathbf{d}}{t} = \mathbf{F} \cdot \bar{\mathbf{v}}$$

$$= (200 \text{ N})(10 \text{ m/s})$$

$$= 2000 \text{ W}$$

7.3   ENERGY

When work is done on an object, the "energy" of the object is changed. *Energy* is a very broad term related to work, and it has a variety of forms. In this chapter we will consider two of these forms, kinetic and potential energy. Together the kinetic energy and the potential (gravitational and elastic) energies are called the *mechanical energy* of the object.

Kinetic Energy

An object is traveling along a frictionless horizontal surface. A constant force is applied to the object in the direction of its displacement. What is the result of the work done on the object?

We know from Newton's second law that $F = ma$, therefore $W = ma \cdot d$. If the force on the object is also constant, its acceleration is constant and we can write

$$v_f^2 = v_i^2 + 2a \cdot d$$

If we solve this relationship for $a \cdot d$, we find that

$$a \cdot d = \frac{v_f^2 - v_i^2}{2}$$

We can substitute this solution for $\mathbf{a} \cdot d$ into our work relationship:

$$W = ma \cdot d$$

$$= m \left(\frac{v_f^2 - v_i^2}{2} \right)$$

$$= \frac{1}{2} mv_f^2 - \frac{1}{2} mv_i^2$$

The work done on the object changes a quantity called kinetic energy that is related to the mass and the square of the speed of the object. We define kinetic energy (KE) to be:

PHYSICS CONCEPTS

$$KE = \frac{1}{2} mv^2$$

Therefore, $W = KE_f - KE_i = \Delta KE$.

PROBLEM

A 10-kilogram object subjected to a 20.-newton force moves across a horizontal, frictionless surface in the direction of the force. Before the force was applied, the speed of the object was 2.0 meters per second. When the force is removed, the object is traveling at 6.0 meters per second. Calculate the following quantities: (a) KE_i, (b) KE_f, (c) ΔKE, (d) W, and (e) d.

SOLUTION

(a) $KE_i = \dfrac{1}{2} mv_i^2$

$= \dfrac{1}{2} (10.\ \text{kg})(2.0\ \text{m/s})^2$

$= 20.\ \text{J}$

(b) $KE_f = \dfrac{1}{2} mv_f^2$

$= \dfrac{1}{2} (10.\ \text{kg})(6.\ \text{m/s})^2$

$= 180\ \text{J}$

(c) $\Delta KE = KE_f - KE_i$

$\qquad = 180 \text{ J} - 20. \text{ J}$

$\qquad = 160 \text{ J}$

(d) $W = \Delta KE$

$\qquad = 160 \text{ J}$

(e) $\quad W = F \cdot d$

$\quad d = \dfrac{W}{F}$

$\qquad = \dfrac{160 \text{ J}}{20. \text{ N}}$

$\qquad = 8.0 \text{ m}$

Gravitational Potential Energy

Darya lifts a textbook vertically off a desk and holds it above her head. It is clear that Darya has done work on the book because she has applied a force through a distance. However, the change in the kinetic energy of the book is zero. What did Darya's work accomplish?

In this case, the work overcame the attraction of the gravitational field, and as a result the position of the book with respect to the Earth changed. We relate this change to a quantity we call **gravitational potential energy**.

To calculate the change in the gravitational potential energy (PE) of the object, we measure the work done on the object. The force needed to overcome gravity is simply $\mathbf{F}_g$, which is equal to mg. Therefore, since $W = \mathbf{F}_g \cdot d$, we define the change in the gravitational potential energy (ΔPE) as:

PHYSICS CONCEPTS

$$\Delta PE = mg\,\Delta h$$

Here we use Δh, rather than d, to represent the change in vertical displacement above the Earth. Since we are dealing with a scalar quantity, we will not consider the algebraic signs of the quantities involved. We will simply agree that an object decreases its gravitational potential energy as it moves closer to the Earth (and vice versa).

PROBLEM

A 2.00-kilogram mass is lifted to a height of 10.0 meters above the surface of the Earth. Calculate the change in the gravitational potential energy of the object.

SOLUTION

$$\Delta PE = mg\Delta h$$
$$= (2.00 \text{ kg}) (9.8 \text{ m/s}^2)(10.0 \text{ m})$$
$$= 196 \text{ J}$$

Since the object has moved *away* from the Earth's surface, its gravitational potential energy has *increased* by 196 joules.

For a change in gravitational energy to occur, there must be a change in the *vertical* displacement of an object; if it is moved only *horizontally*, its gravitational potential energy change is zero. If an object is moved up an inclined plane, its potential energy change is measured by calculating only its *vertical* displacement; the horizontal part of its displacement does not change its potential energy.

Interaction of PE and KE (Conservation of Mechanical Energy)

Ketan tosses an object upward, and it returns to the Earth. Let's analyze the motion of this object using an "energy" point of view. For simplicity we will ignore air resistance.

Ketan's hand does work on the object. The work is transformed into kinetic energy. As the object rises, we observe that its speed decreases to zero. As a result, the kinetic energy of the object is decreasing while its potential energy is increasing. This represents a transformation of energy from kinetic to potential.

On the downward trip, the speed of the object increases. As the potential energy of the object decreases, its kinetic energy increases. This represents a transformation from potential to kinetic energy.

In the system, the sum of potential energy and kinetic energy (the total mechanical energy) has been conserved (i.e., is constant); a change in one is accompanied by an opposite change in the other.

$$\Delta PE = -\Delta KE$$
$$PE_i + KE_i = PE_f + KE_f$$

This can also be expressed as the law of conservation of energy, which states that energy can neither be created nor destroyed. In an ideal mechanical system (a closed system upon which no friction or other external forces act) the total mechanical energy is constant.

═══════════════ **PHYSICS CONCEPTS** ═══════════════

$$E_f = E_i$$
$$E_T = PE + KE$$

*Note that these equations do not appear as seen here on the Reference Tables.

PROBLEM

A 0.50-kilogram ball is projected vertically and rises to a height of 2.0 meters above the ground. Calculate: (a) the increase in the ball's potential energy, (b) the decrease in the ball's kinetic energy, (c) the initial kinetic energy, and (d) the intitial speed of the ball.

SOLUTION

(a) $\Delta PE = mg\Delta h$

$\quad = (0.50 \text{ kg})(9.8 \text{ m/s}^2)(2.0 \text{ m})$

$\quad = 9.8 \text{ J}$

(b) $\Delta KE = -\Delta PE$

$\quad = -9.8 \text{ J}$

(c) At the highest point, the speed of the ball is zero; therefore its kinetic energy is zero. As a result, the initial kinetic energy represents the change in the kinetic energy of the object.

$$\Delta KE = KE_f - KE_i$$
$$-9.8 \text{ J} = 0 - KE_i$$
$$KE_i = 9.8 \text{ J}$$

(d) $KE_i = \dfrac{1}{2} mv_i^2$

Solving for v_i gives

$$v_i = \sqrt{\dfrac{2KE_i}{m}}$$

$$\quad = \sqrt{\dfrac{2(9.8 \text{ J})}{0.50 \text{ kg}}}$$

$$\quad = 6.3 \text{ m/s}$$

An amusement-park roller coaster is an example of the interchange of the kinetic and potential energies of the coaster car. As the car falls, its kinetic energy increases; as it rises, its kinetic energy decreases, as shown in the diagram.

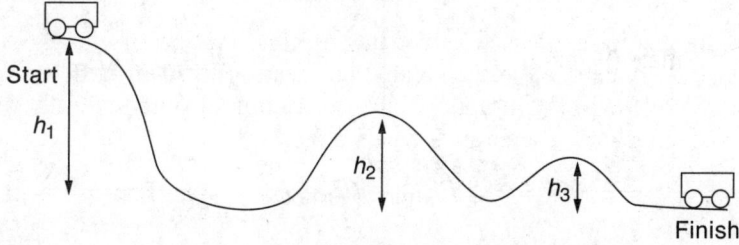

Subsequent hills are made shorter and shorter so that the car will continue to have kinetic energy as it moves along the track.

A simple pendulum, shown in the following diagram, is another device that illustrates the transformation between kinetic and potential energies.

In the absence of friction, the swing of a pendulum back and forth will go on continuously. This type of motion, known as *simple harmonic motion* (SHM), occurs often in nature. For example, the oscillation of a spring and the vibration of a tuning fork are examples of SHM.

The time needed to complete one full swing of the pendulum is known as its period (T). The period of a pendulum (for small angles) is related to the length of the pendulum (ℓ) and to gravitational acceleration (g) according to the relationship

$$\bigstar \quad T = 2\pi \sqrt{\frac{\ell}{g}}$$

Note that the period of a simple pendulum is independent of the mass of the bob.

PROBLEM

A pendulum whose bob weighs 12 newtons is lifted a vertical height of 0.40 meter from its equilibrium position. Calculate: (a) the change in potential energy between maximum height and equilibrium height, (b) the gain in kinetic energy, and (c) the velocity at the equilibrium point.

SOLUTION

PE_{max} KE_{min}

PE_{max} KE_{min}

PE_{min} KE_{max}

$\bigstar$ Note that this equation does not appear on the Reference Tables and is not tested as part of the core curriculum.

(a) We take the potential energy at the lowest point to be zero. Then:
$$\Delta PE = mg\Delta h \quad = F_g\Delta h$$
$$= (12 \text{ N})(-0.40 \text{ m})$$
$$= -4.8 \text{ J}$$

(b) $\quad \Delta KE = -\Delta PE$
$$= -(-4.8 \text{ J})$$
$$= +4.8 \text{ J}$$

(c) First we calculate the mass of the bob:
$$F_g = mg$$
$$m = \frac{F_g}{g}$$
$$= \frac{12 \text{ N}}{9.8 \text{ m/s}^2}$$
$$= 1.2 \text{ kg}$$

Then we calculate the velocity:
$$KE = \frac{1}{2}mv^2$$
$$v = \sqrt{\frac{2KE}{m}}$$
$$= \sqrt{\frac{2(4.8 \text{ J})}{1.2 \text{ kg}}}$$
$$= 2.8 \text{ m/s}^2$$

Elastic Potential Energy and Springs

Consider the arrangement shown in the diagram:

Here a spring is attached to a wall and to a mass resting on a horizontal, frictionless table. If we apply a force on the mass and displace it to the right, we have done work. This work has been converted into the spring's potential energy, a quantity we call **elastic potential energy**.

We can calculate the work done in stretching the spring as follows. We know that springs obey Hooke's law, a graph of which is shown below.

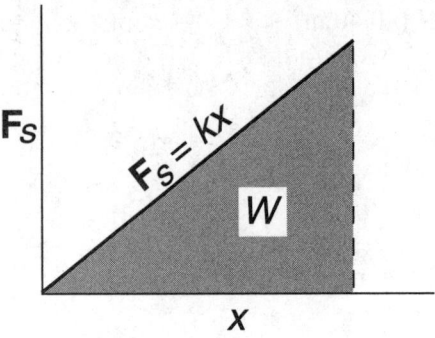

Since the area under this graph equals the work done, and we know that the area of a triangle is equal to ½ (base · height), we use the equations

$$W = \frac{1}{2} \; x \cdot \mathbf{F}_S \text{ and } \mathbf{F}_S = kx$$

Then we have

$$W = \frac{1}{2} \; x \cdot kx = \frac{1}{2} \; kx^2$$

and the potential energy of the spring (PE_s) is given by:

PHYSICS CONCEPTS

$$PE_s = \frac{1}{2} \; kx^2$$

PROBLEM
A spring whose constant is 2.0 newtons per meter is stretched 0.40 meter from its equilibrium position. What is the increase in the elastic potential energy of the spring?

SOLUTION

$$\Delta PE_s = \frac{1}{2} \; kx^2$$

$$= \frac{1}{2} \left(2.0 \; \frac{N}{m} \right) (0.40 \text{ m})^2$$

$$= 0.16 \text{ J}$$

What would happen if we released the spring after stretching it? The force exerted on the mass by the spring (the restoring force) would displace the mass toward the wall (to the left). The mass would then overshoot its equilibrium position and compress the spring. In turn, the spring would exert a force on the mass away from the wall (to the right). Since friction is absent, this back-and-forth motion, namely, SHM would continue indefinitely. As the spring moved back and forth, there would be a continual exchange between kinetic and potential energies, as shown in the diagram.

Elastic and Inelastic Collisions

Imagine a ball bouncing repeatedly on a sidewalk, as illustrated below.

After each successive bounce, the ball's height above the ground diminishes; eventually the ball comes to rest on the ground.

When the ball is on the ground, part of its kinetic energy is lost and there is an incomplete conversion to potential energy, a phenomenon known as a **partially inelastic collision**. If the ball stuck to the ground after its first bounce, the collision would be termed **totally inelastic**.

If, however, the ball rose repeatedly to the same height, the collisions would be termed **elastic**. In an elastic collision both kinetic energy and momentum are conserved, as outlined below:

$$\mathbf{p}_{1_i} + \mathbf{p}_{2_i} = \mathbf{p}_{1_f} + \mathbf{p}_{2_f}$$
$$KE_{1_i} + KE_{2_i} = KE_{1_f} + KE_{2_f}$$

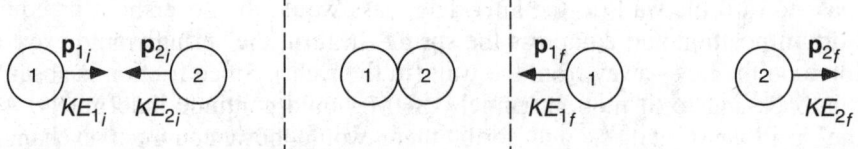

In inelastic collisions, the kinetic energy that is "lost" is converted into internal energy (Q) of the objects by frictional forces. Systems upon which frictional or other external forces act are called nonideal mechanical systems. In a nonideal mechanical system, the total energy is constant. The equation representing the laws of conservation of energy on page 166 can be expanded to include Q, and the total energy (E_T) is given by

========= **PHYSICS CONCEPTS** =========

$$E_T = PE + KE + Q$$

7.4   INTERNAL ENERGY AND WORK

Suppose we use a force to move an object along a horizontal table at constant speed. We know that we have done work on the object, but what has this work accomplished? The kinetic energy has not changed because the speed has been kept constant. The gravitational potential energy has not changed because the table is horizontal. Our work has been used to overcome the friction between the object and the table, and, therefore we say that the *internal energy* of the object-table system has been increased by the work we have done.

Roughly speaking, the **internal energy** (Q) of a system is the total kinetic and potential energies of the atoms and molecules that make up the system. A change in the internal energy of an object is usually accompanied by a change in its *temperature*.

PART A AND B–1 QUESTIONS

1. Which represents a scalar quantity?
   (1) acceleration   (3) energy
   (2) momentum   (4) displacement

2. Work is measured in the same units as
   (1) force   (3) mass
   (2) momentum   (4) energy

3. Which terms represent scalar quantities?
   (1) power and force   (3) time and energy
   (2) work and displacement   (4) distance and velocity

4. Which term is a unit of power?
(1) joule   (2) newton   (3) watt   (4) hertz

5. Which symbolic expression shows how the energy unit (joule) is related to the fundamental units of kilogram, meter, and second?
(1) $kg \cdot m^2/s^2$ (2) $N \cdot m$ (3) $kg \cdot m/s$ (4) $kg \cdot m^2 \cdot s^2$

6. A 2.2-kilogram mass is pulled by a 30.-newton force through a distance of 5.0 meters, as shown in the diagram below.

What amount of work is done?
(1) 11 J   (2) 66 J   (3) 150 J   (4) 330 J

7. Work is being done when a force
(1) acts vertically on a cart that can move only horizontally
(2) is exerted by one team in a tug of war when there is no movement
(3) is exerted on a wagon while pulling it up a hill
(4) of gravitational attraction acts on a person standing on the surface of the earth

8. In the diagram below, 55 joules of work is needed to raise a 10.-newton weight 5.0 meters at a constant speed.

How much work is done to overcome friction as the weight is raised?
(1) 5 J   (2) 5.5 J   (3) 11 J   (4) 50. J

9. A 2.0-kilogram block sliding down a ramp from a height of 3.0 meters above the ground reaches the ground with a kinetic energy of 50. joules. The total work done by friction on the block as it slides down the ramp is approximately
(1) 6 J   (2) 9 J   (3) 18 J   (4) 44 J

10. A person weighing 6.0×10^2 newtons rides an elevator upward at an average speed of 3.0 meters per second for 5.0 seconds. How much does this person's gravitational potential energy increase as a result of this ride?
(1) 3.6×10^2 J   (3) 3.0×10^3 J
(2) 1.8×10^3 J   (4) 9.0×10^3 J

Base your answers to questions 11 and 12 on the information below.

A 2.0-kilogram mass is pushed along a horizontal, frictionless surface by a 3.0-newton force that is parallel to the surface.

11. How much work is done in moving the mass 1.5 meters horizontally?
(1) 4.5 J   (2) 2.0 J   (3) 3.0 J   (4) 30 J

12. How much gravitational potential energy would be gained by the mass if it is moved 2 meters horizontally?
(1) 0 J   (2) 6 J   (3) 40 J   (4) 4 J

13. An object has a mass of 8.0 kilograms. A 2.0-newton force displaces the object a distance of 3.0 meters to the east, and then 4.0 meters to the north. What is the total work done on the object?
(1) 10. J   (2) 14 J   (3) 28 J   (4) 56 J

14. If a 2.0-kilogram mass is raised 0.05 meter vertically, the work done on the mass is approximately
(1) 0.10 J   (2) 0.98 J   (3) 40. J   (4) 100 J

15. A 20-newton block is at rest at the bottom of a frictionless incline as shown in the diagram below.

3.0 m

4.0 m

How much work must be done against gravity to move the block to the top of the incline?
(1) 10 J   (2) 60 J   (3) 80 J   (4) 100 J

16. The work done in raising an object must result in an increase in the object's
(1) gravitational potential energy   (3) internal energy
(2) kinetic energy   (4) heat energy

17. A 1.0-kilogram mass falls a distance of 0.50 meter, causing a 2.0-kilogram mass to slide the same distance along a table top, as represented in the diagram below.

How much work is done by the falling mass?
(1) 1.5 J   (2) 4.9 J   (3) 9.8 J   (4) 14.7 J

18. Which is *not* a unit of power?
(1) joules/second   (3) watts
(2) newton-seconds   (4) calories/second

19. As the time required to lift a mass the same vertical distance is increased, the power developed
(1) decreases   (2) increases   (3) remains the same

20. As the power of a machine is increased, the time required to move an object a fixed distance
(1) decreases   (2) increases   (3) remains the same

21. The rate at which a force does work may be measured in
(1) watts   (2) newtons   (3) joules   (4) kilocalories

22. An electric motor lifts a 10-kilogram mass 100 meters in 10 seconds. The power developed by the motor is
(1) 9.8 W   (2) 98 W   (3) 980 W   (4) 9800 W

23. A horizontal force of 40 newtons pushes a block along a level table at a constant speed of 2 meters per second. How much work is done on the block in 6 seconds?
(1) 80 J (2) 120 J (3) 240 J (4) 480 J

24. The potential energy stored in a compressed spring is to the change in the spring's length as the kinetic energy of a moving body is to the body's
(1) speed (2) mass (3) radius (4) acceleration

Note that question 25 has only three choices.

25. The diagram below shows an ideal simple pendulum.

As the pendulum swings from position A to position B, what happens to its total mechanical energy? [Neglect friction.]
(1) It decreases. (3) It remains the same.
(2) It increases.

26. Car A and car B are of equal mass and travel up a hill. Car A moves up the hill at a constant speed that is twice the constant speed of car B. Compared to the power developed by car B, the power developed by car A is
(1) the same (3) half as great
(2) twice as great (4) 4 times as great

27. A weightlifter lifts a 2000-newton weight a vertical distance of 0.5 meter in 0.1 second. What is the power output?
(1) 1×10^{-4} W (2) 4×10^{-4} W (3) 1×10^{4} W (4) 4×10^{4} W

28. One elevator lifts a mass a given height in 10 seconds, and a second elevator does the same work in 5 seconds. Compared to the power developed by the first elevator, the power developed by the second elevator is
(1) one-half as great (3) the same
(2) twice as great (4) 4 times as great

29. What is the maximum distance that a 60.-watt motor may vertically lift a 90.-newton weight in 7.5 seconds?
(1) 2.3 m (2) 5.0 m (3) 140 m (4) 1100 m

30. If the velocity of an automobile is doubled, its kinetic energy
  (1) decreases to one-half   (3) decreases to one-fourth
  (2) doubles   (4) quadruples

31. If the kinetic energy of a 10-kilogram object is 2000 joules, its velocity is
  (1) 10 m/s   (2) 20 m/s   (3) 100 m/s   (4) 400 m/s

32. Which graph best represents the relationship between the kinetic energy (*KE*) of a moving object as a function of its velocity (**v**)?

    (1)   (2)   (3)   (4)

33. Compared to the kinetic energy of an object that has fallen freely 1 meter, the kinetic energy of the object after falling 2 meters is
  (1) one-half as great   (3) the same
  (2) twice as great   (4) 4 times as great

34. Which cart shown below has the greatest kinetic energy?

    (1)   (3)

    (2)   (4)

35. What is the kinetic energy of a 0.15-kg. baseball if its speed is 10 meters per second?
  (1) 0.75 J   (2) 1.5 J   (3) 7.5 J   (4) 15 J

36. The diagram below represents a cart traveling from left to right along a frictionless surface with an initial speed of **v**.

At which point is the gravitational potential energy of the cart least?
(1) *A*   (2) *B*   (3) *C*   (4) *D*

37. A block raised through a given distance acquires a potential energy of 9.8 joules. A second block with twice the mass is raised through the same distance. The potential energy acquired by the second block is
(1) 4.9 J   (2) 9.8 J   (3) 19.6 J   (4) 39.2 J

38. A mass resting on a shelf 10.0 meters above the floor has a gravitational potential energy of 980. joules with respect to the floor. The mass is moved to a shelf 8.00 meters above the floor. What is the new gravitational potential energy of the mass?
(1) 960. J   (2) 784 J   (3) 490. J   (4) 196 J

39. A 2.0-kilogram rock is raised 5.0 meters above the ground. What is its change in potential energy with respect to the ground?
(1) 7.0 J   (2) 10 J   (3) 98 J   (4) 320 J

40. A 20.-newton block falls freely from rest from a point 3.0 meters above the surface of the earth. With respect to the surface of the Earth, what is the gravitational potential energy of the block-Earth system after the block has fallen 1.5 meters?
(1) 20. J   (2) 30. J   (3) 60. J   (4) 120 J

41. An object gains 10. joules of potential energy as it is lifted vertically 2.0 meters. If a second object with one-half the mass is lifted vertically 2.0 meters, the potential energy gained by the second object will be
(1) 10. J   (2) 20. J   (3) 5.0 J   (4) 2.5 J

42. As a mass is displaced parallel to the surface of the earth, its gravitational potential energy
(1) decreases (2) increases   (3) remains the same

43. A ball is thrown upward from the earth's surface. While the ball is rising, its gravitational potential energy will
(1) decrease (2) increase   (3) remain the same

44. As the velocity of an object falling toward the earth increases, the gravitational potential energy of the object with respect to the Earth (1) decreases (2) increases (3) remains the same

45. A 2.0-newton book falls from a table 1.0 meter high. After falling 0.5 meter, the book's kinetic energy is
(1) 1.0 J   (2) 2.0 J   (3) 10 J   (4) 20 J

46. During an emergency stop, a 1.5×10^3-kilogram car lost a total of 3.0×10^5 joules of kinetic energy. What was the speed of the car at the moment the brakes were applied?
(1) 10. m/s   (2) 14 m/s   (3) 20. m/s   (4) 25 m/s

47. Ten joules of work are done in accelerating a 2.0-kilogram mass from rest across a horizontal, frictionless table. The total kinetic energy gained by the mass is
(1) 3.2 J   (2) 5.0 J   (3) 10. J   (4) 20. J

48. A 10.-kilogram mass falls freely a distance of 6.0 meters near the earth's surface. The total kinetic energy gained by the mass as it falls is approximately
(1) 60. J   (2) 590 J   (3) 720 J   (4) 1200 J

49. As the pendulum swings from position A to position C as shown in the diagram below, what is the relationship of kinetic energy to potential energy? [Neglect friction.]

(1) The kinetic energy decreases more than the potential energy increases.
(2) The kinetic energy increases more than the potential energy decreases.
(3) The kinetic energy decrease is equal to the potential energy increase.
(4) The kinetic energy increase is equal to the potential energy decrease.

50. Mass m is raised to a vertical height h above the ground and then released. After the mass has fallen three-fourths of the way to the ground, its kinetic energy will be equal to
(1) $\dfrac{mgh}{4}$ (2) $\dfrac{3\,mgh}{4}$ (3) mgh (4) $\dfrac{4}{3\,mg}$

51. A car travels at constant speed v up a hill from point A to point B, as shown in the diagram below.

As the car travels from A to B, its gravitational potential energy
(1) increases and its kinetic energy decreases
(2) increases and its kinetic energy remains the same
(3) remains the same and its kinetic energy decreases
(4) remains the same and its kinetic energy remains the same

52. A basketball player who weighs 600 newtons jumps 0.5 meter vertically off the floor. What is her kinetic energy just before hitting the floor?
(1) 30 J   (2) 60 J   (3) 300 J   (4) 600 J

53. In the diagram below, an object starting from rest at A slides down a frictionless hill and then up to B. The object's potential energy at A is 30. joules; its kinetic energy at B is

(1) 3.3 J   (2) 10. J   (3) 13 J   (4) 20. J

54. When a rising baseball encounters air resistance, its total mechanical energy
(1) decreases   (2) increases   (3) remains the same

55. An object is lifted at constant speed a distance h above the surface of the Earth in a time t. The total potential energy gained by the object is equal to the
(1) average force applied to the object
(2) total weight of the object
(3) total work done on the object
(4) total momentum gained by the object

56. As an object falls freely in a vacuum, its total energy
(1) decreases (2) increases   (3) remains the same

57. As a pendulum swings from position A to position B as shown in the diagram, its total mechanical energy (neglecting friction)

(1) decreases (2) increases   (3) remains the same

58. A 2-kilogram mass is thrown vertically upward from the Earth's surface with an initial kinetic energy of 400 newton-meters. The mass will rise to a height of approximately
(1) 10 m   (2) 20 m   (3) 400 m   (4) 800 m

Base your answers to questions 59 through 63 on the diagram below which represents a block with initial velocity $\mathbf{v}_1$ sliding along a frictionless track from point A through point E.

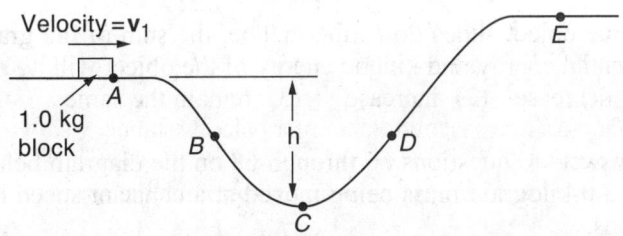

59. At which two points will the kinetic energy of the block be the same?
(1) A and B (2) A and C (3) B and D (4) B and E

60. At which two points will the gravitational potential energies of the block-Earth system be the same?
(1) A and B (2) A and C (3) B and D (4) B and E

61. The kinetic energy of the block will be greatest when it reaches point
(1) A (2) B (3) C (4) D

62. Which expression best represents the kinetic energy of the block at point C?

(1) mgh

(3) $\dfrac{mv_1{}^2}{2} - mgh$

(2) $mgh - \dfrac{mv_1{}^2}{2}$

(4) $\dfrac{mv_1{}^2}{2} + mgh$

63. The velocity of the block will be least at point

(1) A　　　　(2) B　　　　(3) C　　　　(4) E

Base your answers to questions 64 and 65 on the diagram below, which represents a 1.00-kilogram object being held at rest on a frictionless incline.

2.00 meters

1.00 kg

30°

64. The object is released and slides the length of the incline. When it reaches the bottom of the incline, the object's kinetic energy will be closest to

(1) 19.6 J　　(2) 2.00 J　　(3) 9.81 J　　(4) 4.00 J

65. As the object slides down the incline, the sum of the gravitational potential energy and kinetic energy of the object will

(1) decrease　(2) increase　　(3) remain the same

Base your answers to questions 66 through 68 on the diagram below, which represents a 3.0-kilogram mass being moved at a constant speed by a force of 6.0 newtons.

F = 6.0 N　　　　3.0 kg

4 meters

N　　　　　　M

66. What is the change in the kinetic energy of the mass as it moves from point M to point N?

(1) 24 J　　(2) 18 J　　(3) 6 J　　(4) 0 J

67. If energy is supplied at the rate of 10 watts, how much work is done during 2 seconds?
(1) 20 J   (2) 15 J   (3) 10 J   (4) 5 J

68. If the 3.0-kilogram mass were raised 4 meters from the surface, its gravitational potential energy would increase by approximately
(1) 120 J   (2) 40 J   (3) 30 J   (4) 12 J

Base your answers to questions 69 through 71 on the diagram below, which shows an object at A that moves over a frictionless surface from A to E. The object has a mass of m.

69. As the object moves from point A to point D, the sum of its gravitational potential and kinetic energies
(1) decreases, only
(2) decreases and then increases
(3) increases and then decreases
(4) remains the same

70. The object's kinetic energy at point C is less than its kinetic energy at point
(1) A (2) B (3) D (4) E

71. The object's kinetic energy at point D is equal to
(1) mgd (2) $mg(d + h)$ (3) mgh (4) $mg(h - d)$

72. A 0.10-meter spring is stretched from equilibrium to position A and then to position B as shown in the diagram below.

Compared to the spring's potential energy at A, what is its potential energy at B?

(1) the same   (3) one-half as great

(2) twice as great   (4) 4 times as great

73. What is the spring constant of a spring of negligible mass that gained 8 joules of potential energy as a result of being compressed 0.4 meter?

(1) 100 N/m   (2) 50 N/m   (3) 0.3 N/m   (4) 40 N/m

Base your answers to questions 74 and 75 on the figure below. A block of mass M sits on a frictionless surface and is pushed, compressing a spring 0.1 meter. The spring constant is 200 newtons per meter.

74. What force is required to compress the spring 10 meters?

(1) 0.5 N   (2) 20 N   (3) 200 N   (4) 2000 N

75. How much potential energy is stored in the spring?

(1) 1 J   (2) 10 J   (3) 1000 J   (4) 10,000 J

76. A 60.-kilogram student climbs a ladder a vertical distance of 4.0 meters in 8.0 seconds. Approximately how much total work is done against gravity by the student during the climb?

(1) 2.4×10^3 J   (3) 2.4×10^2 J

(2) 2.9×10^2 J   (4) 3.0×10^1 J

77. What is the maximum amount of work that a 6000.-watt motor can do in 10. seconds?

(1) 6.0×10^1 J   (3) 6.0×10^3 J

(2) 6.0×10^2 J   (4) 6.0×10^4 J

78. Which combination of fundamental units can be used to express energy?

(1) kg•m/s   (2) kg•m²/s   (3) kg•m/s²   (4) kg•m²/s²

79. An object is thrown vertically upward. Which pair of graphs best represents the object's kinetic energy and gravitational potential energy as functions of its displacement while it rises?

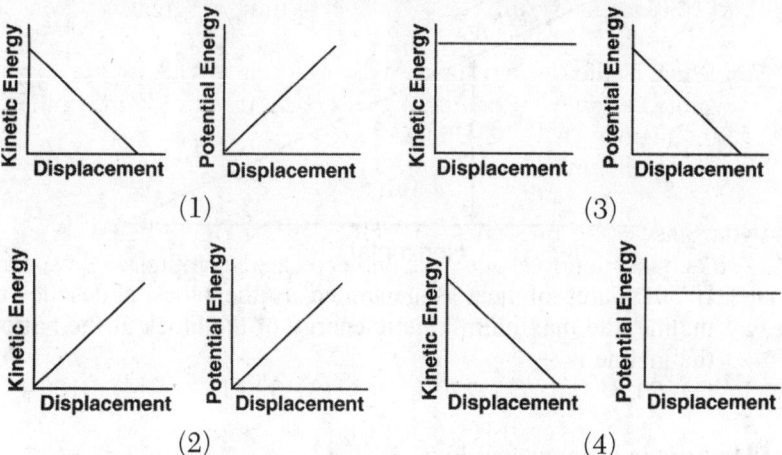

(1)   (3)

(2)   (4)

80. Student *A* lifts a 50.-newton box from the floor to a height of 0.40 meter in 2.0 seconds. Student *B* lifts a 40.-newton box from the floor to a height of 0.50 meter in 1.0 second. Compared to student *A*, student *B* does
    (1) the same work but develops more power
    (2) the same work but develops less power
    (3) more work but develops less power
    (4) less work but develops more power

81. While riding a chairlift, a 55-kilogram skier is raised a vertical distance of 370 meters. What is the total change in the skier's gravitational potential energy?
    (1) 5.4×10^1 J   (3) 2.0×10^4 J
    (2) 5.4×10^2 J   (4) 2.0×10^5 J

82. The work done on a slingshot is 40.0 joules to pull back a 0.10-kilogram stone. If the slingshot projects the stone straight up in the air, what is the maximum height to which the stone will rise? [Neglect friction.]
    (1) 0.41 m   (2) 41 m   (3) 410 m   (4) 4.1 m

83. A block weighing 40. newtons is released from rest on an incline 8.0 meters above the horizontal, as shown in the diagram below.

Horizontal

If 50. joules of heat is generated as the block slides down the incline, the maximum kinetic energy of the block at the bottom of the incline is
(1) 50. J   (2) 270 J   (3) 320 J   (4) 3100 J

84. A joule is equivalent to a
(1) N•m   (2) N•s   (3) N/m   (4) N/s

85. Which graph best represents the relationship between the elastic potential energy stored in a spring and its elongation from equilibrium?

(1)   (2)   (3)   (4)

86. A car with mass m possesses momentum of magnitude p. Which expression correctly represents the kinetic energy, KE, of the car in terms of m and p?

(1) $KE = \dfrac{1}{2}\dfrac{p}{m}$ (3) $KE = \dfrac{1}{2}mp$

(2) $KE = \dfrac{1}{2}mp^2$ (4) $KE = \dfrac{1}{2}\dfrac{p^2}{m}$

87. The table below lists the mass and speed of each of four objects.

Data Table

Objects	Mass (kg)	Speed (m/s)
A	1.0	4.0
B	2.0	2.0
C	0.5	4.0
D	4.0	1.0

Which two objects have the same kinetic energy?
(1) *A* and *D*   (2) *B* and *D*   (3) *A* and *C*   (4) *B* and *C*

88. A horizontal force of 5.0 newtons acts on a 3.0-kilogram mass over a distance of 6.0 meters along a horizontal, frictionless surface. What is the change in kinetic energy of the mass during its movement over the 6.0-meter distance?
(1) 6.0 J   (2) 15 J   (3) 30. J   (4) 90. J

89. Which quantity is a measure of the rate at which work is done?
(1) energy   (2) power   (3) momentum (4) velocity

90. A pendulum is pulled to the side and released from rest. Which graph best represents the relationship between the gravitational potential energy of the pendulum and its displacement from its point of release?

(1)   (2)   (3)   (4)

91. Which graph best represents the relationship between the power required to raise an elevator and the speed at which the elevator rises?

(1)   (2)   (3)   (4)

92. A spring with a spring constant of 80. newtons per meter is displaced 0.30 meter from its equilibrium position. The potential energy stored in the spring is
   (1) 3.6 J   (2) 7.2 J   (3) 12 J   (4) 24 J

93. The work done in accelerating an object along a frictionless horizontal surface is equal to the change in the object's
   (1) momentum   (3) potential energy
   (2) velocity   (4) kinetic energy

94. As a block slides across a table, its speed decreases while its temperature increases. Which two changes occur in the block's energy as it slides?
   (1) a decrease in kinetic energy and an increase in internal energy
   (2) an increase in kinetic energy and a decrease in internal energy
   (3) a decrease in both kinetic energy and internal energy
   (4) an increase in both kinetic energy and internal energy

95. Which graph best represents the relationship between the gravitational potential energy of an object near the surface of Earth and its height above Earth's surface?

(1)

(2)

(3)

(4)

96. A 1.00-kilogram ball is dropped from the top of a building. Just before striking the ground, the ball's speed is 12.0 meters per second. What was the ball's gravitational potential energy, relative to the ground, at the instant it was dropped? [Neglect friction.]
   (1) 6.00 J   (2) 24.0 J   (3) 72.0 J   (4) 144 J

97. A 110-kilogram bodybuilder and his 55-kilogram friend run up identical flights of stairs. The bodybuilder reaches the top in 4.0 seconds while his friend takes 2.0 seconds. Compared to the power developed by the bodybuilder while running up the stairs, the power developed by his friend is
   (1) the same   (3) half as much
   (2) twice as much   (4) four times as much

98. Which two quantities can be expressed using the same units?
(1) energy and force (3) momentum and energy
(2) impulse and force (4) impulse and momentum

99. A box is pushed to the right with a varying horizontal force. The graph below represents the relationship between the applied force and the distance the box moves.

Force vs. Distance

What is the total work done in moving the box 6.0 meters?
(1) 9.0 J (2) 18 J (3) 27 J (4) 36 J

100. A 55.0-kilogram diver falls freely from a diving platform that is 3.00 meters above the surface of the water in a pool. When she is 1.00 meter above the water, what are her gravitational potential energy and kinetic energy with respect to the water's surface?
(1) $PE = 1620$ J and $KE = 0$ J (3) $PE = 810$ J and $KE = 810$ J
(2) $PE = 1080$ J and $KE = 540$ J (4) $PE = 540$ J and $KE = 1080$ J

Answers to Part A and B–1 questions can be found on page 456.

PART B–2 AND C QUESTIONS

1. Base your answers to parts *a* through *d* on the information below.

A 6.0-kilogram concrete block is dropped from the top of a tall building. The block has fallen a distance of 55 meters and has a speed of 30. meters per second when it hits the ground.

a At the instant the block was released, what was its gravitational potential energy with respect to the ground? [Show all calculations, including the equation and substitution with units]

189

b Calculate the kinetic energy of the block at the point of impact. [Show all calculations, including the equation and substitution with units.]

c How much mechanical energy was "lost" by the block as it fell?

d Using one or more complete sentences, explain what happened to the mechanical energy that was "lost" by the block.

Base your answers to questions 2 through 5 on the information and diagram below, which is drawn to a scale of 1.0 centimeter = 3.0 meters.

A 650-kilogram roller coaster car starts from rest at the top of the first hill of its track and glides freely. [Neglect friction.]

2. Using a metric ruler and the scale of 1.0 cm = 3.0 m, determine the height of the first hill.

3. Determine the gravitational potential energy of the car at the top of the first hill. [Show all calculations, including the equation and substitution with units.]

4. Using one or more complete sentences, compare the kinetic energy of the car at the top of the second hill to its kinetic energy at the top of the third hill.

5. Base your answers to parts *a* through *c* on the information at the top of page 191.

A student performs a laboratory activity in which a 15-newton force acts on a 2.0-kilogram mass. The work done over time is summarized in the table below.

DATA TABLE

Time (s)	Work (J)
0	0
1.0	32
2.0	59
3.0	89
4.0	120

a Using the information in the data table, construct a graph on the grid provided, following the directions below.
   (1) Develop an appropriate scale for work, and plot the points for a *work*-versus-*time* graph.
   (2) Draw the best-fit line.

b Calculate the value of the slope of the graph constructed in part a. [Show all calculations, including equations and substitutions with units.]

c Based on your graph, how much time did it take to do 75 joules of work?

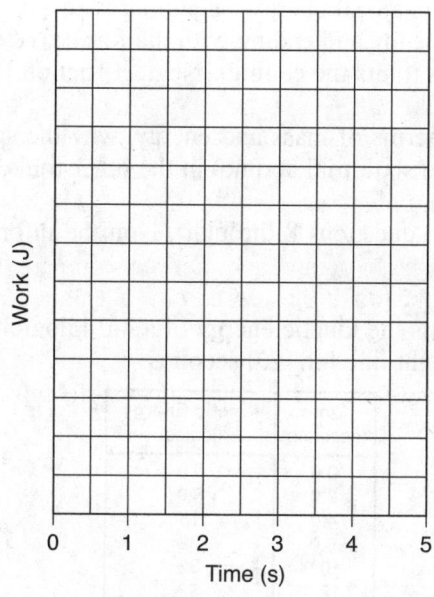

Base your answers to questions 6 and 7 on the information below and on your knowledge of physics.

Using a spring toy like the one shown in the diagram, a physics teacher pushes on the toy, compressing the spring, causing the suction cup to stick to the base of the toy.

When the teacher removes her hand, the toy pops straight up and just brushes against the ceiling. She does this demonstration five times, always with the same result.

When the teacher repeats the demonstration for the sixth time the toy crashes against the ceiling with considerable force. The students notice that in this trial, the spring and toy separated from the base at the moment the spring released.

The teacher puts the toy back together, repeats the demonstration and the toy once again just brushes against the ceiling.

6. Describe the conversions that take place between pairs of the three forms of mechanical energy, beginning with the work done by the teacher on the toy and ending with the form(s) of energy possessed by the toy as it hits the ceiling. [Neglect friction.]

7. Explain, in terms of mass and energy, why the spring toy hits the ceiling in the sixth trial and not in the other trials.

Base your answers to questions 8 through 11 on the information and table below.

The table lists the kinetic energy of a 4.0-kilogram mass as it travels in a straight line for 12.0 seconds.

Time (seconds)	Kinetic Energy (joules)
0.0	0.0
2.0	8.0
4.0	18
6.0	32
10.0	32
12.0	32

Directions (8–9): Using the information in the data table, construct a graph on the grid provided below.

8. Mark an appropriate scale on the axis labeled "Kinetic Energy (J)."

9. Plot the data points for kinetic energy versus time.

Kinetic Energy vs Time

10. Calculate the speed of the mass at 10.0 seconds. [Show all work, including the equation and substitution with units.]

11. Compare the speed of the mass at 6.0 seconds to the speed of the mass at 10.0 seconds.

Base your answers to questions 12 and 13 on the information and diagram below.

A block of mass m starts from rest at height h on a frictionless incline. The block slides down the incline across a frictionless level surface and comes to rest by compressing a spring through distance x, as shown in the diagram below.

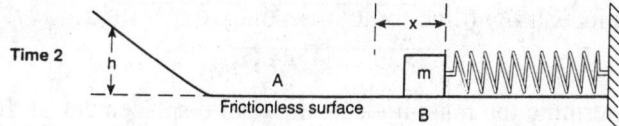

12. Name the forms of mechanical energy possessed by the system when the block is in position A and in position B.

    Position A: _____
    Position B: _____

13. Determine the spring constant, k, in terms of g, h, m, and x. [Show all work, including formulas and an algebraic solution for k.]

A 160.-newton box sits on a 10.-meter-long frictionless plane inclined at an angle of 30.° to the horizontal as shown. Force (F) applied to a rope attached to the box causes the box to move with a constant speed up the incline.

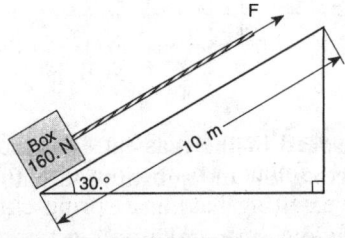

14. Calculate the amount of work done in moving the box from the bottom to the top of the inclined plane. [Show all work, including the equation and substitution with units.]

15. The graph below represents the velocity of an object traveling in a straight line as a function of time.

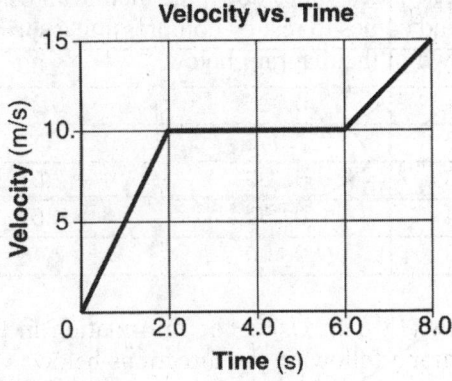

Determine the magnitude of the total displacement of the object at the end of the first 6.0 seconds.

15. _____ m

Base your answers to questions 16 and 17 on the information below.

A 65-kilogram pole vaulter wishes to vault to a height of 5.5 meters.

16. On a separate piece of paper, calculate the *minimum* amount of kinetic energy the vaulter needs to reach this height if air friction is neglected and all the vaulting energy is derived from kinetic energy. [Show all work, including the equations and substitution with units.]

17. On a separate piece of paper, calculate the speed the vaulter must attain to have the necessary kinetic energy. [Show all work, including the equation and substitution with units.]

Base your answers to questions 18 through 21 on the information and data table below.

The spring in a dart launcher has a spring constant of 140 newtons per meter. The launcher has six power settings, 0 through 5, with each successive setting having a spring compression 0.020 meter beyond the previous setting. During testing, the launcher is aligned to the vertical, the spring is compressed, and a dart is fired upward. The maximum vertical displacement of the dart in each test trial is measured. The results of the testing are shown in the table below.

Data Table

Power Setting	Spring Compression (m)	Dart's Maximum Vertical Displacement (m)
0	0.000	0.00
1	0.020	0.29
2	0.040	1.14
3	0.060	2.57
4	0.080	4.57
5	0.100	7.10

Directions (18–19): Using the information in the data table, construct a graph following the directions below.

18. Plot the data points for the dart's maximum vertical displacement versus spring compression.

19. Draw the line or curve of best fit.

20. Using information from your graph, calculate the energy provided by the compressed spring that causes the dart to achieve a maximum vertical displacement of 3.50 meters. [Show all work, including the equation and substitution with units.]

21. Determine the magnitude of the force, in newtons, needed to compress the spring 0.040 meter.

22. A spring in a toy car is compressed a distance, x. When released, the spring returns to its original length, transferring its energy to the car. Consequently, the car having mass m moves with speed v.

    Derive the spring constant, k, of the car's spring in terms of m, x, and v. [Assume an ideal mechanical system with no loss of energy.] [Show all work, including the equations used to derive the spring constant.]

Base your answers to questions 23 and 24 on the information below.

    A 75-kilogram athlete jogs 1.8 kilometers along a straight road in 1.2×10^3 seconds.

23. Determine the average speed of the athlete in meters per second.

24. Calculate the average kinetic energy of the athlete. [Show all work, including the equation and substitution with units.]

25. A car, initially traveling at 30. meters per second, slows uniformly as it skids to a stop after the brakes are applied. On the axes below, sketch a graph showing the relationship between the kinetic energy of the car as it is being brought to a stop and the work done by friction in stopping the car.

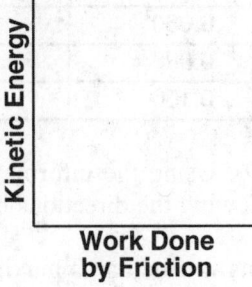

26. A book sliding across a horizontal tabletop slows until it comes to rest. Describe what change, if any, occurs in the book's kinetic energy and internal energy as it slows.

Base your answers to questions 27 and 28 on the information and diagram below.

A pop-up toy has a mass of 0.020 kilogram and a spring constant of 150 newtons per meter. A force is applied to the toy to compress the spring 0.050 meter.

27. Calculate the potential energy stored in the compressed spring. [Show all work, including the equation and substitution with units.]

28. The toy is activated and all the compressed spring's potential energy is converted to gravitational potential energy. Calculate the maximum vertical height to which the toy is propelled. [Show all work, including the equation and substitution with units.]

Base your answers to questions 29 and 30 on the information and diagram below.

A 10.-kilogram block is pushed across a floor by a horizontal force of 50. newtons. The block moves from point A to point B in 3.0 seconds.

29. Using a scale of 1.0 centimeter = 1.0 meter, determine the magnitude of the displacement of the block as it moves from point A to point B.

30. Calculate the power required to move the block from point *A* to point *B* in 3.0 seconds. [Show all work, including the equation and substitution with units.]

Base your answers to questions 31 through 33 on the information and diagram below.

A 3.0-kilogram object is placed on a frictionless track at point *A* and released from rest. (Assume the gravitational potential energy of the system to be zero at point *C*.)

31. Calculate the gravitational potential energy of the object at point *A*. [Show all work, including the equation and substitution with units.]

32. Calculate the kinetic energy of the object at point *B*. [Show all work, including the equation and substitution with units.]

33. Which letter represents the farthest point on the track that the object will reach?

Base your answers to questions 34 through 37 on the graph below, which represents the relationship between vertical height and gravitational potential energy for an object near Earth's surface.

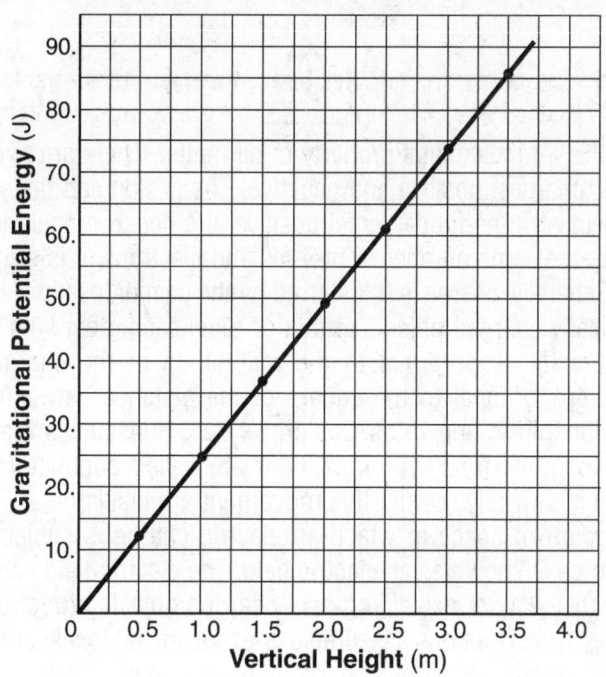

Gravitational Potential Energy vs. Vertical Height

34. Based on the graph, what is the gravitational potential energy of the object when it is 2.25 meters above the surface of Earth?

35. Using the graph, calculate the mass of the object. [Show all work, including the equation and substitution with units.]

36. What physical quantity does the slope of the graph represent?

37. Using a straightedge, draw a line on the graph to represent the relationship between gravitational potential energy and vertical height for an object having a greater mass.

Answers to Part B–2 and C questions can be found on pages 491–501.

**Chapter
Eight**

STATIC ELECTRICITY

KEY IDEAS

Electricity is a fundamental property of all matter. There are two types of electric charges: positive and negative. The proton and the electron are, respectively, the fundamental positive and negative charges. If an object has the same number of protons and electrons, it is electrically neutral. Generally, charge is transferred by the gain or loss of electrons.

Electrically charged objects attract or repel each other with a force that is directly proportional to the magnitude of the charges and inversely proportional to the square of the distance between them. The relationship is known as Coulomb's law. Substances that allow charges to move freely through them are called conductors; substances that severely restrict this movement are insulators.

The region of space in which an electric charge is subject to an electric force is known as an electric field. The electric-field concept is an alternative way of explaining how charged objects attract or repel each other. Constructions called field lines are an aid to visualizing the electric fields around various charge configurations.

The work done in moving a unit charge between two points in an electric field is known as the potential difference between these points. Potential difference describes the electric field in terms of energy and work.

The charge that a conductor acquires is proportional to the potential difference across the conductor. The ratio of charge to potential difference is called capacitance. Devices that make use of this property are known as capacitors; they are used to store electric charge for a variety of applications.

KEY OBJECTIVES

At the conclusion of this chapter you will be able to:
- Define the term *electric charge*, and state the SI unit for charge.
- Relate neutral and charged objects to protons and electrons.
- Explain how neutral objects may become charged by contact.
- Solve problems involving elementary charges.
- Define the terms *conductor, insulator,* and *grounding.*

- Describe the difference between charging by induction and charging by conduction.
- Explain how an electroscope operates.
- State the equation for Coulomb's law, and solve problems using the equation.
- Define the term *electric field*, and describe how an electric field is represented by field lines.
- Draw simple field configurations.
- State the equation for measuring electric field intensity, and solve problems using the equation.
- Define the terms *potential difference* and *electric potential*, and state the SI units for measuring these quantities.
- Define the term *electron-volt* and relate it to the joule.
- Relate the electric field strength between oppositely charged parallel plates to the potential difference across them, and use this relationship to solve problems.
- Describe Millikan's oil drop experiment and its contribution to the understanding of electric charge.

8.1 WHAT IS ELECTRICITY?

We have all experienced the effects of static electricity: A balloon sticks to a wall after being rubbed on a shirt or blouse or a person receives a shock after walking on a carpet and then touching an electrical appliance.

In Chapter 5, we learned that gravitation is a universal attraction between two masses and obeys an *inverse square* law, known as Newton's law of universal gravitation:

$$\mathbf{F}_g = \frac{Gm_1m_2}{r^2}$$

Imagine a force like gravitation that also obeys an inverse square law but is a *billion–billion–billion–billion* times stronger! There are other differences as well. The electric force between two objects depends on their *charges* rather than on their masses. Also, there are two types of **electric charges**, which we call *positive* and *negative*. Like charges (positive–positive and negative–negative) repel each other, and unlike charges (positive–negative) attract.

The phenomenon of electricity was recognized in ancient Greece nearly 5,000 years ago, but it was not understood completely until the twentieth century when the electrical model of the atom was developed.

8.2 ELECTRIC CHARGES

The fundamental positive charge is the *proton,* which is found in the nucleus of the atom along with uncharged neutrons. The fundamental negative charge is the *electron,* which is located outside the nucleus. The properties of these three particles are compared in the table below:

Particle	Relative Charge	Charge (C)	Mass (kg)
Proton	+1	$+1.60 \times 10^{-19}$	1.66×10^{-27}
Electron	−1	-1.60×10^{-19}	9.11×10^{-31}
Neutron	0	0.00	1.67×10^{-27}

As we can see, the proton and the electron have equal, but opposite, charges. Therefore, a *neutral* object has the same number of protons and electrons. The proton is nearly 2,000 times more massive than the electron and is tightly bound in the nucleus (along with the neutrons). As a result, ordinary objects, such as balloons, become electrically charged by gaining or losing electrons. If an object gains electrons, the *excess* of electrons gives the object a *negative* charge. If an object loses electrons, the *deficiency* of electrons gives the object a *positive* charge.

It has long been known that a hard rubber rod becomes negatively charged when rubbed with animal fur. How does this occur? Since electrons are transferred in charging, the rubber rod *gains* electrons and so becomes negatively charged. The fur *loses* an equivalent number of electrons and becomes positively charged. This example illustrates the fact that electric charge is *conserved*; it cannot be created or destroyed. The law of conservation of electric charge is a fundamental law of physics, as are the laws of conservation of energy and momentum.

PROBLEM
A glass rod becomes positively charged when it is rubbed with silk. Explain how this occurs.

SOLUTION
The glass rod loses electrons to the silk, which becomes negatively charged.

As indicated in the preceding table, the unit of charge in the SI system of measurement is the *coulomb* (C). By inspecting the table, we see that the magnitude of the charge on a proton or an electron is 1.60×10^{-19} coulomb. This quantity is known as the *elementary charge* and is denoted by the letter e.

Since electric charges ultimately come from protons and electrons, charge must be some multiple of the elementary charge. We can write this fact as an equation:

===== **PHYSICS CONCEPTS** =====

$$Q = ne$$

*Note that this equation does not appear as seen here on the Reference Tables.

where Q is the charge on the object (in coulombs), n is the number of elementary charges, and e is the elementary charge itself.

PROBLEM
A balloon has acquired a charge of -3.20×10^{-17} coulomb. How many excess electrons does this charge represent?

SOLUTION
We use the equation $Q = ne$ to solve the problem. In this case, n represents the number of *excess* electrons on the balloon (since the charge it has acquired is negative), and e represents the charge on one electron:

$$Q = ne$$
$$-3.20 \times 10^{-17} \text{ C} = n(-1.60 \times 10^{-19} \text{ C})$$
$$n = 200. \text{ excess electrons}$$

PROBLEM
How many elementary charges are present in 1.00 coulomb of charge?

SOLUTION
We solve this problem as we did the one above:

$$Q = ne$$
$$1.00 \text{ C} = n(1.60 \times 10^{-19} \text{ C})$$
$$n = 6.25 \times 10^{18} \text{ elementary charges}$$

Certain substances, such as sodium chloride (common table salt), consist of positive and negative *ions*—atoms that have lost or gained electrons. In a water solution of sodium chloride, charge is transferred by ions, rather than by free electrons.

✪ 8.3   CONDUCTORS AND INSULATORS

Certain materials, for example, metals and solutions of ionic substances, permit charged particles such as electrons or ions to move freely through them; these materials are known as **conductors**. Copper, silver, and a water

✪ indicates material that is not part of the core curriculum.

solution of sodium chloride are examples of electrical conductors. Conductors cannot hold a charge if they are in contact with other materials since the charged particles move easily through them.

Other materials, such as nonmetals, do not readily permit the free movement of charges; these materials are called **insulators**. Rubber, glass, and air are examples of electrical insulators. When an insulator (e.g., a glass rod) is given an electric charge, the charge remains confined to the area where the charge was placed.

The Earth is an electrical conductor and can accept or donate large numbers of electrons. If a charged object is placed in contact with the Earth, it loses its own charge to the Earth. Because of its large size, the Earth remains essentially neutral.

The process of allowing the flow of charge into the Earth is known as *grounding*. Lightning rods are conductors through which dangerously large buildups of atmospheric charge pass harmlessly into the ground.

Human beings, because of the dissolved salts they contain, are also conductors and can act as "grounds" for electric charge. Touching an exposed electrical wire serves to confirm this shocking point.

8.4 CHARGING OBJECTS

The diagram below represents a negatively charged rod that is brought near a neutral object.

Negatively charged rod

Neutral object

Negatively charged rod

Neutral object

As the rod is brought near the object, the excess electrons in the rod repel the electrons in the object. The result is a *redistribution* of charge within the object. It is still neutral, but some of the charges have been separated, as shown in the lower half of the diagram. This phenomenon is known as *induction*. If the rod were removed, the original distribution of charge would return.

If the rod *touched* the object, some of the excess electrons from the rod would be transferred to the neutral object, giving it a permanent negative charge. This process, known as *charging by conduction,* is shown in the following diagram.

Negatively charged rod

Neutral object

Negatively charged rod

Negatively charged object

Another way in which an object can be charged is illustrated in the following diagram:

(1)

Negatively charged rod

Neutral object

(3)

Positively charged object

(2)

Some electrons flow into ground.

Ground

In step (1) the negatively charged rod is brought near the object, causing a redistribution of the object's charges. When the object is grounded in step (2), some of the electrons flow into the ground because they are repelled by the rod. When the ground is removed, the object is then *deficient* in electrons and therefore has acquired a positive charge. This process is known as *charging by induction.*

Note that in all of the examples above, the neutral objects are conductors while the rods are insulators.

8.5 THE ELECTROSCOPE

The *Braun electroscope*, diagramed below, is a device for detecting the presence of electric charge. It consists of a flat plate, a vertical post, and a "leaf," all of which are conductors. In addition, there is a circular shield, which prevents stray charges from affecting the electroscope. The shield is separated from the rest of the device by an insulating collar placed under the plate.

Uncharged Electroscope   Charged Electroscope

When the electroscope is uncharged, the leaf is collapsed; that is, it rests against the vertical post. If, however, the electroscope is brought into contact with a charged object, some of the charge is transferred to the electroscope. Since like charges repel, the leaf moves away from the post as shown in the diagram at the right. The only way to tell whether the charge on a charged electroscope is positive or negative is to test the electroscope by using another object of known charge.

8.6   COULOMB'S LAW

We know that like charges repel and unlike charges attract. Now we will investigate how the force on each charge is calculated. Suppose we have two point charges, q_1 (+) and q_2 (−), separated in space by a distance r. (By "point charges" we mean charged objects separated by a distance that is much larger than the size of either object.) This situation is represented in the diagram.

In Section 8.1 we indicated that electric charges obey a law similar to Newton's law of universal gravitation. This law, known as *Coulomb's law*, takes this form:

‎

PHYSICS CONCEPTS

$$F_e = \frac{kq_1q_2}{r^2}$$

Here, q_1 and q_2 represent the two charges, r represents the distance between their centers, and k, known as the electrostatic constant, has the value 9.0×10^9 N · m²/C². F_e is the magnitude of the force on either charge. In the diagram above, these forces are shown as attractive forces because the charges are unlike.

PROBLEM
(a) Calculate the magnitude of the force between two positive charges, $q_1 = 3.0 \times 10^{-6}$ coulomb and $q_2 = 6.0 \times 10^{-5}$ coulomb, separated by a distance of 9.0 meters.
(b) Draw a diagram representing this situation.

SOLUTION
(a) We solve this problem by substituting the known quantities into Coulomb's law:

$$\mathbf{F}_e = \frac{kq_1q_2}{r^2}$$

$$= \frac{\left(9.0 \times 10^9 \, \frac{\text{N} \cdot \text{m}^2}{\text{C}^2}\right)(3.0 \times 10^{-6} \, \text{C})(6.0 \times 10^{-5} \, \text{C})}{(9.0 \, \text{m})^2}$$

$$= 2.0 \times 10^{-2} \, \text{N}$$

(b)

+3.0 10⁻⁶C   +6.0 10⁻⁵C

9.0 m

8.7 THE ELECTRIC FIELD

An **electric field** exists in a region of space if an electric force is exerted on a charged particle. The idea of an electric field was first developed by the great English scientist Michael Faraday and was perfected by other physicists during the nineteenth century.

Before the concept of the electric field was developed, it was assumed that charges affected each other directly. This idea, known as *action at a distance*, is illustrated in the diagram below:

Action at a Distance

The problem with action at a distance is that there is no way to explain how one charge "knows" that another charge is near it.

The electric field concept assumes that one charge (q_1 in the diagram below) somehow changes the space around it. A second charge (q_2) then interacts with the field with the result that a force is exerted on this charge. One type of interaction is represented in the diagram below.

8.8   ELECTRIC FIELD LINES (LINES OF FORCE)

Field lines, also called *lines of force*, are models that we create in order to visualize an electric field. A field line is the path that a very small *positive* charge (known as a *test charge*) takes while in the field; it is drawn as a line (straight or curved) with an arrow to indicate the proper direction. A *negative* charge would move along the same field line but in the opposite direction.

When a sufficient number of field lines have been drawn, the result is a visual representation of the field. The diagrams that follow represent the electric fields in the vicinity of isolated positive and negative charges:

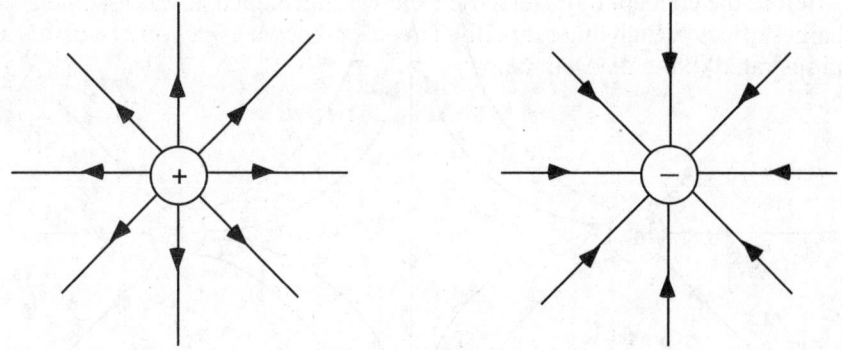

As we can see, the field surrounding each charge has a great deal of *symmetry* associated with it. A positive test charge would move radially *outward* from the isolated positive charge (left diagram) and *inward* toward the isolated negative charge (right diagram). The *number* of lines is an indication of the magnitude of the charge. For example, we might have drawn 16 lines to represent the field of an electric charge 2 times as large as the one shown above.

The *spacing* (or *concentration*) of the field lines indicates the relative *strength* of the electric field. The field is stronger in a region where the lines are more closely spaced and weaker where they are spread farther apart. The diagrams show that each electric field increases in strength as either charge is approached.

Now we will draw the electric field around two equal and opposite charges placed near each other (a configuration known as a *dipole*):

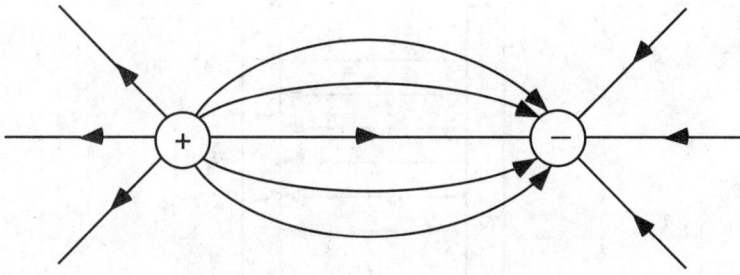

For contrast the electric field in the vicinity of two positive (like) charges is shown in the diagram that follows:

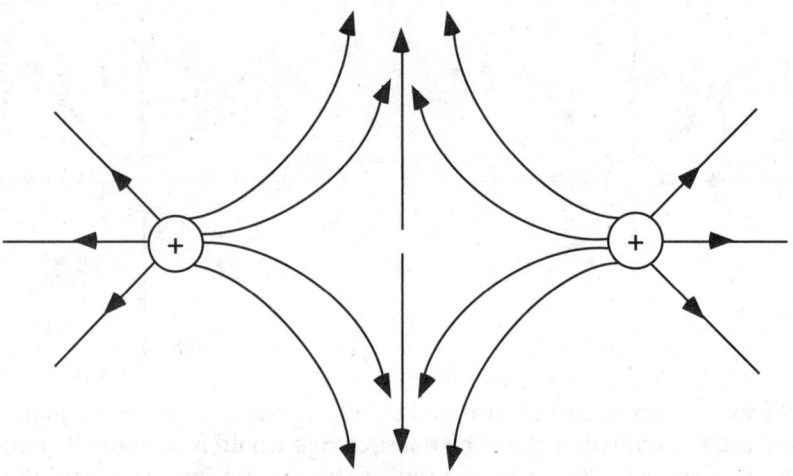

Notice that no field lines pass through the point midway between the charges. If another charge were placed at that point, it would experience no net force; in other words, the field strength is zero at that point. If we had used two *negative* charges, the field lines would have had the same shapes but would have pointed inward toward the charges.

Now let's draw the field lines between two oppositely charged parallel plates. This configuration, known as a *parallel plate capacitor*, is shown in the diagram below. We assume that the plates are very large in size. Actually, this situation can be approximated by placing the plates very close together and considering the field near the center of the plates.

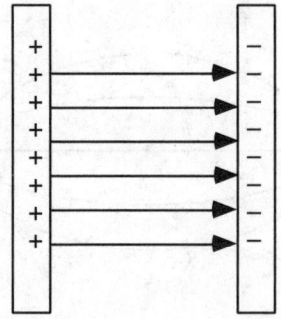

Notice that the field lines are parallel. The result is that the electric field between the plates is uniform; it does not increase or decrease in strength.

Finally, we will consider the electric field in the vicinity of a (positively) charged hollow conductor, as shown in the following diagram:

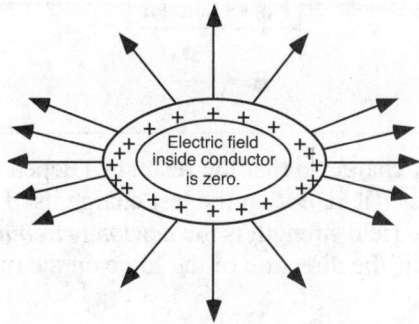

We can see that the field lines are not uniformly distributed: they are more concentrated around the curved parts of the conductor and are less concentrated around the flatter parts. Consequently, pointed objects have a greater buildup of charge. This is the principle upon which the lightning rod operates. The lightning tends to discharge on the pointed rod. Then, since the rod is grounded, any excess charge is passed harmlessly into the earth.

Also note that no electric field exists *inside* the conductor. This statement is true for every hollow conductor, regardless of its shape. As a result, hollow conductors can act as shields against electric charges.

8.9   ELECTRIC FIELD STRENGTH

The electric field is a vector quantity. We can verify this fact by referring to the diagrams in Section 8.8: the arrows point in the direction of the field, and the concentrations of the field lines indicate the magnitude, or strength, of the field.

To *measure* the strength of an electric field, we take a very small positive test charge, place it in the field, and measure the force on it, as the diagram illustrates.

Measuring the Electric Field Due to +q

$$E = \frac{F}{q}$$

We define the strength of the electric field (E) as the ratio of the force (F) to the magnitude of the test charge (q):

=== PHYSICS CONCEPTS ===

$$E = \frac{F}{q}$$

We *divide* by the test charge so that the result (**E**) depends on the charge (or charges) producing the field, *not* on the test charge itself.

The unit of electric field strength is the *newton per coulomb* (N/C), and the direction of the field is the direction of the force on the (positive) test charge.

PROBLEM

A test charge of $+2.0 \times 10^{-6}$ coulomb experiences a force of 2.4×10^{-3} newton [east] when placed in an electric field. Determine the magnitude and the direction of the electric field.

SOLUTION

We solve the problem by using the definition of electric field:

$$E = \frac{F}{q}$$

$$= \frac{2.4 \times 10^{-3} \text{ N [E]}}{2.0 \times 10^{-6} \text{ C}}$$

$$= 1.2 \times 10^3 \text{ N/C [E]}$$

8.10   POTENTIAL DIFFERENCE

We have described the electric field in terms of the force on a charged particle. We can also describe the electric field in terms of work and energy as shown in the diagram.

**Potential Difference Between
Two Points in an Electric Field**

B q A

$$V_{AB} = \frac{W_{AB}}{q}$$

In the diagram, we move a test charge q_0 between two points, A and B, in an electric field. If the charge is repelled by the field, we must do work on the charge to move it between the two points. The work we do against the field (W_{AB}) will increase the potential energy of the test charge. If we were to release the test charge at point B so that the field repels the charge back to point A, we can then say that the field has done work (W_{BA}) on the charge where the work is equivalent to the change in the kinetic energy of the charge. Note that this elementary charge is the charge on a single electron and was also subsequently determined to be the magnitude of the charge on a single proton.

Another way of describing this situation is to say that a **potential difference** exists between points A and B in the electric field. We define this potential difference (V) as follows:

PHYSICS CONCEPTS

$$V = \frac{W}{q}$$

Potential difference is a scalar quantity, as is work. The unit of potential difference is the *joule per coulomb* (J/C) called the *volt* (V) in honor of Alessandro Volta, an Italian scientist.

PROBLEM
When a charge of -4×10^{-3} coulomb is moved between two points in an electric field, 0.8 joule of work is done on the charge. Calculate the potential difference between the two points.

SOLUTION
The *sign* of the charge is not needed to solve the problem. We need know only the magnitude of the charge and the work done on it.

$$V = \frac{W}{q}$$

$$= \frac{0.8 \text{ J}}{4 \times 10^{-3} \text{ C}}$$

$$= 200 \text{ V}$$

PROBLEM
Calculate the work done on an elementary charge that is moved between two points in an electric field with a potential difference of 1 volt.

SOLUTION
We can rearrange our equation as follows:

$$W = q V$$

Then we have

$$W = (1.6 \times 10^{-19} \text{ C})(1.0 \text{ V})$$

$$= 1.6 \times 10^{-19} \text{ J}$$

213

The very small quantity of work in the problem shown above is frequently used as a unit of energy in atomic and nuclear physics. This unit is known as an **electron-volt** (eV). Some multiples of the electron-volt are shown in the table, where the letter M stands for *mega-*; the letter G for *giga-*; and the letter T for *tera*.

Multiple of eV	Abbreviation
10^6	MeV
10^9	GeV
10^{12}	TeV

PROBLEM

A charge, equal to 2×10^7 elementary charges, is moved through a potential difference of 3000 volts. What is the change in the potential energy of the charge?

SOLUTION

The change in the potential energy of the charge is equal to the work done on it. We can calculate this work *directly* in electron-volts by using the unit *elementary charge,* rather than coulomb, for the charge:

$$W = q\,V$$

$$= (2 \times 10^7 \text{ el. ch.})(3000 \text{ V})$$

$$= 6 \times 10^{10} \text{ eV} = 60 \text{ GeV}$$

8.11 ELECTRIC POTENTIAL

We know that it is possible to measure the potential difference between two points, A and B, by dividing the work done on a charge by the magnitude of the charge. Suppose, however, that we wish to know the **electric potential** at just *one* of the points, A or B. How can we accomplish this?

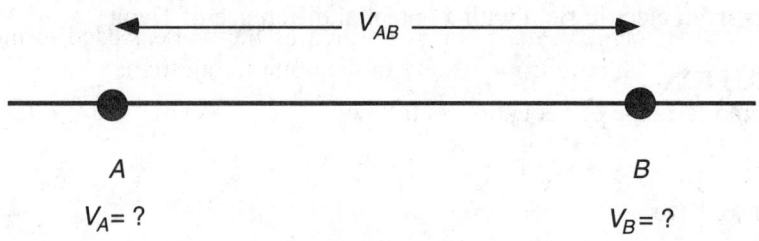

This situation is similar to climbing 525 meters vertically between two points on a hill. It seems only natural to ask how *high* the first and second points are.

The secret is to establish a reference point whose value is zero. With regard to the hill, sea level, with a height of 0 meter, is the reference point, and the height (or altitude) of each point is the distance of that point above sea level. The distance *between* two points is then the *difference* in their altitudes. The diagram below illustrates this concept.

Similarly, to assign electric potentials, we establish a reference point of 0 volt. For an isolated charge, the reference point is taken to be infinitely far from the charge. For other situations, the ground may be taken as a reference point. We then measure the potential difference between the point in question and the reference point, assigning this value as the electric potential of the point. The diagram below, where A represents the point in question, illustrates how this is accomplished.

The electric potential at a point is defined as the work needed to move a charge of +1 coulomb from infinity to the point in question.

✪ 8.12   THE MILLIKAN OIL DROP EXPERIMENT

Robert Millikan, an American physicist, used a modified pair of charged parallel plates to measure the charges on microscopic oil droplets. The diagram shows how the observation was performed.

At equilibrium:
$$F_E = -F_g$$

The charged droplet was injected through the opening into the uniform field between the parallel plates, where it was observed by a microscope. By changing the potential difference between the plates, the electric field strength was varied until the upward electric force on the droplet was balanced by the weight of the droplet. Millikan was then able to calculate the electric charge on each oil droplet that he observed. By measuring thousands of such charges, he determined that the charges were all multiples of 1.60×10^{-19} coulomb and thus concluded that the smallest charge, the *elementary charge*, is equal to 1.60×10^{-19} coulomb.

PART A AND B–1 QUESTIONS

1. A glass rod becomes positively charged when rubbed with silk. The silk becomes charged because it
   (1) loses protons   (3) gains protons
   (2) loses electrons   (4) gains electrons

2. A potential difference of 10.0 volts exists between two points, A and B, within an electric field. What is the magnitude of charge that requires 2.0×10^{-2} joule of work to move it from A to B?
   (1) 5.0×10^2 C   (3) 5.0×10^{-2} C
   (2) 2.0×10^{-1} C   (4) 2.0×10^{-3} C

✪   indicates this concept is not tested as part of the core curriculum.

3. A negatively charged object is brought near the knob of a negatively charged electroscope. The leaves of the electroscope will
   (1) move closer together
   (2) move farther apart
   (3) become positively charged
   (4) become neutral

4. A neutral rubber rod is rubbed with fur and acquires a charge of -2×10^{-6} coulomb. The charge on the fur is
   (1) $+1 \times 10^{-6}$ C
   (2) $+2 \times 10^{-6}$ C
   (3) -1×10^{-6} C
   (4) -2×10^{-6} C

5. Which diagram shows the leaves of an electroscope charged negatively by induction?

| Electroscope | Electroscope | Electroscope | Electroscope |
| (1) | (2) | (3) | (4) |

6. A small, uncharged metal sphere is placed near a larger, negatively charged sphere. Which diagram best represents the charge distribution on the smaller sphere?

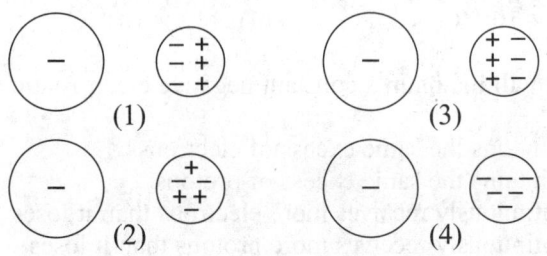

7. When a rubber rod is rubbed with fur, the rod becomes negatively charged because of the transfer of
   (1) electrons to the fur
   (2) protons to the fur
   (3) electrons to the rod
   (4) protons to the rod

8. A charge of 10 elementary charges is equivalent to
   (1) 1.60×10^{-19} C
   (2) 1.60×10^{-18} C
   (3) 6.25×10^{18} C
   (4) 6.25×10^{19} C

9. Sphere A has a charge of +2 units and sphere B, which is identical to sphere A, has a charge of –4 units. If the two spheres are brought together and then separated, the charge on each sphere will be
   (1) –1 unit (2) –2 units   (3) +1 unit   (4) +4 units

10. The unit of charge in the SI system is the
    (1) ohm   (2) ampere   (3) coulomb   (4) volt

11. In the Millikan oil drop experiment, an oil drop is found to have a charge of -4.8×10^{-19} coulomb. How many excess electrons does the oil drop have?
    (1) 1.6×10^{-19} (2) 2   (3) 3   (4) 6.3×10^{18}

12. Two identical metal spheres, charged as shown in the diagram, are brought into contact and then separated.

    A B

    +5 10^{-6} +7 10^{-6}
    coulomb   coulomb

    What will be the charge on sphere A after separation?
    (1) -1×10^{-6} C   (3) $+6 \times 10^{-6}$ C
    (2) $+1 \times 10^{-6}$ C   (4) $+12 \times 10^{-6}$ C

13. A body will maintain a constant negative electrostatic charge if the body
    (1) maintains the same excess of electrons
    (2) maintains the same excess of protons
    (3) continuously receives more electrons than it loses
    (4) continuously receives more protons than it loses

14. A positively charged body must have
    (1) an excess of neutrons
    (2) an excess of electrons
    (3) a deficiency of protons
    (4) a deficiency of electrons

15. A neutral object is attracted by a charged object because the neutral object's
    (1) charges are redistributed
    (2) charge is lost to the surroundings
    (3) net charge is changed by induction
    (4) net charge is changed by conduction

16. An electron is located in the electric field between two parallel metal plates as shown in the diagram below.

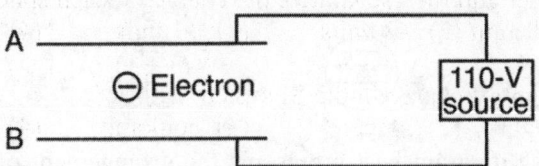

If the electron is attracted to plate *A*, then plate *A* is charged
(1) positively, and the electric field is directed from plate *A* toward plate *B*
(2) positively, and the electric field is directed from plate *B* toward plate *A*
(3) negatively, and the electric field is directed from plate *A* toward plate *B*
(4) negatively, and the electric field is directed from plate *B* toward plate *A*

17. A rod is rubbed with wool. Immediately after the rod and wool have been separated, the net charge of the rod-wool system
(1) decreases (2) increases   (3) remains the same

18. As shown in the diagram below, a charged rod is held near, but does not touch, a neutral electroscope.

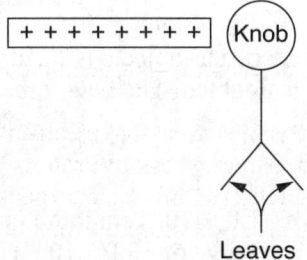

The charge on the knob becomes
(1) positive and the leaves become positive
(2) positive and the leaves become negative
(3) negative and the leaves become positive
(4) negative and the leaves become negative

19. An uncharged metal sphere is placed midway between spheres A and B, represented in the diagram below.

Which diagram best represents the arrangement of the charges in the uncharged sphere?

   (1)   (2)   (3)   (4)

20. The electrostatic force of attraction between two small spheres that are 1.0 meter apart is **F**. If the distance between the spheres is decreased to 0.5 meter, the electrostatic force will then be
(1) $\dfrac{\mathbf{F}}{2}$ (2) 2**F**   (3) $\dfrac{\mathbf{F}}{4}$ (4) 4**F**

21. The distance between two point charges is tripled. Compared to the original force, the new electrostatic force between the charges is
(1) decreased to one-ninth   (3) increased by a factor of 3
(2) decreased to one-third   (4) increased by a factor of 9

22. The electrical force of attraction between two point charges is **F**. The charge on one of the objects is quadrupled, and the charge on the other object is doubled. The new force between the objects is

(1) 8**F**   (2) 2**F**   (3) $\dfrac{1}{2}$ **F**   (4) 4**F**

23. A point charge of -1.0×10^{-9} coulomb is located 5.0×10^{-2} meter from another point charge of $+3.0 \times 10^{-9}$ coulomb. What is the magnitude of the electric force between the two point charges?
(1) 6.0×10^{-17} N   (3) 5.4×10^{-7} N
(2) 1.2×10^{-15} N   (4) 1.1×10^{-5} N

24. Two point charges 1 meter apart repel each other with a force of 9 newtons. What is the force of repulsion when these two charges are 3 meters apart?
(1) 1 N   (2) 27 N   (3) 3 N   (4) 81 N

25. If the magnitude of the charge on each of two positively charged objects is halved, the electrostatic force between the objects will
(1) decrease to one-half
(2) decrease to one-quarter
(3) decrease to one-sixteenth
(4) remain the same

26. Which procedure will double the force between two point charges?
(1) doubling the distance between the charges
(2) doubling the magnitude of one charge
(3) halving the distance between the charges
(4) halving the magnitude of one charge

27. The force between two fixed, charged spheres is **F**. If the charge on each sphere is halved, the force between them will be
(1) **F**   (2) $\dfrac{F}{2}$ (3) $\dfrac{F}{4}$ (4) **4F**

28. The diagram below represents two charges with a separation of d. Which step would produce the greatest increase in the force between the two charges?

(1) doubling charge q_1, only
(2) doubling d, only
(3) doubling charge q_1, only, and d
(4) doubling both charges and d

29. What is the magnitude of the electric field intensity at a point in space where a charge of 100 coulombs experiences a force with a magnitude of 10 newtons?
(1) 1 N/C   (2) 10 N/C   (3) 0.1 N/C   (4) 100 N/C

30. Which is a vector quantity?
(1) electric energy
(2) electric charge
(3) electric power
(4) electric field intensity

31. The electric field intensity at a given distance from a point charge is **E**. If the charge is doubled and the distance remains fixed, the electric field intensity will be
(1) $\dfrac{E}{2}$ (2) 2E   (3) $\dfrac{E}{4}$ (4) 4E

32. The electric field around a point charge is
(1) radial   (2) elliptical   (3) parabolic   (4) circular

33. Gravitational force is to mass as electrical force is to
(1) weight   (2) charge   (3) gravity   (4) electricity

34. As the electric field intensity at a point in space decreases, the electrostatic force on a unit charge at this point
(1) decreases (2) increases (3) remains the same

35. The diagram below shows some of the lines of electric force around a positive point charge.

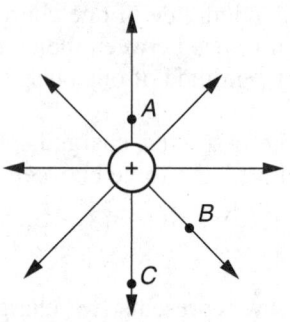

The strength of the electric field is
(1) greatest at point A (3) greatest at point C
(2) greatest at point B (4) equal at points A, B, and C

36. The diagram below represents a uniformly charged rod.

Which graph best represents the relationship between the magnitude of the electric field intensity (E) and the distance from the rod as measured along line AB?

37. The diagram below represents an electron within an electric field between two parallel plates that are charged with a potential difference of 40.0 volts.

If the magnitude of the electric force on the electron is 2.00×10^{-15} newton, the magnitude of the electric field strength between the charged plates is
(1) 3.20×10^{-34} N/C   (3) 1.25×10^4 N/C
(2) 2.00×10^{-14} N/C   (4) 2.00×10^{16} N/C

38. What force will a proton experience in a uniform electric field whose strength is 2.00×10^5 newtons per coulomb?
(1) 8.35×10^{24} N   (3) 3.20×10^{-14} N
(2) 1.67×10^5 N   (4) 1.67×10^{-14} N

39. A proton experiences an electrical force **F** when placed at a point in an electric field. If the proton is replaced by an electron, what force will the electron experience?

(1) $-\mathbf{F}$ (2) $-\mathbf{F}\dfrac{m_p}{m_e}$ (3) $\mathbf{F}\dfrac{m_p}{m_e}$ (4) $\mathbf{F}$

Base your answers to questions 40 through 41 on the diagram below, which represents two large parallel metal plates with a small charged sphere between them.

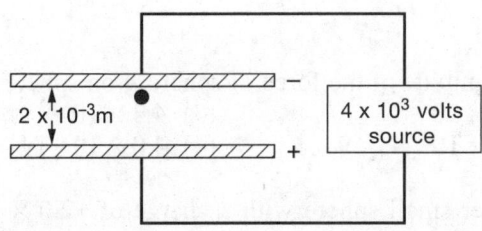

40. The energy gained by the charged sphere as it moves from the negative plate to the positive plate can be measured in
(1) electron-volts   (3) coulombs/volt
(2) volt·meters   (4) volts/meter

41. As the sphere moves from the negative plate to the positive plate, the force on the sphere
(1) decreases   (2) increases   (3) remains the same

42. Gravitational forces differ from electrostatic forces in that gravitational forces are
(1) attractive, only   (3) neither attractive nor repulsive
(2) repulsive, only   (4) both attractive and repulsive

Base your answers to questions 43 through 47 on the diagram below, which represents two small charged spheres, A and B, 3 meters apart. Each sphere has a charge of $+2.0 \times 10^{-6}$ coulomb.

43. Which diagram best illustrates the electric field between charges A and B?

44. The magnitude of the force of sphere A on sphere B is
(1) 1.2×10^{-2} N   (3) 4.4×10^{-13} N
(2) 2.0×10^{3} N   (4) 4.0×10^{-3} N

45. If another small sphere with a charge of $+2.0 \times 10^{-6}$ coulomb is placed at point y, the net force on this sphere will be
(1) 0 N   (2) 40. N   (3) 80. N   (4) 240 N

46. If a positive charge is placed at point x, the direction of the net force on the charge will be
(1) into the page   (3) toward the left
(2) out of the page   (4) toward the right

47. If sphere A is moved toward sphere B, the electric field intensity at point x will
(1) decrease   (2) increase   (3) remain the same

Base your answers to questions 48 and 49 on the information and diagram below.

Two small metallic spheres, A and B, are separated by a distance of 4.0×10^{-1} meter, as shown. The charge on each sphere is $+1.0 \times 10^{-6}$ coulomb. Point P is located near the spheres.

48. What is the magnitude of the electrostatic force between the two charged spheres?
(1) 2.2×10^{-2} N   (3) 2.2×10^{4} N
(2) 5.6×10^{-2} N   (4) 5.6×10^{4} N

49. Which arrow best represents the direction of the resultant electric field at point P due to the charges on spheres A and B?

50. Two charged spheres are shown in the diagram.

Which polarities will produce the electric field shown?
(1) A and B both negative   (3) A positive and B negative
(2) A and B both positive   (4) A negative and B positive

51. If 1.0 joule of work is required to move 1.0 coulomb of charge between two points in an electric field, the potential difference between the two points is
(1) 1.0×10^0 V   (3) 6.3×10^{18} V
(2) 9.0×10^9 V   (4) 1.6×10^{-19} V

52. Which quantity of excess electric charge could be found on an object?
(1) 6.25×10^{-19} C   (3) 6.25 elementary charges
(2) 4.80×10^{-19} C   (4) 1.60 elementary charges

53. The diagram below represents two electrically charged identical-sized metal spheres, A and B.

+2.0 × 10⁻⁷ C   +1.0 × 10⁻⁷ C

If the spheres are brought into contact, which sphere will have a net gain of electrons?
(1) A, only   (3) both A and B
(2) B, only   (4) neither A nor B

Base your answers to questions 54 through 57 on the diagram below which shows a positive point charge placed at A.

54. The electric field intensity at point B is E. At point D the field intensity will be equal to
(1) $\dfrac{1}{9} E$ (2) $\dfrac{1}{3} E$ (3) $3E$ (4) $9E$

55. If a positive charge is placed at point B, the force exerted on this charge by charge A will be directed toward
(1) the top of the page   (3) A
(2) the bottom of the page   (4) C

56. If the charge is moved from point B to point C, the force between the two charges will
(1) decrease   (2) increase   (3) remain the same

57. The electric field surrounding charge *A* is best represented by which diagram?

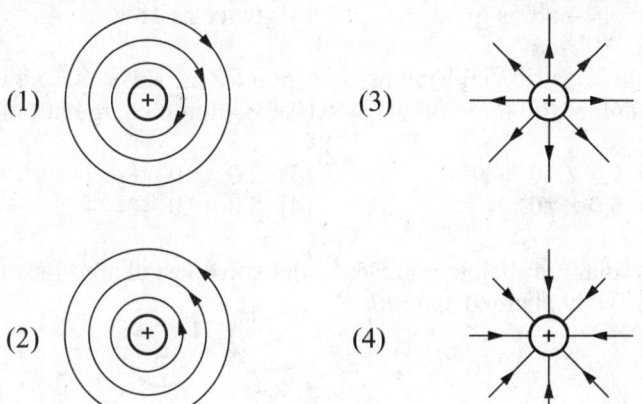

(1)　　　　　　　　　　　　　(3)

(2)　　　　　　　　　　　　　(4)

Base your answers to questions 58 through 62 on the diagram below which shows four point charges, *A*, *B*, *C*, and *D*, and three points, *X*, *Y*, and *Z*, in the electric field of these point charges.

58. Between which pair of charges is the electric force greatest?
(1) *A* and *B*　　(2) *A* and *D*　　(3) *B* and *C*　　(4) *B* and *D*

59. The direction of the electric field of charge *A* at point *X* is
(1) upward　　　　　　　　　(3) toward charge *A*
(2) downward　　　　　　　　(4) toward charge *B*

60. Compared to the intensity of the field of charge *A* at point *X*, the intensity of the field of charge *A* at point *Y* is
(1) one-fourth as great　　　　(3) twice as great
(2) the same　　　　　　　　(4) 4 times as great

61. Compared to the intensity of the field of charge A at point X, the intensity of the field of charge B at point X is
(1) one-fourth as great   (3) the same
(2) one-half as great   (4) twice as great

62. If the electric field intensity at point Z is 1.0×10^5 newtons per coulomb, the force on a 2.0×10^{-6} coulomb charge at point Z will be
(1) 2.0×10^{-1} N   (3) 2.0×10^5 N
(2) 5.0×10^4 N   (4) 5.0×10^{10} N

63. The diagram below represents a positive test charge located near a positively charged sphere.

The greatest increase in the electric potential energy of the test charge relative to the sphere would be caused by moving the charge to point
(1) A (2) B (3) C (4) D

64. It takes 15 joules of work to bring 3.0 coulombs of positive charge from infinity to a point. What is the electric potential at this point in an electric field?
(1) 45. V   (2) 5.0 V   (3) 0.20 V   (4) 0 V

65. A positive test charge is moving between two oppositely charged plates. As the charge moves toward the negative plate, its potential energy
(1) decreases (2) increases (3) remains the same

66. A volt is defined as a
(1) joule/coulomb   (3) coulomb/second
(2) joule/second   (4) joule·second/coulomb

67. The work required to move 2 coulombs of charge through a potential difference of 5 volts is
(1) 10 J   (2) 2 J   (3) 25 J   (4) 50 J

68. If 8.0 joules of work is required to transfer 4.0 coulombs of charge between two points, the potential difference between the two points is
    (1) 6.4 V   (2) 2.0 V   (3) 32 V   (4) 40. V

69. Two similar metal spheres, *A* and *B*, have charges of $+2.0 \times 10^{-6}$ coulomb and $+1.0 \times 10^{-6}$ coulomb, respectively, as shown in the diagram below.

$$+2.0 \times 10^{-6} \text{ C} \qquad +1.0 \times 10^{-6} \text{ C}$$

The magnitude of the electrostatic force on *A* due to *B* is 2.4 newtons. What is the magnitude of the electrostatic force on *B* due to *A*?
    (1) 1.2 N   (2) 2.4 N   (3) 4.8 N   (4) 9.6 N

70. If 1.6×10^{-12} joule of energy is needed to move a charge through a potential difference of 1×10^7 volts then the magnitude of this charge is
    (1) 1.6×10^{-19} C   (3) 1.6×10^5 C
    (2) 1.6×10^{-5} C   (4) 1.6×10^{19} C

71. An electron-volt is a unit of
    (1) potential difference   (3) current
    (2) charge   (4) energy

72. An electron gains 2 electron-volts of energy as it is transferred from point *A* to point *B*. The potential difference between points *A* and *B* is
    (1) 3.2×10^{-19} V   (3) 32 V
    (2) 2 V   (4) 1.25×10^{19} V

73. An energy of 13.6 electron-volts is equivalent to
    (1) 1.60×10^{-19} J   (3) 6.25×10^{-19} J
    (2) 2.18×10^{-18} J   (4) 6.63×10^{-18} J

74. Which quantity is equivalent to 3.2×10^{-17} joule?
    (1) 8.00×10^{-3} eV   (3) 3.20 eV
    (2) 3.20×10^{-17} eV   (4) 200 eV

75. What is the maximum amount of kinetic energy that may be gained by a proton accelerated through a potential difference of 50 volts?
    (1) 1 eV   (2) 10 eV   (3) 50 eV   (4) 100 eV

Base your answers to questions 76 through 78 on the diagram below. The diagram shows a negatively charged pith ball with a mass of 10^{-3} kilogram that is held suspended in the air by the attractive force of a positively charged pith ball. The distance between the centers of the two pith balls is 0.01 meter.

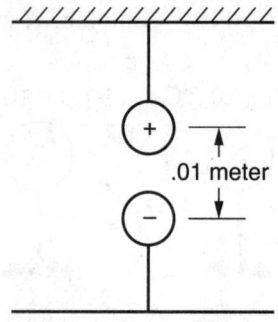

.01 meter

76. The minimum force needed to keep the negatively charged pith ball suspended in the air is approximately
    (1) 10^{-5} N   (2) 10^{-4} N   (3) 10^{-3} N   (4) 10^{-2} N

77. The electrostatic field between the two pith balls is best represented by

(1)   (3)

(2)   (4)

78. If the magnitude of the charge on each pith ball is 1×10^{-7} coulomb, the attractive force between them is
    (1) 9×10^{-3} N   (2) 9×10^{-2} N   (3) 9×10^{-1} N   (4) 9.0 N

Base your answers to questions 79 through 81 on the information below.

Charge $+q$ is located a distance r from charge $+Q$. Each charge is 1 coulomb.

79. The magnitude of the electric field due to charge $+Q$ at distance r is equal to

(1) $\dfrac{kQ}{F}$ (2) $\dfrac{kQq}{r}$ (3) $\dfrac{Q}{r^2}$ (4) $\dfrac{kQ}{r^2}$

80. If 200 joules of work was required to move $+q$ through distance r to $+Q$, the potential difference between the two charges would be

(1) 100 V   (2) 200 V   (3) 800 V   (4) 50 V

81. If distance r is doubled, the force that $+Q$ exerts on $+q$ is

(1) quartered (2) halved   (3) unchanged (4) doubled

82. If the distance separating an electron and a proton is halved, the magnitude of the electrostatic force between these charged particles will be

(1) unchanged   (3) quartered

(2) doubled   (4) quadrupled

83. In the diagram below, P is a point near a negatively charged sphere.

Which vector best represents the direction of the electrical field at point P?

84. Metal sphere A has a charge of -2 units and an identical metal sphere, B, has a charge of -4 units. If the spheres are brought into contact with each other and then separated, the charge on sphere B will be

(1) 0 units   (2) -2 units   (3) -3 units   (4) $+4$ units

Base your answers to questions 85 through 88 on the diagram below, which shows two identical metal spheres. Sphere *A* has a charge of +12 coulombs, and sphere *B* is a neutral sphere.

+ 12 coulombs

85. When spheres *A* and *B* come into contact, sphere *B* will
    (1) gain 6 C of protons   (3) gain 6 C of electrons
    (2) lose 6 C of protons   (4) lose 6 C of electrons

86. When spheres *A* and *B* are in contact, the total charge of the system is
    (1) neutral   (2) +6 C   (3) +12 C   (4) +24 C

87. When spheres *A* and *B* are separated, the charge on sphere *A* is
    (1) +12 C
    (2) one-fourth of the original amount
    (3) one-half of the original amount
    (4) 4 times the original amount

88. After contact, the spheres are moved apart. As the distance between the spheres is increased, the electric potential energy of the system
    (1) decreases   (2) increases   (3) remains the same

Base your answers to questions 89 through 93 on the diagram below, which shows three small metal spheres with different charges.

89. Which vector best represents the net force on sphere *B*?

90. Compared to the force between spheres *A* and *B*, the force between spheres *B* and *C* is
(1) one-quarter as great   (3) one-half as great
(2) twice as great   (4) 4 times as great

91. If sphere *A* is moved further to the left, the magnitude of the net force on sphere *B* will
(1) decrease   (2) increase   (3) remain the same

92. If the charge on sphere *B* were decreased, the magnitude of the net force on sphere *B* would
(1) decrease   (2) increase   (3) remain the same

93. If sphere *B* were removed, the force of sphere *C* on sphere *A* would
(1) decrease   (2) increase   (3) remain the same

Base your answers to questions 94 through 98 on the diagram below, which represents two charged spheres, *X* and *Y*.

94. At which point is the magnitude of the electric field equal to zero?
(1) *A*   (2) *B*   (3) *C*   (4) *D*

95. Which arrow best represents the direction of the electric field at point *A*?

(1)   (2)   (3)   (4)

233

96. If a unit positive charge moves directly from point B to point D, the potential energy of the charge will
    (1) decrease, only   (3) decrease, then increase
    (2) increase, only   (4) increase, then decrease

97. Moving the two spheres toward each other would cause their electric potential energy to
    (1) decrease   (2) increase   (3) remain the same

98. Compared to the force of the electric field of sphere X on sphere Y, the force of the electric field of sphere Y on sphere X is
    (1) less   (2) greater   (3) the same

99. Which graph best represents the relationship between the strength of an electric field and distance from a point charge?

(1)   (2)   (3)   (4)

100. An electron placed between oppositely charged parallel plates A and B moves toward plate A, as represented in the diagram below.

    What is the direction of the electric field between the plates?
    (1) toward plate A (3) into the page
    (2) toward plate B (4) out of the page

Answers to Part A and B–1 questions can be found on page 457.

PART B–2 AND C QUESTIONS

Base your answers to questions 1 and 2 on the information and diagram below.

Two parallel plates separated by a distance of 2.0×10^{-2} meter are charged to a potential difference of 1.0×10^2 volts. Points A, B, and C are located in the region between the plates.

1. *On the diagram*, sketch the electric field lines between the oppositely charged parallel plates through points A, B, and C. [Draw lines with arrowheads in the proper direction.]

2. Calculate the magnitude of the electric field strength between the plates. [Show all calculations, including the equation and substitutions with units.]

Base your answer to question 3 on the information below.

A scientist set up an experiment to collect data about lightning. In one lightning flash, a charge of 25 coulombs was transferred from the base of a cloud to the ground. The scientist measured a potential difference of 1.8×10^6 volts between the cloud and the ground and an average current of 2.0×10^4 amperes.

3. Determine the amount of energy, in joules, involved in the transfer of the electrons from the cloud to the ground. [Show all calculations, including the equation and substitution with units.]

Base your answers to questions 4 through 6 on the information and diagram below.

Two small charged spheres, A and B, are separated by a distance of 0.50 meter. The charge on sphere A is $+2.4 \times 10^{-6}$ coulomb and the charge on sphere B is -2.4×10^{-6} coulomb.

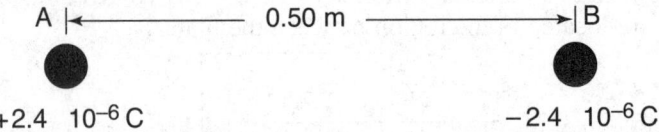

4. On the diagram, sketch *three* electric field lines to represent the electric field in the region between sphere A and sphere B. [Draw an arrowhead on each field line to show the proper direction.]

5. Calculate the magnitude of the electrostatic force that sphere A exerts on sphere B. [Show all calculations, including the equation and substitution with units.]

6. Using the axes, sketch the general shape of the graph that shows the relationship between the magnitude of the electrostatic force between the two charged spheres and the distance separating them. The charge on each sphere remains constant as the distance separating them is varied.

7. Four small metal spheres R, S, T, and U on insulating stands act on each other by means of electrostatic forces.

It was known that sphere S is negatively charged. The following observations were made:

• Sphere S attracts all the other spheres.
• Spheres T and U repel each other.
• Sphere R attracts all the other spheres.

Determine the charge on each sphere and complete the table below noting for each sphere if it is positive (+), negative (−), or neutral (0).

Sphere	Charge
R	
T	
U	

8. Two small identical metal spheres, A and B, on insulated stands, are each given a charge of $+2.0 \times 10^{-6}$ coulomb. The distance between the spheres is 2.0×10^{-1} meter. On a separate piece of paper, calculate the magnitude of the electrostatic force that the charge on sphere A exerts on the charge on sphere B. [Show all work, including the equation and substitution with units.]

Base your answers to questions 9 and 10 on the information below.

The magnitude of the electric field strength between two oppositely charged parallel metal plates is 2.0×10^3 newtons per coulomb. Point P is located midway between the plates.

9. On the diagram below, sketch *at least five* electric field lines to represent the field between the two oppositely charged plates. [Draw an arrowhead on each field line to show the proper direction.]

10. An electron is located at point P between the plates. Calculate the magnitude of the force exerted on the electron by the electric field. [Show all work, including the equation and substitution with units.]

Answers to Part B–2 and C questions can be found on pages 501–506.

Chapter Nine

ELECTRIC CURRENT AND CIRCUITS

KEY IDEAS

Electric current is defined as the time rate at which charge flows through a substance. Currents are established and maintained through a conductor by the application of a potential difference across the conductor.

The resistance of a substance is the ratio of the potential difference across the substance to the current through it. Resistance depends on the nature of the substance, its length, its cross-sectional area, and the temperature. In metallic conductors, the resistance at a given temperature is relatively constant, a fact that is the basis for a relationship known as Ohm's law. Certain substances lose all resistance at low temperatures, a phenomenon known as superconductivity.

Current and potential difference are usually measured in an electric circuit that is a closed path for the flow of charge. A circuit usually contains a source of potential difference and one or more resistances, and may include other devices as well. A circuit that has only one current path is known as a series circuit; a circuit with more than one current path, as a parallel circuit.

Complex circuits may consist of both series and parallel branches or be even more complicated. Such circuits need to be solved by means of Kirchhoff's two laws, which are mathematical statements of the laws of conservation of energy and of electric charge.

KEY OBJECTIVES
At the conclusion of this chapter you will be able to:
- Define the term *electric current*, and state the SI unit for it.
- Solve problems involving current, charge, and time.
- Distinguish between conventional current and electron flow.
- Define the term *resistance*, and state the SI unit for it.
- Solve problems that relate current, potential difference, and resistance.
- Relate the resistance of a material to its length, cross section, resistivity, and temperature.
- Solve problems that relate resistivity, cross section, and length.

- Define the terms *superconductor, series circuit,* and *parallel circuit.*
- State Ohm's law, relate it to metallic conductors, and solve Ohm's law problems.
- Define the term *electric circuit*, and list the components in a simple circuit.
- Draw the circuit symbols for a resistor and a source of potential difference.
- State the equations for determining power and energy output in electric circuits, and solve problems using these equations.
- State the relationships in a series circuit, and solve problems using these relationships.
- State the relationships in a parallel circuit, and solve problems using these relationships.
- Compare and contrast series and parallel circuits.
- State Kirchhoff's rules as they apply to electric circuits.

9.1 INTRODUCTION

In Chapter 8, we learned about static charges. In this chapter we learn about charges in motion and the way they function in electric circuits. Without electric circuits it would be impossible to operate most of the devices that we take for granted, such as televisions, telephones, blenders, and vacuum cleaners.

9.2 ELECTRIC CURRENT

The word *current* means "flow" and **electric current** means "flow of charge." When we use the term *electric current*, we are referring, not to the speed of the charged particles, but to the *quantity* of charge that passes a single point in time.

As an analogy, consider the movement of cars on a highway. We measure the speed of a single car by calculating the distance it travels in a given amount of time. We measure the flow of traffic by counting the number of cars passing a given point in a given amount of time. During rush hour, the flow of traffic will be high, but the speed of any individual car may be quite small. At 4 A.M., however, the reverse is likely to be true.

The SI unit of current is the *ampere* (A) which is equivalent to 1 *coulomb per second*. The symbol used to represent current is I, and we can write

PHYSICS CONCEPTS

$$I = \frac{\Delta q}{t}$$

PROBLEM
What is the electric current in a conductor if 240 coulombs of charge pass through it in 1.0 minute?

SOLUTION
We solve the problem by using the relation $I = \dfrac{\Delta q}{t}$ and remembering that time must be measured in seconds.

$$I = \frac{\Delta q}{t}$$

$$= \frac{240\ \text{C}}{60.\ \text{s}} = 4.0\ \text{A}$$

In metallic conductors, the current is carried by freely moving electrons; in solutions, the current is carried by ions; in gases, the current is carried by both electrons and ions.

9.3   CURRENT AND POTENTIAL DIFFERENCE

How do we establish a current in a conductor? To initiate the flow of charge, we find that a *potential difference* is necessary:

We can see from the diagram that the potential difference is oriented so that positive charges will flow toward the right (the negative terminal) while negative charges (including electrons) will flow toward the left (the positive terminal).

We call the apparent flow of positive charges a *conventional current,* and its direction is always opposite to the direction of electron flow. A conventional current is equivalent to an *electron flow* of the same magnitude in every respect *except direction.* In this book, when we speak of the direction of the current, we are referring to conventional current.

9.4 RESISTANCE

Suppose we removed the source of potential difference in the diagram in Section 9.3. What would happen to the flow of charge in the conductor? We know what would happen: the current would cease. A real-life situation of this type is turning off the switch of a lit lamp.

The reason the current ceases is that, as charges travel through the conductor, they lose energy because of collisions with the atoms of the conductor. This situation is roughly analogous to an object's coming to rest on a surface because of friction. We call the electrical analog of friction *resistance*. An effect of electrical resistance is the almost total conversion of the lost energy into heat. Resistance is the reason that wires become hot as they conduct a current.

We measure the resistance of a material by placing a potential difference across it and then measuring the amount of current that passes through the material. **Resistance** is defined as the ratio of potential difference to current:

PHYSICS CONCEPTS

$$R = \frac{V}{I}$$

The SI unit of resistance is the *volt per ampere* (V/A), which is called the *ohm* (Ω) in honor of German physicist Georg Ohm.

PROBLEM

When a conductor has a potential difference of 110 volts placed across it, the current through it is 0.50 ampere. What is the resistance of the conductor?

SOLUTION

$$R = \frac{V}{I}$$

$$= \frac{110 \text{ V}}{0.50 \text{ A}}$$

$$= 220 \ \Omega$$

The resistance of a material depends on (1) the nature of the material, (2) the geometry of the conductor, and (3) the temperature at which the resistance is measured.

1. Metallic substances are good conductors; that is, they have low resistances. Silver and copper are the best metallic conductors. The quantity that measures how well a substance resists carrying a current is known as the *resistivity* (ρ); its unit is the *ohm·meter* ($\Omega \cdot$ m). The table that follows lists the resistivities of some common materials.

Substance	Resistivity ($\Omega \cdot$ m at 20°C)
Silver	1.59×10^{-8}
Copper	1.68×10^{-8}
Aluminum	2.65×10^{-8}
Iron	9.71×10^{-8}
Silicon (semiconductor)	$0.1 - 60$
Glass (insulator)	$10^9 - 10^{12}$

The greater the resistivity of a material, the greater its resistance will be.

2. The resistance of a regularly shaped conductor is directly proportional to its length and inversely proportional to its cross-sectional area. This fact makes sense because making a conductor longer increases the likelihood that the electron will collide with the atoms of the conductor, thereby increasing the resistance. Making a conductor wider increases the number of *paths* that the electrons can take and, therefore, decreases the resistance.

3. Generally, the resistance of a metallic conductor increases with rising temperature. This fact also makes sense because increasing the temperature of a conductor increases the vibrational kinetic energy of its atoms, making collisions with electrons more likely. In other materials, such as semiconductors, resistance actually decreases with increasing temperature. In practice, a temperature such as 20°C is chosen as a standard temperature for comparing resistances.

We can combine all of these factors into one relationship:

PHYSICS CONCEPTS

$$R = \rho \cdot \frac{L}{A} \text{ (at a specified temperature)}$$

where L represents the length of the conductor and A is its cross-sectional area.

PROBLEM
Calculate the resistance at 20°C of an aluminum wire that is 0.200 meter long and has a cross-sectional area of 1.00×10^{-3} square meter.

SOLUTION
We know from the table given above that aluminum has a resistivity of 2.65×10^{-8} $\Omega \cdot$ m. Therefore:

$$R = \rho \cdot \frac{L}{A} \text{ (at a specified temperature)}$$

$$= 2.65 \times 10^{-8} \ \Omega \cdot m \cdot \frac{0.200 \ m}{1.00 \times 10^{-3} \ m^2}$$

$$= 5.30 \times 10^{-6} \ \Omega$$

Certain substances, called **superconductors**, lose all of their resistance when cooled to very low temperatures. It is also interesting that superconductors do not conduct very well at higher temperatures. If a substance could be made to be superconducting at or near atmospheric temperatures, vast quantities of energy could be conserved because the heat lost by superconducting wires would be minimal.

9.5 ELECTRIC CIRCUITS AND OHM'S LAW

The word *circuit* means "closed path." By an **electric circuit** we mean an arrangement where electric charges can flow in a closed path. The simplest electric circuit consists of a source of potential difference (a battery or a power source), a single resistance, and connecting wires (which are assumed to have negligible resistance).

A device that provides resistance to a circuit is called a *resistor*, and its symbol is as follows:

The symbol for a source of potential difference is

The diagram below represents the completed circuit:

As the electrons begin to flow from the negative terminal, their energy is provided by the source of potential difference. When the electrons reach the positive terminal, all of their energy has been expended (i.e., converted to heat), and the power source must provide them with additional energy for the next trip around the circuit.

Suppose a student performs an experiment using the circuit diagrammed above. The student varies the potential difference across the circuit and measures the current in the circuit with each change. Her results are recorded in the table below.

Potential Difference (V)	Current (A)
0.0	0.0
1.0	1.8
2.0	4.3
3.0	5.7
4.0	8.2
5.0	9.9
6.0	11.8
7.0	14.3
8.0	16.4

To analyze the data, the student finds it useful to plot the data points and then draw a graph, as shown here.

We can see that the graph of the data points, within experimental error, is a straight line that passes through the origin. The *slope* of the graph is the ratio of potential difference to current (which we know as *resistance*). Since the slope is constant, it follows that the resistance of the material is also constant.

PROBLEM
Calculate the resistance of the conductor whose graph is given above.

SOLUTION
To calculate the resistance, we need only calculate the slope of the straight line on the graph.

As a result of this calculation ($\Delta V / \Delta I$), we find that the resistance of the conductor is 2.0 ohms.

Materials that have constant resistances are said to obey *Ohm's law*. The mathematical statement of this relationship is as follows:

PHYSICS CONCEPTS

$$V = IR$$

Ohm's law is a fundamental relationship in electric circuits. It describes how much potential difference is required to move charges through a resistance at a given current.

9.6 POWER AND ENERGY IN ELECTRIC CIRCUITS

Suppose we wish to calculate the *rate* at which energy is supplied to the simple circuit drawn in Section 9.5. We know that potential difference is related to work and energy and that current is related to time. If we multiply the potential difference across the circuit by the current in the circuit and examine the *units*, we find that

Potential difference · current = volts · amperes

$$= \frac{\text{joules}}{\text{coulomb}} \cdot \frac{\text{coulombs}}{\text{second}}$$

$$= \frac{\text{joules}}{\text{second}} = \text{watts} \equiv \text{power}$$

Therefore, the *power* supplied to a circuit by the source is given by this relationship:

PHYSICS CONCEPTS

$$P_{\text{source}} = VI$$

PROBLEM

Calculate the rate at which energy is supplied by a 120-volt source to a circuit if the current in the circuit is 5.5 amperes.

SOLUTION

$$P_{source} = VI$$

$$= (120 \text{ V})(5.5 \text{ A}) = 660 \text{ W}$$

Since our "simple" circuit contains only resistance (i.e., no other device, such as a motor, is present), all of the energy is dissipated as heat. We can calculate the rate at which heat energy is produced by using the power relationship ($P = VI$) and Ohm's law to substitute the term IR for V:

PHYSICS CONCEPTS

$$P_{heat} = VI = IR(I) = I^2R$$

This relationship can be used to calculate the rate at which any resistance in a circuit produces heat energy.

PROBLEM
A 150-ohm resistor carries a current of 2.0 amperes. Calculate the rate at which heat energy is produced by the resistor.

SOLUTION

$$P_{heat} = I^2R$$

$$= (2.0 \text{ A})^2(150 \text{ }\Omega)$$

$$= 600 \text{ W}$$

If, however, another device is present in the circuit, the term VI will not be equal to the term I^2R because not all of the electric energy is converted to heat energy.

There is a third relationship for calculating power: $P = \dfrac{V^2}{R}$. See whether you can derive this relationship using Ohm's law and the power equation.

If we know how much power is developed by a circuit, it is an easy matter to calculate the amount of energy produced: we need only to multiply the power (in watts) by the time the circuit operates (in seconds). The result will be the amount of energy (in joules).

PHYSICS CONCEPTS

$$\text{Energy } (W) = \text{Power } (P) \cdot \text{time } (t)$$

$$= VIt = I^2Rt = \dfrac{V^2t}{R}$$

PROBLEM

How much energy is produced by 50.-volt source that generates a current of 5.0 amperes for 2.0 minutes?

SOLUTION

$$Energy = VIt$$
$$= (50. \text{ V})(5.0 \text{ A})(120 \text{ s})$$
$$= 30,000 \text{ J}$$

9.7 SERIES CIRCUITS

At one time, small holiday lights were arranged so that, if one bulb burned out, the entire string of lights remained unlit. We call the type of electric circuit that produced this effect a series circuit. A **series circuit** has only one current path and if that path is interrupted, the *entire* circuit ceases to operate.

The diagram represents a circuit containing three resistors arranged in series. In addition a number of *meters* have been placed in order to measure various characteristics of the circuit.

Each symbol ─Ⓥ─ represents a *voltmeter*; this very-high-resistance device measures the potential difference across two points in a circuit. The symbol V_t represents the total potential difference across the circuit. The symbol ─Ⓐ─ represents an *ammeter*; this very-low-resistance device measures the current passing through any part of the circuit. The subscript t indicates that the ammeter in the diagram is measuring the total current through the circuit.

Since a series circuit contains only one current path, the current throughout the circuit is constant; therefore, an ammeter placed at any other position in the circuit would record the same value.

The situation is not the same with potential difference: The potential difference across two points depends on the work the source must do in order to move the charge between these two points. In the circuit shown in the diagram, the resistance across any two points will determine how much work needs to be done to transport the charges through the circuit. Our aim is to calculate the readings on all of the meters in the diagram.

We can solve this problem by being aware of the following relationships, which hold true for *any series circuit*:

PHYSICS CONCEPTS

$$I = I_1 = I_2 = I_3 = \ldots$$
$$V = V_1 + V_2 + V_3 + \ldots$$
$$V_n = I_n R_n$$

The first relationship states that the current through any resistance in a series circuit is constant throughout the circuit. The second relationship states that the potential difference across the entire circuit (V_t), supplied by the power source, is equal to the sum of the potential differences (V_1, V_2, . . .) across all the resistances. (This is really a statement of the law of conservation of energy and is known, in honor of German physicist Gustav Kirchhoff, as *Kirchhoff's first rule* or is called, more simply, the *loop rule*.) The third relationship states that Ohm's law holds for each resistance.

If we combine the three statements, we can develop a means of finding the resistance of the circuit as a whole:

$$V = V_1 + V_2 + V_3 + \ldots$$
$$IR = I_1 R_1 + I_2 R_2 + I_3 R_3 + \ldots$$
$$= I_t R_1 + I_t R_2 + I_t R_3 + \ldots$$
$$= I_t (R_1 + R_2 + R_3 + \ldots)$$

PHYSICS CONCEPTS

$$R_{eq} = R_1 + R_2 + R_3 + \ldots$$

R_{eq} is known as the *equivalent resistance* of the circuit.

Now let us examine the series circuit diagram again and calculate all of the meter readings.

The equivalent resistance (R_{eq}) of the circuit is found from the relationship

$$R_{eq} = R_1 + R_2 + R_3 + \ldots$$
$$= 3\,\Omega + 6\,\Omega + 9\,\Omega = 18\,\Omega$$

The current through the circuit (I) is found from the relationship

$$V = IR_{eq}$$

We know that V_t equals 24 volts since the source supplies the entire circuit:

$$24\text{ V} = I(18\,\Omega)$$
$$I = 1.33\text{ A}$$

The potential difference across each resistance can be found by using Ohm's law:

$$V_1 = (1.33\text{ A})(3\,\Omega) = 4\text{ V}$$
$$V_2 = (1.33\text{ A})(6\,\Omega) = 8\text{ V}$$
$$V_3 = (1.33\text{ A})(9\,\Omega) = 12\text{ V}$$

What does voltmeter V_4 read? Since we neglect the resistance of the connecting wires, we assume their resistance to be (nearly) 0 ohm, so the potential difference needed to move the charges across that section of the wire is (nearly) 0 volt.

Another fact about series circuits is important: As the *number* of resistances in a series circuit increases, the equivalent resistance of the circuit increases and the current through the circuit *decreases*. This effect is roughly equivalent to that obtained by increasing the length of a conductor. If the resistances were light bulbs, the bulbs would get dimmer as more were added to the circuit. The next problem illustrates this point.

PROBLEM
Suppose a fourth resistance of 18 ohms is added to the series circuit we have been considering. Calculate (a) the equivalent resistance of the circuit and (b) the current through the circuit.

SOLUTION

(a) The equivalent resistance (R_{eq}) is now:

$$R_{eq} = 3\ \Omega + 6\ \Omega + 9\ \Omega + 18\ \Omega = 36\ \Omega$$

(b) The current through the circuit is found, once again, from Ohm's law:

$$V = IR_{eq}$$
$$24\ \text{V} = I(36\ \Omega)$$
$$I = 0.67\ \text{A}$$

As stated above, the equivalent resistance has increased and the current through the circuit has decreased in comparison to the original circuit.

9.8 PARALLEL CIRCUITS

In contrast to a series circuit, a **parallel circuit** has more than one current path. If a segment of a parallel circuit is interrupted, the result will not necessarily be that the entire circuit ceases to operate. In a home, for example, the burning out of a single bulb does not usually darken the entire house.

The diagram below represents a parallel circuit containing two resistances and a number of suitably placed meters.

In this type of circuit, the current separates into more than one path. The point (or points) where this separation occurs is known as a *junction*. As a consequence of the law of conservation of electric charge, the sum of the currents *entering* a junction must be equal to the sum of the currents *leaving* the junction. This statement is known as *Kirchhoff's second rule* or, more simply, as the *junction rule*.

PROBLEM
In the diagram below, what are the magnitude and the direction of the current in wire X?

SOLUTION

There is *more than one* answer to this problem! The value of the current in X depends on the directions of the other currents. We know that, in each case, the sum of the currents *entering* the junction must be equal to the sum of the currents *leaving* it.

Two solutions are shown in the diagrams. Can you find other solutions?

We can analyze the circuit shown on page 251 by using the following relationships, which are valid for *any parallel circuit*:

PHYSICS CONCEPTS

$$V = V_1 = V_2 = V_3 = \ldots = V_n$$
$$I = I_1 + I_2 + I_3 + \ldots + I_n$$
$$V_n = I_n R_n$$

The first relationship states that the potential difference across a parallel circuit is constant and equal to the battery voltage. This relationship follows from the fact that each resistance comprises an independent path for the flowing charges and that, if one resistance ceases to operate, the others can continue to function. The second relationship states that the current through

252

the entire circuit is equal to the sum of the currents through all the resistances. (This is really an application of Kirchhoff's second rule.) Once again, the third relationship states that Ohm's law holds for each resistance.

If we combine the three statements, we can develop a means of finding the resistance of the parallel circuit as a whole:

$$V_n = I_n R_n \Rightarrow I_n = \frac{V_n}{R_n}$$

$$I = I_1 + I_2 + I_3 + \dots$$

$$\frac{V}{R_{eq}} = \frac{V_1}{R_1} + \frac{V_2}{R_2} + \frac{V_3}{R_3} + \dots$$

$$= \frac{V}{R_1} + \frac{V}{R_2} + \frac{V}{R_3} + \dots$$

$$= V \left(\frac{1}{R_{eq}} = \frac{1}{R_1} + \frac{1}{R_2} + \frac{1}{R_3} + \dots \right)$$

PHYSICS CONCEPTS

$$\frac{1}{R_{eq}} = \frac{1}{R_1} + \frac{1}{R_2} + \frac{1}{R_3} + \dots + \frac{1}{R_n}$$

Here, R_{eq} is the equivalent resistance of the parallel circuit. We will use these relationships to calculate the meter readings in the diagram of the parallel circuit on page 251.

First, we can use Ohm's law $\left(I = \frac{V}{R} \right)$ to calculate currents I_1 and I_2:

$$I_1 = \frac{V}{R_1} = \frac{24 \text{ V}}{3 \text{ }\Omega} = 8 \text{ A}$$

$$I_2 = \frac{V}{R_2} = \frac{24 \text{ V}}{6 \text{ }\Omega} = 4 \text{ A}$$

Next, we calculate the total current I by adding I_1 and I_2:

$$I = 8 \text{ A} + 4 \text{ A} = 12 \text{ A}$$

We find the equivalent resistance R_{eq} from the relationship

$$\frac{1}{R_{eq}} = \frac{1}{R_1} = \frac{1}{R_2}$$

$$= \frac{1}{3\ \Omega} + \frac{1}{6\ \Omega} = \frac{1}{2\ \Omega}$$

$$R_{eq} = 2\ \Omega$$

We could also have used Ohm's law ($V = IR_{eq}$) to calculate the equivalent resistance of this circuit.

We note that the equivalent resistance is less than any single resistance in the circuit. This is characteristic of parallel circuits in general.

If more resistance is added in parallel, the equivalent resistance *decreases* and the total current *increases* because each new parallel resistance creates another independent path in which charges can flow. The result is roughly equivalent to that obtained by increasing the cross-sectional area of a conductor. For this reason, overloading a household circuit by connecting too many electrical appliances is dangerous. As the current in the house wires increases, the amount of heat energy also increases, a situation that may lead to fires in unprotected circuits. Fortunately, fuses and circuit-breakers are designed to prevent such fires from occurring. The next problem illustrates this effect.

PROBLEM
A 6-ohm resistor is added in parallel to the parallel circuit shown at the beginning of this section. Calculate (a) the equivalent resistance and (b) the total current of the altered circuit.

SOLUTION
The diagram of the modified circuit is as follows:

254

(a)

$$\frac{1}{R_{eq}} = \frac{1}{R_1} + \frac{1}{R_2} + \frac{1}{R_3}$$

$$= \frac{1}{3\ \Omega} + \frac{1}{6\ \Omega} + \frac{1}{6\ \Omega} = \frac{1}{1.5\ \Omega}$$

$$R_{eq} = 1.5\ \Omega$$

(b)

$$I = \frac{V}{R_{eq}} = \frac{24\ V}{1.5\ \Omega} = 16\ A$$

As we can see, the equivalent resistance has decreased to 1.5 Ω and the total current has increased to 36 A.

Most circuits represent more complex combinations of series and parallel arrangements than are shown in this chapter. In addition, they may include additional power sources, current loops, and junctions. These complex circuits will not be analyzed in this book. You should be aware, however, that Kirchhoff's rules and some fancy algebra can be used for these analyses. Ask your physics teacher to show you how to analyze one of these complex circuits. Here is a situation where your teacher earns his or her richly deserved pay!

PART A AND B–1 QUESTIONS

1. An ampere can be defined as 1
   (1) C/s   (3) J/C
   (2) Ω/V   (4) N/C

2. Electrical conductivity in liquid solutions depends on the presence of free
   (1) neutrons   (2) protons   (3) molecules   (4) ions

3. As the temperature of a coil of copper wire increases, its electrical resistance
   (1) decreases   (2) increases   (3) remains the same

4. If the cross-sectional area of a metallic conductor is halved and the length of the conductor is doubled, the resistance of the conductor will be
   (1) halved   (2) doubled   (3) unchanged   (4) quadrupled

5. Which segment of copper wire has the highest resistance at room temperature?
   (1) 1.0 m length, 1.0×10^{-6} m^2 cross-sectional area
   (2) 2.0 m length, 1.0×10^{-6} m^2 cross-sectional area
   (3) 1.0 m length, 3.0×10^{-6} m^2 cross-sectional area
   (4) 2.0 m length, 3.0×10^{-6} m^2 cross-sectional area

6. The ratio of the potential difference across a conductor to the current in the conductor is called
   (1) conductivity   (3) charge
   (2) resistance   (4) power

7. Most metals are good electrical conductors because
   (1) their molecules are close together
   (2) they have high melting points
   (3) they have many intermolecular spaces through which the current can flow
   (4) they have a large number of free electrons

8. Which graph represents a circuit element at constant temperature that obeys Ohm's law?

   (1)   (2)   (3)   (4)

9. Which graph best represents the relationship between the resistance (R) of a solid conductor of constant cross section and its length (L)?

   (1)   (2)   (3)   (4)

10. The resistance of a wire at constant temperature depends on the wire's
    (1) length, only
    (2) type of metal, only
    (3) length and cross-sectional area, only
    (4) length, cross-sectional area, and type of metal

11. The current in a circuit is supplied by a generator. If the resistance in the circuit is increased, the force required to keep the generator turning at the same speed is
    (1) decreased (2) increased (3) the same

12. The ratio of the potential difference across a conductor to the current in the conductor is called
    (1) energy   (2) charge   (3) resistance   (4) power

13. The graph below shows how the voltage and current are related in a simple electric circuit.

    For any point on the line, what does the ratio of V to I represent?
    (1) work, in joules   (3) resistance, in ohms
    (2) power, in watts   (4) charge, in coulombs

14. The graph below represents the relationship between potential difference and current for four different resistors. Which resistor has the greatest resistance?

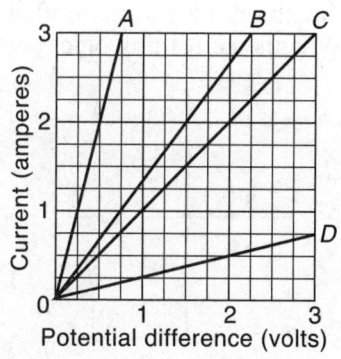

    (1) A   (2) B   (3) C   (4) D

15. Three ammeters are located near junction P in a direct current circuit as shown in the diagram below. If ammeter A_1 reads 3 amperes and ammeter A_2 reads 4 amperes, what does ammeter A_3 read?

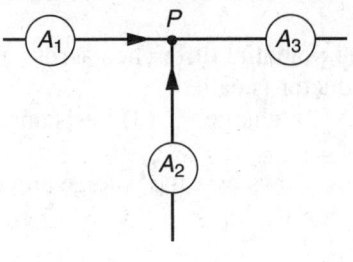

(1) 5 A   (2) 7 A   (3) 3 A   (4) 8 A

16. The diagram below represents currents flowing in branches of an electric circuit. What is the reading on ammeter A?

(1) 13 A   (2) 17 A   (3) 3 A   (4) 33 A

17. The diagram below represents currents flowing in branches of an electric circuit. What is the reading on ammeter A?

(1) 13 A   (2) 17 A   (3) 3 A   (4) 33 A

18. If the voltage across a 4-ohm resistor is 12 volts, the current through the resistor is

(1) 0.33 A   (2) 48 A   (3) 3.0 A   (4) 4.0 A

19. A charge of 5.0 coulombs moves through a circuit in 0.50 second. How much current is flowing through the circuit?
    (1) 2.5 A　　(2) 5.0 A　　(3) 7.0 A　　　(4) 10. A

20. What is the current in a conductor if 6.25×10^{18} electrons pass a given point each second?
    (1) 1 A　(2) 1.6×10^{-19} A　(3) 2.6 A　(4) 6.25×10^{18} A

21. A resistor carries a current of 0.10 ampere when the potential difference across it is 5.0 volts. The resistance of the resistor is
    (1) 0.020 Ω　(2) 0.50 Ω　(3) 5.0 Ω　　(4) 50. Ω

22. If the potential difference across a 12-ohm resistor is 6 volts, the current through the resistor is
    (1) 0.33 A　　(2) 0.50 A　　(3) 3.0 A　　　(4) 4.0 A

23. Three resistors, 4 ohms, 6 ohms, and 8 ohms, are connected in parallel in an electric circuit. The equivalent resistance of the circuit is
    (1) less than 4 Ω
    (2) between 4 Ω and 8 Ω
    (3) between 10. Ω and 18 Ω
    (4) 18 Ω

24. Which combination of current and electromotive force would use energy at the greatest rate?
    (1) 10 A at 110 V　　　(3) 3 A at 220 V
    (2) 8 A at 110 V　　　(4) 5 A at 110 V

25. A 10-volt potential difference maintains a 2-ampere current in a resistor. The total energy expended by this resistor in 5 seconds is
    (1) 10 J　　　(2) 20 J　　　(3) 50 J　　　(4) 100 J

26. What is the current in a normally operating 60-watt, 120-volt lamp?
    (1) 1.0 A　　　(2) 2.0 A　　(3) 0.50 A　　　(4) 4.0 A

27. How much electric energy will be used by a 115-volt, 60-watt light bulb in 1 minute?
    (1) 60 J　　　(2) 115 J　　(3) 3,600 J　　(4) 6,900 J

28. If the current and the resistance of an electric circuit are each doubled, the power will
    (1) remain the same　　(3) be 8 times as large
    (2) be doubled　　　　(4) be quadrupled

29. An ampere-volt is a unit of
   (1) work   (2) resistance   (3) energy   (4) power

30. An electric heater raises the temperature of a measured quantity of water. The water absorbs 6000 joules of energy from the heater in 30.0 seconds. What is the minimum power supplied to the heater?
   (1) 5.00×10^2 W   (3) 1.80×10^5 W
   (2) 2.00×10^2 W   (4) 2.00×10^3 W

31. A circuit contains a rheostat (variable resistor) connected to a source of constant voltage. As the resistance of the rheostat increases, the power dissipated in the circuit
   (1) decreases   (2) increases   (3) remains the same

32. In the circuit represented below, which switches must be closed to produce a current in conductor *AB*?

   (1) 1 and 4   (2) 2 and 3   (3) 1, 2, and 3   (4) 2, 3, and 4

Note that question 33 has only three choices.

33. An electric circuit contains a variable resistor connected to a source of constant voltage. As the resistance of the variable resistor is increased, the power dissipated in the circuit
   (1) decreases   (2) increases   (3) remains the same

34. A circuit consists of a resistor and a battery. Increasing the voltage of the battery while keeping the temperature of the circuit constant would result in an increase in
   (1) current, only   (3) both current and resistance
   (2) resistance, only   (4) neither current nor resistance

35. Charge flowing at the rate of 2.50×10^{16} elementary charges per second is equivalent to a current of

(1) 2.50×10^{13} A   (3) 4.00×10^{-3} A

(2) 6.25×10^{5} A   (4) 2.50×10^{-3} A

36. An electric drill operating at 120. volts draws a current of 3.00 amperes. What is the total amount of electrical energy used by the drill during 1.00 minutes of operation?

(1) 2.16×10^{4} J   (3) 3.60×10^{2} J

(2) 2.40×10^{3} J   (4) 4.00×10^{1} J

37. A 0.686-meter-long wire has a cross-sectional area of 8.23×10^{-6} meter^2 and a resistance of 0.125 ohm at 20° Celsius. This wire should be made of

(1) aluminum   (3) nichrome

(2) copper   (4) tungsten

38. The algebraic sum of all the potential drops and applied voltages around a complete circuit is equal to zero. This is an application of the law of conservation of

(1) mass   (2) energy   (3) charge   (4) momentum

Base your answers to questions 39 through 43 on the information below.

An electric heater rated at 4800 watts is operated on 120 volts.

39. What is the resistance of the heater?

(1) 576,000 Ω   (2) 120 Ω   (3) 3.0 Ω   (4) 40. Ω

40. How much energy is used by this heater in 10.0 seconds?

(1) 1.15 J   (2) 40. J   (3) 4.8×10^{3} J   (4) 4.8×10^{4} J

41. If the heater were replaced by one having a greater resistance, the amount of heat produced each second would

(1) decrease   (2) increase   (3) remain the same

42. If another heater is connected in parallel with the first one and both operate at 120 volts, the current in the first heater will

(1) decrease   (2) increase   (3) remain the same

43. If the original heater were operated at fewer than 120 volts, the amount of heat produced would

(1) decrease   (2) increase   (3) remain the same

Base your answers to questions 44 through 49 on the diagram below, which represents an electric circuit. Charge is transferred from point A to point B at the rate of 5 coulombs per second for 120 seconds. Six joules of work are done in transferring each coulomb of charge.

44. What is the current in the wire?
(1) 1.2 A   (2) 5 A   (3) 6 A   (4) 24 A

45. What is the potential difference between points A and B?
(1) 6 V   (2) 30 V   (3) 100 V   (4) 600 V

46. What is the rate at which work is done in this circuit?
(1) 6 W   (2) 30 W   (3) 144 W   (4) 720 W

47. What is the total work done in this circuit?
(1) 6 J   (2) 30 J   (3) 720 J   (4) 3600 J

48. What is the rate at which electrons are transferred in this circuit?
(1) 8×10^{-19} electrons/s   (3) 1.1×10^{18} electrons/s
(2) 5 electrons/s   (4) 3.1×10^{19} electrons/s

49. What force moves 1 coulomb of charge from point A to point B?
(1) 1 N   (2) 5 N   (3) 30 C   (4) 120 C

50. If V_2 in the diagram reads 24 volts, V_1 will read

(1) 8 V   (2) 24 V   (3) 48 V   (4) 72 V

Base your answers to questions 51 through 54 on the diagram below, which shows three resistors connected to a 15-volt source.

51. The equivalent resistance of the circuit is
    (1) 10 Ω   (2) 20 Ω   (3) 30 Ω   (4) 40 Ω

52. The total power developed in the circuit is
    (1) 2.5 W   (2) 5.0 W   (3) 7.5 W   (4) 10 W

53. Compared to the heat developed in resistor R_1, the heat developed in resistor R_3 is
    (1) one-third as great   (3) 3 times as great
    (2) two times as great   (4) one-fourth as great

54. If resistor R_3 is removed and replaced by a resistor of lower value, the resistance of the circuit will
    (1) decrease   (2) increase   (3) remain the same

Base your answers to questions 55 through 57 on the diagram below.

55. The potential difference across R_1 is
    (1) 0 V   (2) 8.00 V   (3) 12.0 V   (4) 24.0 V

56. What power is supplied by the source?
    (1) 24.0 W   (2) 90.0 W   (3) 3.00 W   · (4) 192 W

57. What is the current in resistor R_2?
    (1) 8.00 A   (2) 2.00 A   (3) 16.0 A   (4) 4.00 A

58. The diagram below represents an electric circuit.

If the voltage between *A* and *B* is 10 volts, the voltage between *B* and *C* is

(1) 5 V   (2) 10 V   (3) 15 V   (4) 20 V

59. Compared to the potential drop across the 10-ohm resistor shown in the diagram, the potential difference across the 5-ohm resistor is

(1) the same   (3) one-half as great
(2) twice as great   (4) 4 times as great

60. Ammeter *A* in the diagram below will read 2 amperes when switch *B* makes contact with which terminal?

(1) 1   (2) 2   (3) 3   (4) 4

61. As more resistors are added in series across a battery, the potential drop across each resistor
(1) decreases   (2) increases   (3) remains the same

264

62. A 5-ohm resistor and a 10-ohm resistor are connected in series. If the current in the 5-ohm resistor is doubled, the current in the 10-ohm resistor will
(1) be halved
(2) remain the same
(3) be doubled
(4) be quadrupled

Base your answers to questions 63 through 66 on the circuit diagram below. The reading of ammeter A_1 is 2.0 amperes. Neglect the resistance of the connecting wires and the battery.

63. The potential difference supplied by battery B is
(1) 1.0 V   (2) 2.0 V   (3) 6.0 V   (4) 12 V

64. How much heat energy will be produced by resistor R_1 in 3.0 seconds?
(1) 72 J   (2) 2.0 J   (3) 12 J   (4) 24 J

65. How much charge will pass through resistor R_1 in 3.0 seconds?
(1) 1.0 C   (2) 2.0 C   (3) 6.0 C   (4) 12 C

66. Compared to the potential difference across resistor R_3, the potential difference across resistor R_2 is
(1) one-half as much
(2) the same
(3) twice as much
(4) 3 times as much

67. If the current in the 10-ohm resistor in the diagram is 1 ampere, then the current in the 40-ohm resistor is

(1) 1 A   (2) 0.25 A   (3) 5 A   (4) 4 A

68. As additional resistors are connected in parallel to a source of constant voltage, the current in the circuit
(1) decreases   (2) increases   (3) remains the same

Base your answers to questions 69 through 72 on the diagram below, which shows three resistors connected in parallel to a 300-volt direct-current source.

69. A voltmeter connected between points A and B will have a reading of
(1) 20 V   (2) 50 V   (3) 100 V   (4) 300 V

70. What is the reading of ammeter A_3?
(1) 5.0 A   (2) 15 A   (3) 30. A   (4) 55 A

71. The greatest amount of current is through ammeter
(1) A_1 (2) A_2 (3) A_3 (4) A_4

72. Heat is produced by the 10-ohm resistor at a rate of
(1) 30. W   (2) 3.0×10^2 W   (3) 9.0×10^3 W   (4) 3.0×10^3 W

73. Which circuit below would have the lowest voltmeter reading?

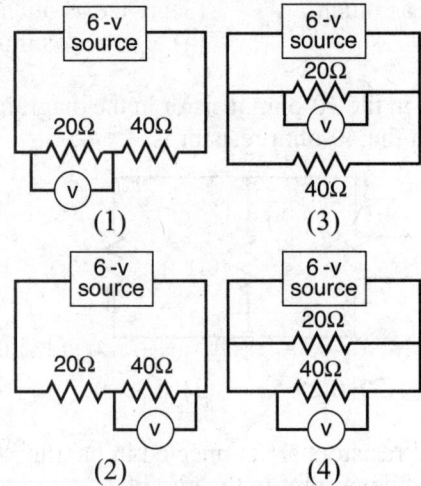

74. Which circuit segment has an equivalent resistance of 6 ohms?

75. A circuit consists of a 10.0-ohm resistor, a 15.0-ohm resistor, and a 20.0-ohm resistor connected in parallel across a 9.00-volt battery. What is the equivalent resistance of this circuit?
(1) $0.200\ \Omega$ (2) $1.95\ \Omega$ (3) $4.62\ \Omega$ (4) $45.0\ \Omega$

76. An electric circuit contains a variable resistor connected to a source of constant potential difference. Which graph best represents the relationship between current and resistance in this circuit?

77. In the circuit diagram below, two 4.0-ohm resistors are connected to a 16-volt battery as shown.

The rate at which electrical energy is expended in this circuit is
(1) 8.0 W   (2) 16 W   (3) 32 W   (4) 64 W

78. The current through a 10.-ohm resistor is 1.2 amperes. What is the potential difference across the resistor?
(1) 8.3 V   (2) 12 V   (3) 14 V   (4) 120 V

79. A copper wire of length L and cross-sectional area A has resistance R. A second copper wire at the same temperature has a length of $2L$ and a cross-sectional area of $\frac{1}{2}A$. What is the resistance of the second copper wire?

(1) R (2) $2R$ (3) $\frac{1}{2}R$ (4) $4R$

80. A 6.0-ohm lamp requires 0.25 ampere of current to operate. In which circuit below would the lamp operate correctly when switch S is closed?

81. What is the total current in a circuit consisting of six operating 100-watt lamps connected in parallel to a 120-volt source?

(1) 5 A   (2) 20 A   (3) 600 A   (4) 12,000 A

82. A 4.50-volt personal stereo uses 1950 joules of electrical energy in one hour. What is the electrical resistance of the personal stereo?

(1) 433 Ω (2) 96.3 Ω (3) 37.4 Ω (4) 0.623 Ω

83. The diagram below represents a simple circuit consisting of a variable resistor, a battery, an ammeter, and a voltmeter.

What is the effect of increasing the resistance of the variable resistor from 1000 Ω to 10,000 Ω? [Assume constant temperature.]
(1) The ammeter reading decreases.
(2) The ammeter reading increases.
(3) The voltmeter reading decreases.
(4) The voltmeter reading increases.

Base your answers to questions 84 through 86 on the diagram below, which represents an electric circuit consisting of four resistors and a 12-volt battery.

84. What is the current measured by ammeter A?
    (1) 0.50 A   (2) 2.0 A   (3) 72 A   (4) 4.0 A

85. What is the equivalent resistance of this circuit?
    (1) 72 Ω   (2) 18 Ω   (3) 3.0 Ω   (4) 0.33 Ω

86. How much power is dissipated in the 36-ohm resistor?
    (1) 110 W   (2) 48 W   (3) 3.0 W   (4) 4.0 W

87. Which quantity and unit are correctly paired?
    (1) resistivity and $\dfrac{\Omega}{m}$

    (2) potential difference and eV

    (3) current and C•s

    (4) electric field strength and $\dfrac{N}{C}$

88. Which wavelength is in the infrared range of the electromagnetic spectrum?
    (1) 100 nm   (2) 100 mm   (3) 100 m   (4) 100 µm

89. What is the resistance at 20.°C of a 2.0-meter length of tungsten wire with a cross-sectional area of 7.9×10^{-7} meter^2?
    (1) 5.7×10^{-1} Ω   (3) 7.1×10^{-2} Ω
    (2) 1.4×10^{-1} Ω   (4) 4.0×10^{-2} Ω

90. A 6.0-ohm resistor that obeys Ohm's law is connected to a source of variable potential difference. When the applied voltage is decreased from 12 V to 6.0 V, the current passing through the resistor
(1) remains the same   (3) is halved
(2) is doubled   (4) is quadrupled

91. In which circuit represented below are meters properly connected to measure the current through resistor R_1 and the potential difference across resistor R_2?

      (1)   (3)

      (2)   (4)

92. A 50-watt lightbulb and a 100-watt lightbulb are each operated at 110 volts. Compared to the resistance of the 50-watt bulb, the resistance of the 100-watt bulb is
(1) half as great   (3) one-fourth as great
(2) twice as great   (4) four times as great

93. A device operating at a potential difference of 1.5 volts draws a current of 0.20 ampere. How much energy is used by the device in 60. seconds?
(1) 4.5 J   (2) 8.0 J   (3) 12 J   (4) 18 J

94. As the number of resistors in a parallel circuit is increased, what happens to the equivalent resistance of the circuit and total current in the circuit?
    (1) Both equivalent resistance and total current decrease.
    (2) Both equivalent resistance and total current increase.
    (3) Equivalent resistance decreases and total current increases.
    (4) Equivalent resistance increases and total current decreases.

95. Pieces of aluminum, copper, gold, and silver wire each have the same length and the same cross-sectional area. Which wire has the *lowest* resistance at 20°C?
    (1) aluminum (2) copper (3) gold (4) silver

96. The graph below represents the relationship between the potential difference (V) across a resistor and the current (I) through the resistor.

    Through which entire interval does the resistor obey Ohm's law?
    (1) *AB* (2) *BC* (3) *CD* (4) *AD*

97. How much electrical energy is required to move a 4.00-microcoulomb charge through a potential difference of 36.0 volts?
    (1) 9.00×10^6 J   (3) 1.44×10^{-4} J
    (2) 144 J   (4) 1.11×10^{-7} J

98. What must be inserted between points A and B to establish a steady electric current in the incomplete circuit represented in the diagram below?

    (1) switch   (3) magnetic field source
    (2) voltmeter   (4) source of potential difference

99. In a series circuit containing two lamps, the battery supplies a potential difference of 1.5 volts. If the current in the circuit is 0.10 ampere, at what rate does the circuit use energy?
(1) 0.015 W   (2) 0.15 W   (3) 1.5 W   (4) 15 W

100. Which changes would cause the greatest increase in the rate of flow of charge through a conducting wire?
(1) increasing the applied potential difference and decreasing the length of wire
(2) increasing the applied potential difference and increasing the length of wire
(3) decreasing the applied potential difference and decreasing the length of wire
(4) decreasing the applied potential difference and increasing the length of wire

Answers to Part A and B–1 questions can be found on page 457.

PART B–2 AND C QUESTIONS

1. Base your answers to parts *a* through *d* on the diagram below, which represents a circuit containing a 120-volt power supply with switches S_1 and S_2 and two 60.-ohm resistors.

a If switch S_1 is kept open and switch S_2 is closed, what is the circuit resistance?
b If switch S_2 is kept open and switch S_1 is closed, how much current will flow through the circuit? [Show all calculations, including equations and substitutions with units.]
c When both switches are closed, what is the current in the ammeter?
d When both switches are closed, what is the reading of the voltmeter?

2. Base your answers to parts *a* and *b* on the information below.

Two resistors are connected in parallel to a 12-volt battery. One resistor, R_1, has a value of 18 ohms. The other resistor, R_2, has a value of 9 ohms. The total current in the circuit is 2 amperes. A student wishes to measure the current through R_1 and the potential difference across R_2.

a Using the symbols below for a battery, an ammeter, a voltmeter, and resistors, draw and label a circuit diagram that will enable the student to make the desired measurements.

Symbols: —|||— Battery

—(A)— Ammeter

—(V)— Voltmeter

—/\/\/— Resistor

b Calculate the value of the current in resistor R_1. [Show all calculations, including equations and substitutions with units.]

3. Base your answers to parts *a* through *c* on the information and data table below.

A resistor was held at constant temperature in an operating electric circuit. A student measured the current through the resistor and the potential difference across it. The measurements are shown in the data table below.

Data Table

Current (A)	Potential Difference (V)
0.010	2.3
0.020	5.2
0.030	7.4
0.040	9.9
0.050	12.7

a Using the information in the data table, construct a graph on the grid provided, following the directions below.

(1) Mark an appropriate scale on the axis labeled "Current (A)."
(2) Plot the data points for potential difference versus current.
(3) Draw the best-fit line.

Potential Difference vs. Current

b Using your graph, find the slope of the best-fit line. [Show all calculations, including the equation and substitution with units.]
c What physical quantity does the slope of the graph represent?

Base your answers to questions 4 through 6 on the information below.

A 5.0-ohm resistor, a 20.0-ohm resistor, and a 24-volt source of potential difference are connected in parallel. A single ammeter is placed in the circuit to read the total current.

4. In the space provided, draw a diagram of this circuit, using the symbols with labels given below. [Assume availability of any number of wires of neglible resistance.]

5.0 Ω

20.0 Ω

24-volt source

(A)

5. Determine the total circuit resistance. [Show all calculations, including the equation and substitution with units.]

6. Determine the total circuit current. [Show all calculations, including the equation and substitution with units.]

Base your answers to questions 7 and 8 on the information and diagram below.

A 5.0-ohm resistor, a 15.0-ohm resistor, and an unknown resistor, R, are connected as shown with a 15-volt source. The ammeter reads a current of 0.50 ampere.

7. Determine the reading of the voltmeter connected across the 5.0-ohm resistor. [Show all calculations, including the equation and substitution with units.]

8. Determine the total electrical energy used in the circuit in 600. seconds. [Show all calculations, including the equation and substitution with units.]

Base your answers to questions 9 through 13 on the information and data table below.

A variable resistor was connected to a battery. As the resistance was adjusted, the current and power in the circuit were determined. The data are recorded in the table below.

Current (amperes)	Power (watts)
0.75	2.27
1.25	3.72
2.25	6.75
3.00	9.05
4.00	11.9

9–10. Using the information in the data table, construct a line graph on the grid provided below.

9. Plot the data points for power versus current.

10. Draw the best-fit line.

11. Using your graph, determine the power delivered to the circuit at a current of 3.5 amperes.

12. Calculate the slope of the graph. [Show all calculations, including the equation and substitution with units.]

13. What is the physical significance of the slope of the graph?

14. Your school's physics laboratory has the following equipment available for conducting experiments:

accelerometers	lasers	stopwatches
ammeters	light bulbs	thermometers
bar magnets	meter sticks	voltmeters
batteries	power supplies	wires
electromagnets	spark timers	

Explain how you would find the resistance of an unknown resistor in the laboratory. Your explanation must include:

a Measurements required

b Equipment needed

c Complete circuit diagram

d Any equation(s) needed to calculate the resistance

Base your answers to questions 15 and 16 on the information below.

An electric circuit contains two 3.0-ohm resistors connected in parallel with a battery. The circuit also contains a voltmeter that reads the potential difference across one of the resistors.

15. Draw a diagram of this circuit, using the symbols from the *Reference Tables for Physical Setting/Physics*. [Assume availability of any number of wires of negligible resistance.]

16. Calculate the total resistance of the circuit. [Show all work, including the equation and substitution with units.]

Base your answers to questions 17 and 18 on the information below.

A toaster having a power rating of 1050 watts is operated at 120. volts.

17. Calculate the resistance of the toaster. [Show all work, including the equation and substitution with units.]

18. The toaster is connected in a circuit protected by a 15-ampere fuse. (The fuse will shut down the circuit if it carries more than 15 amperes.) Is it possible to simultaneously operate the toaster and a microwave oven that requires a current of 10.0 amperes on this circuit? Justify your answer mathematically.

Base your answers to questions 19 through 21 on the information below.

You are given a 12-volt battery, ammeter A, voltmeter V, resistor R_1, and resistor R_2
Resistor R_2 has a value of 3.0 ohms.

19. Using appropriate symbols from the *Reference Tables for Physical Setting/Physics*, draw and label a complete circuit showing:

   • resistors R_1 and R_2 connected in parallel with the battery
   • the ammeter connected to measure the current through resistor R_1, only
   • the voltmeter connected to measure the potential drop across resistor R_1

20. If the total current in the circuit is 6.0 amperes, determine the equivalent resistance of the circuit.

21. If the total current in the circuit is 6.0 amperes, determine the resistance of resistor R_1. [Show all calculations, including the equation and substitution with units.]

Base your answers to questions 22 and 23 on the information and diagram below.

A 10.0-meter length of copper wire is at 20°C. The radius of the wire is 1.0×10^{-3} meter.

Cross Section of Copper Wire

$r = 1.0 \times 10^{-3}$ m

22. Determine the cross-sectional area of the wire.

23. Calculate the resistance of the wire. [Show all work, including the equation and substitution with units.]

Base your answers to questions 24 through 26 on the information and diagram below.

A 15-ohm resistor, R_1, and a 30.-ohm resistor, R_2, are to be connected in parallel between points A and B in a circuit containing a 90.-volt battery.

90. V

24. Complete the diagram above to show the two resistors connected in parallel between points A and B.

25. Determine the potential difference across resistor R_1.

26. Calculate the current in resistor R_1. [Show all work, including the equation and substitution with units.]

Base your answers to questions 27 and 28 on the information below.

A copper wire at 20°C has a length of 10.0 meters and a cross-sectional area of 1.00×10^{-3} meter^2. The wire is stretched, becomes longer and thinner, and returns to 20°C.

27. What effect does this stretching have on the wire's resistance?

28. What effect does this stretching have on the wire's resistivity?

29. An electric circuit contains a source of potential difference and 5-ohm resistors that combine to give the circuit an equivalent resistance of 15 ohms. Draw a diagram of this circuit using circuit symbols given in the *Reference Tables for Physical Setting/Physics*. [Assume the availability of any number of 5-ohm resistors and wires of negligible resistance.]

Base your answers to questions 30 through 32 on the information and diagram below.

A 3.0-ohm resistor, an unknown resistor, R, and two ammeters, A_1 and A_2, are connected as shown with a 12-volt source. Ammeter A_2 reads a current of 5.0 amperes.

30. Determine the equivalent resistance of the circuit.

    _____ Ω

31. Calculate the current measured by ammeter A_1. [Show all work, including the equation and substitution with units.]

32. Calculate the resistance of the unknown resistor, R. [Show all work, including the equation and substitution with units.]

33. Calculate the resistance of a 1.00-kilometer length of nichrome wire with a cross-sectional area of 3.50×10^{-6} meter^2 at 20°C. [Show all work, including the equation and substitution with units.]

34. A generator produces a 115-volt potential difference and a maximum of 20.0 amperes of current. Calculate the total electrical energy the generator produces operating at maximum capacity for 60. seconds. [Show all work, including the equation and substitution with units.]

Base your answers to questions 35 through 37 on the information and diagram below.

A 50.-ohm resistor, an unknown resistor R, a 120-volt source, and an ammeter are connected in a complete circuit. The ammeter reads 0.50 ampere.

35. Calculate the equivalent resistance of the circuit. [Show all work, including the equation and substitution with units.]

36. Determine the resistance of resistor R.

    _____ Ω

37. Calculate the power dissipated by the 50.-ohm resistor. [Show all work, including the equation and substitution with units.]

38. The diagram below shows two resistors, R_1 and R_2, connected in parallel in a circuit having a 120-volt power source. Resistor R_1 develops 150 watts and resistor R_2 develops an unknown power. Ammeter A in the circuit reads 0.50 ampere.

Calculate the amount of charge passing through resistor R_2 in 60. seconds. [Show all work, including the equation and substitution with units.]

Base your answers to questions 39 through 41 on the information below.

An 18-ohm resistor and a 36-ohm resistor are connected in parallel with a 24-volt battery. A single ammeter is placed in the circuit to read its total current.

39. Draw a diagram of this circuit using symbols from the *Reference Tables for Physical Setting/Physics*. [Assume the availability of any number of wires of negligible resistance.]

40. Calculate the equivalent resistance of the circuit. [Show all work, including the equation and substitution with units.]

41. Calculate the total power dissipated in the circuit. [Show all work, including the equation and substitution with units.]

Answers to Part B–2 and C questions can be found on pages 506–522.

MAGNETISM; ELECTROMAGNETISM AND ITS APPLICATIONS
Chapter
Ten

KEY IDEAS

One of the phenomena associated with all substances is magnetism. Magnets have north and south poles, named for the way they orient themselves in the Earth's magnetic field. Unlike electric charges, single magnetic poles have not been discovered: A north pole is found in conjunction with a south pole. Like poles repel each other and unlike poles attract, a property similar to that observed in electric charges. The magnetic field around a magnetic configuration may be mapped using field lines in the same way an electric field is mapped.

Magnetic fields exist whenever an electric current is present. The direction of the magnetic field is always perpendicular to the electric field. Current-carrying wires and charges moving perpendicularly through a magnetic field experience forces that are perpendicular to both the direction of the magnetic field and the motion of the charges.

If a conductor is moved perpendicularly through a magnetic field, a potential difference is established across the conductor.

If a charge is accelerated, the changing electric and magnetic fields will give rise to electromagnetic waves that are carriers of energy.

KEY OBJECTIVES
At the conclusion of this chapter you will be able to:
- Define the terms *magnet, north pole, south pole, temporary magnet, permanent magnet.*
- Define the term *domain*, and describe how domains contribute to the magnetic properties of a metal such as iron.
- State the conventions for drawing magnetic field lines, and draw simple magnetic field configurations.
- Define the term *magnetic induction*, and state the SI unit for magnetic induction (field strength).
- Use an appropriate hand rule to describe the magnetic field around a current-carrying wire.
- Use an appropriate hand rule to determine the magnetic polarity of a current-carrying coil (solenoid).
- State the factors that influence the magnetic induction in a straight wire and in a solenoid.

- Use an appropriate hand rule to determine the force on a current-carrying wire in an external magnetic field.
- Describe how a potential difference may be induced across a conductor moving in a magnetic field.
- Describe how electromagnetic waves may be produced from accelerating charges.

10.1   INTRODUCTION

The phenomenon of magnetism was known in ancient times, when it was observed that certain rocks (called lodestones) attracted iron. It was also observed that, when pieces of iron were rubbed with lodestones, the iron became magnetized and that, if a very thin magnet was floated on water, one end of the magnet always pointed in the northern direction. As a result of these discoveries, the Chinese used magnets to create compasses with which to navigate their waters.

✪ 10.2   GENERAL PROPERTIES OF MAGNETS

A **magnet**, then, is any substance that possesses the properties discussed in Section 10.1. One common shape of a magnet is a rectangular bar, as shown in the diagram:

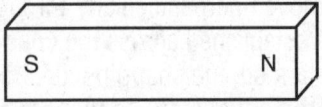

We know that magnets have "polarity." The end that points northward is the **north pole** of the magnet, and the end that points southward is the **south pole**. Also, when two magnets are brought near one another, it is observed that *like poles repel* and *unlike poles attract*. For these reasons we can conclude that the Earth itself behaves as a giant magnet. Earth's "North Pole" behaves like a magnetic south pole since it attracts the north pole of compass needles. If a piece of metal is placed in the vicinity of a magnet, the metal itself will become magnetized. If the metal retains its magnetism after the original magnet is removed, it is called a **permanent magnet**; otherwise it is a **temporary magnet**. Alloys such as ALNICO make good permanent magnets, while soft iron produces excellent temporary magnets.

Why do certain substances, such as iron, have magnetic properties? It has been discovered that groups of atoms align their unpaired electrons so that they spin in the same direction, giving rise to microscopic magnets called **domains**. An external magnetic field causes all the domains to align in the same direction, as shown in the diagram on the right.

✪ indicates material that is not part of the core curriculum.

10.3 MAGNETIC FIELDS

In the same way that electrostatic and gravitational forces can be explained by electric and gravitational fields, the existence of magnetic forces can be explained by the presence of magnetic fields. Also, just as field lines are used to visualize electric and gravitational fields, magnetic field lines (called *flux lines*) are used to visualize a magnetic field.

The magnetic fields between various poles of two adjacent magnets are shown in the diagrams, as well as the field around a horseshoe magnet, which is simply a bar magnet that has been bent so that the north and south poles are near each other. By agreement the field lines point away from the north and toward the south.

We can also draw the magnetic field around a single bar magnet, as shown below.

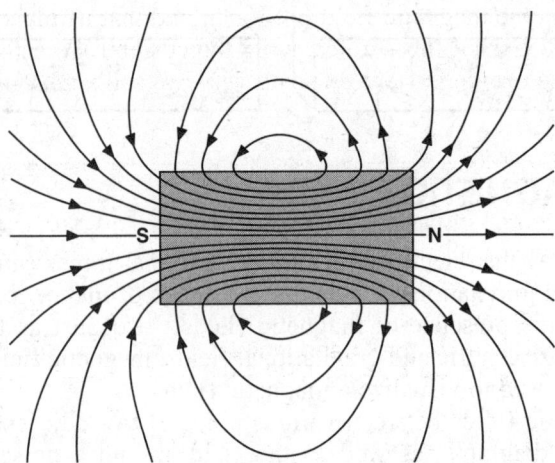

Note that the field lines form closed loops inside the magnet. As a consequence of this fact, magnetic "monopoles" are not believed to exist: a north pole of a magnet is always accompanied by a corresponding south pole.

✪ Each magnetic flux line has been standardized and has the SI unit of 1 weber (Wb). The strength of the magnetic field, known as the **magnetic induction**, is given by the concentration of these flux lines, that is, the number of flux lines per unit area. We represent magnetic induction by the letter **B,** and its unit is the *weber per square meter* (Wb/m^2), also known as the tesla (T). Magnetic induction is a vector quantity because it has both magnitude and direction. A weak field, such as the Earth's magnetic field, has a magnetic induction of approximately 5×10^{-5} tesla. A field of 1 tesla is extremely strong and is used in applications such as magnetic resonance imaging (MRI) and nuclear particle accelerators.

10.4   ELECTROMAGNETISM

In 1820, the Danish physicist Hans Oersted discovered that a wire carrying a current produced a magnetic field as shown in the diagram.

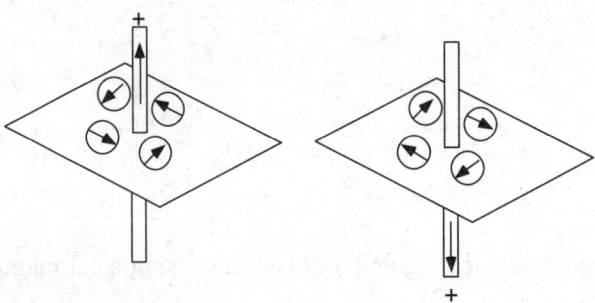

✪ indicates material that is not part of the core curriculum.

We note that the magnetic field is circular and that its plane is perpendicular to the direction of the wire carrying the current. We can determine the direction of the magnetic field by using what we call *right-hand rule 1:*

PHYSICS CONCEPTS

✪ The thumb of the right hand is pointed in the direction of the conventional current. The fingers of the right hand (from wrist to fingertips) will curl in the direction of the magnetic field. (*Note*: If you use electron flow instead of conventional current, use your *left* hand instead of your right hand.)

To represent the direction of a magnetic field in two dimensions, we use dots (•) to indicate that the direction is out of the plane of the paper and X's (X) to indicate that the direction is into the plane of the paper, as shown in the diagram.

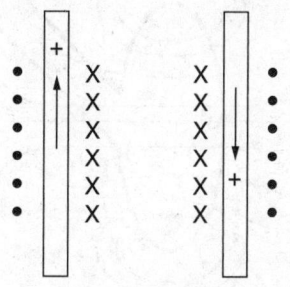

✪ indicates updated Physics concepts, but not tested on the Regents exam.

The magnetic induction in the vicinity of a straight wire is directly proportional to the current in the wire and is inversely proportional to the distance from the wire. A uniform magnetic field into or out of the plane of the paper can be represented as shown below:

Uniform field out of the page   Uniform field into the page

Magnetic Field Around a Coil

If we bend our wire into a single loop, the magnetic field will appear as illustrated.

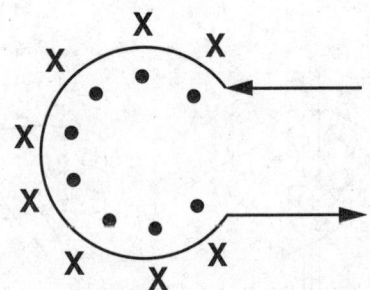

We note below that the magnetic field around the loop has the appearance of a very thin bar magnet. Each face of the single loop is a magnetic pole.

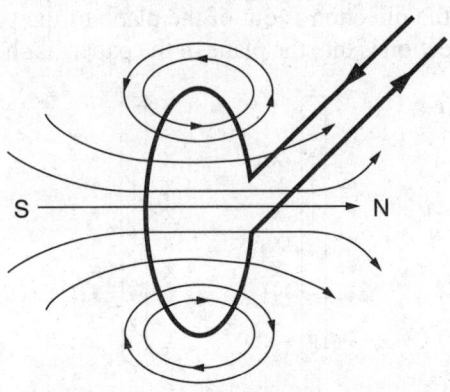

If we link a number of loops together, we produce a coil (or a *solenoid*) whose magnetic field is the result of the fields of the individual loops. The magnetic field around a solenoid is shown below.

We note that the magnetic field of the solenoid is nearly identical to the magnetic field of a bar magnet. We can determine the north pole of our coil by using what we call *right-hand rule 2:*

PHYSICS CONCEPTS

✪ The fingers of the right hand are wrapped in the direction of the conventional current (electron-flow users, be sure to use the left hand). The thumb will point to the end of the coil, which is the north pole.

We note that the magnetic field is strongest inside the coil and is also uniform because the lines are closely spaced and are parallel. The magnetic induction depends on the current in the coil, the number of turns per unit length of the coil, and the nature of the core. If a *ferromagnetic* core such as soft iron is placed inside the coil, the magnetic induction can be increased thousands of times—a fact that is applied in making commercial electromagnets.

Forces on a Current Carrier in a Magnetic Field

If a wire carrying a current is placed in a magnetic field, so that the direction of the current is perpendicular to the direction of the magnetic field, the magnetic

---

✪ indicates updated Physics concepts, but not tested on the Regents exam.

field of the wire will interact with the magnetic field of the magnet to produce a force on the wire. This is illustrated in the diagram.

If, however, the current is parallel to the magnetic field, no force will be present on the wire.

To determine the direction of the force on the wire, we use what we call *right-hand rule 3*, illustrated in the diagram below.

====== **PHYSICS CONCEPTS** ======

✪ The thumb points in the direction of conventional current, the fingers point in the direction of the magnetic field, and the direction of the force points away from the palm (electron-flow users, left hand, please!).

Suppose the wire was neither perpendicular nor parallel to the magnetic field. In that case, there would be a force on the wire but it would be less than the maximum force present when the wire is perpendicular, but greater than zero when the wire is parallel.

✪ 10.5   ELECTROMAGNETIC INDUCTION

A motor uses a magnetic field to convert electrical energy into mechanical energy. It is also possible to accomplish the reverse process, that is, to use a magnetic field to convert mechanical energy into electrical energy. Devices that accomplish this purpose are known as *generators*.

Let's begin by moving a wire perpendicularly through a magnetic field.

If we focus on one electron in the wire, indicated in the diagram below:

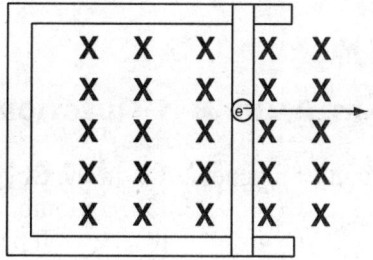

we can apply *right-hand rule 3,* but we must use the *left* hand because we are dealing with a negatively charged particle, to show that there is a downward force on the electron. This is true for all electrons passing through the moving segment of the wire. These moving electrons constitute an electric current.

Since work has been done in moving the electrons through the wire, a potential difference has been induced across the ends of the wire. This potential difference depends on the strength of the magnetic field, the length of the wire in the magnetic field, and the speed with which the wire is moved. It is important to note that in order to induce maximum voltage the wire needs to "cut" through the magnetic flux lines at a 90° angle. As the angle decreases, the voltage induced decreases. If the wire moves parallel to the magnetic flux lines, it doesn't "cut" through any flux lines, and so zero potential difference is induced.

Electromagnetic Waves

A changing electric field produces a changing magnetic field and vice versa. If the changing electric field is produced by an *accelerating* charge, energy will be radiated away from the charge in the form of *electromagnetic waves*. In an electromagnetic wave, both the electric field and the magnetic field vary as a sine wave and the two fields are perpendicular to each other and to their direction of motion. All electromagnetic waves (collectively known as the *electromagnetic spectrum*) travel in space at the speed of light.

PART A AND B–1 QUESTIONS

✪ **1.** Which is the unit of magnetic flux in the SI system?
   (1) weber
   (2) joule
   (3) coulomb
   (4) newton per ampere-meter

2. The presence of a uniform magnetic field may be detected by using a
   (1) stationary charge
   (2) small mass
   (3) beam of neutrons
   (4) magnetic compass

3. In the diagram below, what is the direction of the magnetic field at point *P*?

Poles of Permanent Magnet

   (1) toward *A*   (2) toward *B*   (3) toward *C*   (4) toward *D*

4. Which diagram best illustrates the direction of the magnetic field between the unlike poles of two bar magnets?

5. Which diagram best represents the magnetic field near the poles of a horseshoe magnet?

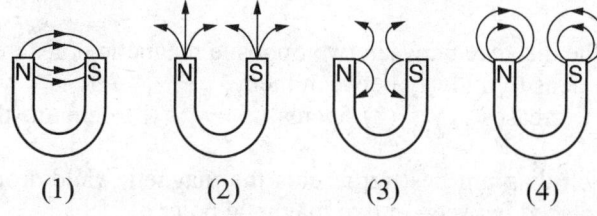

6. Which vector best represents the direction of the magnetic field at point *A* near the two north magnetic poles shown in the diagram?

7. Which arrow in the diagram below represents the direction of the flux *inside* the bar magnet?

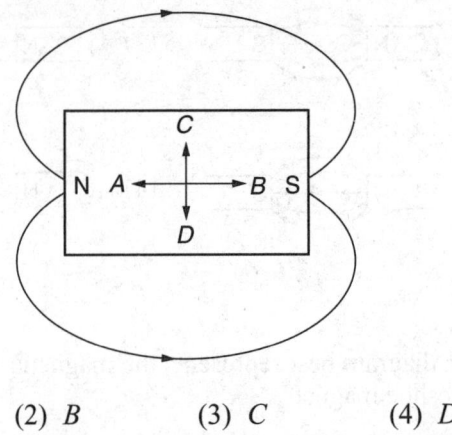

(1) *A*   (2) *B*   (3) *C*   (4) *D*

8. As the distance between two opposite magnetic poles increases, the flux density midway between them
(1) decreases   (2) increases   (3) remains the same

9. Which diagram best represents the magnetic field around an iron bar placed between unlike magnetic poles?

Base your answers to questions 10 through 14 on the diagram below, which represents a cross section of an operating solenoid. A compass is located at point *C*.

✪ **10.** Which diagram best represents the shape of the magnetic field around the solenoid?

✪ **11.** Which shows the direction of the compass needle at point *C*?

✪ **12.** If *B* is the north pole of the solenoid, which diagram best represents the direction of *electron flow* in one of the wire loops?

✪ **13.** If an iron rod were inserted into the solenoid, the strength of the magnetic field inside the solenoid would
   (1) decrease   (2) increase   (3) remain the same

✪ indicates material that is not part of the core curriculum.

14. Each diagram below represents a cross section of a long, straight, current-carrying wire with the electron flow into the page. Which diagram best represents the magnetic field near the wire?

(1)   (3)

(2)   (4)

15. A magnetic field will be produced by
(1) moving electrons   (3) stationary protons
(2) moving neutrons   (4) stationary ions

16. What is the direction of the magnetic field near the center of the current-carrying wire loop shown in the diagram below?

(1) into the page   (3) to the left
(2) out of the page   (4) to the right

17. The diagram represents a wire with electrons moving in the direction shown. At point A, the magnetic field is directed

(1) out of the page   (3) from left to right
(2) into the page   (4) from right to left

18. Which diagram best represents the direction of the magnetic field around a wire conductor in which the electrons are moving as indicated? [The X's indicate that the field is directed into the paper, and the dots indicate that the field is directed out of the page.]

(1)   (2)   (3)   (4)

✪ 19. As the permeability of a substance in a magnetic field decreases, the flux density within the substance
(1) decreases   (2) increases   (3) remains the same

✪ 20. In the diagram below, electron current is passed through a solenoid. The north pole of the solenoid is nearest to point

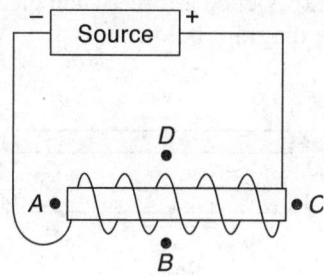

(1) *A*   (2) *B*   (3) *C*   (4) *D*

✪ 21. If the current through a solenoid increases, the magnetic field strength of the solenoid
(1) decreases   (2) increases   (3) remains the same

✪ 22. As materials of increasing permeability are placed at position *X* in the diagram, the flux density at position *X*

(1) decreases   (2) increases   (3) remains the same

✪ indicates material that is not part of the core curriculum.

✪ **23.** The diagram below shows an end view of a current-carrying wire between the poles of a magnet. The wire is perpendicular to the magnetic field.

If the direction of the electron flow is out of the page, which arrow correctly shows the direction of the magnetic force **F** acting on the wire?

    (1)   (2)   (3)   (4)

✪ **24.** Wires x and y experience a force when electrons pass through them as shown in the diagram below.

The force on wire y will be toward
(1) A (2) B (3) C (4) D

✪ **25.** When electrons flow from point A to point B in the wire shown in the diagram, a force will be produced on the wire

(1) toward N   (3) into the page
(2) toward S   (4) out of the page

✪ indicates material that is not part of the core curriculum.

26. The diagram below shows conductor *C* between two opposite magnetic poles. Which procedure will produce the greatest induced potential difference in the conductor?

(1) holding the conductor stationary between the poles
(2) moving the conductor out of the page
(3) moving the conductor toward the right side of the page
(4) moving the conductor toward the north pole

The diagram below represents a U-shaped wire conductor positioned perpendicular to a uniform magnetic field that acts into the page. *AB* represents a second wire, which is free to slide along the U-shaped wire. The length of wire *AB* is 1 meter, and the magnitude of the magnetic field is 8.0 webers per meter^2.

27. If wire *AB* is moved to the left at a constant speed, the direction of the induced electron motion in wire *AB* will be
(1) toward *A*, only   (3) first toward *A* and then toward *B*
(2) toward *B*, only   (4) first toward *B* and then toward *A*

The diagram below represents wires A and B both carrying a flow of electrons out of the paper. The wires are 1 meter apart.

⚙ 28. As a result of the magnetic fields associated with the wires, wire A will experience a force directed toward
(1) 1   (2) 2   (3) 3   (4) 4

A stream of electrons from heated filament F is accelerated by a potential difference of 80.0 volts toward plate P. Some of the electrons in the beam pass through the hole in plate P and follow the path shown in the diagram. The magnetic and electric fields are uniform.

⚙ 29. With the magnetic field removed and the electric field turned on, the beam would be deflected
(1) upward   (3) into the page
(2) downward   (4) out of the page

30. Which statement best describes a proton that is being accelerated?
(1) It produces electromagnetic radiation.
(2) The magnitude of its charge increases.
(3) It absorbs a neutron to become an electron.
(4) It is attracted to other protons.

Answers to Part A and B–1 questions can be found on page 458.

⚙ indicates material that is not part of the core curriculum.

Part B–2 and C Questions

1. The diagram below shows two compasses located near the ends of a bar magnet. The north pole of compass X points toward end A of the magnet.

On the diagram provided draw the correct orientation of the needle of compass Y and label its polarity.

2. On the following diagram of a bar magnet draw a minimum of four field lines to show the magnitude and direction of the magnetic field in the region surrounding the bar magnet.

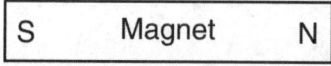

Answers to Part B–2 and C questions can be found on page 523.

WAVES AND SOUND

Chapter
Eleven

KEY IDEAS

Waves transfer energy without the transfer of mass. Mechanical waves, such as sound, require a medium for transmission, whereas electromagnetic waves, such as visible light, do not.

A wave may be longitudinal, transverse, or a combination of both, depending on the direction in which the medium vibrates in relation to the movement of the wave's energy. Longitudinal waves exhibit parallel vibrations; in transverse waves the vibrations are perpendicular.

The characteristics of a periodic wave include speed, wavelength, frequency and period, and amplitude. Among the properties of periodic waves are reflection, refraction (the change in the direction of a wave that enters a medium at an angle), interference (the combination of two or more waves simultaneously in a medium), diffraction (the apparent "bending" of a wave around an obstacle), and the Doppler effect (the apparent change in the frequency of a wave as perceived by an observer because of the relative motion between the wave source and the observer).

KEY OBJECTIVES

At the conclusion of this chapter you will be able to:

- Define the terms *periodic wave, wave motion, transverse wave, longitudinal wave,* and *surface wave,* and provide examples of each.
- Compare and contrast mechanical waves with electromagnetic waves.
- Define the terms *period, frequency, amplitude,* and *wavelength,* and solve problems that relate these quantities to wave speed.
- Use a diagram of a periodic wave to identify the following: crest, trough, amplitude, phase, and wavelength.
- Define the term *reflection*, and apply the law of reflection.
- Define the term *ray*, and apply it to various types of periodic waves.
- Define the term *refraction*, and apply Snell's law.
- Define the terms *constructive interference, destructive interference, resonance,* and *diffraction.*
- Explain how interference can produce standing waves and beats.
- Define the term *Doppler effect*, and explain this phenomenon.

11.1   DEFINITION OF WAVE MOTION

The diagram represents a coiled spring held between two people; a handkerchief is tied to the spring. If one person quickly jerks the end of the spring up and down, there will be a disturbance in the spring. When the disturbance reaches the other person's hand, it will cause the hand to jerk. Therefore, the disturbance, or *wave pulse,* transfers energy. A moving particle can also transfer energy, but its mass is transferred as well.

If we look at the handkerchief tied to the spring in the diagrams below, however, we can see that it has vibrated about the spring's rest position, that is, it has moved up and down, but it has not moved along with the energy. A wave pulse, or a series of identical, repeating, evenly spaced pulses (called a **periodic wave**), transfers energy, but not mass.

11.2 TYPES OF WAVES

A wave is a vibratory disturbance that is transmitted through a material or through space. Water waves, sound waves, and waves that travel along a spring are examples of *mechanical* waves. Mechanical waves require a material medium for transmission. The energy disturbance is propagated by the molecules of water, air, or, as in the example in Section 11.1, the metal atoms of a spring.

Light waves, microwaves, and radio waves are examples of *electromagnetic* waves. Electromagnetic waves do not need a material medium; they are the result of changes in the field strengths of electric and magnetic fields and can travel in space (a vacuum). Since electromagnetic waves cannot be observed directly, we will use mechanical wave models to explore wave properties and behavior. Light will be studied in greater detail in Chapter 12.

Mechanical waves can be divided into three different types: transverse, longitudinal, and surface waves. In a **transverse wave**, the particles of the medium vibrate or exhibit simple harmonic motion (SHM) about a rest position *perpendicular* to the direction of motion of the wave. Waves on a string, as shown below, are an example of a transverse wave.

In a **longitudinal wave,** diagramed below, a disturbance causes the particles of the material to vibrate in SHM in a direction *parallel* to the direction of motion of the wave.

Sound is an example of a longitudinal wave. Fluids usually transmit only longitudinal waves.

On the surface of water, the motions of particles are both parallel and perpendicular to the direction of motion of the wave. In the diagram, the combination of transverse and longitudinal waves produces what is known as a **surface wave**.

Wave motion

11.3 CHARACTERISTICS OF PERIODIC WAVES

A periodic disturbance (e.g., water dripping from a faucet into a sink of water) will produce a traveling wave. As a result each particle in the medium will vibrate in simple harmonic motion in response to the disturbance.

If we look at a single particle in the medium (e.g., the handkerchief on the spring in Section 11.1), we see that it moves up and down as we send a traveling wave through the spring. The time it takes for its motion to repeat itself is called the **period** (*T*), and it is measured in seconds. The number of times the motion repeats itself in a time interval of one unit of time is known as the **frequency** (*f*) of the wave. Frequency is measured in hertz (Hz), which is equivalent to cycles per second or reciprocal seconds (s^{-1}).

Frequency and period are inversely proportional to each other and are related by this equation:

PHYSICS CONCEPTS

$$T = \frac{1}{f}$$

In other words, frequency and period are reciprocals of each other.

The maximum displacement of any particle in the medium relative to its rest position is called the **amplitude** of the wave. In a transverse wave the maximum upward displacement is known as a *crest*; the maximum downward displacement, as a *trough*. In a longitudinal wave, the particles in the medium produce areas of maximum compression called ✪ *condensations* and areas of maximum separation called *rarefactions*. Condensations are analogous to crests, and rarefactions to troughs. Areas of condensation and rarefaction are shown in the diagram below.

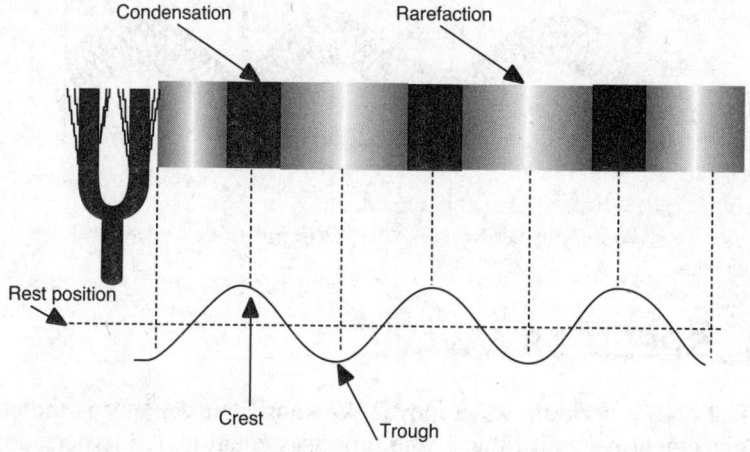

✪ Note: The Regents exam will simply refer to condensation as compression.

305

Amplitude is related to the energy carried by the wave. In sound waves, amplitude corresponds to loudness; in light, to brightness.

Points on a periodic wave that are at equal displacements from their rest position and are experiencing identical movements, that is, are moving in *the same direction* toward or away from the rest position, are said to be *in phase*. Points on a periodic wave that are at equal displacements from their rest position but are experiencing motion in *opposite directions* from each other are described as being *180° out of phase* or completely out of phase. In the diagram, points 1 and 5, 2 and 6, 3 and 7, 4 and 8 are in phase. Points 1 and 3, 2 and 4, 3 and 5, 4 and 6, 5 and 7, 6 and 8 are completely out of phase.

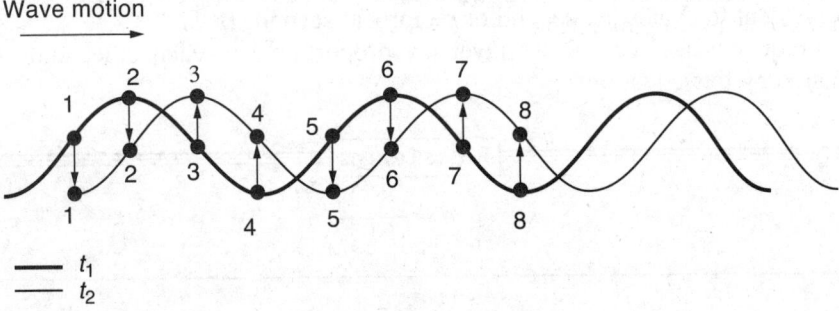

The distance between two successive points on a periodic wave that are in phase is called the **wavelength** (λ). Wavelength is measured in meters. Successive points that are 180° out of phase are therefore separated by a distance of one-half wavelength.

11.4 SPEED OF A WAVE

How fast does a periodic wave move? We know that velocity is the change in displacement per unit time. If the time was equal to T, the period of the wave motion, the wave would move a distance of one wavelength (λ). Therefore, $v = \dfrac{\lambda}{T}$. Since period and frequency are reciprocals of each other, we can rewrite the equation as follows:

PHYSICS CONCEPTS

$$v = f\lambda$$

PROBLEM

What is the wavelength of a sound wave whose speed is 330 meters per second and whose frequency is 990 hertz?

SOLUTION

$$v = f\lambda$$

$$\lambda = \frac{v}{f}$$

$$= \frac{330 \text{ m/s}}{990 \text{ s}^{-1}}$$

$$= 0.33 \text{ m}$$

All electromagnetic waves travel in space at the speed of light, which is denoted by the letter c and approximately equal to 3.0×10^8 meters per second. Generally, the speed of a mechanical wave depends only on the properties of the medium, not on the amplitude or frequency of the wave. A wave with large amplitude transmits more energy than a wave with low amplitude, but both travel at the same speed through a given medium. If two waves have the same speed, the wave with the higher frequency will have a shorter wavelength than the wave with the lower frequency. This is a direct result of the equation $v = f\lambda$.

11.5   REFLECTION

When a wave travels from one medium to another, part of the energy of the wave is transmitted into the new medium with the same frequency, part is absorbed, and part moves back into the original medium, that is, it is *reflected*, with the same frequency.

If the difference between the media is small, most of the wave's energy or amplitude will be transmitted and very little will be reflected. If, however, the two media are very different, very little energy will be transmitted and most will be reflected.

If the wave travels from a less dense to a more dense medium, the reflected wave will be inverted, or 180° out of phase.

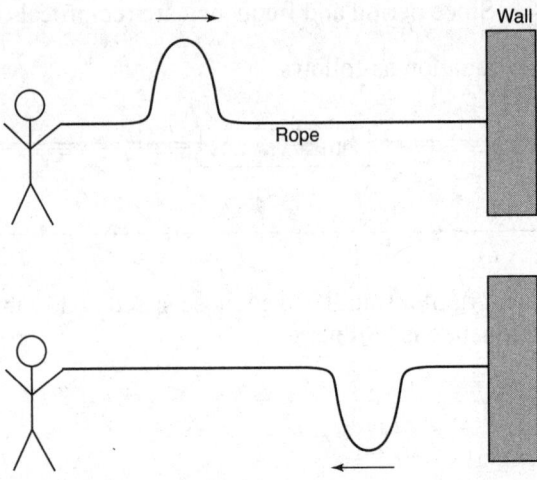

If, on the other hand, the wave travels from a more dense to a less dense medium, the reflected wave will *not* undergo a phase change.

Although transverse waves are used in the two diagrams above to illustrate this property, longitudinal waves behave in the same manner.

Wave Shape

A wave can have various shapes, depending on the source that produces it. If a person drops pebbles into a pond and then views the result from above, the diagram below shows what is seen.

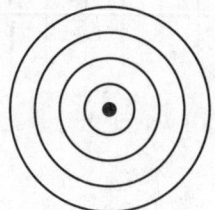

The pebbles are known as a *point source,* and the wave is circular, that is, it spreads out evenly in all directions. The circles in the diagram represent the crests of the wave and are called *wave fronts*.

A line that indicates the direction of motion is called a **ray**. The rays of circular waves are radial lines.

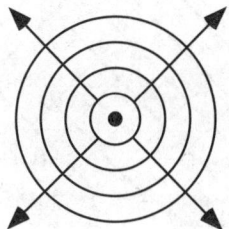

If the source of the waves is broad, such as a wooden plate bobbing in the water, the result will be a series of plane waves, as shown in the diagram below.

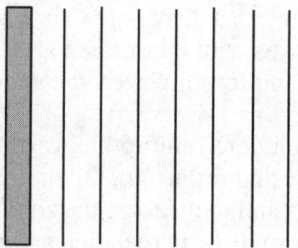

In the next diagram the wave fronts are straight lines. The rays all point in the same direction, that is, they are parallel to one another. At distances very far from a point source, waves become nearly plane in shape.

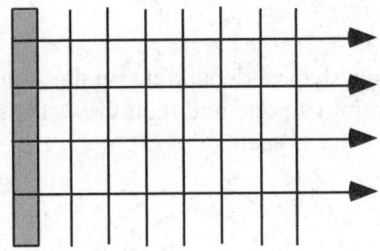

The diagrams show that in all waves the rays are perpendicular to the wave fronts.

Incident and Reflected Waves

When a plane wave strikes a reflecting surface at an angle, it is reflected at the same angle, as shown in the diagram below.

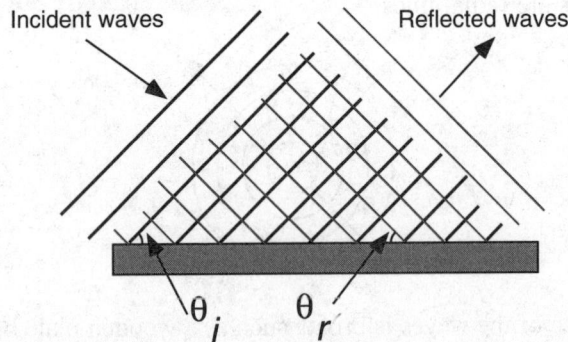

The wave that strikes the surface is called the *incident* wave. The angle it makes with the surface is the angle of incidence (represented by θ_i in the diagram). Similarly, the wave that leaves the surface is termed the *reflected* wave, and the angle it makes with the surface is the reflected angle (θ_r in the diagram). In every case, and for all waves, the angle of incidence is equal to the angle of reflection.

It is not always convenient to refer to the wave fronts themselves in measuring angles of incidence and reflection. Often it is easier to refer to the rays associated with the waves and to measure the angles of incidence and reflection with respect to a normal (a perpendicular line drawn to the surface). This relationship is known as the law of reflection.

PHYSICS CONCEPTS

$$\theta_i = \theta_r$$

This situation is diagrammed below.

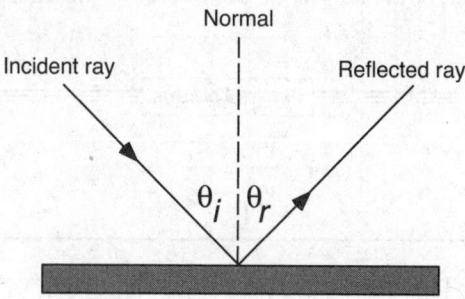

11.6 REFRACTION

If a wave passes from one medium to another at an angle to the boundary, and its speed changes, its direction in the second medium will also change, as shown in the diagram below.

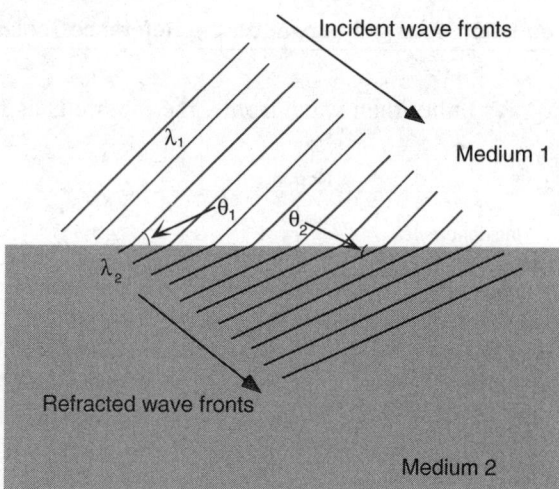

The wave that strikes the boundary in medium 1 is the *incident* wave, and the angle it makes with the boundary surface (θ_1) is the angle of incidence. The wave in medium 2 is called the *refracted* wave, and the angle it makes with the boundary surface (θ_2) is the angle of refraction.

As the wave enters the second medium, its frequency does not change; therefore, its change in velocity is accompanied by a change in its wavelength. In the diagram, the wavelength in medium 2 is less than the wavelength in medium 1 (λ_1); consequently, the speed of the wave in medium 2 is also less than the speed of the wave in medium 1. This happens when the wave is traveling from a less dense medium to a more dense medium. The opposite would be true if the wave were traveling from a more dense to a less dense medium. The wavelength of the wave would increase as would the speed of the wave.

PHYSICS CONCEPTS

$$\frac{v_1}{v_2} = \frac{\lambda_1}{\lambda_2}$$

We could prove this statement using the relationship $v = f\lambda$. Using simple trigonometry, it can be shown that the ratio of the speeds in the two media is related to the ratio of the sines of the angles in these media. This relationship is known as Snell's law, after the Dutch astronomer and mathematician Willebrord Snell.

PHYSICS CONCEPTS

$$\frac{\sin \theta_2}{\sin \theta_1} = \frac{v_2}{v_1}$$

***Note that this equation does not appear on the Reference Tables.**

If we refer to rays rather than wave fronts, the picture is as follows:

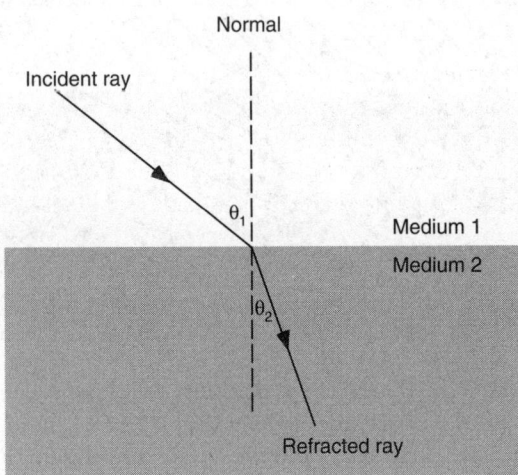

Once again, we measure the angles of incidence and refraction with respect to a normal.

By examining the sizes of the angles, we can draw conclusions about the relative speeds of the waves in the two media. The ray representing a slower wave is positioned closer to the normal than is the ray representing a faster wave. We will explore this phenomenon further in Chapter 12.

11.7 INTERFERENCE

Two or more waves passing simultaneously through the same area of a medium affect the medium independently but do not affect each other. The resultant displacement of any point in the medium is the algebraic sum of the displacements of all the individual waves; this is known as the principle of *superposition*, and the result is called *interference*.

There exist two kinds of interference, constructive and destructive. **Constructive interference** occurs when the individual wave displacements, A and B in the diagram below, are in the same direction. In this case, the resulting amplitude, $A + B$, is greater than any individual wave amplitude.

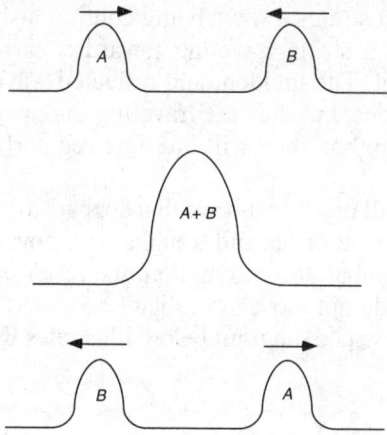

Destructive interference occurs when displacements A and B are in opposite directions, as illustrated below. In this case, the resulting amplitude is less than any individual wave amplitude. If the displacements are equal in magnitude, complete or maximum destructive interference occurs. If the displacements are not equal in magnitude, the result is partial destructive interference.

11.8 STANDING WAVES

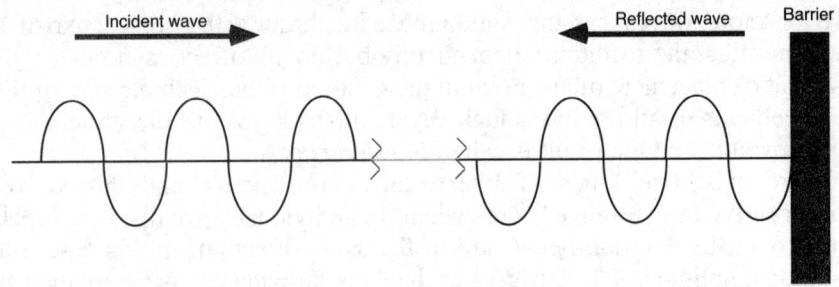

The diagram above illustrates a wave being continuously produced by an up-and-down motion on a string traveling toward a barrier and the reflected wave emerging from it. The incident and reflected waves have the same frequencies and amplitudes, but they are traveling in opposite directions. When the waves pass one another, they will interfere regularly, both constructively and destructively.

This interference will produce a wave that appears to "stand still" in the horizontal direction. Adjacent crests and troughs will move vertically in opposite directions about points that have no motion; the result is known as a *standing wave*. The points that do not move are called *nodes,* and the crest-trough combinations are *antinodes*. The diagram below illustrates this phenomenon.

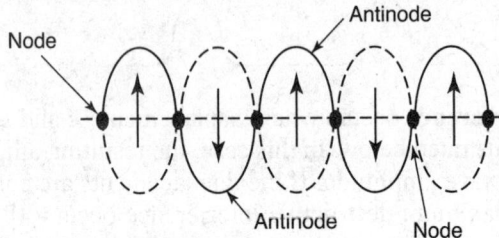

11.9 RESONANCE

If a person were to pluck a stretched guitar string that was not connected to a guitar, the sound would not be heard. What exactly does the guitar *body* contribute to the production of audible sounds?

When a string is plucked, a standing wave pattern is established in the string. The guitar box spontaneously vibrates at the frequencies produced by the strings, and a standing-wave pattern is established in the guitar itself. This phenomenon is known as **resonance**. The *amplitudes* of the standing waves (antinodes) in the guitar are much larger than those in the string, and therefore we hear the sound. In general, musical instruments act as resonance devices.

Sometimes, resonance can be an unwanted phenomenon. Years ago, a gale-force wind caused a bridge in Tacoma, Washington, to resonate at its *natural frequency of vibration*. The energy produced by the standing-wave pattern was great enough to cause the bridge to collapse. When bridges and like structures are built today, devices are incorporated to prevent the production of these destructive standing-wave patterns.

11.10 DIFFRACTION

Diffraction is the bending of a wave around an obstacle. If a person stands beside an open door, he or she can usually hear conversation taking place in the room. As the sound waves emerge through the door, they are able to "bend around" the doorway. Similarly, water waves seem to be able to pass through pier barriers as though no obstruction were present.

The requirement for diffraction is that the size of the opening be on the order of the length of the wave being diffracted. For this reason, light will *not* diffract through a doorway because the opening is far too large in comparison to the wavelength of light. The diagram below illustrates the process of diffraction.

11.11 DOPPLER EFFECT

All of us are familiar with the sound of a siren on a moving vehicle—a fire engine, for example. As the vehicle approaches, the *apparent* pitch of the siren is increased; as the vehicle passes us and then recedes, the apparent pitch is decreased.

This phenomenon, known as the **Doppler effect**, occurs with all types of waves, including light. It is the result of relative motion between a source of waves and an observer. As the distance between the source and the observer decreases, the frequency of the source, as perceived by the observer, is increased; as the distance increases, the apparent frequency is decreased.

Effect on Mechanical and Electromagnetic Waves

For mechanical waves, such as sound and water, the effect produced by a source in motion is different from the effect experienced by an observer in motion, even though the general outcome for both is similar. For example, if an observer is moving toward a stationary source of sound, his or her ear drum receives more waves than if the observer were at rest, and the apparent frequency of the sound is increased. If, however, the source is moving toward a stationary observer, the result is a series of sound waves that are crowded together on the side nearest the observer. The result is that the observer's ear drum receives more waves than if the source were at rest, and the frequency of the sound appears to be increased. The following diagram illustrates the situation in which a source of sound is in motion and the observers are stationary.

For electromagnetic waves, such as visible light, the frequency change is recorded as a *color shift*, a phenomenon important in astronomy and astrophysics. An increase in frequency exhibits a blue shift. A decrease in frequency exhibits a red shift.

Bow Waves, Shock Waves, and Sonic Booms

If you have ever seen a duck swimming on a lake or pond, you may have observed a V-shaped wave produced by the duck. This phenomenon, known as a *bow wave,* is also produced by a boat in motion on a body of water. A bow wave is a special case of the Doppler effect. As the duck (or boat) travels on the water, it produces water waves. If the speed of the traveler is *greater* than the speed of the water waves, a bow wave results.

When planes exceed the speed of sound, they produce *shock waves,* which are exactly analogous to bow waves. The diagram below illustrates how shock waves are formed. A shock wave is accompanied by an explosionlike sound known as a *sonic boom.*

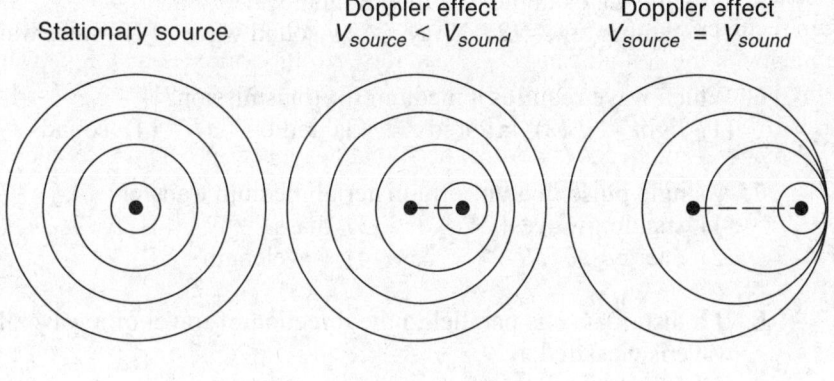

Stationary source

Doppler effect
$V_{source} < V_{sound}$

Doppler effect
$V_{source} = V_{sound}$

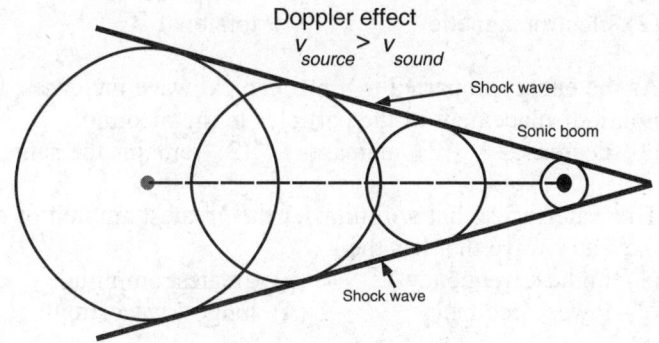

Doppler effect
$v_{source} > v_{sound}$

Shock wave

Sonic boom

Shock wave

PART A AND B-1 QUESTIONS

1. A series of pulses generated at regular time intervals in an elastic medium will produce
(1) nodes
(2) antinodes
(3) a polarized wave
(4) a periodic wave

2. Waves transfer energy between two points with no transfer of
(1) work
(2) momentum
(3) mass
(4) force

3. Compression waves in a spring are an example of
(1) longitudinal waves
(2) transverse waves
(3) polarized waves
(4) torsional waves

4. A wave in which the vibration is at right angles to the wave's direction of motion is called a
(1) longitudinal wave
(2) compressional wave
(3) transverse wave
(4) torsional wave

317

5. Which is an example of a longitudinal wave?
   (1) gamma ray   (2) X ray   (3) sound wave   (4) water wave

6. Which wave requires a medium for transmission?
   (1) light   (2) infrared   (3) radio   (4) sound

7. A single pulse in a uniform material medium transfers
   (1) standing waves   (3) mass
   (2) energy   (4) wavelength

8. If a disturbance is parallel to the direction of travel of a wave, the wave is classifed as
   (1) longitudinal   (3) transverse
   (2) electromagnetic   (4) torsional

9. As the energy imparted to a mechanical wave increases, the maximum displacement of the particles in the medium
   (1) decreases   (2) increases   (3) remains the same

10. The water wave that will transfer the greatest amount of energy is the water wave that has the
    (1) highest frequency   (3) greatest amplitude
    (2) lowest frequency   (4) longest wavelength

11. A wave is generated in a rope, which is represented by the solid line in the diagram below. As the wave moves to the right, point P on the rope is moving toward which position?

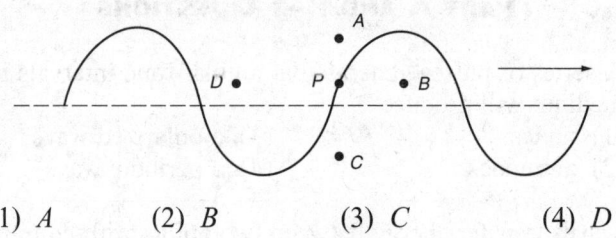

    (1) A (2) B (3) C (4) D

12. What is the amplitude of the wave represented in the diagram?

    (1) 1 m   (2) 2 m   (3) 3 m   (4) 6 m

13. Which distance represents the wavelength of the wave shown below?

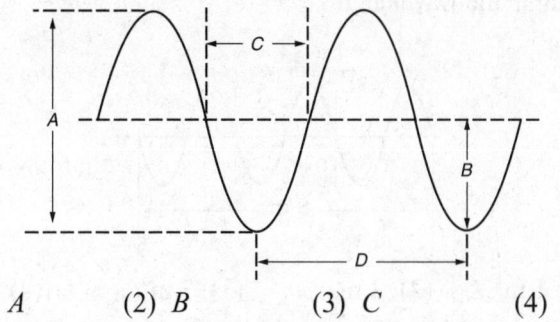

(1) *A*   (2) *B*   (3) *C*   (4) *D*

14. Which two wave representations in the diagram below have the same amplitude?

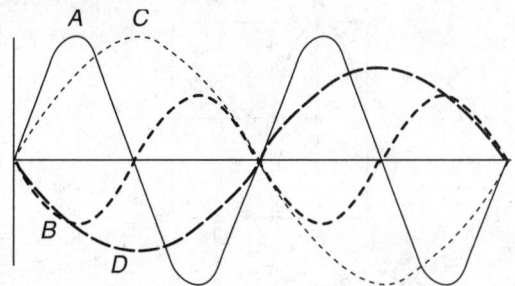

(1) *A* and *C*   (2) *A* and *B*   (3) *B* and *C*   (4) *B* and *D*

15. The graph below represents the displacement of a point in a medium as a function of time as a wave passes through the medium. What is the frequency of the wave?

(1) 1 Hz   (2) 2 Hz   (3) $\frac{1}{4}$ Hz   (4) 4 Hz

16. In the diagram below, a train of waves is moving along a string. What is the wavelength?

(1) 1 m   (2) 2 m   (3) 3 m   (4) 6 m

17. Which distance on the diagram below identifies the amplitude of the given wave?

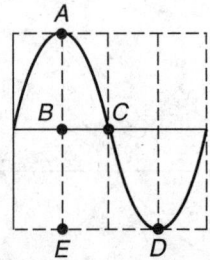

(1) *AE*   (2) *AB*   (3) *AC*   (4) *AD*

18. The number of water waves passing a given point each second is the wave's
(1) frequency (2) amplitude (3) wavelength (4) velocity

19. The wavelength of the periodic wave shown in the diagram below is 4.0 meters. What is the distance from point *B* to point *C*?

(1) 1.0 m   (2) 2.0 m   (3) 3.0 m   (4) 4.0 m

20. An electromagnetic wave traveling through a vacuum has a wavelength of 1.5×10^{-1} meter. What is the period of this electromagnetic wave?
(1) 5.0×10^{-10} s   (3) 4.5×10^{7} s
(2) 1.5×10^{-1} s   (4) 2.0×10^{9} s

21. As the period of a wave decreases, the wave's frequency
(1) decreases   (2) increases   (3) remains the same

22. Waves are traveling with a speed of 3 meters per second toward P, as shown in the diagram.

P

If three crests pass P in 1 second, the wavelength is
(1) 1 m   (2) 6 m   (3) 3 m   (4) 9 m

23. A source produces periodic waves with a frequency of f and a speed of v. The distance traveled by the wave in a time interval equal to one period of the wave is equal to

(1) $\dfrac{v}{f}$ (2) $\dfrac{f}{v}$ (3) $\dfrac{v}{2}$ (4) fv

24. A sound wave takes 1 second to travel from a source to observer A. How long does the same sound wave take to travel in the same medium to observer B, who is located twice as far from the source as observer A?

(1) $\dfrac{1}{4}$ s   (2) 2 s   (3) $\dfrac{1}{2}$ s   (4) 4 s

25. What is the period of a wave with a frequency of 250 hertz?
(1) 1.2×10^{-3} s   (3) 9.0×10^{-3} s
(2) 2.5×10^{-3} s   (4) 4.0×10^{-3} s

26. A wave has a frequency of 2.0 hertz and a velocity of 3.0 meters per second. The distance covered by the wave in 5.0 seconds is
(1) 30. m   (2) 15 m   (3) 7.5 m   (4) 6.0 m

27. As the amplitude of a periodic wave increases, its wavelength
(1) decreases   (2) increases   (3) remains the same

28. The number of water waves passing a given point each second is a measure of the wave's
(1) wavelength (2) amplitude (3) frequency (4) velocity

29. If the period of a wave is doubled, its wavelength will be
(1) halved   (2) doubled   (3) unchanged   (4) quartered

30. The speed of a transverse wave in a string is 10. meters per second. If the frequency of the source producing this wave is 2.5 hertz, what is its wavelength?
(1) 0.25 m   (2) 2.0 m   (3) 25 m   (4) 4.0 m

31. The frequency of a water wave is 6.0 hertz. If its wavelength is 2.0 meters, the speed of the wave is
    (1) 0.33 m/s   (2) 2.0 m/s   (3) 6.0 m/s   (4) 12 m/s

32. A radio station transmits waves with a wavelength of 30 meters. The frequency of the transmitted waves is
    (1) 1×10^7 Hz   (3) 3×10^9 Hz
    (2) 1×10^9 Hz   (4) 9×10^9 Hz

33. If the frequency of a light wave in a vacuum is 5.1×10^{14} hertz, what is its wavelength?
    (1) 5.9×10^{-7} m   (3) 1.5×10^{-7} m
    (2) 1.7×10^{-7} m   (4) 8.1×10^{-7} m

34. What is the distance between two consecutive points in phase on a wave called?
    (1) frequency   (2) period   (3) amplitude   (4) wavelength

35. As a wave travels into a medium in which its speed increases, its wavelength
    (1) decreases   (2) increases   (3) remains the same

36. If a wave has a frequency of 110. hertz, its period is
    (1) 9.09×10^{-4} s   (3) 1.00×10^{-1} s
    (2) 9.09×10^{-3} s   (4) 1.00×10^{1} s

Base your answers to questions 37 and 38 on the information below.

The frequency of a wave is 2.0 hertz, and its speed is 0.04 meter per second.

37. The period of the wave is
    (1) 0.005 s   (2) 2.0 s   (3) 0.50 s   (4) 0.02 s

38. The wavelength of the wave is
    (1) 1.0 m   (2) 0.02 m   (3) 0.08 m   (4) 4.0 m

39. Sound waves with a constant frequency of 250 hertz are traveling through air at STP. What is the wavelength of the sound waves?
    (1) 0.76 m   (2) 1.3 m   (3) 250 m   (4) 83,000 m

40. Wave X travels eastward with frequency f and amplitude A. Wave Y, traveling in the same medium, interacts with wave X and produces a standing wave. Which statement about wave Y is correct?
  (1) Wave Y must have a frequency of f, an amplitude of A, and be traveling eastward.
  (2) Wave Y must have a frequency of $2f$, an amplitude of $3A$, and be traveling eastward.
  (3) Wave Y must have a frequency of $3f$, an amplitude of $2A$, and be traveling westward.
  (4) Wave Y must have a frequency of f, an amplitude of A, and be traveling westward.

41. Maximum constructive interference will occur at points where the phase difference between two waves is
  (1) 0°   (2) 90°   (3) 180°   (4) 270°

42. Two pulses approach each other as shown.

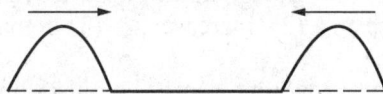

Which diagram best represents the wave formed when the two pulses meet?

(1)   (3)

(2)   (4)

43. The diagram below represents two waves traveling simultaneously in the same medium.

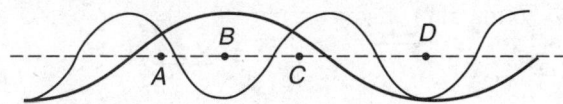

At which point will maximum constructive interference occur?
  (1) A (2) B (3) C (4) D

44. Two pulses are traveling along a string toward each other as represented in the diagram below.

Which phenomenon will occur as the pulses meet?
(1) reflection   (3) interference
(2) polarization   (4) refraction

45. Which phenomenon must occur when two or more waves pass simultaneously through the same region in a medium?
(1) refraction   (3) dispersion
(2) interference   (4) reflection

46. As the phase difference between two superposed waves changes from 180° to 90°, the amount of destructive interference
(1) decreases   (2) increases   (3) remains the same

47. The diagram below represents two pulses approaching each other from opposite directions in the same medium.

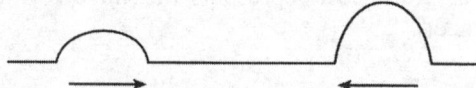

Which diagram best represents the medium after the pulses have passed through each other?

48. Maximum constructive interference between two waves of the same frequency could occur when their phase difference is

(1) λ (2) $\dfrac{\lambda}{2}$ (3) $\dfrac{3\lambda}{2}$ (4) $\dfrac{\lambda}{4}$

49. In the diagram below which point is in phase with point X?

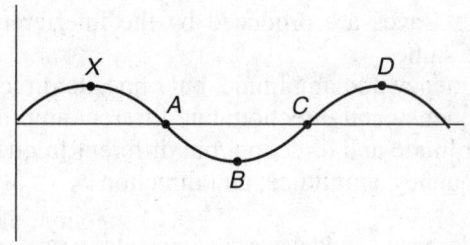

(1) A (2) B (3) C (4) D

50. The diagram below shows a rope with two waves moving along it in the directions shown.

What will be the resultant wave pattern at the instant when the maximum displacement of both pulses is at point O on the rope?

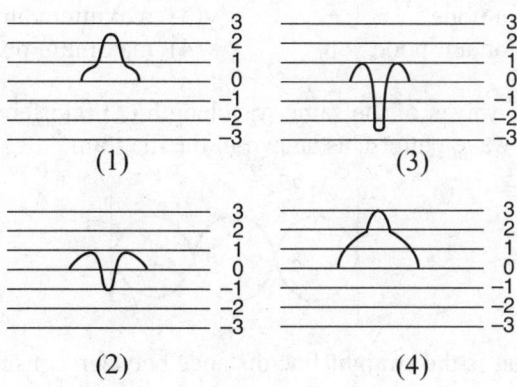

51. Maximum destructive interference will occur between two waves where their phase difference is

(1) 0°   (2) 90°   (3) 180°   (4) 270°

52. Two points on a periodic wave in a medium are said to be in phase
   if they
   (1) have the same amplitude, only
   (2) are moving in the same direction, only
   (3) have the same period
   (4) have the same amplitude and are moving in the same direction

53. Standing waves are produced by the interference of two waves
   with the same
   (1) frequency and amplitude, but opposite directions
   (2) frequency and direction, but different amplitudes
   (3) amplitude and direction, but different frequencies
   (4) frequency, amplitude, and direction

54. If two identical sound waves arriving at the same point are in phase,
   the resulting wave, compared to the original waves, will have
   (1) an increase in speed   (3) a larger amplitude
   (2) an increase in frequency   (4) a longer period

55. When the stretched string of the apparatus represented below is
   made to vibrate, point P does not move.

Point P is most probably at the location of
   (1) a node   (3) maximum amplitude
   (2) an antinode   (4) maximum pulse

56. Two waves of the same wavelength (λ) interfere to form a standing wave pattern, as shown in the diagram.

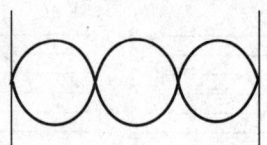

What is the straight-line distance between consecutive nodes?
   (1) 1λ (2) 2λ (3) $\frac{1}{2}\lambda$ (4) $\frac{1}{4}\lambda$

57. A car's horn is producing a sound wave having a constant frequency of 350 hertz. If the car moves toward a stationary observer at constant speed, the frequency of the car's horn detected by this observer may be
(1) 320 Hz   (2) 330 Hz   (3) 350 Hz   (4) 380 Hz

58. Which characteristic of a wave is always changed whenever a wave is reflected, refracted, or diffracted?
(1) wavelength   (3) speed
(2) period   (4) direction of travel

59. Which characteristic of a wave changes as the wave travels across a boundary between two different media?
(1) frequency   (2) period   (3) phase   (4) speed

60. When a pulse traveling in a medium strikes the boundary of a different medium, the energy of the pulse will be
(1) completely absorbed by the boundary
(2) entirely transmitted into the new medium
(3) entirely reflected back into the original medium
(4) partly reflected back into the original medium and partly transmitted or absorbed into the new medium

61. Standing waves can be produced in a vibrating rope because of the phenomenon of
(1) reflection   (2) refraction   (3) dispersion   (4) diffraction

62. Compared to the frequency of a source wave, the frequency of the echo as the wave reflects from a stationary object is
(1) smaller   (2) larger   (3) the same

63. As a wave enters a different medium with no change in velocity, the wave will be
(1) reflected but not refracted   (3) both reflected and refracted
(2) refracted but not reflected   (4) neither reflected nor refracted

64. A pulse traveling along a stretched spring is reflected from the fixed end. Compared to the pulse's speed before reflection, its speed after reflection is
(1) less   (2) greater   (3) the same

65. Which diagram best illustrates wave refraction?

66. In the diagram below, ray *AB* is incident on surface *XY* at point *B*. If the corresponding wave is traveling more rapidly in medium 2 than in medium 1, through which point will the ray most likely pass?

(1) *E*   (2) *F*   (3) *C*   (4) *D*

67. As a wave travels from one medium to another, its speed decreases. The ratio of the angle of incidence to the angle of refraction is
(1) equal to 0   (3) equal to 1
(2) less than 1, but greater than 0   (4) greater than 1

68. A car traveling at 70 kilometers per hour accelerates to pass a truck. When the car reaches a speed of 90 kilometers per hour the driver hears the glove compartment door start to vibrate. By the time the speed of the car is 100 kilometers per hour, the glove compartment door has stopped vibrating. This vibrating phenomenon is an example of
(1) the Doppler effect   (3) resonance
(2) diffraction   (4) destructive interference

69. Which wave phenomenon occurs when vibrations in one object cause vibrations in a second object?
(1) reflection   (3) intensity
(2) resonance   (4) tuning

328

70. The diagram below represents straight wave fronts passing from deep water into shallow water, with a change in speed and direction.

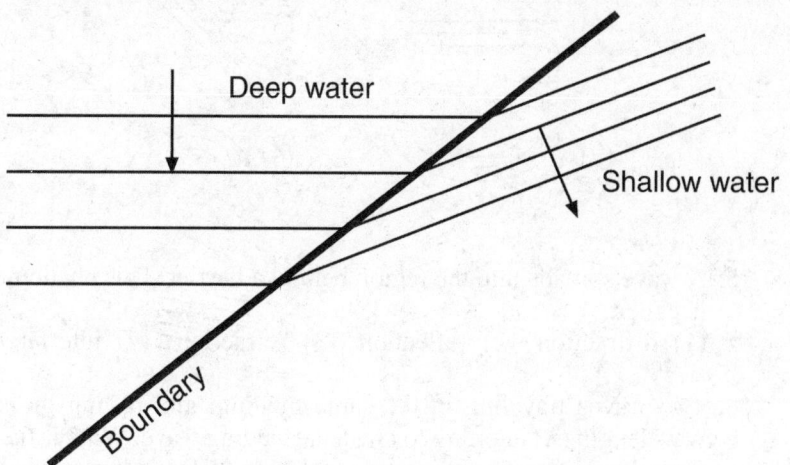

Which phenomenon is illustrated in the diagram?
(1) reflection   (3) diffraction
(2) refraction   (4) interference

71. A girl on a swing may increase the amplitude of the swing's oscillations if she moves her legs at the natural frequency of the swing. This is an example of
(1) the Doppler effect   (3) wave transmission
(2) destructive interference (4) resonance

72. If a ray is bent away from the normal when entering a new medium, the speed of its wave has
(1) decreased (2) increased (3) remained the same

73. Refraction of a wave is caused by a change in the wave's
(1) amplitude (2) frequency (3) phase   (4) speed

74. The diagram below represents straight wave fronts approaching an opening in a barrier.

Which diagram best represents the shape of the waves after passing through the opening?

75. A wave spreads into the region behind a barrier. This phenomenon is called
(1) diffraction (2) reflection (3) refraction (4) interference

76. Two waves traveling in the same medium and having the same wavelength (λ) interfere to create a standing wave. What is the distance between two consecutive nodes on this standing wave?
(1) λ　　　　(2) $\dfrac{3\lambda}{4}$　　　　(3) $\dfrac{\lambda}{2}$　　　　(4) $\dfrac{\lambda}{4}$

77. An observer detects an apparent change in the frequency of sound waves produced by an airplane passing overhead. This phenomenon illustrates
(1) the Doppler effect　　　　(3) an increase in wave amplitude
(2) the refraction of sound waves (4) an increase in wave intensity

78. A girl moves away from a source of sound at a constant speed. Compared to the frequency of the sound wave produced by the source, the frequency of the sound wave heard by the girl is
(1) lower　　(2) higher　　(3) the same

79. An Earth satellite in orbit emits a radio signal of constant frequency. Compared to the emitted frequency, the frequency of the signal received by a stationary observer will appear to be
(1) higher as the satellite approaches
(2) higher as the satellite moves away
(3) lower as the satellite approaches
(4) unaffected by the satellite's motion

Base your answers to questions 80 through 83 on the diagram below which represents the wave pattern produced by a vibrating source moving linearly in a shallow tank of water. The pattern is viewed from above, and the lines represent wave crests.

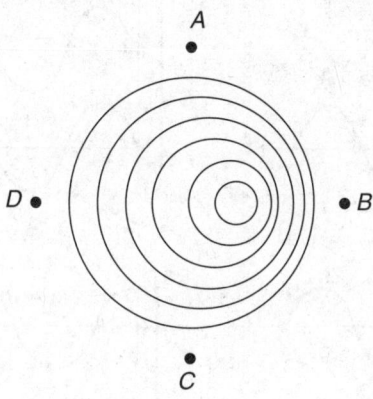

80. The source is moving toward point

(1) *A*   (2) *B*   (3) *C*   (4) *D*

81. The wave pattern is an illustration of
   (1) diffraction   (3) dispersion
   (2) interference   (4) the Doppler effect

82. Compared to the frequency of the waves observed at point *D*, the frequency of the waves observed at point *B* is
   (1) lower   (2) higher   (3) the same

83. The velocity of the source is increased. The wavelength of the waves observed at point *D* will
   (1) decrease   (2) increase   (3) remain the same

The answers to questions 84 through 88 are to be chosen from the four sets of wave diagrams below. Each diagram represents a graph of the amplitude of a periodic wave as a function of time. The amplitude and time scales for all graphs are identical.

A

C

B

D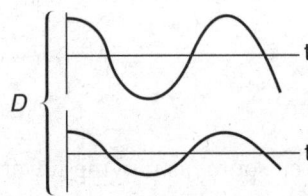

84. Which set shows two waves of unequal frequency?
 (1) *A*   (2) *B*   (3) *C*   (4) *D*

85. Which set of waves would produce the greatest constructive interference?
 (1) *A*   (2) *B*   (3) *C*   (4) *D*

86. Which set of waves shows equal wavelength but different phase?
 (1) *A*   (2) *B*   (3) *C*   (4) *D*

87. Which set shows two waves of equal amplitude that are in phase?
 (1) *A*   (2) *B*   (3) *C*   (4) *D*

88. Which set of waves shows unequal amplitude but the same period?
 (1) *A*   (2) *B*   (3) *C*   (4) *D*

Base your answers to questions 89 through 92 on your knowledge of physics and on the diagram below, which shows the water waves produced by two sources, X and Y. Three waves are produced by each source every second, and the semicircles represent wave crests.

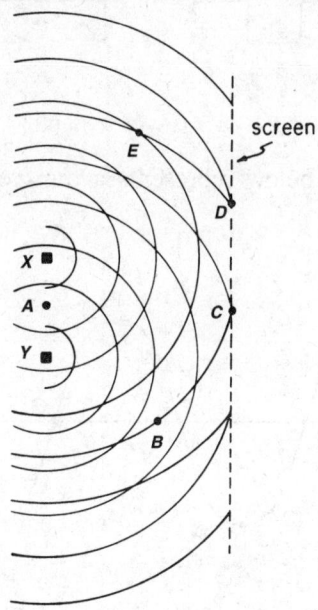

89. A point where destructive interference occurs is
(1) *E*   (2) *B*   (3) *C*   (4) *D*

90. The distance between points *X* and *D* differs from the distance between points *Y* and *D* by how many wavelengths?

(1) 1   (2) 2   (3) $\frac{1}{2}$ (4) $1\frac{1}{2}$

91. The period of the waves is
(1) 0.33 s   (2) 2.5 s   (3) 3.0 s   (4) 4.0 s

92. If the wave travels into an area of different water depth, there will be a change in the wave's
(1) velocity, only   (3) frequency and velocity
(2) wavelength, only   (4) wavelength and velocity

93. The diagram below represents two pulses approaching each other.

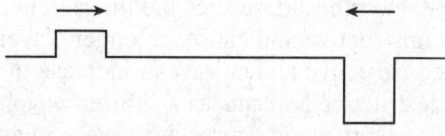

Which diagram best represents the resultant pulse at the instant the pulses are passing through each other?

(1)   (2)   (3)   (4)

94. The diagram below represents a transverse wave traveling in a string.

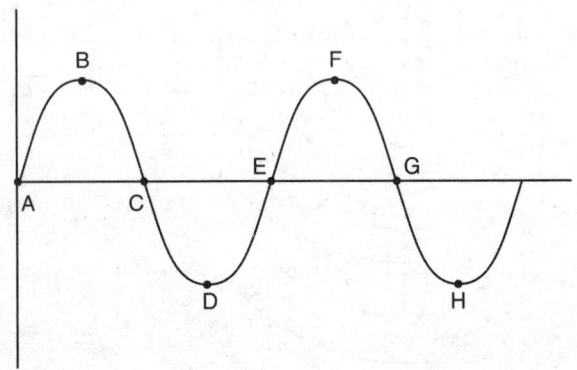

Which two labeled points are 180° out of phase?
(1) *A* and *D*   (2) *B* and *F*   (3) *D* and *F*   (4) *D* and *H*

95. The diagram below represents shallow water waves of constant wavelength passing through two small openings, *A* and *B*, in a barrier.

——— Crest
- - - - Trough

Which statement best describes the interference at point *P*?
(1) It is constructive and causes a longer wavelength.
(2) It is constructive and causes an increase in amplitude.
(3) It is destructive and causes a shorter wavelength.
(4) It is destructive and causes a decrease in amplitude.

Base your answers to questions 96 through 98 on the diagram below, which represents periodic water waves in a ripple tank. The speed of a wave decreases as it moves from the deep to the shallow portion of the tank.

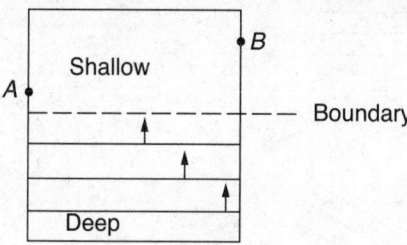

96. Which drawing best represents the waves after they enter the shallow section?

97. Which diagram best represents the pattern produced when the waves are reflected from the boundary between the deep and shallow sections of the ripple tank?

98. If a barrier is placed in the ripple tank connecting points A and B, which drawing best represents the waves after reflection from the barrier?

Base your answers to questions 99 through 103 on the diagram and information below.

Two speakers are arranged as shown so that initially they will emit tones that are in phase, equal in volume, and equal in frequency. A microphone is placed at position A, which is equidistant from both speakers, and then is moved along a line parallel to the line joining the speakers until no sound is heard. [This is position B.] The microphone is then moved to position C, where sound is again picked up by the microphone.

99. Which phenomenon caused the sound to be louder at position C than at position B?
(1) reflection (2) dispersion (3) polarization (4) interference

100. Distance D_2 is shorter than distance D_1 by an amount equal to
(1) the wavelength of the emitted sound
(2) one-half the wavelength of the emitted sound
(3) twice the wavelength of the emitted sound
(4) the distance between the two speakers

336

101. If the sound waves emitted by D_1 and D_2 have a frequency of 660 hertz and a speed of 330 meters per second, their wavelength is
(1) 1.0 m   (2) 2.0 m   (3) 0.25 m   (4) .50 m

102. As the first speaker is adjusted so that the sound that it emits is 180° out of phase with the sound emitted by the second speaker, the loudness of the sound received at A is
(1) greater   (2) less   (3) the same

103. If speaker D_1 were removed and speaker D_2 were accelerated toward microphone B, the frequency of the waves detected at B would
(1) decrease   (2) increase   (3) remain the same

Base your answers to questions 104 through 108 on the information and diagram below. The diagram represents two sound waves that are produced in air by two tuning forks. The frequency of wave A is 400 hertz.

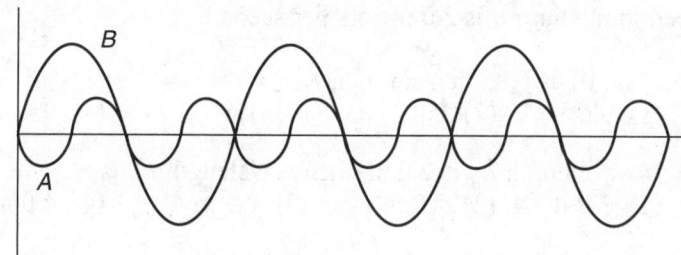

104. Under standard conditions of temperature and pressure, the wavelength in air of A is
(1) 2.5 m   (2) 12 m   (3) $\dfrac{331}{400}$ m   (4) 331×400 m

105. The frequency of wave B is
(1) 200 Hz   (2) 400 Hz   (3) 600 Hz   (4) 800 Hz

106. Sound waves produced by tuning forks are
(1) longitudinal   (2) hyperbolic   (3) torsional   (4) elliptical

107. Compared to the amplitude of wave B, the amplitude of wave A is
(1) less   (2) greater   (3) the same

108. Compared to the speed of wave A, the speed of wave B is
(1) less   (2) greater   (3) the same

Base your answers to questions 109 and 110 on the diagram and information below.

The diagram represents wave fronts traveling from medium A to medium B at boundary XY. The wave speed in medium A is 4.0 meters per second. The wave speed in medium B is 2.0 meters per second.

109. If $\sin \theta_A$ is 0.8, then $\sin \theta_B$ is
   (1) 0.5   (2) 0.8   (3) 1.6   (4) 0.4

110. If wavelength λ_A is 2.0 m, then wavelength λ_B is
   (1) 1.0 m   (2) 2.0 m   (3) 0.5 m   (4) 4.0 m

Answers to Part A and B–1 questions can be found on page 458.

PART B–2 AND C QUESTIONS

1. Base your answers to parts *a* and *b* on the information and diagram below.

   The sonar of a stationary ship sends a signal with a frequency of 5.0 $\times 10^3$ hertz down through water. The speed of the signal is 1.5×10^3 meters per second. The echo from the bottom is detected 4.0 seconds later.

KEY
〰 Incident Waves
╌ Reflected Waves

a What is the wavelength of the sonar wave? [Show all calculations, including the equation and substitution with units.]

b What is the depth of the water under the ship? [Show all calculations, including the equation and substitution with units.]

Base your answers to questions 2 through 4 on the information and diagram below.

A barrier is placed in the ripple tank as shown in the diagram below.

2. *On the diagram,* use a protractor and a straightedge to construct an arrow to represent the direction of the velocity of the reflected waves.

3. Determine the speed of the waves in the ripple tank. [Show all calculations, including the equation and substitution with units.]

4. Using one or more complete sentences, state the Law of Reflection.

Base your answers to questions 5 through 7 on the information and diagram below.

Two waves, *A* and *B*, travel in the same direction in the same medium at the same time.

5. On the grid provided, draw the resultant wave produced by the superposition of waves *A* and *B*.

Position in meters

6. What is the amplitude of the resultant wave?

7. What is the wavelength of the resultant wave?

Base your answers to questions 8 and 9 on the information below.

A periodic transverse wave has an amplitude of 0.20 meter and a wavelength of 3.0 meters.

8. On the grid provided below, draw at least one cycle of this periodic wave.

```
0.20 m  ─ ─ ─ ─ ─ ─ ─ ─ ─ ─ ─ ─ ─ ─ ─ ─ ─ ─ ─ ─

0.10 m  ─ ─ ─ ─ ─ ─ ─ ─ ─ ─ ─ ─ ─ ─ ─ ─ ─ ─ ─ ─

0.00 m  ─ ─ ─ ─ ─ ─ ─ ─ ─ ─ ─ ─ ─ ─ ─ ─ ─ ─ ─ ─
           1.0 m      2.0 m      3.0 m      4.0 m
-0.10 m  ─ ─ ─ ─ ─ ─ ─ ─ ─ ─ ─ ─ ─ ─ ─ ─ ─ ─ ─

-0.20 m  ─ ─ ─ ─ ─ ─ ─ ─ ─ ─ ─ ─ ─ ─ ─ ─ ─ ─ ─
```

9. If the frequency of this wave is 12 Hz, what is its speed?
   (1) 0.25 m/s   (2) 12 m/s   (3) 36 m/s   (4) 4.0 m/s

Base your answers to questions 10 through 13 on the information and diagram below.

A wave generator having a constant frequency of 15 hertz produces a standing wave pattern in a stretched string.

Fixed support

Wave generator

340

10. Using a ruler, measure the amplitude of the wave shown. Record the value to the *nearest tenth of a centimeter*.

11. Using a ruler, measure the wavelength of the wave shown. Record the value to the *nearest tenth of a centimeter*.

12. State what would happen to the wavelength of the wave if the frequency of the wave were increased.

13. How many antinodes are shown in the diagram?

14. The diagram below represents a transverse wave moving on a uniform rope with point *A* labeled as shown. On the diagram below, mark an **X** at the point on the wave that is 180° out of phase with point *A*.

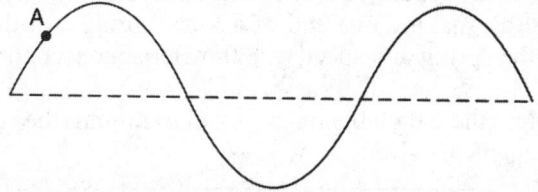

15. The diagram below represents a periodic transverse wave traveling in a uniform medium.

On the diagram above, draw a wave having *both* a smaller amplitude and the same wavelength as the given wave.

Base your answers to questions 16 and 17 on the information and diagram at the top of page 342.

A 1.50×10^{-6}-meter-long segment of an electromagnetic wave having a frequency of 6.00×10^{14} hertz is represented as follows.

— 1.50 × 10⁻⁶ m —

16. On the diagram above, mark *two* points on the wave that are in phase with each other. Label each point with the letter *P*.

17. According to the *Reference Tables for Physical Setting/Physics*, which type of electromagnetic wave does the segment in the diagram represent?

Base your answers to questions 18 through 19 on the information below.

A student generates a series of transverse waves of varying frequency by shaking one end of a loose spring. All the waves move along the spring at a speed of 6.0 meters per second.

18. Complete the data table below, by determining the wavelengths for the frequencies given.

Data Table	
Frequency (Hz)	Wavelength (m)
1.0	
2.0	
3.0	
6.0	

19. Plot the data points for wavelength versus frequency.

20. Draw the best-fit line or curve.

21. The graph below represents the relationship between wavelength and frequency of waves created by two students shaking the ends of a loose spring.

Wavelength vs. Frequency

Calculate the speed of the waves generated in the spring. [Show all work, including the equation and substitution with units.]

22. The diagram below represents a transverse wave, *A*, traveling through a uniform medium. On the diagram, draw a wave traveling through the same medium as wave *A* with twice the amplitude and twice the frequency of wave *A*.

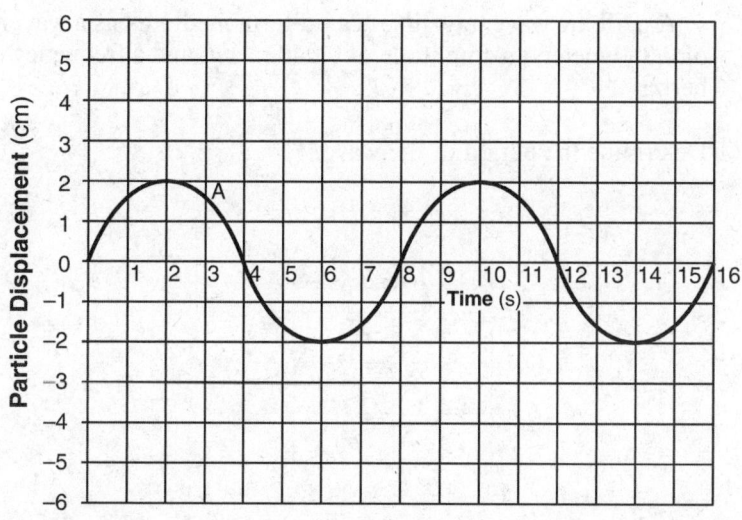

Base your answers to questions 23 and 24 on the information and diagram below.

A student standing on a dock observes a piece of wood floating on the water as shown below. As a water wave passes, the wood moves up and down, rising to the top of the wave crest every 5.0 seconds.

(Not drawn to scale)

23. Calculate the frequency of the passing water waves. [Show all work, including the equation and substitution with units.]

24. Calculate the speed of the water waves. [Show all work, including the equation and substitution with units.]

25. An FM radio station broadcasts its signal at a frequency of 9.15×10^7 hertz. Determine the wavelength of the signal in the air.

Base your answers to questions 26 through 28 on the information below.

A periodic wave traveling in a uniform medium has a wavelength of 0.080 meter, an amplitude of 0.040 meter, and a frequency of 5.0 hertz.

26. Determine the period of the wave.

27. On the grid below, starting at point *A*, sketch a graph of *at least one* complete cycle of the wave showing its amplitude and period.

Particle Displacement vs. Time

28. Calculate the speed of the wave. [Show all work, including the equation and substitution with units.]

Answers to Part B–2 and C questions can be found on pages 523–531.

LIGHT AND GEOMETRIC OPTICS

Chapter Twelve

KEY IDEAS

Visible light is part of the electromagnetic spectrum of waves. Electromagnetic waves are transverse and have a constant speed in space. Because light is a periodic wave, it possesses the characteristics and properties of all periodic waves: reflection, refraction, interference, and diffraction, and it exhibits the Doppler effect.

The principal applications of reflected light involve the use of plane and curved mirrors. Lenses, prisms, and fiber-optic bundles are applications of refraction of light.

Diffraction and interference can be demonstrated by passing light through a single- or double-slit arrangement. These devices can be used to measure the wavelength of light. Interference also occurs with the reflected light from thin films and is responsible for the colors seen on soap bubbles and oil slicks.

If monochromatic light is generated so that all of the waves have a constant phase relationship, the light is said to be coherent. Lasers produce intense beams of coherent light.

KEY OBJECTIVES
At the conclusion of this chapter you will be able to:
- Define the term *polarization*, and explain why polarization distinguishes between transverse and longitudinal waves.
- Explain how reflection of light produces an image in a plane mirror, and describe the characteristics of such an image.
- Explain how light refracts as it passes from one medium to another.
- Define the term *absolute index of refraction*, and solve problems using this concept.
- State Snell's law in terms of absolute indices of refraction, and solve problems using this equation.
- Define the terms *critical angle* and *total internal reflection,* and relate them to Snell's law.
- Define the term *dispersion*.
- Describe the patterns produced when monochromatic light passes through a double-slit arrangement, and explain how these patterns are formed.

- Explain the difference between the pattern produced by a double-slit arrangement and that produced by a single-slit arrangement.
- Define the term *laser*, and explain how laser light differs from ordinary light.

12.1 INTRODUCTION

Light is an electromagnetic wave. The different forms of light constitute the *electromagnetic spectrum*, diagramed below, of which visible light is only a very small part.

Electromagnetic waves differ in their frequency and in the sources used to produce them.

✪ 12.2 POLARIZATION OF LIGHT

Polarization is the separation of a beam of light so that the vibrations are in one plane. It is an exclusive property of transverse waves. We know that light is a transverse wave because it can be *polarized*. When a light wave is produced, it vibrates in many directions, as shown below.

Unpolarized Light Beam

✪ indicates material that is not part of the core curriculum.

If, however, a beam of light passes through a polarizing filter, the beam that emerges will vibrate in one plane only and is said to be *plane polarized*.

Unpolarized Light Beam

Plane-polarized
Light Beam

Since longitudinal waves, such as sound, vibrate parallel to the direction of motion, they cannot be polarized; therefore, polarization distinguishes between transverse and longitudinal waves. If a second polarizing filter is placed at a right angle to the plane of polarized light it will block nearly all of the light, as shown in the diagram.

Unpolarized Light Beam

Plane-polarized
Light Beam

No Light

When light is reflected from a nonmetallic surface, it is polarized. Polarized sunglasses can be used to eliminate the glare associated with this type of reflection.

12.3 SPEED OF LIGHT

Anyone who has observed lightning or fireworks from a distance knows that the flash of light appears nearly instantaneously, while the explosion is heard somewhat later. We know that sound travels in air at approximately 330 meters per second (about 750 mi/h), but how fast does light travel?

This question was first answered in 1675 when the Dutch astronomer Olaus Roemer used his observations of Jupiter and the eclipse of one of its moons to measure the speed of light. In the nineteenth century the American physicist Albert Michaelson used sunlight and rotating mirrors to obtain more precise measurements.

As a result of Einstein's special theory of relativity, discussed in Chapter 15, we know that the speed of light in a vacuum is constant under all circumstances. It has been set at the value 2.99792458×10^8 meters per second (approximately 186,000 mi/s). The letter c is used to represent the speed of light in a vacuum.

The speed of light is less in a material medium than in a vacuum and depends on the nature of the medium and the frequency of the light.

12.4 VISIBLE LIGHT

The visible portion of the electromagnetic spectrum ranges from red to violet. The following table indicates the approximate ranges for visible light in a vacuum.

Wavelengths of Light in a Vacuum	
Violet	$4.0 - 4.2 \times 10^{-7}$ m
Blue	$4.2 - 4.9 \times 10^{-7}$ m
Green	$4.9 - 5.7 \times 10^{-7}$ m
Yellow	$5.7 - 5.9 \times 10^{-7}$ m
Orange	$5.9 - 6.5 \times 10^{-7}$ m
Red	$6.5 - 7.0 \times 10^{-7}$ m

Monochromatic light consists of light of a single color, that is, light of a single wavelength (or frequency). If all the colors of visible light are mixed together, the result is white light. Black is the complete absence of visible light.

12.5 REFLECTION

Law of Reflection

When light is reflected from a surface, the angle that the incident ray makes with the normal to the surface is equal to the angle that the reflected ray makes with the normal to the surface. This statement is known as the *law of reflection*.

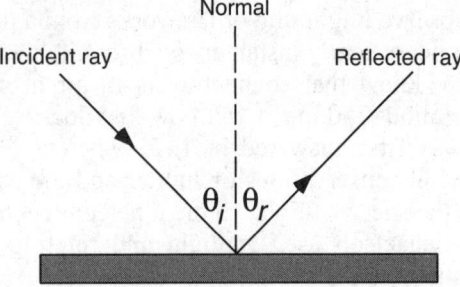

Polished surfaces such as plane mirrors produce *regular* reflection.

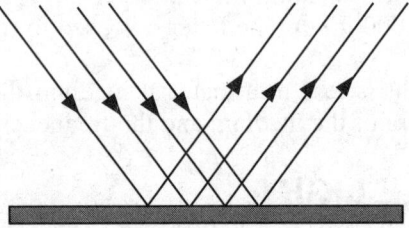

An irregular surface such as a windblown water surface or the paper on which this book is printed produces *diffuse* reflection.

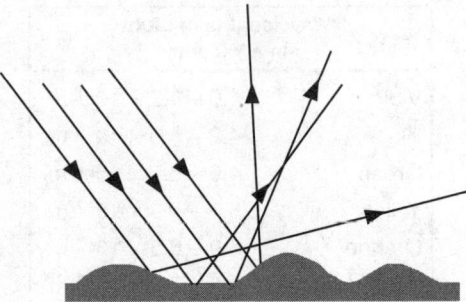

We note that, while the reflected rays emerge at different angles in the diagram above, the law of reflection holds for each individual pair of incident and reflected rays.

Mirrors

One primary application of reflected light is the image formed by a mirror. Mirrors may be plane or curved.

PLANE MIRRORS

When an object is viewed in a plane mirror, the image that is formed is erect (upright), left-right reversed, and the same size as the object. The object and image distances from the mirror are equal. These relationships can be proved by using the law of reflection and simple geometry.

The image produced by a plane mirror is *virtual* because light does not actually pass through the mirror to form the image; it only appears to be located inside the mirror.

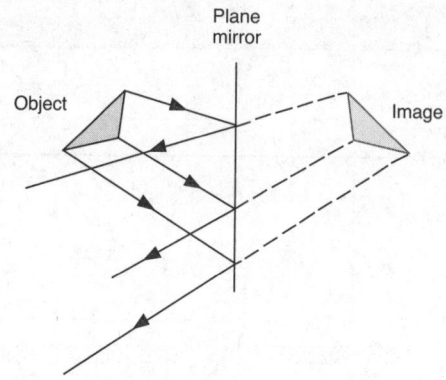

This phenomenon occurs because light appears to travel in straight lines, and a device such as a human eye searches for the apparent origin of the light.

PROBLEM

A beam of light enters and exits a hollow rectangular box. How could plane mirrors be placed in the box to produce the following patterns?

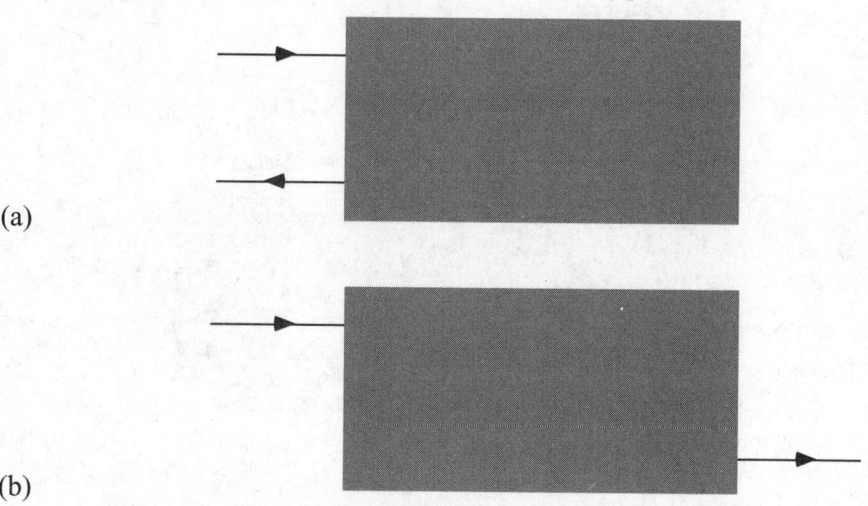

(a)

(b)

SOLUTION
The problem in each case is solved by orienting two plane mirrors at 45° angles in the upper and lower right corners of the box. Since the angles of incidence are all 45°, the incident and reflected rays will be perpendicular to one another, as illustrated in the diagrams below.

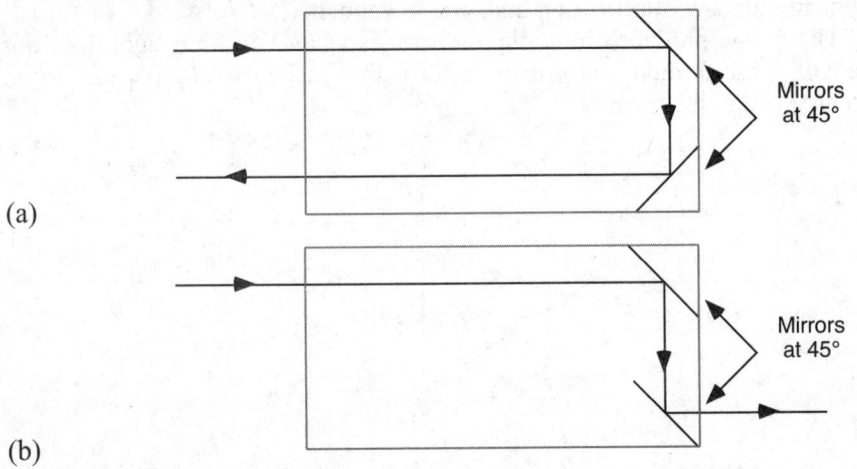

(a)

(b)

12.6 REFRACTION

When monochromatic light travels between two media, there is a change in the speed of the light wave. If the light enters at an oblique angle, it will change direction as it passes into the second medium.

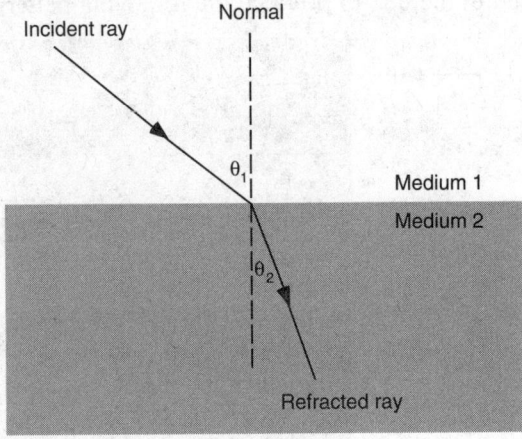

$$\frac{\sin \theta_1}{\sin \theta_2} = \frac{v_1}{v_2}$$

*Note that this equation does not appear as seen here on the Reference Tables.

A result of the relationship expressed in this equation (Snell's law) is that the slower the speed, the smaller the angle.

PROBLEM
Monochromatic light passes between two media. If the angle in medium 1 is 45° and the angle in medium 2 is 30.°, calculate the ratio of the light speeds between media 1 and 2.

SOLUTION

$$\frac{\sin \theta_1}{\sin \theta_2} = \frac{v_1}{v_2}$$

$$\frac{v_1}{v_2} = \frac{\sin 45°}{\sin 30.°} = 1.4$$

$$v_1 = 1.4v_2$$

Absolute Index of Refraction

To simplify refraction problems, a quantity known as the **absolute index of refraction** is defined as follows:

$$n = \frac{c}{v}$$

The equation states that the absolute index of refraction of a medium (n) is the ratio of the speed of light in a vacuum (c) to the speed of light in a medium (v). The absolute index of refraction is always greater than or equal to 1. The larger the index of refraction, the slower the speed of light in a medium.

PROBLEM
The speed of light in a medium is 2.4×10^8 meters per second. What is the absolute index of refraction of the medium?

SOLUTION

$$n = \frac{c}{v}$$

$$= \frac{3.0 \times 10^8 \text{ m/s}}{2.4 \times 10^8 \text{ m/s}}$$

$$= 1.25$$

The table below lists the indices of refraction for some common materials.

Absolute Indices of Refraction ($\lambda = 5.9 \times 10^{-7}$ m)	
Air	1.00
Alcohol (Ethyl)	1.36
Benzene	1.50
Canada Balsam	1.53
Carbon Tetrachloride	1.46
Corn Oil	1.47
Diamond	2.42
Glass, Crown	1.52
Glass, Flint	1.66
Glycerol	1.47
Lucite	1.50
Quartz, Fused	1.46
Sodium Chloride	1.54
Water	1.33
Zircon	1.92

Snell's Law

In terms of the index of refraction, the relationship

$$\frac{\sin \theta_1}{\sin \theta_2} = \frac{v_1}{v_2}$$

becomes

$$\frac{\sin \theta_1}{\sin \theta_2} = \frac{\dfrac{c}{n_1}}{\dfrac{c}{n_2}} = \frac{n_2}{n_1}$$

We write this as follows:

$$n_1 \sin \theta_1 = n_2 \sin \theta_2$$

This relationship is an alternate form of Snell's law.

From the equations above, and the proportional relationship between the velocity of each wave and its wavelength, we can also write an equation to express the relationship between the speed of light and index of refraction in two mediums.

$$\frac{n_2}{n_1} = \frac{v_1}{v_2} = \frac{\lambda_1}{\lambda_2}$$

PROBLEM

A monochromatic light ray is incident on a surface boundary between air and corn oil at an angle of 60.° to the normal. Calculate the refracted angle of the ray in the corn oil.

SOLUTION

$$n_1 \sin \theta_1 = n_2 \sin \theta_2$$

$$\sin \theta_2 = \frac{n_1 \sin \theta_1}{n_2}$$

$$= \frac{(1)(\sin 60.°)}{1.47} = 0.59$$

$$\theta_2 = \sin^{-1}(0.59) = 36°$$

PROBLEM

A ray of monochromatic light in air is incident on a transparent block of material whose absolute index of refraction is 1.50, as shown in the diagram below.

If the angle of incidence is 30°, trace the path of the light ray through the block and back into the air.

SOLUTION

We begin the solution by applying Snell's law in order to calculate the angle of refraction:

$$n_{air} \sin \theta_{air} = n_{block} \sin \theta_{block}$$

$$(1.00)(\sin 30°) = (1.50)(\sin \theta_{block})$$

$$\sin \theta_{block} = 0.333$$

$$\theta_{block} = 19.5°$$

We now extend the ray into the block at 19.5°. Using geometry (alternate interior angles), we find that the ray reaches the second surface at an angle of 19.5°.

If we applied Snell's law again, we would conclude that the ray would emerge in air at the original angle of 30°, as illustrated below.

The tendency of light to emerge back into the air at the same angle to the normal as it entered into the medium as in the example above is referred to as **parallelism of light** and occurs whenever light passes from air into a more dense medium with parallel sides and comes back out into air.

☢ Critical Angle

If monochromatic light passes from medium 1 in which its speed is slower to medium 2 in which its speed is faster, we can draw a ray diagram as follows:

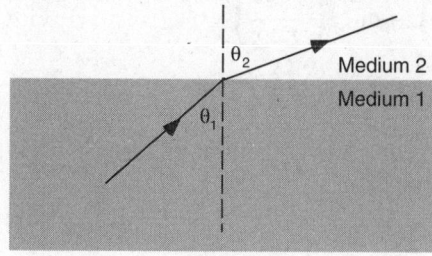

As angle θ_1 is made larger, angle θ_2 also becomes larger ($n_1 \sin \theta_1 = n_2 \sin \theta_2$).

There is one unique angle in medium 1 that will produce an angle of 90° in medium 2, as illustrated.

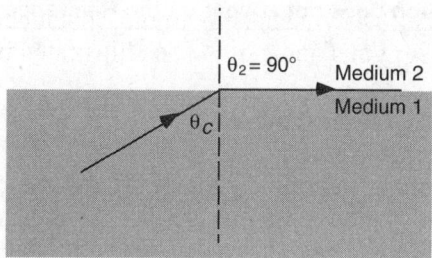

This unique angle is called the **critical angle** (θ_c). We can derive an equation for the critical angle using Snell's law:

$$n_1 \sin \theta_c = n_2 \sin 90°$$

Then, since $\sin 90° = 1$,

$$n_1 \sin \theta_c = n_2$$

=== **PHYSICS CONCEPTS** ===

$$\sin \theta_c = \frac{n_2}{n_1}$$

Note that this equation does not appear on the Reference Tables.

PROBLEM
Calculate the critical angle between diamond and Lucite.

☢ indicates material that is not part of the core curriculum.

SOLUTION

$$\sin \theta_c = \frac{n_2}{n_1}$$

$$\theta_c = \sin^{-1}\left(\frac{n_2}{n_1}\right) = \sin^{-1}\left(\frac{1.50}{2.42}\right)$$

$$= 38°$$

In the event that medium 2 is a vacuum or air ($n_2 = 1$), the above equation simplifies to this expression:

PHYSICS CONCEPTS

$$\sin \theta_c = \frac{1}{n_1}$$

---

*Note that this equation does not appear on the Reference Tables.

○ Total Internal Reflection

What is so special about the critical angle? At 90° the maximum refracted angle possible, light would skim the surface boundary between the two media. If the critical angle were exceeded, the light could no longer escape but would be reflected back into the medium. In this case, the law of reflection would hold as with any other pair of incident and reflected rays.

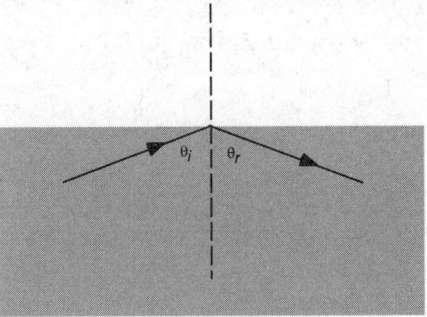

We note that the equation $\sin \theta_c = 1/n_1$, given above, implies that larger values of n yield smaller critical angles. A material such as diamond ($n = 2.42$) has a critical angle of only 24°. Because of this small value, much of the light inside a diamond is totally internally reflected. This **total internal reflection** is responsible for much of the diamond's sparkle.

○ indicates material that is not part of the core curriculum.

PROBLEM

A monochromatic ray of light in air enters a triangular prism whose absolute index of refraction is 1.50. The prism is in the shape of an isosceles right triangle, and the incident ray is perpendicular to the surface as shown in the diagram.

Complete the path of the light ray as it enters the prism.

SOLUTION

We begin by noting that the angle of incidence is 0°; consequently the angle of refraction will also be 0°. The ray will enter the prism without bending and will make an angle of 45° as it strikes the second surface (hypotenuse) of the prism.

We now must ask: Will the critical angle of the block-air combination be exceeded?

$$\sin \theta_c = \frac{n_{air}}{n_{block}}$$

$$= \frac{1.00}{1.50} = 0.667$$

$$\theta_c = 41.8°$$

Since the angle of the ray is 45°, the critical angle is exceeded and total internal reflection results. The ray continues through the prism and exits into the air as shown below.

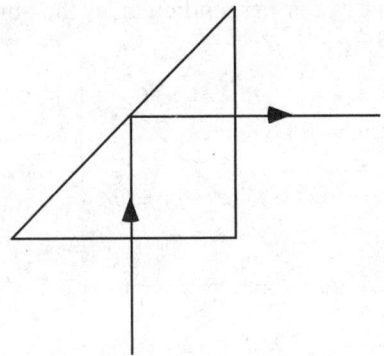

✪ Dispersion

If white (polychromatic) light is passed through a prism, it is separated into its component colors as shown in the diagram.

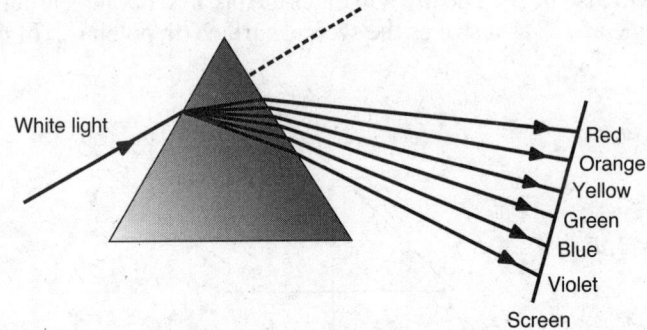

This phenomenon is known as **dispersion**. When water droplets in the air disperse light, rainbows may be produced.

Dispersion occurs because the speed of light inside the prism depends on the color of the light. The index of refraction is different for each component color of the light. From the diagram we can see that red light travels fastest because it has the largest refracted angle (that is, it is bent the least).

12.7 DIFFRACTION AND INTERFERENCE OF LIGHT

The Double-Slit Experiment

If light is passed through a pair of closely spaced slits, an alternating pattern of bright and dark bands will appear on a distant screen, as shown below.

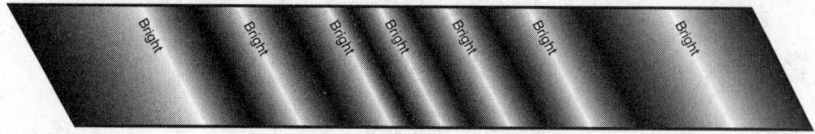

To explain this phenomenon we must first consider how a wave can travel from one point to another. Danish scientist Christian Huygens developed a principle that states that every point on a wave front can be considered a point source of secondary spherical waves (called *wavelets*). After a time, the new location of the wave is found by drawing a common tangent to these wavelets.

Source
•

In the double-slit arrangement illustrated on page 362, each slit serves as a point source of waves. We see that, as the waves grow, the wave fronts interfere with one another. Points that interfere constructively will ultimately produce the bright bands, while points that interfere destructively will produce the dark bands, as shown in the diagram on the next page.

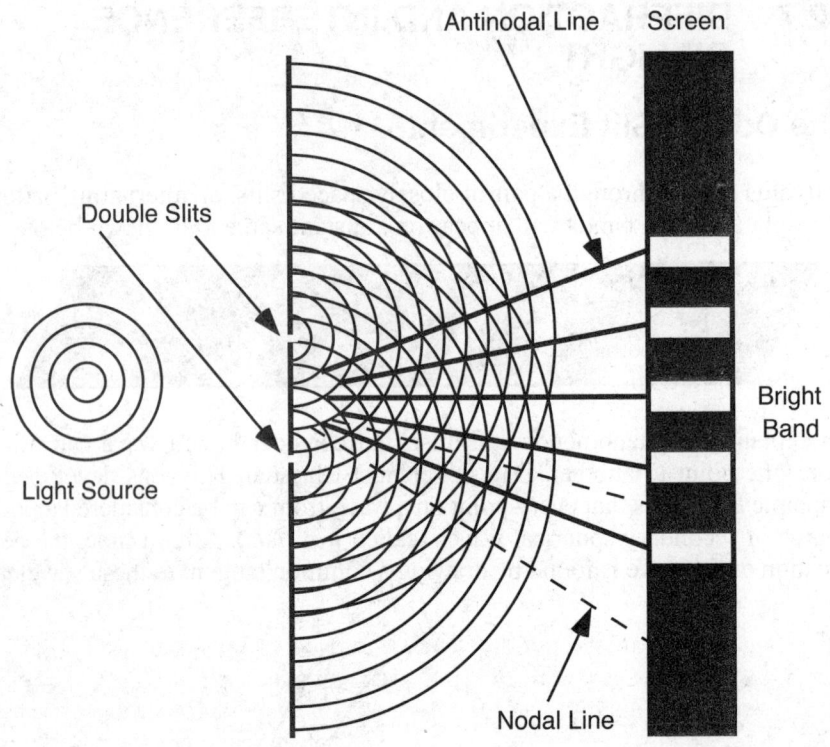

○ Single-Slit Diffraction

A single slit can also be used to produce a diffraction pattern. The width of the slit must have the same order of magnitude as the wavelength of the light used. The pattern produced by single-slit diffraction is different from the pattern obtained with double-slit diffraction in two respects: (1) the central bright band is much wider than any of the other bright bands, and (2) the intensity of the central band is greater than the intensity of any of the other bright bands.

★ indicates material that is not part of the core curriculum.

✪ Lasers

When light (even monochromatic light) is emitted by a source, the light waves have no special (phase) relationship with one another.

As a result, the light beam spreads and loses its intensity (brightness) quickly. This light is said to be *incoherent*.

A **laser** is a device that produces monochromatic light in which nearly all of the waves are in phase.

The word *laser* is an acronym for light amplification by stimulated emission of radiation. The beam produced by a laser spreads very little and is extraordinarily intense. This light is called *coherent*. Because of its properties, laser light can be used for a variety of applications, from bar-code scanning to eye surgery.

The color of laser light depends on the sources that produce the light. Helium-neon lasers, for example, produce red light.

PART A AND B–1 QUESTIONS

✪ 1. Which phenomenon is associated only with transverse waves?
(1) interference (2) dispersion (3) refraction (4) polarization

✪ 2. Which characterizes a polarized wave?
(1) transverse and vibrating in one plane
(2) transverse and vibrating in all directions
(3) circular and vibrating at random
(4) longitudinal and vibrating at random

✪ 3. A longitudinal wave cannot be
(1) polarized   (2) diffracted   (3) refracted   (4) reflected

✪ indicates material that is not part of the core curriculum.

4. If the electromagnetic waves received on Earth from a source in outer space appear to be increasing in frequency, the distance between the source and the Earth is probably
   (1) decreasing   (2) increasing   (3) remaining the same

5. A star recedes rapidly from the Earth. The frequencies observed from the Earth compared to the frequencies of light emitted by the star are
   (1) lower   (2) higher   (3) the same

6. The wavelength of a spectral line emitted from a distant star is shifted toward a longer wavelength. The star is assumed to be
   (1) stationary relative to the Earth
   (2) moving in a circle around the Earth
   (3) moving away from the Earth
   (4) approaching the Earth

7. Which electromagnetic wave has the highest frequency?
   (1) radio   (2) infrared   (3) X-ray   (4) visible

8. Which electromagnetic radiation has a wavelength shorter than that of visible light?
   (1) ultraviolet waves   (3) radio waves
   (2) infrared waves   (4) microwaves

9. Which is not in the electromagnetic spectrum?
   (1) light waves (2) radio waves (3) sound waves   (4) X-rays

10. Which of the following electromagnetic waves has the lowest frequency?
   (1) violet light (2) green light (3) yellow light   (4) red light

11. In a vacuum, the wavelength of ultraviolet light is greater than that of
   (1) X-rays   (3) blue light
   (2) radio waves   (4) infrared light

12. In a vacuum, light waves and radio waves have the same
   (1) frequency   (2) period   (3) speed   (4) wavelength

13. If the frequency of a light wave in a vacuum is increased, its wavelength
   (1) decreases   (2) increases   (3) remains the same

14. Light of frequency 5.0×10^{14} hertz has a wavelength of 4.0×10^{-7} meter while traveling in a certain material. The speed of light in the material is
    (1) 1.3×10^7 m/s   (3) 3.0×10^8 m/s
    (2) 2.0×10^8 m/s   (4) 1.3×10^{21} m/s

15. What is the wavelength of X-rays with a frequency of 1.5×10^{18} hertz traveling in a vacuum?
    (1) 4.5×10^{26} m   (3) 5.0×10^{-10} m
    (2) 2.0×10^{-10} m   (4) 5.0×10^9 m

16. What is the color of a light wave with a frequency of 5.65×10^{14} hertz?
    (1) red   (2) yellow   (3) green   (4) blue

17. Which formula represents a constant for light waves of different frequencies in a vacuum?

    (1) $f\lambda$ (2) $\dfrac{f}{\lambda}$ (3) $\dfrac{\lambda}{f}$ (4) $f + \lambda$

18. A light beam from Earth is reflected by an object in space. If the round trip takes 2.0 seconds, then the distance of the object from Earth is
    (1) 6.7×10^7 m (2) 1.5×10^8 m (3) 3.0×10^8 m (4) 6.0×10^8 m

19. Which set of electromagnetic radiations is arranged in order of increasing frequency?
    (1) radio, ultraviolet, visible, gamma
    (2) gamma, radio, visible, ultraviolet
    (3) radio, visible, ultraviolet, gamma
    (4) visible, ultraviolet, gamma, radio

20. The time required for light to travel a distance of 1.5×10^{11} meters is closest to
    (1) 5.0×10^2 s (2) 2.0×10^{-3} s (3) 5.0×10^{-1} s (4) 4.5×10^{19} s

21. The observed color of light depends on the light's
    (1) speed   (2) amplitude   (3) intensity   (4) frequency

22. An observer at point *O* sees the reflected light ray as shown in the diagram below.

The image of point *P* that the observer sees is located at
(1) *A*   (2) *B*   (3) *C*   (4) *D*

23. The diagram below shows two rays of light striking a plane mirror.

Which diagram below best represents the reflected rays?

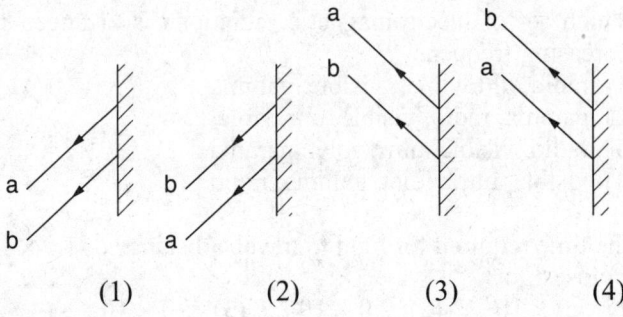

(1)   (2)   (3)   (4)

24. Which property of light is illustrated by the diagram below?

(1) diffraction   (2) dispersion   (3) refraction   (4) reflection

25. Which phenomenon of light is illustrated by the diagram below?

(1) refraction
(2) dispersion
(3) regular reflection
(4) diffuse reflection

26. Which diagram below best represents the image of the object formed by the plane mirror in the diagram below?

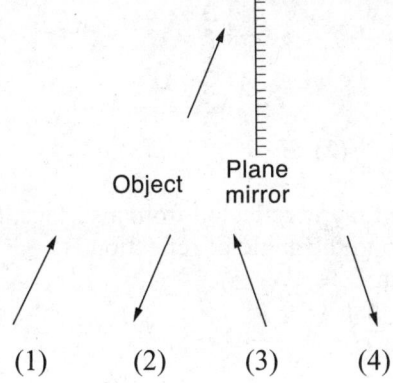

27. Which graph best represents the relationship between the image size and the object size for an object reflected in a plane mirror?

28. The diagram below shows a light ray being reflected from a plane mirror. What is the angle of reflection?

(1) 90°   (2) 80°   (3) 50°   (4) 40°

29. When a light wave enters a new medium and is refracted, there must be a change in the light wave's
    (1) color   (2) frequency   (3) period   (4) speed

30. The speed of light in a piece of plastic is 2.00×10^8 meters per second. What is the absolute index of refraction of this plastic?
    (1) 1.00   (2) 0.670   (3) 1.33   (4) 1.50

31. When a ray of light strikes a mirror perpendicularly to its surface, the angle of reflection is
    (1) 0°   (2) 45°   (3) 60°   (4) 90°

32. If an observer at point O looks at the mirror, M, fixed in position as shown in the diagram below, which point will be seen?

    (1) A (2) B (3) C (4) D

33. When a light ray is reflected from a surface, the ratio of the angle of incidence to the angle of reflection is
    (1) less than 1
    (2) greater than 1
    (3) equal to 1

34. A light ray is incident upon a plane mirror. As the angle of incidence increases, the angle between the incident ray and the mirror's surface
    (1) decreases   (2) increases   (3) remains the same

35. As a person approaches a plane mirror, the size of his or her image
    (1) decreases   (2) increases   (3) remains the same

36. A ray of light ($f = 5.09 \times 10^{14}$ Hz) traveling in air strikes a block of sodium chloride at an angle of incidence of 30.°. What is the angle of refraction for the light ray in the sodium chloride?
    (1) 19°   (2) 25°   (3) 40.°   (4) 49°

37. The speed of a ray of light traveling through a substance having an absolute index of refraction of 1.1 is
    (1) 1.1×10^8 m/s   (3) 3.0×10^8 m/s
    (2) 2.7×10^8 m/s   (4) 3.3×10^8 m/s

38. Which diagram best represents the behavior of a ray of monochromatic light in air incident on a block of crown glass?

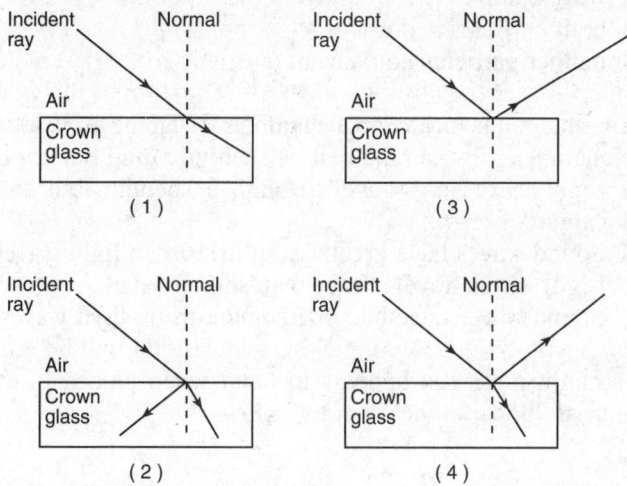

Note that question 39 has only three choices.

39. As yellow light ($f = 5.09 \times 10^{14}$ Hz) travels from zircon into diamond, the speed of the light
   (1) decreases   (2) increases   (3) remains the same

40. A beam of monochromatic light approaches a barrier having four openings, A, B, C, and D, of different sizes as shown below.

Which opening will cause the greatest diffraction?
   (1) A (2) B (3) C (4) D

41. Light demonstrates the characteristics of
  (1) particles, only
  (2) waves, only
  (3) both particles and waves
  (4) neither particles nor waves

42. A ringing bell is located in a chamber. When the air is removed from the chamber, why can the bell be seen vibrating but *not* be heard?
  (1) Light waves can travel through a vacuum, but sound waves cannot.
  (2) Sound waves have greater amplitude than light waves.
  (3) Light waves travel slower than sound waves.
  (4) Sound waves have higher frequency than light waves.

43. A straight glass rod appears to bend when placed in a beaker of water, as shown in the diagram below.

What is the best explanation for this phenomenon?
  (1) The water is warmer than the air.
  (2) Light travels faster in water than in air.
  (3) Light is reflected at the air–water interface.
  (4) Light is refracted as it crosses the air–water interface.

44. What happens to the speed and frequency of a light ray when it passes from air into water?
  (1) The speed decreases and the frequency increases.
  (2) The speed decreases and the frequency remains the same.
  (3) The speed increases and the frequency increases.
  (4) The speed increases and the frequency remains the same.

45. Parallel wave fronts incident on an opening in a barrier are diffracted. For which combination of wavelength and size of opening will diffraction effects be greatest?
    (1) short wavelength and narrow opening
    (2) short wavelength and wide opening
    (3) long wavelength and narrow opening
    (4) long wavelength and wide opening

46. What is the wavelength of a light ray with frequency 5.09×10^{14} hertz as it travels through Lucite?
    (1) 3.93×10^{-7} m   (3) 3.39×10^{14} m
    (2) 5.89×10^{-7} m   (4) 7.64×10^{14} m

47. A person observes a fireworks display from a safe distance of 0.750 kilometer. Assuming that sound travels at 340. meters per second in air, what is the time between the person seeing and hearing a fireworks explosion?
    (1) 0.453 s   (2) 2.21 s   (3) 410. s   (4) 2.55×10^5 s

48. Electromagnetic radiation having a wavelength of 1.3×10^{-7} meter would be classified as
    (1) infrared   (2) orange   (3) blue   (4) ultraviolet

49. Which diagram best represents the path taken by a ray of monochromatic light as it passes from air through the materials shown?

(1)

(3)

(2)

(4)

50. What is the speed of a ray of light ($f = 5.09 \times 10^{14}$ hertz) traveling through a block of sodium chloride?
(1) 1.54×10^8 m/s   (3) 3.00×10^8 m/s
(2) 1.95×10^8 m/s   (4) 4.62×10^8 m/s

Note that question 51 has only three choices.

51. Compared to the speed of microwaves in a vacuum, the speed of x-rays in a vacuum is
(1) less   (2) greater   (3) the same

52. Which ray diagram illustrates refraction?

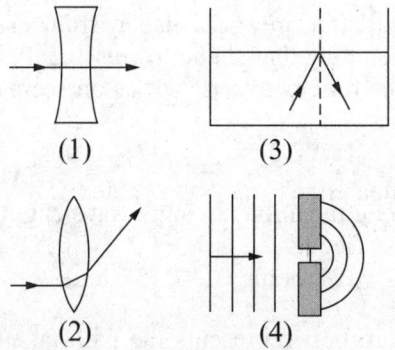

53. The diagram below shows light ray R entering air from water.

Through which point is the ray most likely to pass?
(1) A (2) B (3) C (4) D

54. The ray R of monochromatic yellow light shown in the diagram is incident upon a glass surface at an angle of θ.

Which resulting ray is *not* possible?
(1) A (2) B (3) C (4) D

55. Light travels from medium A, where its speed is 2.5×10^8 meters per second, into medium B, where its speed is 2.0×10^8 meters per second. Compared to the absolute index of refraction of medium A, the absolute index of refraction of medium B is
(1) less (2) greater (3) the same

56. If the speed of light in a medium is 1.5×10^8 meters per second, the index of refraction of the medium is
(1) 0.5 (2) 2.0 (3) 0.67 (4) 1.5

57. The diagram shows a transparent cube made of a substance that has an index of refraction (n) of 1.73 in air.

$n = 1.73$

If a ray of light strikes a side of the cube with an angle of incidence of 60°, the angle of refraction will be
(1) smaller than 30° (3) 45°
(2) 30° (4) greater than 45°

58. Which graph best shows the relationship between the sine of the angle of incidence (sin i) and the sine of the angle of refraction (sin r) for light moving from air to water?

      (1)   (2)   (3)   (4)

59. As a light wave passes from Lucite into flint glass, the frequency of the wave will
(1) decrease   (2) increase   (3) remain the same

60. For a given medium, the index of refraction (n) for a given light wave is equal to

(1) $\dfrac{c}{v}$ (2) cf (3) $\dfrac{\lambda}{c}$ (4) cv

Base your answers to questions 61 through 66 on the diagram below, which represents two media with parallel surfaces in air and a ray of light passing through them.

61. What is the approximate speed of light in water?
(1) 4.4×10^7 m/s   (3) 3.0×10^8 m/s
(2) 2.3×10^8 m/s   (4) 4.0×10^8 m/s

62. If the angle of refraction in air is 45°, what is the sine of the angle of incidence in water?
(1) 0.53   (2) 0.71   (3) 0.94   (4) 1.33

63. Which line best represents the incident ray in the air?
    (1) *AE*   (2) *BE*   (3) *CE*   (4) *DE*

64. What is the sine of the critical angle for light passing from water into air?
    (1) 0.50   (2) 0.75   (3) 0.87   (4) 1.33

65. Compared to the speed of light in water, the speed of light in medium *X* is
    (1) lower   (2) higher   (3) the same

66. Ray *EFG* would be a straight line if the index of refraction for medium *X* were
    (1) less than 1.33
    (2) greater than 1.33
    (3) equal to 1.33

Base your answers to questions 67 through 70 on the information and diagram below.

This diagram represents three transparent media arranged one on top of the other. A light ray in air is incident on the upper surface of layer *A*.

67. If layers *B* and *C* both have the same index of refraction, in which direction will the light ray travel after reaching the boundary between layers *B* and *C*?
    (1) *W*   (2) *X*   (3) *Y*   (4) *Z*

68. If angle *E* is 60° and layer *A* has an index of refraction of 1.61, the sine of angle *F* will be closest to
    (1) 1.00   (2) 0.866   (3) 0.538   (4) 0.400

69. If angle *E* were increased, angle *F* would
    (1) decrease   (2) increase   (3) remain the same

70. Compared to the apparent speed of light in layer *A*, the apparent speed of light in layer *B* is
    (1) greater   (2) the same   (3) less

✪ 71. As the index of refraction of an alcohol-water mixture increases, the critical angle for the mixture
    (1) decreases   (2) increases   (3) remains the same

✪ 72. If the critical angle of glass is 45°, the light ray shown in the diagram will be

    (1) reflected   (2) refracted   (3) dispersed   (4) diffracted

✪ 73. The critical angle for light traveling from a given medium to air is 38°. If light is traveling from another medium with a greater index of refraction to air, the critical angle is
    (1) less than 38°
    (2) greater than 38°
    (3) equal to 38°

74. The diagram shows a ray of light (*R*) incident upon a surface at an angle greater than the critical angle. Through which point is the ray most likely to pass?

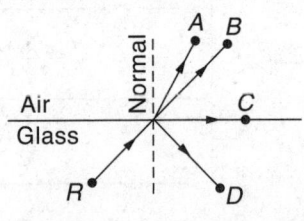

    (1) *A*   (2) *B*   (3) *C*   (4) *D*

75. As a beam of monochromatic light passes from water into air, the frequency of the light
    (1) decreases   (2) increases   (3) remains the same

✪ 76. The critical angle is the angle of incidence that produces an angle of refraction of
    (1) 0°   (2) 45°   (3) 60°   (4) 90°

✪ indicates material that is not part of the core curriculum.

77. The index of refraction of a transparent material is 2.0. Compared to the speed of light in air, the speed of light in this material is
(1) less   (2) greater   (3) the same

✪ **78.** As the index of refraction of a medium increases, the critical angle between the medium and air θ_c
(1) decreases   (2) increases   (3) remains the same

✪ **79.** The critical angle for a monochromatic light ray traveling from a dispersive material into air is 45°. What is the index of refraction for the material?
(1) 2.41   (2) 1.71   (3) 1.41   (4) 0.707

✪ **80.** If a ray of light in glass is incident upon an air surface at an angle greater than the critical angle, the ray will
(1) reflect, only   (3) partly refract and partly reflect
(2) refract, only   (4) partly refract and partly diffract

✪ **81.** A light ray travels from an unknown material into air. If the sine of the critical angle is 0.685, the substance could be
(1) alcohol   (3) flint glass
(2) diamond   (4) carbon tetrachloride

✪ **82.** The critical angle of glass is 43°. In which diagram below will the ray of light be refracted?

(1)   (2)   (3)   (4)

✪ **83.** In which diagram will the ray of light most likely undergo total internal reflection?

(1)   (2)   (3)   (4)

✪ indicates material that is not part of the core curriculum.

Base your answers to questions 84 through 87 on the accompanying diagram, which represents a ray of monochromatic light incident upon the surface of plate X. The values of n in the diagram represent absolute indices of refraction.

84. What is the relative index of refraction of the light going from plate X to plate Y?
    (1) $\dfrac{1.0}{1.7}$ (2) $\dfrac{1.0}{1.4}$ (3) $\dfrac{1.7}{1.4}$ (4) $\dfrac{1.4}{1.7}$

85. The speed of the light ray in plate X is approximately
    (1) 1.8×10^8 m/s   (3) 2.5×10^8 m/s
    (2) 2.1×10^8 m/s   (4) 2.9×10^8 m/s

86. Compared to angle θ, the angle of refraction of the light ray in plate X is
    (1) smaller   (2) greater   (3) the same

87. Compared to angle θ, the angle of refraction of the ray emerging from plate Y into air will be
    (1) smaller   (2) greater   (3) the same

✪ 88. In the diagram below, a monochromatic light ray is passing from medium A into medium B. The angle of incidence, θ, is varied by moving the light source, S.

When angle θ becomes the critical angle, the angle of refraction will be
    (1) $0°$ (3) greater than θ, but less than $90°$
    (2) θ (4) $90°$

✪ 89. When a ray of white light is refracted, which component color has the greatest change in direction?
    (1) orange   (2) red   (3) green   (4) violet

✪ indicates material that is not part of the core curriculum.

378

90. Compared to the wavelength of a wave of green light in air, the wavelength of the same wave of green light in Lucite is
(1) less   (2) greater   (3) the same

✪ **91.** Monochromatic light cannot be
(1) dispersed   (2) absorbed   (3) reflected   (4) refracted

✪ **92.** Compared to the speed of light in a vacuum, the speed of light in a dispersive medium is
(1) less   (2) greater   (3) the same

✪ **93.** The separation of white light into its component colors as it passes through a prism is called
(1) convergence   (3) diffraction
(2) dispersion   (4) total internal reflection

✪ **94.** Which terms best describe the phenomenon illustrated by the diagram below?

(1) scattering and diffraction   (3) transmission and Doppler effect
(2) reflection and interference   (4) refraction and dispersion

Base your answers to questions 95 through 99 on the diagram below, which represents a ray of monochromatic green light incident upon the surface of a glass prism.

95. The index of refraction of the glass prism for green light equals
(1) $\dfrac{\theta_1}{\theta_2}$ (2) $\dfrac{\theta_2}{\theta_1}$ (3) $\dfrac{\sin \theta_2}{\sin \theta_1}$ (4) $\dfrac{\sin \theta_1}{\sin \theta_2}$

✪ indicates material that is not part of the core curriculum.

96. After the ray leaves the prism, it will most likely pass through point
   (1) *A*   (2) *B*   (3) *C*   (4) *D*

✪ **97.** If the monochromatic green ray is replaced by a monochromatic red ray, θ_2 will
   (1) decrease   (2) increase   (3) remain the same

✪ **98.** Compared to the speed of the monochromatic green light in the prism, the speed of the monochromatic red light in the prism is
   (1) less   (2) greater   (3) the same

✪ **99.** Compared to the frequency of the green light in the prism, the frequency of the red light in the prism is
   (1) less   (2) greater   (3) the same

Base your answers to questions 100 through 102 on the diagram below, which shows a ray of white light incident upon a surface of a triangular glass prism in air.

100. The index of refraction for the violet light ray is approximately
   (1) 0.577   (2) 2.00   (3) 1.24   (4) 1.73

101. If the index of refraction for the red light is approximately 1.22, the speed of the red light in glass is approximately
   (1) 1.22×10^8 m/s   (3) 3.00×10^8 m/s
   (2) 2.46×10^8 m/s   (4) 4.07×10^8 m/s

102. Which color of light has the greatest speed in the prism?
   (1) red   (2) green   (3) blue   (4) violet

✪ indicates material that is not part of the core curriculum.

380

✪ **103.** Which pattern of bright and dark bands could have been produced by a beam of monochromatic light passing through a single narrow slit?

104. Interference experiments demonstrate the
(1) particle nature of light
(2) polarization of light
(3) intensity of light
(4) wave nature of light

105. Which diagram best represents the phenomenon of diffraction?

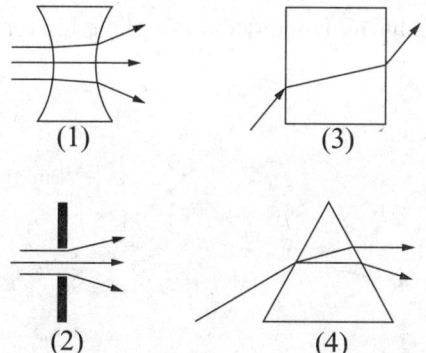

106. Which phenomenon is the best evidence for the wave nature of light?
(1) reflection
(2) photoelectric emission
(3) diffusion
(4) interference

✪ **107.** Which term best describes the light generated by a laser?
(1) diffused   (2) coherent   (3) dispersive   (4) longitudinal

✪ indicates material that is not part of the core curriculum.

⭐ **108.** The diagram below represents a group of light waves emitted simultaneously from a single source.

These light waves would be classified as
(1) coherent, but not monochromatic
(2) monochromatic, but not coherent
(3) both monochromatic and coherent
(4) neither monochromatic nor coherent

⭐ **109.** Two sources are coherent if the waves that they produce
(1) have the same amplitude   (3) are both transverse in nature
(2) travel at the same speed   (4) have a constant phase relation

Answers to Part A and B–1 questions can be found on page 459.

PART B–2 AND C QUESTIONS

1. Base your answers to parts *a* through *c* on the information and diagram below.

A ray of light *AO* is incident on a plane mirror as shown.

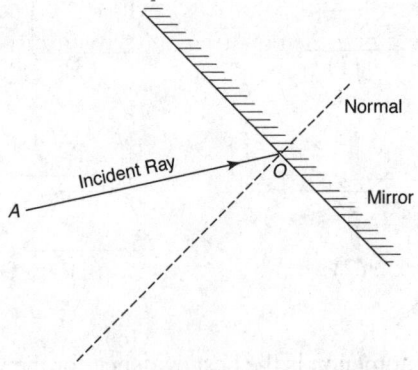

a Using a protractor, measure the angle of incidence for light ray *AO* and record the value on your answer sheet.

b What is the angle of reflection of the light ray?

c Using a protractor and straightedge, construct the reflected ray on the diagram.

⭐ indicates material that is not part of the core curriculum.

Base your answers to questions 2 through 4 on the diagram below, which shows light ray *AO* in Lucite. The light ray strikes the boundary between Lucite and air at point *O* with an angle of incidence of 30.°. The dotted line represents the normal to the boundary at point *O*.

2. Calculate the angle of refraction for incident ray *AO*. [Show all calculations, including the equation and substitution with units.]

3. *On the diagram*, using your answer from question 2, construct an arrow with a protractor and straightedge, to represent the refracted ray.

✪ 4. Calculate the critical angle for a Lucite-air boundary. [Show all calculations, including the equation and substitution with units.]

✪ indicates material that is not part of the core curriculum.

Base your answers to questions 5 and 6 on the diagram below, which shows a light ray in water incident at an angle of 60.° on a boundary with plastic.

5. Using a protractor, measure the angle of refraction to the *nearest degree*.

6. Determine the absolute index of refraction for the plastic. [Show all calculations, including the equation and substitution with units.]

7. A ray of light traveling in air is incident on an air-water boundary as shown below.

On the diagram provided, draw the path of the ray in the water.

Base your answers to questions 8 through 10 on the diagram below which shows a ray of monochromatic light ($f = 5.09 \times 10^{14}$ hertz) passing through a flint glass prism.

8. Calculate the angle of refraction (in degrees) of the light ray as it enters the air from the flint glass prism. [Show all calculations, including the equation and substitution with units.]

9. Using a protractor and a straightedge, construct the refracted light ray in the air on the diagram above.

10. What is the speed of the light ray in flint glass?
    (1) 5.53×10^{-9} m/s   (3) 3.00×10^8 m/s
    (2) 1.81×10^8 m/s   (4) 4.98×10^8 m/s

11. Determine the color of a ray of light with a wavelength of 6.21×10^{-7} meter.

Base your answers to questions 12 through 14 on the information and diagram below.

A monochromatic beam of yellow light, AB, is incident upon a Lucite block in air at an angle of 33°.

12. Calculate the angle of refraction for incident beam *AB*. [Show all work, including the equation and substitution with units.]

13. Using a straightedge, a protractor, and your answer from question 12, draw an arrow to represent the path of the refracted beam.

14. Compare the speed of the yellow light in air to the speed of the yellow light in Lucite.

Base your answers to questions 15 through 17 on the information and diagram below.

A ray of monochromatic light having a frequency of 5.09×10^{14} hertz is incident on an interface of air and corn oil at an angle of $35°$ as shown. The ray is transmitted through parallel layers of corn oil and glycerol and is then reflected from the surface of a plane mirror, located below and parallel to the glycerol layer. The ray then emerges from the corn oil back into the air at point *P*.

15. Calculate the angle of refraction of the light ray as it enters the corn oil from air. [Show all work, including the equation and the substitution with units.]

16. Explain why the ray does *not* bend at the corn oil–glycerol interface.

17. On the diagram above, use a protractor and straightedge to construct the refracted ray representing the light emerging at point *P* into air.

Base your answers to questions 18 and 19 on the information and diagram below.

Two plane mirrors are positioned perpendicular to each other as shown. A ray of monochromatic red light is incident on mirror 1 at an angle of 55°. This ray is reflected from mirror 1 and then strikes mirror 2.

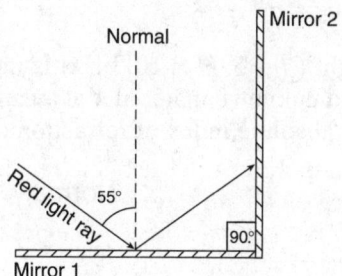

18. Determine the angle at which the ray is incident on mirror 2.

19. On the diagram above, use a protractor and a straightedge to draw the ray of light as it is reflected from mirror 2.

Base your answers to questions 20 through 23 on the diagram below, which shows a light ray ($f = 5.09 \times 10^{14}$ Hz) in air, incident on a boundary with fused quartz. At the boundary, part of the light is refracted and part of the light is reflected.

20. Using a protractor, measure the angle of incidence of the light ray at the air-fused quartz boundary.

21. Calculate the angle of refraction of the incident light ray. [Show all work, including the equation and substitution with units.]

22. Using a protractor and straightedge, construct the refracted light ray in the fused quartz on the diagram on page 387.

23. Using a protractor and straightedge, construct the reflected light ray on the diagram on page 387.

Base your answers to questions 24 through 27 on the information and diagram below.

A ray of light ($f = 5.09 \times 10^{14}$ Hz) is incident on the boundary between air and unknown material X at an angle of incidence of 55°, as shown. The absolute index of refraction of material X is 1.66.

24. Identify a substance of which material X may be composed.

25. Determine the speed of this ray of light in material X.

_____ m/s

26. Calculate the angle of refraction of the ray of light in material X. [Show all work, including the equation and substitution with units.]

27. On the diagram above, use a straightedge and protractor to draw the refracted ray of light in material X.

Base your answers to questions 28 and 29 on the information and diagram below.

> A ray of monochromatic light of frequency 5.09×10^{14} hertz is traveling from water into medium X. The angle of incidence in water is 45° and the angle of refraction in medium X is 29°, as shown.

28. Calculate the absolute index of refraction of medium X. [Show all work, including the equation and substitution with units.]

29. Medium X is most likely what material?

Base your answers to questions 30 through 33 on the diagram below, which represents a ray of monochromatic light ($f = 5.09 \times 10^{14}$ hertz) in air incident on flint glass.

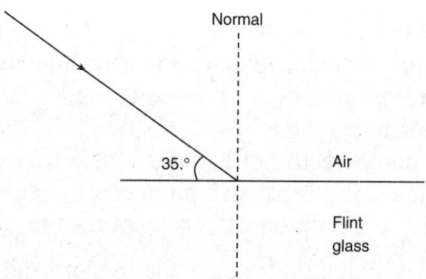

30. Determine the angle of incidence of the light ray in air.

31. Calculate the angle of refraction of the light ray in the flint glass. [Show all work, including the equation and substitution with units.]

32. Using a protractor and straightedge, draw the refracted ray on the diagram.

33. What happens to the light from the incident ray that is *not* refracted or absorbed?

Answers to Part B–2 and C questions can be found on pages 531–541.

MODERN PHYSICS

KEY IDEAS

Energy is not continuous; it occurs in a series of discrete bundles called quanta. Einstein's explanation of the photoelectric effect introduced the concept that light energy is quantized in units called photons. The energy of a photon is directly related to its frequency. Compton's experiments verified the photon nature of light. De Broglie proposed that matter such as electrons possess wavelike characteristics, and diffraction experiments verified this theory.

Rutherford established the nuclear model of the atom by analyzing the results when alpha particles were scattered by metallic foils. Bohr refined the model of hydrogen by proposing that the energy levels in the atom are quantized and that electrons can make only discrete transitions within these levels. The currently accepted model of the atom is based on the idea of probability, and each electron is viewed as a cloud rather than as a specific point in space.

KEY OBJECTIVES

At the conclusion of this chapter you will be able to:

- Define the term *quantum of energy*, and relate this term to Planck's constant.
- Describe the photoelectric effect and Einstein's explanation of it.
- Solve problems using Einstein's photoelectric equation.
- Explain how the Compton effect supports the photon theory of light.
- Calculate the momentum of a photon, given its frequency or wavelength.
- Describe Rutherford's experiments involving the scattering of alpha particles by metallic foils and the model of the atom he proposed as a result of those experiments.
- Explain why the Rutherford model did *not* provide a complete picture of the atom.
- State the hypotheses that Bohr used in developing his model of the hydrogen atom.
- Define the terms *ground state, excited state,* and *stationary state* as they apply to the Bohr model.

- Describe how Bohr was able to explain the existence of line spectra.
- Define the term *ionization*, and use the energy level diagrams for hydrogen and mercury to calculate the energies involved in various electron transitions.
- Define the term *electron cloud*, and state why the cloud model was needed to provide a more nearly complete picture of the atom.

13.1 INTRODUCTION

The title of this chapter is not completely accurate since the theories and discoveries we describe originated between 1895 and 1930. In this time period, two major scientific theories were advanced: quantum physics and Einstein's theory of relativity. Although both theories revolutionized the sciences and technology, they presented various aspects of nature in ways that defied the familiar, "commonsense" view of the world.

13.2 BLACK-BODY RADIATION AND PLANCK'S HYPOTHESIS

When a solid is heated, it emits a variety of electromagnetic radiation. As the temperature of the solid is increased, the radiation shifts toward shorter wavelengths, a change that explains why heated solids begin to glow red, then orange, and finally white. To study this radiation effectively, physicists used a device called a *cavity radiator*, also known as an *ideal black body*. This device is a hollow solid with a small opening drilled in one of the walls, as shown in the following diagram.

Emitted radiation

When the solid was heated, it was found that that the radiation emitted *through the opening* depended only on the temperature of the solid, not on the material of which the cavity radiator was made. The origin of the radiation was presumed to be the atoms of the solid. The heating caused the atoms to oscillate (just as springs do), and then the energy was released into the cavity as electromagnetic radiation. However, there was serious disagreement between the theory and the experimental results.

In 1900, German physicist Max Planck reported a startling discovery: the experimental results could be explained precisely if it was assumed that the atomic "oscillators" can have only certain energies given by this equation:

=== **PHYSICS CONCEPTS** ===

$$E = nhf$$

*Note that this equation does not appear on the Reference Tables.

---

where E is the energy of the atomic oscillator, f is its frequency of oscillation, n is an integer (1, 2, 3, . . .) and h, known as *Planck's constant*, has the value of 6.6×10^{-34} joule $\cdot$ second. Each of these discrete values of energy is known as a *quantum* of energy. As a result, the radiation emitted from the cavity opening is also restricted to certain values.

This idea was revolutionary because physicists had always assumed that a particle could take on or emit any value of energy. (In fact, Planck himself was not comfortable with the concept of quantized energy!) As we will see in the next section, Planck's idea was reinforced in 1905 by German-American physicist Albert Einstein, who used it to explain the nature of light in an experiment involving the *photoelectric effect*.

13.3 THE PHOTOELECTRIC EFFECT

The photoelectric effect is based on a very simple premise: When light energy strikes the surface of certain materials, electrons are ejected, creating a small but measurable current. Today, photoelectric devices are in wide use in products such as cameras and automatic door openers. The diagram below illustrates how the photoelectric effect can be studied in a laboratory.

As monochromatic light strikes a metal surface, a potential difference causes the ejected photoelectrons (e^-) to move onto a collector. The galvanometer (G) measures the current that is produced as a result. The arrangement illustrated above allows for changes in the (1) metal surface, (2) potential difference, (3) light intensity, and (4) color of the light used.

Such photoelectric experiments yielded results that did not agree with the idea that light behaves as a wave. For example, certain colors of light did not eject electrons, regardless of the intensity of the incident light. According to wave theory, very bright light (of any color) should possess a great deal of energy and be able to eject electrons from a metal surface. It was found that, when electrons are ejected, the *color* of the light determines how energetic the electrons are; the intensity determines the *rate* at which the electrons are ejected (i.e., the current).

The potential difference can also be made to *oppose* the movement of the electrons. There is a value, known as the *stopping potential*, that will stop even the most energetic electrons from reaching the collector. If we measure the value of the stopping potential (V_0), we can calculate the kinetic energy of the fastest electrons (KE_{max}). Since energy (work) is equal to potential difference multiplied by charge, it follows that:

PHYSICS CONCEPTS

$$KE_{max} = V_0 e$$

*Note that this equation does not appear on the Reference Tables.

PROBLEM

In a photoelectric experiment, the stopping potential is 5.0 volts. Calculate the maximum kinetic energy of the photoelectrons.

SOLUTION

$KE_{max} = V_0 e$

$\quad\quad = (5.0 \text{ V})(1.6 \times 10^{-19} \text{ C})$

$\quad\quad = 8.0 \times 10^{-19} \text{ J } [5.0 \text{ eV}]$

The graph below represents an experiment that relates the color (i.e., the frequency) of the incident light to the maximum kinetic energy of the photoelectrons.

For any photoemissive material tested, the graph is a straight line whose slope is Planck's constant. There is a frequency, called the *threshold frequency* (f_0) at which the maximum kinetic energy of the photoelectrons is zero. The energy of a photon at this frequency is just sufficient to break the bonds holding the electron to the surface of the material. This energy, known as the *work* function (W), is the y-intercept of the graph. At frequencies *less* than the threshold value, photons cannot produce a photoelectric effect.

Einstein predicted these results before there was any experimental verification. He theorized that light behaved as if it were a collection of particles called *photons*. The energy of each photon depends on the frequency of the light. The energy is given by the following equation:

PHYSICS CONCEPTS

$$E = hf \text{ or substituting } \frac{c}{\lambda} \text{ for } f$$

$$E = \frac{hc}{\lambda}$$

394

Einstein summarized his theory by means of his *photoelectric equation*:

$$KE_{max} = hf - W$$

*Note that this equation does not appear on the Reference Tables.

When a photon with energy hf strikes a surface, a part of its energy (the work function, W) frees the electron from its bonds. The remainder of the energy ($hf - W$) gives the electron its kinetic energy (KE_{max}).

PROBLEM
A photon with a frequency of 8.0×10^{14} hertz strikes a photoemissive surface whose work function is 1.7×10^{-19} joule. Calculate (a) the maximum kinetic energy of the ejected photoelectrons and (b) the threshold frequency of the surface.

SOLUTION
(a) $KE_{max} = hf - W$
$\qquad\quad = (6.6 \times 10^{-34} \text{ J} \cdot \text{s})(8.0 \times 10^{14} \text{ Hz}) - 1.7 \times 10^{-19} \text{ J}$
$\qquad\quad = 3.6 \times 10^{-19} \text{ J}$

(b) The threshold frequency (f_0) is the frequency at which the maximum kinetic energy of the photoelectrons is just zero. Applying the photoelectric equation gives

$KE_{max} = hf - W$
$\quad\ 0 = hf_0 - W$
$\quad\ W = hf_0$
$1.7 \times 10^{-19} \text{ J} = (6.6 \times 10^{-34} \text{ J} \cdot \text{s})f_0$
$\qquad f_0 = 2.6 \times 10^{14} \text{ Hz}$

✪ 13.4   THE COMPTON EFFECT

The photon theory of light was strongly supported by the experiments of Arthur Compton, a U.S. physicist. When Compton bombarded a block of graphite with X-rays of known frequency, he discovered that both electrons and X-rays emerged from the block, as shown in the diagram.

✪ indicates material that is not part of the core curriculum.

Compton observed that the scattered X-rays had lower frequencies than the incident X-rays, and he recognized that both energy and momentum were conserved in these collisions. Compton used the following relationship for the magnitude of the momentum of a photon:

=== **PHYSICS CONCEPTS** ===

$$p = \frac{E}{c} = \frac{hf}{c} = \frac{h}{\lambda}$$

Note that this equation does not appear on the Reference Tables.

PROBLEM
Calculate the momentum of an X-ray photon whose wavelength is 1.0×10^{-10} meter.

SOLUTION
$$p = \frac{h}{\lambda} = \frac{6.6 \times 10^{-34} \text{ J·s}}{1.0 \times 10^{-10} \text{ m}} = 6.6 \times 10^{-24} \text{ kg·m/s}$$

This relationship demonstrates that momentum (p), the particle property of light, wavelength (λ), and the wave property of light, are inversely related: High-frequency light (ultraviolet, X-rays, gamma radiation) behaves more like particles and less like waves; low-frequency light (radio waves, microwaves, infrared radiation) behaves more like waves and less like particles.

13.5   MODELS OF THE ATOM

The impact of modern physics is most evident in the development of the atomic model of matter. Though the concept of the atom goes back to ancient Greece and Rome, it was not until the twentieth century that the structure of the atom was reasonably well understood. We use the term *atomic model* to indicate that we are trying to describe the key features of the atom. We really do not know what an atom "looks like" because we have no instruments for its direct observation. A model is like a road map: it gives us information to get from place to place, but it does *not* describe the actual landscape.

Early Models

The first scientific model of the atom was conceived by English chemist and physicist John Dalton at the beginning of the nineteenth century. His model explained the mathematics of chemical combinations, but the internal structure of the atom remained a mystery. At the beginning of the twentieth century, the discoveries of Antoine Henri Becquerel, the Curies, Frédéric Joliot-Curie,

Iréne Joliot-Curie, and J.J. Thomson led to the idea that an atom is constructed of positively and negatively charged particles, most notably the electron.

Rutherford Model

British physicist Ernest Rutherford and his two assistants, Hans Geiger and Ernest Marsden, bombarded thin metallic foils, such as gold, with massive positively charged particles known as *alpha particles*. They counted scintillations of scattered alpha particles on a zinc sulfide screen. They observed the following:
1. Most of the particles passed through the foils without being deflected.
2. A very small number of particles were deflected through large angles, some approaching 180°.

The diagrams below represent the results of Rutherford's experiments.

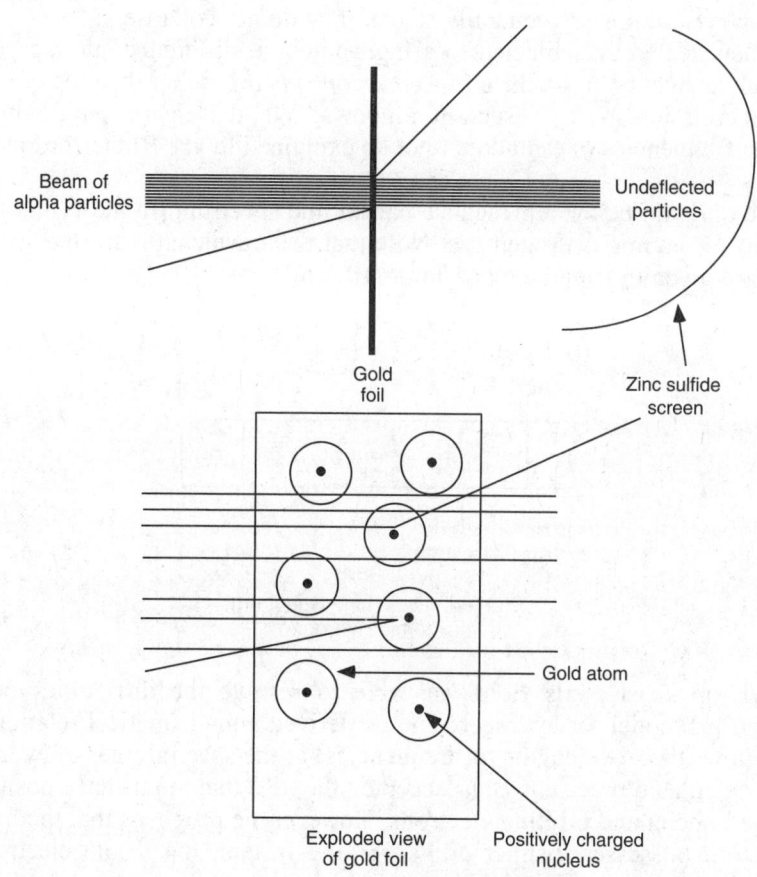

On the basis of these experiments, Rutherford drew the following conclusions:
1. Most of the atom is empty space.
2. Most of the mass of the atom is concentrated in a dense, positively charged *nucleus*. The negative electrons orbit the nucleus much as the planets orbit the Sun.
3. According to calculations using Coulomb's law, the interaction of alpha particles with a very small, massive nucleus would result in it following a hyberbolic path and produce the observed angular pattern of scattering.

Bohr Model

Although Rutherford's nuclear model was an important advance in our knowledge of atomic structure, it had two serious failings:
1. Orbiting electrons are *accelerating* charges and, as such, should produce electromagnetic radiation. This release of energy should cause any electron's orbit to decrease, leading to the collapse of any and all atoms. However, atoms are generally stable; they do not collapse.
2. When heated or subjected to a high potential difference, atomic gases, such as hydrogen, produce *line emission spectra,* rather than the continuous emission spectrum seen in rainbows. Why these gases emit only certain frequencies of light could not be explained by the Rutherford model.

The diagram below represents a partial line spectrum (in the visible-light region) for atomic hydrogen gas. Note that the wavelengths for the spectral lines are given in nanometers (1 nm = 10^{-9} m).

(Partial) Visible Line Spectrum
for Atomic Hydrogen

Danish physicist Neils Bohr was able to resolve the difficulties of the Rutherford model for hydrogen, and he derived a mathematical relationship that allows the wavelengths (or frequencies) of the spectral lines of hydrogen to be calculated precisely. Bohr accepted the idea that atoms have positively charged nuclei and orbiting electrons. However, he proposed that the hydrogen atom possesses distinct orbits. At any instant, the single electron of hydrogen can be associated with one and only one of these orbits. Each orbit gives the hydrogen atom a specific amount of energy. (For this reason, Bohr called the orbits *energy levels.*) Energy levels are identified by integers: 1, 2, 3, . . , *n*. Today these integers are known as *quantum numbers*.

If the electron is on the first energy level, the atom is said to be in the **ground state**. On any other level, the atom is in an **excited state**. No energy is either emitted or absorbed by the atom when it is in any single state, known as a *stationary state*. Rather, energy is *emitted* (as photons) when the atom goes from a higher energy to a lower energy state, and energy is *absorbed* (as photons) when the atom goes from a lower energy to a higher energy state.

Bohr's relationship for the energy levels in a hydrogen atom is as follows:

===| PHYSICS CONCEPTS |===

$$E_n = -\frac{13.6 \text{ eV}}{n^2}$$

*Note that this equation does not appear on the Reference Tables.

In the ground state ($n = 1$), the energy of the hydrogen atom is -13.6 eV. Then, for example, in the excited state where $n = 3$, the energy of the hydrogen atom is

$$E_3 = -\frac{13.6 \text{ eV}}{3^2} = -1.51 \text{ eV}$$

If *n* is considered infinitely large, the energy of the hydrogen atom is taken to be zero because the electron is no longer considered to be associated with the atom, a condition known as **ionization**.

✪ PROBLEM
Calculate the energy of a hydrogen atom when its electron is associated with energy level 2.

SOLUTION

$$E_n = -\frac{13.6 \text{ eV}}{n^2}$$

$$E_2 = -\frac{13.6 \text{ eV}}{2^2} = -3.40 \text{ eV}$$

✪ indicates material that is not part of the core curriculum.

The diagram below is a graphic representation of the energy levels of the hydrogen atom:

Energy Levels of the Hydrogen Atom

When a hydrogen atom changes from energy state $E_{initial}$ to energy state E_{final}, the amount of energy emitted or absorbed by the atom is determined as follows:

PHYSICS CONCEPTS

$$\Delta E = E_{final} - E_{initial} = \frac{-13.6 \text{ eV}}{n^2_{final}} - \frac{-13.6 \text{ eV}}{n^2_{initial}}$$

which can be simplied to

$$\Delta E = 13.6 \text{ eV} \left(\frac{1}{n^2_{initial}} - \frac{1}{n^2_{final}} \right)$$

*Note that this equation does not appear on the Reference Tables.

If ΔE is a *negative* number, the energy is emitted; if it is positive, the energy is absorbed. The frequency of the photon that is emitted (or absorbed) can be calculated from the familiar relationship $\Delta E = hf$.

PROBLEM
Calculate the energy of the photon that is emitted when a hydrogen atom changes from energy state $n = 3$ to $n = 2$.

SOLUTION

We have already calculated the two energy states: $E_3 = -1.51$ eV; $E_2 = -3.40$ eV.

$$\Delta E = E_{\text{final}} - E_{\text{initial}} = E_2 - E_3$$
$$= (-3.40 \text{ eV}) - (-1.51 \text{ eV}) = -1.89 \text{ eV}$$

A 1.89–eV photon is emitted by the hydrogen atom in changing from $n = 3$ to $n = 2$. This corresponds to the red line of the hydrogen spectrum shown above.

The visible line spectrum for atomic hydrogen is known as the *Balmer series*. It consists of electron transitions to energy level 2 from higher levels (3, 4, 5, . . .). Transitions down to energy level 1 yield a line spectrum in the ultraviolet region, while transitions to level 3, 4, or 5 yield line spectra in the infrared region.

PROBLEM

How much energy is needed to ionize a hydrogen atom in the ground state?

SOLUTION

We must raise the energy level of the hydrogen atom from level 1 to "infinity."

$$\Delta E = E_\infty - E_1 = 0.0 \text{ eV} - (-13.6 \text{ eV}) = +13.6 \text{ eV}$$

Thus, 13.6 eV must be absorbed by a hydrogen atom in the ground state in order to ionize it. This energy is known as the *ionization energy* (or *ionization potential*) of the hydrogen atom.

Unfortunately, the Bohr model is not successful in explaining atoms that have more than one electron.

The diagram on the following page represents a portion of the line spectrum of mercury. Notice that the spacing of the lines is more complex and less regular than for hydrogen. The Bohr model *cannot* be applied to this kind of atom. Nevertheless, we can still calculate the energy of an electron transition between states because we have been given the energy values (labeled *a* through *j*) associated with them.

A Few Energy Levels for the Mercury Atom

The equation we use in this case is simplified to the following:

PHYSICS CONCEPTS

$$E = E_i - Ef$$

Cloud Model

The current model of the atom is able to explain the structure of all of the atoms in the Periodic Table. This model is based on a mathematical area of physics known as *quantum mechanics* or *wave mechanics*. Quantum mechanics does not place electrons in specific orbits; rather, it indicates the *probability* that an electron will be in a region of space near the nucleus. The most probable regions of the electron's location define what is known as the **electron cloud**. The energy levels of the Bohr model are subdivided into other levels termed *sublevels* and *orbitals*. Together, these define the energy and structure of a particular atom.

13.6   ENERGY AND MASS

The term mc^2 is the total energy of an object (most familiarly stated as $E = mc^2$); it is the sum of the object's kinetic energy and rest energy. We state, without showing the mathematics, that at speeds that are low compared to the speed of light the kinetic energy of an object reduces to the familiar form:

$$KE = \frac{1}{2} mv^2$$

PART A AND B–1 QUESTIONS

1. The ratio of the energy of a quantum of electromagnetic radiation to its frequency is
   (1) the electrostatic constant   (3) the gravitational constant
   (2) the electron-volt   (4) Planck's constant

2. The energy of a photon varies directly with its
   (1) frequency   (2) wavelength   (3) speed   (4) rest mass

3. According to the quantum theory of light, the energy of light is carried in discrete units called
   (1) alpha particles   (3) photons
   (2) protons   (4) photoelectrons

4. Which graph best represents the energy of a photon as a function of its frequency?

(1)

(3)

(2)

(4)

5. Which graph best represents the relationship between the energy of a photon and its wavelength?

(1)   (2)   (3)   (4)

6. Light demonstrates the characteristics of
   (1) particles, only   (3) both particles and waves
   (2) waves, only   (4) neither particles nor waves

7. Which color of light has the greatest energy per photon?
   (1) red   (2) green   (3) blue   (4) violet

8. What is the energy of a photon of blue light whose frequency is
   6×10^{14} hertz?
   (1) 4×10^{-6} J (2) 4×10^{-10} J (3) 4×10^{-14} J (4) 4×10^{-19} J

9. The wavelength of photon A is greater than that of photon B.
   Compared to the energy of photon A, the energy of photon B is
   (1) less   (2) greater   (3) the same

10. Which phenomenon best supports the particle theory of light?
    (1) photoelectric effect   (3) interference
    (2) diffraction   (4) polarization

11. A mercury atom in the ground state absorbs 20.00 electronvolts of
    energy and is ionized by losing an electron. How much kinetic
    energy does this electron have after the ionization?
    (1) 6.40 eV   (2) 9.62 eV   (3) 10.38 eV   (4) 13.60 eV

12. Which graph best represents the relationship between photon
    energy and photon frequency?

    (1)   (2)   (3)   (4)

13. A variable-frequency light source emits a series of photons. As the
    frequency of the photon increases, what happens to the energy and
    wavelength of the photon?
    (1) The energy decreases and the wavelength decreases.
    (2) The energy decreases and the wavelength increases.
    (3) The energy increases and the wavelength decreases.
    (4) The energy increases and the wavelength increases.

14. The slope of a graph of photon energy versus photon frequency rep-
    resents
    (1) Planck's constant   (3) the speed of light
    (2) the mass of a photon   (4) the speed of light squared

15. A photon having an energy of 9.40 electronvolts strikes a hydrogen atom in the ground state. Why is the photon *not* absorbed by the hydrogen atom?
    (1) The atom's orbital electron is moving too fast.
    (2) The photon striking the atom is moving too fast.
    (3) The photon's energy is too small.
    (4) The photon is being repelled by electrostatic force.

16. A photon of light traveling through space with a wavelength of 6.0 $\times 10^{-7}$ meter has an energy of
    (1) 4.0×10^{-40} J   (3) 5.4×10^{10} J
    (2) 3.3×10^{-19} J   (4) 5.0×10^{14} J

17. What is the net electrical charge on a magnesium ion that is formed when a neutral magnesium atom loses two electrons?
    (1) -3.2×10^{-19} C   (3) $+1.6 \times 10^{-19}$ C
    (2) -1.6×10^{-19} C   (4) $+3.2 \times 10^{-19}$ C

18. Which type of photon is emitted when an electron in a hydrogen atom drops from $n = 2$ to the $n = 1$ energy level?
    (1) ultraviolet (2) visible light (3) infrared   (4) radio wave

19. Oil droplets may gain electrical charges as they are projected through a nozzle. Which quantity of charge is *not* possible on an oil droplet?
    (1) 8.0×10^{-19} C   (3) 3.2×10^{-19} C
    (2) 4.8×10^{-19} C   (4) 2.6×10^{-19} C

20. All photons in a vacuum have the same
    (1) speed   (2) wavelength (3) energy   (4) frequency

21. Which phenomenon best supports the theory that matter has a wave nature?
    (1) electron momentum   (3) photon momentum
    (2) electron diffraction   (4) photon diffraction

22. The diagram below represents the bright-line spectra of four elements, *A*, *B*, *C*, and *D*, and the spectrum of an unknown gaseous sample.

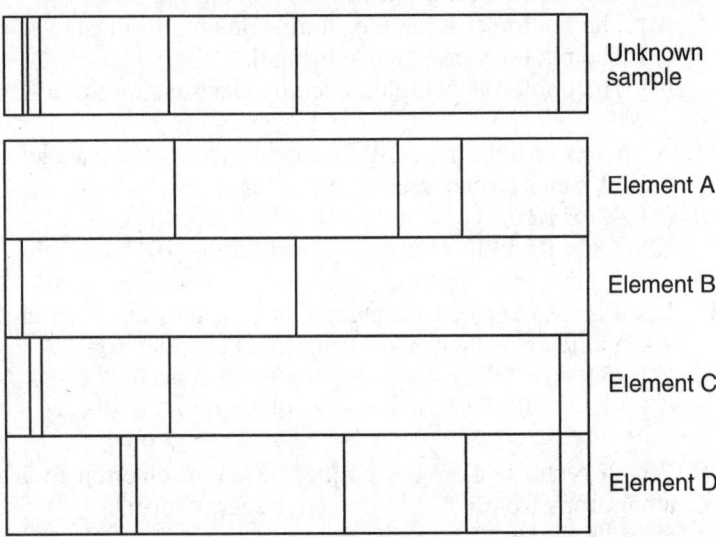

Based on comparisons of these spectra, which two elements are found in the unknown sample?

(1) *A* and *B*   (2) *A* and *D*   (3) *B* and *C*   (4) *C* and *D*

Base your answers to questions 23 through 25 on the diagram below, which represents monochromatic light incident upon photoemissive surface *A*. Each photon has 8.0×10^{-19} joule of energy. *B* represents the particle emitted when a photon strikes surface *A*.

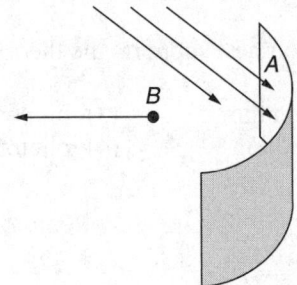

✪ **23.** What is particle *B*?

    (1) an alpha particle   (3) a neutron

    (2) an electron   (4) a proton

✪ indicates material that is not part of the core curriculum.

✪ **24.** If the work function of metal A is 3.2×10^{-19} joule, the energy of particle B is
(1) 3.0×10^{-19} J   (3) 8.0×10^{-19} J
(2) 4.8×10^{-19} J   (4) 11×10^{-19} J

25. The frequency of the incident light is approximately
(1) 1.2×10^{15} Hz   (3) 3.7×10^{15} Hz
(2) 5.3×10^{15} Hz   (4) 8.3×10^{15} Hz

26. Which occurs when a photon and a free electron collide?
(1) Momentum is conserved.
(2) Only the kinetic energy of the photon is conserved.
(3) The momentum of the photon is increased.
(4) The wavelength of the photon is unchanged.

27. A gamma photon makes a collision with an electron at rest. During the interaction, the momentum of the photon will
(1) decrease   (2) increase   (3) remain the same

28. If an X-ray photon collides with an electron, the frequency of the photon will
(1) decrease   (2) increase   (3) remain the same

Base your answers to questions 29 through 32 on the information and diagram below.

An incident photon with a frequency of 5.0×10^{16} hertz strikes a stationary electron. The scattered photon rebounds at an angle of 90°, and the electron moves away with a kinetic energy of 100. electron-volts.

29. Which vector best represents the direction of motion of the electron after the collision?

✪ indicates material that is not part of the core curriculum.

407

30. What is the energy of the incident photon?
(1) 1.3×10^{-18} J   (3) 5.0×10^{16} J
(2) 5.0×10^{-17} J   (4) 3.3×10^{-17} J

31. Compared to the speed of the incident photon, the speed of the scattered photon is
(1) smaller   (2) greater   (3) the same

32. Compared to the wavelength of the incident photon, the wavelength of the scattered photon is
(1) shorter   (2) longer   (3) the same

33. As the speed of an electron increases, its wavelength
(1) decreases   (2) increases   (3) remains the same

✪ 34. In the Rutherford experiment, a beam of alpha particles was directed at a thin gold foil. The deflection pattern of the alpha particles showed that
(1) the electrons of gold atoms have waves
(2) the nuclear volume is a small part of the atomic volume
(3) the energy levels of a gold atom are quantized
(4) gold atoms can emit photons under bombardment

✪ 35. According to the Rutherford model of the atom, the volume of an atom is composed mainly of
(1) electrons   (2) protons   (3) neutrons   (4) empty space

✪ 36. Which type of force causes the hyperbolic trajectory of alpha particles in Rutherford's scattering experiment?
(1) gravitational (2) electrostatic (3) magnetic   (4) nuclear

✪ 37. A scattering experiment is performed in which alpha particles from a single source are deflected by the nuclei of various atomic elements. The nuclei causing the greatest amount of alpha-particle scattering are those having the
(1) smallest neutron number   (3) smallest mass number
(2) greatest photon number   (4) greatest atomic number

✪ 38. Rutherford observed that most of the alpha particles directed at a metallic foil appear to pass through unhindered, with only a few deflected at large angles. What did he conclude?
(1) Alpha particles behave like waves when they interact with atoms.
(2) Atoms have most of their mass distributed loosely in an electron cloud.
(3) Atoms can easily absorb and reemit alpha particles.
(4) Atoms consist mainly of empty space and have small, dense nuclei.

✪ indicates material that is not part of the core curriculum.

✪ **39.** High-speed alpha particles strike a metal foil. Which element, when used in the foil, will tend to scatter the alpha particles through the greatest angles?
(1) platinum, with 78 elementary charges per nucleus
(2) silver, with 47 elementary charges per nucleus
(3) copper, with 29 elementary charges per nucleus
(4) vanadium, with 23 elementary charges per nucleus

Base your answers to questions 40 through 44 on Rutherford's experiments in which alpha particles were allowed to pass into a thin gold foil. All alpha particles had the same speed.

✪ **40.** The paths of the scattered alpha particles were
(1) hyperbolic (2) circular (3) parabolic (4) elliptical

✪ **41.** Some of the alpha particles were deflected. The explanation for this phenomenon is that
(1) electrons have a small mass
(2) electrons have a small charge
(3) the gold leaf was only a few atoms thick
(4) the nuclear charge and mass are concentrated in a small volume

✪ **42.** The alpha particles were scattered because of
(1) gravitational forces   (3) magnetic forces
(2) coulomb forces   (4) nuclear forces

✪ **43.** As the distance between the nuclei of the gold atoms and the paths of the alpha particles increases, the angle of scattering of the alpha particles
(1) decreases (2) increases (3) remains the same

✪ **44.** If a foil were used whose nuclei had a greater atomic number, the angle of scattering of the alpha particles would
(1) decrease (2) increase (3) remain the same

✪ **45.** In alpha-particle scattering, the nucleus produces an effect on the scattering angles. This is due primarily to the fact that the nucleus
(1) has a small total charge
(2) has a mass close to that of the alpha particles
(3) exerts coulomb forces
(4) is widely dispersed throughout the atom

✪ indicates material that is not part of the core curriculum.

✪ 46. Which diagram best represents the path of a positively charged particle as it passes near the nucleus of an atom?

(1)   (2)   (3)   (4)

47. As excited hydrogen atoms return to the ground state, they emit
(1) electrons   (2) protons   (3) photons   (4) neutrons

48. As an electron orbits a nucleus in the same energy level, the energy of the electron
(1) decreases   (2) increases   (3) remains the same

✪ 49. As the radius of an electron orbit in a Bohr atom increases, the magnitude of the energy of the atom
(1) decreases   (2) increases   (3) remains the same

✪ 50. Which is a characteristic of both the Bohr and the Rutherford atomic model?
(1) Neutrons exist in the nuclei of all atoms.
(2) Only a limited number of specified orbits is permitted.
(3) The nucleus is concentrated in a small, dense core.
(4) Electron energy level changes are in discrete amounts.

✪ 51. When a hydrogen atom changes from one energy level (E_1) to a lower energy level (E_2), the expression for the frequency of the emitted photon is
(1) $\dfrac{\lambda}{E}$ (2) $\dfrac{h\lambda}{E_1 - E_2}$ (3) $\dfrac{E_1 - E_2}{h}$ (4) $(E_1 - E_2)\dfrac{\lambda}{2}$

52. The lowest energy state of an atom is called its
(1) ground state   (3) initial energy state
(2) ionized state   (4) final energy state

53. When an excited atom emits a photon, the total energy of the atom
(1) decreases   (2) increases   (3) remains the same

54. When an electron changes from a higher energy state to a lower energy state within an atom, a quantum of energy is
(1) fissioned   (2) fused   (3) emitted   (4) absorbed

✪ indicates material that is not part of the core curriculum.

✪ 55. According to the Bohr model of the atom, an electron in a stable orbit does *not*
   (1) have potential energy   (3) undergo acceleration
   (2) emit radiation   (4) have kinetic energy

56. Which are emitted as atoms of a given element return to the ground state?
   (1) electrons   (3) alpha particles
   (2) photons   (4) neutrons

✪ 57. In his model of the atom, Bohr assumed that the electrons
   (1) are distributed evenly throughout the atom
   (2) are located only in the nucleus of the atom
   (3) are located only in a limited number of specified orbits
   (4) emit energy while in orbit

58. A hydrogen atom undergoes a transition from the $n = 3$ state to the ground state. The total number of different possible photon energies that may be emitted is
   (1) 1   (2) 2   (3) 3   (4) 4

59. A photon with an energy of 10.2 electron-volts is absorbed by a hydrogen atom. This may cause the energy state of the hydrogen atom to move from
   (1) $n = 1$ to $n = 2$ (3) $n = 1$ to $n = 4$
   (2) $n = 1$ to $n = 3$ (4) $n = 1$ to $n = 5$

60. The energy needed to ionize a hydrogen atom in the ground state is
   (1) 2.9 eV   (2) 3.2 eV   (3) 13.06 eV (4) 13.6 eV

61. What is the minimum amount of energy required to ionize a hydrogen atom in the $n = 2$ state?
   (1) 13.6 eV   (2) 10.2 eV   (3) 3.40 eV   (4) 0 eV

62. An atom changing from an energy state of –0.54 eV to an energy state of –0.85 eV will emit a photon whose energy is
   (1) 0.31 eV   (2) 0.54 eV   (3) 0.85 eV   (4) 1.39 eV

63. A hydrogen atom can be raised from the $n = 2$ state to the $n = 3$ state by a photon with an energy of
   (1) 1.89 eV   (2) 10.2 eV   (3) 12.1 eV   (4) 22.3 eV

64. A hydrogen atom in the ground state receives 10.2 electron-volts of energy. To which energy level may the atom become excited?
   (1) $n = 5$ (2) $n = 2$ (3) $n = 3$ (4) $n = 4$

✪ indicates material that is not part of the core curriculum.

65. A photon having an energy of 15.5 electron-volts is incident upon a hydrogen atom in the ground state. The photon may be absorbed by the atom and
   (1) ionize the atom
   (2) excite the atom to $n = 2$
   (3) excite the atom to $n = 3$
   (4) excite the atom to $n = 4$

66. A hydrogen atom is in the $n = 5$ energy state after having absorbed a 0.97-eV photon. What was the original energy state of the hydrogen atom?
   (1) $n = 1$
   (2) $n = 2$
   (3) $n = 3$
   (4) $n = 4$

67. A model of the atom in which the electrons can exist only in specified orbits was suggested by
   (1) Bohr
   (2) Planck
   (3) Einstein
   (4) Rutherford

68. In a hydrogen atom the electron makes the following successive transitions:

$$n = 5 \rightarrow n = 4 \rightarrow n = 3 \rightarrow n = 2 \rightarrow n = 1$$

The emitted photon energies for the successive transitions
   (1) decrease
   (2) increase
   (3) remain the same

Base your answers to questions 69 and 70 on the Energy Levels for Hydrogen chart in the Physics Reference Tables.

69. Which photon could be absorbed by a hydrogen atom in the ground state?
   (1) a 11.0-eV photon
   (2) a 10.2-eV photon
   (3) a 3.4-eV photon
   (4) a 0.54-eV photon

70. Which energy-level jump would show as a bright line in the visible spectrum of hydrogen?
   (1) 1 to 2
   (2) 2 to 3
   (3) 3 to 2
   (4) 4 to 7

71. A hydrogen atom, in the ground state, is bombarded by an 11-electron-volt photon. Which statement best describes the interaction that occurs?
   (1) The photon collides elastically and leaves the atom with an energy of 11 electron-volts.
   (2) The photon collides inelastically and retains an energy of 0.8 electron-volt.
   (3) The photon collides inelastically and disappears.
   (4) The atom is completely ionized to a +1 ion, and the photon disappears.

72. A photon with an energy of 20. electron-volts is completely absorbed by a hydrogen atom in the ground state, ionizing the atom. What is the approximate energy of the incoming photon?
(1) 8.0×10^{-20} J
(3) 3.2×10^{-18} J
(2) 1.6×10^{-19} J
(4) 20 J

73. A hydrogen atom undergoes a transition from the $n = 4$ state to the $n = 1$ state. The energy of the single photon emitted during the transition is approximately
(1) 13.6 eV   (2) 12.75 eV   (3) 2.55 eV   (4) 0.85 eV

Base your answers to questions 74 through 77 on the information and diagram below and on the Physics Reference Tables.

The diagram represents the model of the Bohr hydrogen atom in the ground state. The speed of the electron is 2.17×10^6 meters per second.

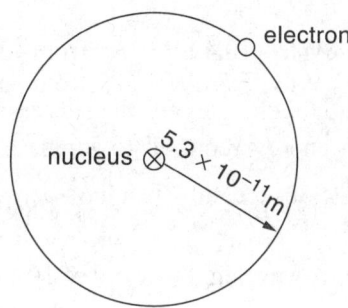

74. What is the kinetic energy of the electron in the ground state?
(1) 4.36×10^{-18} J
(3) 9.87×10^{-24} J
(2) 2.14×10^{-18} J
(4) 0

75. Compared to the electrostatic force between the proton and the electron, the centripetal force on the electron is
(1) one-fourth as much
(3) the same
(2) one-half as much
(4) twice as much

76. The electrostatic force between the electron and the proton is approximately
(1) 4.3×10^{-18} N
(3) 8.2×10^{-8} N
(2) 4.1×10^{-8} N
(4) 5.1×10^1 N

77. Compared to the electrostatic force between the proton and the electron, the gravitational force between them is
(1) less   (2) greater   (3) the same

Base your answers to questions 78 and 79 on the diagram below, which shows some of the energy levels of a mercury atom. The range of the energies of visible photons is approximately 1.5 eV to 3.0 eV.

78. Which energy level transition would cause the emission of visible photons?
    (1) from −5.5 eV to −10.4 eV   (3) from −3.7 eV to −1.6 eV
    (2) from −3.7 eV to −5.5 eV   (4) from −5.5 eV to −1.6 eV

79. The minimum energy required to ionize a mercury atom in the ground state is
    (1) 4.9 eV   (2) 6.7 eV   (3) 8.8 eV   (4) 10.4 eV

80. A photon with a wavelength of 6.0×10^{-7} meter would have an energy of
    (1) 9.4×10^{-20} J   (3) 6.0×10^{-7} J
    (2) 3.3×10^{-19} J   (4) 9.4×10^{18} J

81. A 3.0-electron-volt photon would have a wavelength of approximately
    (1) 4×10^{-7} m   (3) 6×10^{-7} m
    (2) 5×10^{-7} m   (4) 7×10^{-7} m

Base your answer to question 82 on the cartoon below and your knowledge of physics.

82. In the cartoon, Einstein is contemplating the equation for the principle that
(1) the fundamental source of all energy is the conversion of mass into energy
(2) energy is emitted or absorbed in discrete packets called photons
(3) mass always travels at the speed of light in a vacuum
(4) the energy of a photon is proportional to its frequency

Answers to Part A and B–1 questions can be found on page 459.

PART B-2 AND C QUESTIONS

Base your answers to questions 1 and 2 on the diagram below, which shows some energy levels for an atom of an unknown substance.

1. Determine the minimum energy necessary for an electron to change from the *B* energy level to the *F* energy level.

2. Calculate the frequency of the photon emitted when an electron in this atom changes from the *F* energy level to the *B* energy level. [Show all work, including the equation and substitution with units.]

Base your answers to questions 3 through 6 on the information below.

When an electron in an excited hydrogen atom falls from a higher to a lower energy level, a photon having a wavelength of 6.58×10^{-7} meter is emitted.

3. Calculate the energy of a photon of this light wave in joules. [Show all calculations, including the equation and substitution with units.]

4. Convert the energy of the photon to electronvolts.

5. Determine which *two* energy levels the electron has fallen between to emit this photon.

6. Is this photon an X-ray photon? Justify your answer.

Base your answers to questions 7 and 8 on the information below.

In a mercury atom, as an electron moves from energy level i to energy level a, a single photon is emitted.

7. Determine the energy, in electronvolts, of this emitted photon.

8. Determine this photon's energy, in joules.

Base your answers to questions 9 through 11 on the information below.

A photon with a frequency of 5.02×10^{14} hertz is absorbed by an excited hydrogen atom. This causes the electron to be ejected from the atom, forming an ion.

9. Calculate the energy of this photon in joules. [Show all work, including the equation and substitution with units.]

10. Determine the energy of this photon in electronvolts.

_____ eV

11. What is the number of the *lowest* energy level (closest to the ground state) of a hydrogen atom that contains an electron that would be ejected by the absorption of this photon?

$n =$ _____

Base your answers to questions 12 and 13 on the statement below.

The spectrum of visible light emitted during transitions in excited hydrogen atoms is composed of blue, green, red, and violet lines.

12. What characteristic of light determines the amount of energy carried by a photon of that light?
(1) amplitude (2) frequency (3) phase (4) velocity

13. Which color of light in the visible hydrogen spectrum has photons of the shortest wavelength?
(1) blue (2) green (3) red (4) violet

14. Determine the frequency of a photon whose energy is 3.00×10^{-19} joule.

Base your answers to questions 15 through 18 on the Energy Level Diagram for Hydrogen in the *Reference Tables for Physical Settings/Physics*.

15. Determine the energy, in electronvolts, of a photon emitted by an electron as it moves from the $n = 6$ to the $n = 2$ energy level in a hydrogen atom.

16. Convert the energy of the photon to joules.

17. Calculate the frequency of the emitted photon. [Show all work, including the equation and substitution with units.]

18. Is this the only energy and/or frequency that an electron in the $n = 6$ energy level of a hydrogen atom could emit? Explain your answer.

Answers to Part B–2 and C questions can be found on pages 542–544.

NUCLEAR ENERGY

KEY IDEAS

The main components of the atomic nucleus are protons and neutrons, also called nucleons. The atomic number of an atom is the number of protons its nucleus contains; the mass number is the sum of the numbers of protons and neutrons in the nucleus.

Isotopes are atoms with the same atomic numbers but differing mass numbers. The unit of atomic mass is based on the isotope carbon-12. One atomic mass unit is equivalent to 931 million electron-volts of energy. The mass of a nucleus is always less than the mass of its individual nucleons. The energy equivalent of this lost mass is related to the stability of the nucleus. Inside the nucleus, the existence of short-range forces contribute to this stability.

A nuclear reaction involves changes in one or more atomic nuclei. Nuclear equations represent nuclear reactions. A nuclear equation is balanced when the atomic and mass numbers agree on both sides of the arrow.

Radioactive decay is a series of naturally occurring nuclear reactions, which occur in order to increase the stability of the resulting nuclei. In alpha decay, alpha particles are ejected from the nucleus. In beta decay and positron emission, negative and positive beta particles, respectively, are ejected from their parent nuclei. In electron capture, a nucleus absorbs one of the inner electrons surrounding it. In gamma decay, a nucleus ejects a gamma photon, thereby lowering the nuclear energy.

Radioactive decay occurs according to strict mathematical laws. As a result, the half life of a radioactive substance, that is, the amount of time required to reduce a sample of the substance to one-half its initial value, is constant under all conditions.

Nuclear reactions can by induced by bombarding nuclei with subatomic particles such as alpha particles. The energies of the bombarding particles needed to produce a reaction are achieved by introducing the particles into a particle accelerator such as a cyclotron. Only charged particles may be accelerated in this way.

It is now known that the proton and the neutron are not funda-mental particles. The existence of these nucleons and other nuclear particles has been explained by the presence of six varieties of parti-cles called *quarks*. The electric charge on these particles is either $-\frac{1}{3}$ or $+\frac{2}{3}$ of the elementary charge.

KEY OBJECTIVES

At the conclusion of this chapter you will be able to:

* Define the term *nucleon,* and distinguish between the two nucleons.
* Interpret the parts of nuclear symbol, and define the terms *atomic number, mass number,* and *isotope.*
* Define the terms *mass defect* and *binding energy*, and explain how they contribute to the stability of the nucleus.
* Explain how nuclear forces differ from gravitational and electro-magnetic forces.
* Balance a nuclear equation.
* List examples of nuclear reactions that occur naturally.
* Describe various types of induced nuclear reactions.
* Describe how quarks are involved in nuclear structure.

14.1   INTRODUCTION

After British physicist Ernest Rutherford proposed his nuclear model in the early part of the twentieth century, physicists began to question whether the nucleus had a structure of its own. At present, there is still no complete answer to this question. By the mid-1930s, however, a simple nuclear model was in place and we will examine that model in this chapter.

14.2   NUCLEONS

The components of the nucleus are called **nucleons** and are described by a number of properties, including electric charge. Nuclear charge is usually measured in terms of the elementary charge (e) rather than the coulomb.

Two of the principal nucleons are the *proton* and the *neutron*. The proton has a charge of $+1e$, and the neutron is uncharged.

14.3 NUCLEAR SYMBOLS

All atomic nuclei (also called *nuclides*)—and their component nucleons—may be represented by the same general symbol:

$$_{Z}^{A}\text{X}$$

The letter X represents the letter(s) used to identify the particle; the letter Z, representing the **atomic number**, indicates the number of elementary charges present (assumed to be positive unless a negative sign is written); and the letter A, representing the **mass number**, is equal to the sum of neutrons and protons present.

The number of neutrons (N) present in an atomic nucleus is given by this expression:

================= **PHYSICS CONCEPTS** =================

$$N = A - Z$$

*Note that this equation does not appear this same way on the Reference Tables.

Using this representation, we can write the symbols for the proton, neutron, and electron, respectively, as follows:

$$_{1}^{1}\text{H (proton)} \quad _{0}^{1}\text{n (neutron)} \quad _{-1}^{0}\text{e (electron)}$$

The symbol for the proton is a result of the fact that a proton is the nucleus of the simplest hydrogen atom. While the electron is not normally considered a nuclear particle, there are occasions when it is produced in the nucleus.

PROBLEM
Identify the nucleons present in the atomic nucleus whose symbol is $_{10}^{22}\text{Ne}$.

SOLUTION

The atomic number of neon (Ne) is 10; therefore, the nucleus contains 10 protons.

The mass number of this nucleus is 22; therefore, the nucleus contains 22 protons *and* neutrons.

The number of neutrons is found by subtracting the atomic number from the mass number:

22 protons and neutrons – 10 protons = 12 neutrons.

PROBLEM

Calculate the number of protons and neutrons in these nuclei: (a) ^2_1H and (b) ^3_1H.

SOLUTION

(a) ^2_1H contains 1 proton and 1 neutron.

(b) ^3_1H contains 1 proton and 2 neutrons.

✪ 14.4   ISOTOPES

All the nuclei in a sample of a given element contain the same number of protons. They may, however, contain different numbers of neutrons. Nuclei that have the same atomic number but have different mass numbers are called **isotopes**. For example, ^1_1H, ^2_1H, and ^3_1H are all isotopes of the element hydrogen. Sometimes an isotope is written without its atomic number (^{22}Ne) or with the name of its element and its mass number (hydrogen-2).

14.5   NUCLEAR MASSES

Since nuclear particles have very small masses, they are usually measured in terms of the *atomic mass unit* (u) rather than the kilogram. The proton and the neutron each have an approximate mass of 1 atomic mass unit, although the neutron is slightly more massive than the proton. The basis of the nuclear-mass scale is the isotope carbon-12, and an *atom* of this isotope is assigned an exact mass of 12 atomic mass units. One atomic mass unit is approximately equal to 1.66×10^{-27} kilogram.

✪ indicates material that is not part of the core curriculum.

Equivalents of Nuclear Masses

Nuclear masses may also be expressed in terms of their *energy equivalents*. Using Einstein's famous mass-energy relationship ($E = mc^2$), it can be shown (see the next problem) that 1 atomic mass unit of mass is equivalent to 931 mega-electron-volts of energy.

PROBLEM
Calculate the energy equivalent (in MeV) of 1 atomic mass unit of mass.

SOLUTION
We need the following relationships in order to solve the problem:

$$1 \text{ eV} = 1.60 \times 10^{-19} \text{ J}$$

$$1 \text{ MeV} = 10^6 \text{ eV}$$

$$1 \text{ u} = 1.66 \times 10^{-27} \text{ kg}$$

$$c = 3.00 \times 10^8 \text{ m/s}$$

First, we calculate the energy equivalent of 1 u in *joules,* using $E = mc^2$:

$$E = (1.66 \times 10^{-27} \text{ kg})(3.00 \times 10^8 \text{ m/s})^2 = 1.49 \times 10^{-10} \text{ J}$$

Next, we convert this number to electron-volts and then to mega-electron-volts:

$$1.49 \times 10^{-10} \text{ J} \left(\frac{1 \text{ eV}}{1.60 \times 10^{-19} \text{ J}} \right) = 9.31 \times 10^8 \text{ eV}$$

$$9.31 \times 10^8 \text{ eV} \left(\frac{1 \text{ MeV}}{10^6 \text{ eV}} \right) = 931 \text{ MeV}$$

The following table lists the masses and energy equivalents of some nuclear particles:

Particle	Mass (u)	Energy Equivalent (MeV)
Electron	0.0005486	0.5110
Proton	1.007276	938.3
Hydrogen (1 atom)	1.007825	938.8
Neutron	1.008665	939.6

Mass Defect and Binding Energy

A nucleus such as $^{56}_{26}$Fe contains 26 (positive) protons concentrated in an extremely small space. We might suppose that the repulsion of the positive charges ought to tear the nucleus apart. However, this nucleus is quite stable. To explain the reason, we need to compare two quantities: the mass of the nucleus itself and the total mass of its nucleons.

PROBLEM
Compare the mass of a $^{56}_{26}$Fe nucleus (mass = 55.9206 u) with the total mass of its nucleons.

SOLUTION
Using the preceding table and the fact that this nucleus contains 26 protons and 30 neutrons (why?), we can calculate the total mass of the nucleons:

Mass of 26 protons   = (26)(1.007276 u) = 26.1892 u
Mass of 30 neutrons = (30)(1.008665 u) = 30.2600 u
Total mass of nucleons   = 56.4492 u

Subtracting the two values (56.4992 u – 55.9206 u), we see that the mass of the nucleus is 0.5286 u *less* than the mass of its nucleons!

The mass difference that we just calculated is known as the **mass defect** of the nucleus. Its energy equivalent is approximately 492 mega-electron-volts (How can we calculate this quantity?) and is called the *binding energy* of the nucleus. The **binding energy** of a nucleus is the amount of energy that must be added in order to separate the nucleus into its component nucleons.

✪ 14.6   AVERAGE BINDING ENERGY PER NUCLEON

One way of estimating the stability of a nucleus is by referring to a quantity known as the average *binding energy per nucleon*. It is calculated by dividing the total binding energy of the nucleus by the number of nucleons present (i.e., the mass number). In the problem in Section 14.5, the binding energy of 492 mega-electron-volts is divided by 56 nucleons to yield 8.79 mega-electron-volts per nucleon. In general, the larger the binding energy per nucleon, the more stable is the nucleus.

The graph below illustrates how the binding energy per nucleon varies with the number of nucleons in a nucleus. The nuclei located toward the center of the graph (e.g., ^{56}Fe) have larger values than the nuclei located at either

✪ indicates material that is not part of the core curriculum.

end (e.g., ^{2}H and ^{238}U). In Section 14.10, we will see how some of the less stable nuclei can be used in the production of energy.

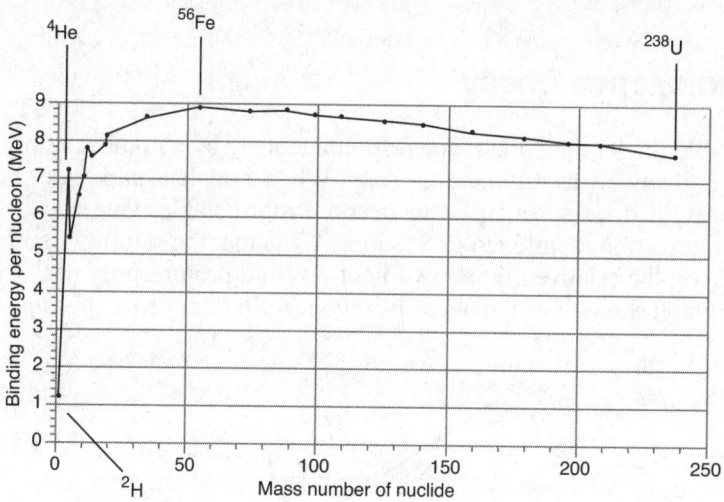

14.7   NUCLEAR FORCES

The stability of a nucleus is tied to the existence of two *nuclear forces*. These forces, called the *strong* and *weak interactions*, are much more powerful at the very small distances present within the nucleus than are gravitational or electromagnetic forces. At larger distances, however, the strong and weak interactions lose their effectiveness, and for this reason they are called *short-range forces*.

14.8   NUCLEAR REACTIONS

Nuclear Equations

A nuclear reaction is a change that occurs within or among atomic nuclei and is represented by a *nuclear equation*, such as the following:

$$^{15}_{7}N + ^{1}_{1}H \rightarrow ^{12}_{6}C + ^{4}_{2}He$$

If we examine this equation carefully, we note that the sum of the atomic numbers on the left side (7 + 1) equals the sum of the atomic numbers on the right side (6 + 2). This equality demonstrates the fact that electric charge must be conserved in a nuclear reaction. Similarly, the sum of the mass numbers on the left side of the equation (15 + 1) equals the sum of the mass numbers on the right side (12 + 4).

This nuclear equation is considered to be *balanced* because both charge and mass number are conserved. All of the nuclear equations we write will be balanced equations.

✪ Radioactive Decay

Less stable nuclei may break down spontaneously by a number of processes known collectively as *radioactive decay*. When a nucleus undergoes radioactive decay, it does so in order to become more stable. One aspect of the strong interaction mentioned in Section 14.7 is that the stability of a nucleus depends on the relative numbers of neutrons and protons present.

The graph shows how the *neutron-proton* (N/P) ratio varies in *stable* nuclei.

As the number of protons increases, the N/P ratio rises from 1 to 1.5. The larger number of neutrons reduces the repulsion among the positively charged protons. The graph does not continue beyond 83 protons because there are no stable isotopes with atomic numbers greater than 83.

Radioactive decay is an attempt to "correct" the ratio of neutrons to protons. However, stability may not occur immediately; a series of decay reactions may be required before a stable nucleus is finally produced.

✪ indicates material that is not part of the core curriculum.

ALPHA DECAY

The following nuclear reaction:

$$^{238}_{92}U \rightarrow\ ^{234}_{90}Th +\ ^4_2He$$

is an example of *alpha decay*. The uranium-238 nucleus (the *parent nucleus*) breaks down to produce a thorium-234 nucleus (the *daughter nucleus*) and a helium-4 nucleus, also known as an *alpha particle*. A symbol for the alpha particle is α.

Whenever alpha decay occurs, the atomic number of the daughter nucleus (as compared with its parent) is *decreased* by 2 and its mass number is *decreased* by 4. Many heavier radioactive nuclei, especially those with atomic numbers greater than 83, undergo alpha decay as a way of reducing the number of protons and neutrons present.

PROBLEM

The nuclide $^{222}_{86}Rn$ undergoes alpha decay and forms an isotope of polonium (Po). Write the nuclear equation for this process.

SOLUTION

Our problem is to find the atomic and mass numbers of the Po nuclide. We write the alpha decay as follows:

$$^{222}_{86}Rn \rightarrow\ ^A_ZPo +\ ^4_2He$$

Since the atomic and mass numbers must be equal on both sides of the equation, the atomic number of polonium must be 84 and the mass number of this nuclide must be 218. Therefore, we can now write the complete equation:

$$^{222}_{86}Rn \rightarrow\ ^{218}_{84}Po +\ ^4_2He$$

BETA (–) DECAY

Certain nuclei undergo radioactive decay and produce an *electron*, also known as a *beta (–) particle*, in the reaction. A symbol for the beta (–) particle is β^-.

Beta decay is governed by the weak interaction mentioned in Section 14.7. The following equation illustrates the process of beta (–) decay:

$$^{214}_{82}Pb \rightarrow\ ^{214}_{83}Bi +\ ^0_{-1}e$$

In beta (–) decay, the atomic number of the daughter is increased by 1 while its mass number remains unchanged. Beta (–) decay occurs in nuclei whose *N/P* ratios lie *above* the band of stability illustrated in the graph on page 425.

Actually, another subatomic particle, called an *antineutrino* is also produced in beta (–) decay. Even though charge and mass number are conserved,
an antineutrino must be produced in order to conserve momentum and
energy as well.

The antineutrino is an example of an *antiparticle*. Every subatomic particle is associated with its own unique antiparticle. Particles and their antiparticles have the same mass, but their electric charges are *opposite* in sign. If a
particle and its antiparticle are brought into contact, all of the mass is transformed entirely into electromagnetic energy in the form of two gamma-ray
photons.

PROBLEM

The nuclide $^{14}_{6}C$ undergoes beta (–) decay and forms an isotope of nitrogen
(N). Write the nuclear equation for this process.

SOLUTION

Our problem is to find the atomic and mass numbers of the N nuclide. We
will write the beta (–) decay as follows:

$$^{14}_{6}C \rightarrow {}^{A}_{Z}N + {}^{0}_{-1}e$$

Since the atomic and mass numbers must be equal on both sides of the equation, the atomic number of nitrogen must be 7 and the mass number of this
nuclide must be 14. Therefore, we can now write the complete equation:

$$^{14}_{6}C \rightarrow {}^{14}_{7}N + {}^{0}_{-1}e$$

Note that the *N/P* ratio of the parent nucleus (^{14}C) is 1.33 (8/6), while the *N/P*
ratio of the daughter nucleus (^{14}N) is 1.00 (7/7). Therefore, beta (–) decay is
a process that *decreases* the *N/P* ratio.

POSITRON DECAY

The name *positron* is a combination of the word parts <u>positive</u> elec<u>tron</u>. The
positron is the *antiparticle* of the electron. The symbol for the positron
is ${}^{0}_{+1}e$; another symbol for the positron is β^+.

The following equation is an example of positron decay:

$$^{19}_{10}Ne \rightarrow {}^{19}_{9}F + {}^{0}_{+1}e$$

In positron decay, the atomic number of the daughter nucleus is *decreased*
by 1 while its mass number remains *unchanged*. An additional particle
(called the *neutrino* and symbolized as ν) is also produced in positron decay
in order to conserve energy and momentum.

If we examine the *N/P* ratios of the parent and daughter nuclei in the equation given above, we see that positron decay *increases* the *N/P* ratio.

ELECTRON CAPTURE

This type of reaction occurs when a nucleus captures one of the inner electrons orbiting the atom. In this reaction, a neutrino is also produced. The following equation is an example of electron capture:

$$^{7}_{4}\text{Be} + ^{0}_{-1}\text{e} \rightarrow ^{7}_{3}\text{Li}$$

Electron capture also serves to *increase* the *N/P* ratio of nuclei.

GAMMA DECAY

Like the electrons in an atom, the nucleus contains energy levels. Occasionally, a nucleus will enter an excited state, known as a *metastable* state. We symbolize a metastable nucleus by adding an *m* to its mass number (e.g., 99m). Eventually, the nucleus will return to its normal state and then a high-energy gamma-ray photon will be emitted.

The symbol for the gamma-ray photon is γ. The following equation is an example of gamma decay:

$$^{99m}_{43}\text{Tc} \rightarrow ^{99}_{43}\text{Tc} + \gamma$$

✪ 14.9 INDUCED NUCLEAR REACTIONS

All of the nuclear reactions we have studied so far have been *natural* processes. We now turn our attention to nuclear changes that have been produced artificially or *induced*. To induce a nuclear reaction, a target nucleus is bombarded with a nuclear particle.

Particle Accelerators

A *particle accelerator* is a device that uses electric and magnetic fields to provide a charged bombarding nuclear particle with sufficient kinetic energy to induce the desired nuclear reaction. As an analogy, consider a bullet fired at a wall at a speed of 10 miles per hour. At this slow speed, the kinetic energy of the bullet would have hardly any effect on the wall. If the bullet were fired at 600 miles per hour, however, its effect on the wall would be devestating!

Examples of modern particle accelerators include the *Van de Graaff accelerator*, the *linear accelerator*, the *cyclotron*, the *synchrotron,* and the *large electron-positron (LEP) collider*. These devices can supply bombarding particles with kinetic energies ranging from 10^3 to 10^{12} electron-volts.

✪ indicates material that is not part of the core curriculum.

☼ Artificial Transmutation

The first induced nuclear reactions used alpha particles (because of their large masses) as the bombarding particles. In 1919, Rutherford bombarded nitrogen-14 nuclei with alpha particles and noted that protons were emitted. In this reaction:

$$^{14}_{7}N + ^{4}_{2}He \rightarrow ^{1}_{1}H + ^{17}_{8}O$$

nitrogen-14 was artificially changed or *transmuted* into oxygen-17. In 1932, English physicist Sir James Chadwick bombarded beryllium-9 and identified a stream of uncharged particles that we now call *neutrons*. The reaction is shown below:

$$^{9}_{4}Be + ^{4}_{2}He \rightarrow ^{1}_{0}n + ^{12}_{6}C$$

In 1934, French physicists Frédéric Joliot-Curie and Irène Joliot-Curie bombarded aluminum-27 and produced the first artificially radioactive isotope, phosphorus-30:

$$^{27}_{13}Al + ^{4}_{2}He \rightarrow ^{1}_{0}n + ^{30}_{15}P$$

The phosphorus-30 undergoes positron decay (see page 428) and forms silicon-30.

14.10 FUNDAMENTAL PARTICLES AND INTERACTIONS

The model of nuclear structure continues to evolve. Over the years, as physicists probed the nucleus, a host of particles were discovered whose functions were largely unknown. In an attempt to explain the existence of these particles, a number of theories were developed.

One of the most successful models is known as the *standard model*, which assumes that four fundamental interactions (also known as fundamental forces) operate in the universe: *electromagnetic*, *weak*, *strong*, and *gravitational*.

FUNDAMENTAL FORCES

Interaction	Relative Strength	Range	Mediating Particle
Strong	1	Short	Gluon
Electromagnetic	0.0073	Long	Photon
Weak	10^{-9}	Very short	W,Z Boson
Gravitational	10^{-38}	Long	Graviton

☆ indicates material that is not part of the core curriculum.

As can be ascertained from the table above, the four fundamental forces in our present Universe are distinct and have very different characteristics. The current thinking in theoretical physics is that this was not always the case. There is a strong belief (yet to be confirmed experimentally) that in the very early Universe when temperatures were very high compared with today, the weak, electromagnetic, and strong forces were unified into a single force. Only when the temperature dropped did these forces separate from each other, with the strong force separating first and then at a still lower temperature the electromagnetic and weak forces separating, leaving us with the four distinct forces we see in our present Universe. The process of the forces separating from each other is called **spontaneous symmetry breaking**.

There is further speculation, which is even less firm than that above, that at even higher temperatures all four forces were unified into a single force.

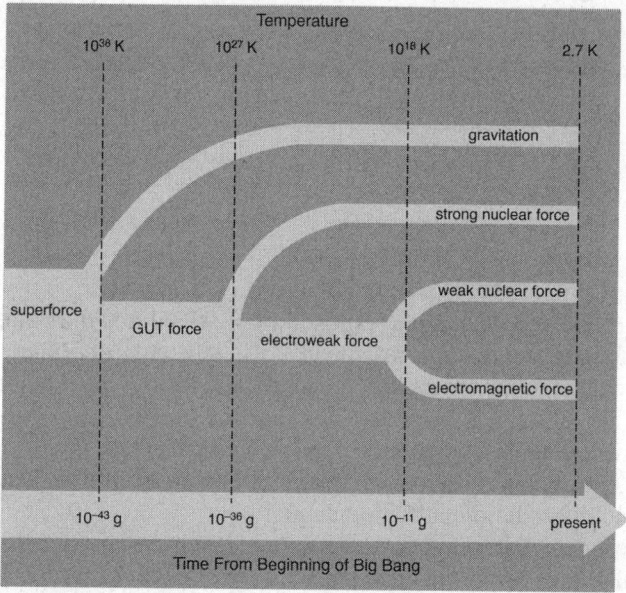

Unified and separate forces: If current theories are correct, the individual forces separated out one by one from a single unified superforce as the very early universe expanded and cooled down.

The standard model describes the behavior and relationships among the first three of these interactions. At present, attempts to incorporate gravitational force within the standard model have been unsuccessful. (At this writing, experiments on a nuclear particle known as the *muon* suggest that the standard model might need to be reformulated in order to explain the results of the experiments.)

An important feature of the standard model is the recognition that there are fundamental particles, known as quarks, which have *fractional elementary charges* and are the building blocks of protons, neutrons, and other nuclear particles.

Particles that interact by the strong interaction are called hadrons. This general classification includes mesons and baryons, but specifically excludes leptons, which do not interact by the strong force. The weak interaction acts on both hadrons and leptons.

The chart below, which is included on the Reference Tables, summarizes the relationship between these different particles of matter.

Classification of Matter

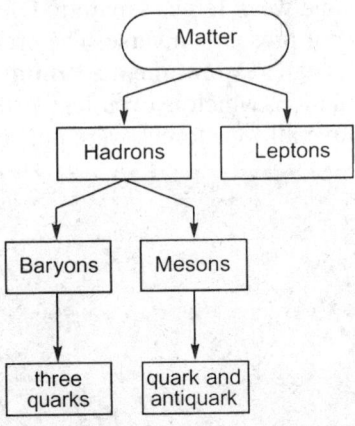

Hadrons are viewed as being composed of quarks, known as mesons (which are intermediate mass particles made up of a quark–antiquark pair) or as baryons (which are massive particles made up of three quarks in the standard model). This class of particles includes the proton and neutron. Other baryons are the lambda, sigma, xi, and omega particles.

Recent experimental evidence shows the existence of five-quark combinations, which are being called pentaquarks.

There are six varieties of quarks (and their antimatter counterparts), whimsically named *up*, *down*, *charm*, *strange*, *top*, and *bottom*. The property with which the names of the quarks are associated is known as *flavor*. (Who says scientists don't have a sense of humor?) The fractional elementary charges of the six quarks are shown in the table:

Flavor	Charge
Up (u)	$+2/3\ e$
Down (d)	$-1/3\ e$
Charm (c)	$+2/3\ e$
Strange (s)	$-1/3\ e$
Top (t)	$+2/3\ e$
Bottom (b)	$-1/3\ e$

According to the standard model scheme, the proton has the structure *uud* and the neutron has the structure *udd*. Using the table, we can calculate the charges on the proton and neutron:

$$\text{Proton} = uud = [(+2/3 \; e) + (+2/3 \; e) + (-1/3 \; e)] = e$$
$$\text{Neutron} = udd = [(+2/3 \; e) + (-1/3 \; e) + (-1/3 \; e)] = 0$$

It is important to note that in the above results and in all combinations of quarks that form mesons and baryons, the sum of the fractional charges always adds up to a whole integer value multiple of *e*, the elementary charge that Robert Millikan found on an electron (see Chapter 8). This is in complete agreement with all well-known scientific observations.

The lepton has a mass less than that of a proton. Electrons and neutrinos are classified as leptons. The chart below, which also appears on the Physics Reference Tables, summarizes the properties of each of the six members of the lepton family.

Leptons

electron	muon	tau
e	μ	τ
–1e	–1e	–1e

electron neutrino	muon neutrino	tau neutrino
ν_e	ν_μ	ν_τ
0	0	0

Note: For each particle there is a corresponding antiparticle with a charge opposite that of its associated particle.

Antiparticles are particles that have the identical mass, lifetime, and spin but opposite charge (if charged) and opposite sign magnetic moment. An antiparticle is denoted by a bar over the symbol for the particle. As previously defined in Section 14.8, the positron is the antiparticle to the electron, and it is denoted by the symbol $\bar{e}$. Antimatter is material consisting of atoms that are comprised of antiprotons, antineutrons, and positrons.

The energies needed to reduce protons, neutrons, and other nuclear particles into their constituent quarks are so high that these quarks cannot be isolated as separate particles. Therefore, their existence has been demonstrated only by indirect means. Whether quarks represent the ultimate structure of matter or whether there are even smaller subunits continues to be the subject of research.

PART A AND B-1 QUESTIONS

✪ 1. Neutral atoms always have equal numbers of
  (1) protons and neutrons
  (2) electrons and neutrons
  (3) protons and electrons
  (4) protons and positrons

✪ 2. What is the relationship between the atomic number Z, the mass number A, and the number of neutrons N in a nucleus?
  (1) $A = Z + N$
  (2) $A = Z - N$
  (3) $A = N/Z$
  (4) $A = NZ$

✪ 3. The ratio of the magnitude of charge on an electron to the magnitude of charge on a proton is
  (1) $1 : 2$
  (2) $1 : 1$
  (3) $1 : 6.25 \times 10^{18}$
  (4) $1 : 1840$

✪ 4. What is the mass number of an atom with 9 protons, 11 neutrons, and 9 electrons?
  (1) 9   (2) 18   (3) 20   (4) 29

✪ 5. An atom consists of 9 protons, 9 electrons, and 10 neutrons. The number of nucleons in this atom is
  (1) 0   (2) 9   (3) 19   (4) 28

✪ 6. A neutral atom could be composed of
  (1) 4 electrons, 5 protons, 6 neutrons
  (2) 5 electrons, 5 protons, 6 neutrons
  (3) 6 electrons, 3 protons, 6 neutrons
  (4) 0 electrons, 5 protons, 5 neutrons

✪ 7. Isotopes of the same element have the same number of
  (1) neutrons and protons, only
  (2) neutrons and electrons, only
  (3) protons and electrons, only
  (4) electrons, protons, and neutrons

✪ 8. If the number of neutrons in an atom increases, the atomic number of the atom
  (1) decreases   (2) increases   (3) remains the same

✪ 9. As the mass number of an isotope increases, its atomic number
  (1) decreases   (2) increases   (3) remains the same

✪ 10. As the number of protons in a nucleus increases, its atomic number
  (1) decreases   (2) increases   (3) remains the same

✪ indicates material that is not part of the core curriculum.

✪ 11. A lithium nucleus contains 3 protons and 4 neutrons. What is the atomic number of the nucleus?
   (1) 1   (2) 7   (3) 3   (4) 4

✪ 12. What is the number of neutrons in the nucleus of $^{222}_{86}Rn$?
   (1) 86   (2) 136   (3) 222   (4) 308

✪ 13. A neutral atom has 24 neutrons and 20 protons. The number of electrons in the atom is
   (1) 24   (2) 20   (3) 44   (4) 4

✪ 14. Which atom is an isotope of $^{238}_{92}U$?
   (1) $^{238}_{91}X$ (2) $^{235}_{92}X$ (3) $^{238}_{93}X$ (4) $^{235}_{93}X$

15. What type of particle has a charge of 1.6×10^{-19} coulomb and a rest mass of 1.67×10^{-27} kilogram?
   (1) a proton   (3) a neutron
   (2) an electron   (4) an alpha particle

16. An atomic mass unit (u) is approximately equal to the mass of
   (1) an alpha particle   (3) a photon
   (2) an electron   (4) a proton

✪ 17. The mass of a nucleus is less than the total mass of its nucleons. This fact indicates that some of the mass has been converted to
   (1) radioactivity   (3) binding energy
   (2) photoelectric effect   (4) thermal energy

18. What is the energy equivalent of a mass of 1 kilogram?
   (1) 9×10^{16} J   (2) 9×10^{13} J   (3) 9×10^{10} J   (4) 9×10^{7} J

19. As the binding energy of a nucleus increases, the energy required to separate the nucleus into nucleons
   (1) decreases   (2) increases   (3) remains the same

20. Which fundamental force is primarily responsible for the attraction between protons and electrons?
   (1) strong   (3) gravitational
   (2) weak   (4) electromagnetic

21. The total conversion of 1.00 kilograms of the Sun's mass into energy yields
   (1) 9.31×10^{2} MeV   (3) 3.00×10^{8} J
   (2) 8.38×10^{19} MeV   (4) 9.00×10^{16} J

✪ indicates material that is not part of the core curriculum.

22. Which graph best represents the relationship between energy and mass in the mass-energy equation?

(1)   (2)   (3)   (4)

23. What is the energy equivalent of a mass of 0.026 kilogram?
(1) 2.34×10^{15} J   (3) 2.34×10^{17} J
(2) 2.3×10^{15} J   (4) 2.3×10^{17} J

24. For a particular nuclear decay, the mass of the products is 0.01 atomic mass unit less than the original nucleus. The total energy released during this decay is
(1) 1.07×10^{-4} Mev   (3) 9.31 Mev
(2) 1.07 Mev   (4) 9.31×10^{6} Mev

25. If the mass of one proton is totally converted into energy, it will yield a total energy of
(1) 5.1×10^{-19} J   (3) 9.3×10^{8} J
(2) 1.5×10^{-10} J   (4) 9.0×10^{16} J

26. The nuclear force that binds nucleons together in the atom is relatively
(1) strong and of long range   (3) weak and of short range
(2) strong and of short range   (4) weak and of long range

Base your answers to questions 27 and 28 on the table below, which shows data about various subatomic particles.

Subatomic Particle Table

Symbol	Name	Quark Content	Electric Charge	Mass (GeV/c^2)
p	proton	uud	+1	0.938
$\bar{p}$	antiproton	$\bar{u}\bar{u}\bar{d}$	−1	0.938
n	neutron	udd	0	0.940
λ	lambda	uds	0	1.116
Ω⁻	omega	sss	−1	1.672

27. Which particle listed on the table has the opposite charge of, and is more massive than, a proton?
   (1) antiproton
   (2) neutron
   (3) lambda
   (4) omega

28. All the particles listed on the table are classified as
   (1) mesons   (2) hadrons   (3) antimatter   (4) leptons

29. A subatomic particle could have a charge of
   (1) 5.0×10^{-20} C
   (2) 8.0×10^{-20} C
   (3) 3.2×10^{-19} C
   (4) 5.0×10^{-19} C

30. The diagram below represents the sequence of events (steps 1 through 10) resulting in the production of a D^- meson and a D^+ meson. An electron and a positron (antielectron) collide (step 1), annihilate each other (step 2), and become energy (step 3). This energy produces an anticharm quark and a charm quark (step 4), which then split apart (steps 5 through 7). As they split, a down quark and an antidown quark are formed, leading to the final production of a D^- meson and a D^+ meson (steps 8 through 10).

Adapted from: Electon/Positron Annihilation http:/www.particleadventure.org/frameless/eedd.html 7/23/2007

Which statement best describes the changes that occur in this sequence of events?
   (1) Energy is converted into matter and then matter is converted into energy.
   (2) Matter is converted into energy and then energy is converted into matter.
   (3) Isolated quarks are being formed from baryons.
   (4) Hadrons are being converted into leptons.

31. A particle unaffected by an electric field could have a quark composition of
   (1) *css*   (2) *bbb*   (3) *udc*   (4) *uud*

Chapter Fourteen NUCLEAR ENERGY

437

32. Which graph best represents the relationship between energy and mass when matter is converted into energy?

(1) (2) (3) (4)

33. The energy produced by the complete conversion of 2.0×10^{-5} kilogram of mass into energy is
    (1) 1.8 TJ (2) 6.0 GJ (3) 1.8 MJ (4) 6.0 kJ

⭐ 34. In the reaction $_4^9Be + _2^4He \rightarrow _6^{12}C + X$, particle X is
    (1) an electron (2) a neutron (3) a positron (4) a proton

⭐ 35. When a gamma ray is emitted by a nucleus, the atomic number of the nucleus
    (1) decreases (2) increases (3) remains the same

36. Baryons may have charges of
    (1) $+1e$ and $+\frac{4}{3}e$ (3) $-1e$ and $+1e$
    (2) $+2e$ and $+3e$ (4) $-2e$ and $-\frac{2}{3}e$

37. The charge of an antistrange quark is approximately
    (1) $+5.33 \times 10^{-20}$ C   (3) $+5.33 \times 10^{20}$ C
    (2) -5.33×10^{-20} C   (4) -5.33×10^{20} C

⭐ 38. If, after beta decay, a nucleus is $_{91}^{234}Pa$, what was the nucleus just before the release of the beta particle?
    (1) $_{90}^{234}Th$ (2) $_{91}^{235}Pa$ (3) $_{91}^{238}Pa$ (4) $_{92}^{234}U$

39. Given the equation $_{13}^{27}Al + _2^4He \rightarrow _{15}^{30}P + X$. The correct symbol for X is
    (1) $_{+1}^{0}e$ (2) $_{-1}^{0}e$ (3) $_2^4He$ (4) $_0^1n$

⭐ 40. When lead $_{82}^{214}Pb$ emits a beta (−) particle, the resultant nucleus will be
    (1) $_{81}^{214}Tl$ (2) $_{82}^{213}Pb$ (3) $_{83}^{214}Bi$ (4) $_{84}^{214}Po$

⭐ 41. In the equation $_{92}^{239}U \rightarrow _{93}^{239}Np + X$, particle X is
    (1) a proton   (3) an alpha particle
    (2) a neutron   (4) a beta (−) particle

⭐ indicates material that is not part of the core curriculum.

✪ 42. When a nucleus captures an electron, the atomic number of the nucleus
(1) decreases (2) increases   (3) remains the same

✪ 43. When a radioactive nucleus emits a beta particle, the mass number of the nucleus will
(1) decrease   (2) increase   (3) remain the same

44. What fundamental force holds quarks together to form particles such as protons and neutrons?
(1) electromagnetic force   (3) strong force
(2) gravitational force   (4) weak force

45. What is the total number of quarks in a helium nucleus consisting of 2 protons and 2 neutrons?
(1) 16   (2) 12   (3) 8   (4) 4

46. A top quark has an approximate charge of
(1) -1.07×10^{-19} C   (3) $+1.07 \times 10^{-19}$ C
(2) -2.40×10^{-19} C   (4) $+2.40 \times 10^{-19}$ C

✪ 47. In the reaction $^{24}_{11}\text{Na} \rightarrow \,^{24}_{12}\text{Mg} + X$, what does X represent?

(1) an alpha particle   (3) a neutron
(2) a beta (–) particle   (4) a positron

48. A tritium nucleus is formed by combining two neutrons and a proton. The mass of this nucleus is 9.106×10^{-3} universal mass unit less than the combined mass of the particles from which it is formed. Approximately how much energy is released when this nucleus is formed?
(1) 8.48×10^{-2} MeV   (3) 8.48 MeV
(2) 2.73 MeV   (4) 273 MeV

49. A lithium atom consists of 3 protons, 4 neutrons, and 3 electrons. This atom contains a total of
(1) 9 quarks and 7 leptons   (3) 14 quarks and 3 leptons
(2) 12 quarks and 6 leptons   (4) 21 quarks and 3 leptons

50. According to the Standard Model of Particle Physics, a meson is composed of
(1) a quark and a muon neutrino
(2) a quark and an antiquark
(3) three quarks
(4) a lepton and an antilepton

✪ indicates material that is not part of the core curriculum.

Base your answers to questions 51 through 53 on the information below. The following equations represent a two-stage nuclear reaction.

$$^{27}_{13}\text{Al} + ^4_2\text{He} \rightarrow ^{30}_{15}\text{P} + \text{X}$$

$$^{30}_{15}\text{P} \rightarrow ^{30}_{14}\text{Si} + \text{Y} + \text{energy}$$

✪ 51. Which nucleus in the two equations has the greatest number of neutrons?

(1) $^{27}_{13}\text{Al}$ (2) ^4_2He (3) $^{30}_{15}\text{P}$ (4) $^{30}_{14}\text{Si}$

✪ 52. What is particle X?
(1) a positron   (2) an electron   (3) a proton   (4) a neutron

✪ 53. Particle Y represents

(1) ^1_0n (2) ^1_1H (3) $^0_{+1}\text{e}$ (4) $^0_{-1}\text{e}$

Base your answers to questions 54 through 57 on the following equation.

$$^{29}_{15}\text{A} + ^1_0\text{n} \rightarrow ^0_{-1}\text{D} + ^w_t\text{E}$$

✪ 54. What type of particle is represented by D?
(1) an electron   (3) an alpha particle
(2) a positron   (4) a gamma ray

55. For particle E, the value of w is
(1) 26   (2) 29   (3) 30   (4) 31

✪ 56. As particle E emits gamma radiation, its atomic number will
(1) decrease by 4   (3) remain the same
(2) increase by 1   (4) decrease by 2

57. For particle E, the value of t is
(1) 13   (2) 14   (3) 15   (4) 16

Base your answers to questions 58 through 61 on the information below.

In the equation $^{221}_{87}\text{Fr} \rightarrow \text{X} + \gamma + Q$, the letter X represents the nucleus produced by the reaction, γ represents a gamma photon, and Q represents additional energy released in the reaction.

✪ 58. Which nucleus is represented by X?

(1) $^{217}_{85}\text{X}$ (2) $^{221}_{87}\text{X}$ (3) $^{220}_{87}\text{X}$ (4) $^{217}_{85}\text{X}$

✪ indicates material that is not part of the core curriculum.

✪ 59. The rest mass of the gamma-ray photon is approximately
  (1) one atomic mass unit   (3) the mass of a neutron
  (2) the mass of a proton   (4) zero

✪ 60. If energy Q equals 9.9×10^{-13} joule, the mass equivalent of this energy is
  (1) 0 kg   (3) 1.1×10^{-29} kg
  (2) 9.1×10^{-31} kg   (4) 3.3×10^{-21} kg

✪ 61. The sample of $^{221}_{87}\text{Fr}$ (half-life = 4.8 min) will decay to one-fourth of its original amount in
  (1) 4.8 min   (2) 9.6 min   (3) 14.4 min (4) 19.2 min

✪ 62. One atomic mass unit is defined as 1/12 of the mass of an isotope of the element
  (1) hydrogen (2) oxygen   (3) uranium   (4) carbon

Base your answers to questions 63 through 66 on the information below:

A photon strikes a stationary deuterium nucleus, $^{2}_{1}\text{H}$, producing two separated, stationary nucleons. The masses of the particles are as follows:

$$\text{proton} = 1.0076 \text{ u}$$
$$\text{neutron} = 1.0090 \text{ u}$$
$$^{2}_{1}\text{H} = 2.0142 \text{ u}$$

✪ 63. If Q represents energy in atomic mass units, then the equation for this nuclear reaction is
  (1) $Q + {}^{2}_{1}\text{H} \rightarrow {}^{1}_{1}\text{H} + {}^{1}_{0}\text{n}$ (3) $^{1}_{1}\text{H} + {}^{1}_{1}\text{H} \rightarrow {}^{2}_{1}\text{H} + Q$
  (2) $Q + {}^{2}_{1}\text{H} \rightarrow {}^{1}_{1}\text{H} + {}^{1}_{1}\text{H}$ (4) $^{1}_{1}\text{H} + {}^{1}_{0}\text{n} \rightarrow {}^{2}_{1}\text{H} + Q$

✪ 64. For this nuclear reaction to occur, the minimum photon energy needed is
  (1) 0 u   (2) 0.0024 u   (3) 2.0142 u   (4) 2.0166 u

✪ 65. What is the binding energy of the deuterium nucleus?
  (1) 2.6×10^{-6} MeV   (3) 2.2 MeV
  (2) 2.2×10^{-3} MeV   (4) 1.9×10^{3} MeV

✪ 66. As the binding energy per nucleon of a nucleus increases, the stability of the nucleus
  (1) decreases (2) increases   (3) remains the same

✪ indicates material that is not part of the core curriculum.

67. A baryon may have a charge of

(1) $-\frac{1}{3}e$ (3) $+\frac{2}{3}e$

(2) $0e$ (4) $+\frac{4}{3}e$

Answers to Part A and B–1 questions can be found on page 460.

PART B–2 AND C QUESTIONS

Base your answers to questions 1 and 2 on the information below.

When an electron and its antiparticle (positron) combine, they annihilate each other and become energy in the form of gamma rays.

1. The positron has the same mass as the electron. Calculate how many joules of energy are released when they annihilate. [Show all work, including the equation and substitution with units.]

2. What conservation law prevents this from happening with two electrons?

Base your answers to questions 3 through 5 on the passage below and on your knowledge of physics.

Forces of Nature

Our understanding of fundamental forces has evolved along with our growing knowledge of the particles of matter. Many everyday phenomena seemed to be governed by a long list of unique forces. Observations identified the gravitational, electric, and magnetic forces as distinct. A large step toward simplification came in the mid-19th century with Maxwell's unification of the electric and magnetic forces into a single electromagnetic force. Fifty years later came the recognition that the electromagnetic force also governed atoms. By the late 1800s, all commonly observed phenomena could be understood with only the electromagnetic and gravitational forces.

Particle Physics–Perspectives and Opportunities
(adapted)

A hydrogen atom, consisting of an electron in orbit about a proton, has an approximate radius of 10^{-10} meter.

442

3. Determine the order of magnitude of the electrostatic force between the electron and the proton.

4. Determine the order of magnitude of the gravitational force between the electron and the proton.

5. In the above passage there is an apparent contradiction. The author stated that "the electromagnetic force also governed atoms." He concluded with "all commonly observed phenomena could be understood with only the electromagnetic and gravitational forces."

   Use your responses to questions 3 and 4 to explain why the gravitational interaction is negligible for the hydrogen atom.

Base your answers to questions 6 and 7 on the information and data table below.

   In the first nuclear reaction using a particle accelerator, accelerated protons bombarded lithium atoms, producing alpha particles and energy. The energy resulted from the conversion of mass into energy. The reaction can be written as shown below.

$$^1_1H + \,^7_3Li \rightarrow \,^4_2He + \,^4_2He + energy$$

Data Table

Particle	Symbol	Mass (u)
proton	^1_1H	1.007 83
lithium atom	^7_3Li	7.016 00
alpha particle	^4_2He	4.002 60

6. Determine the difference between the total mass of a proton plus a lithium atom, $^1_1H + \,^7_3Li$, and the total mass of two alpha particles, $^4_2He + \,^4_2He$, in universal mass units.

7. Determine the energy in megaelectronvolts produced in the reaction of a proton with a lithium atom.

443

8. A tau lepton decays into an electron, an electron antineutrino, and a tau neutrino, as represented in the reaction below.

$$\tau \rightarrow e + \bar{v}_e + v_\tau$$

On the equation below, show how this reaction obeys the Law of Conservation of Charge by indicating the amount of charge on each particle.

_____ e $\rightarrow$ _____ e + _____ e + _____ e

Base your answers to questions 9 through 12 on the passage below and on your knowledge of physics.

More Sci- Than Fi, Physicists Create Antimatter

Physicists working in Europe announced yesterday that they had passed through nature's looking glass and had created atoms made of antimatter, or antiatoms, opening up the possibility of experiments in a realm once reserved for science fiction writers. Such experiments, theorists say, could test some of the basic tenets of modern physics and light the way to a deeper understanding of nature.

By corralling [holding together in groups] clouds of antimatter particles in a cylindrical chamber laced with detectors and electric and magnetic fields, the physicists assembled antihydrogen atoms, the looking glass equivalent of hydrogen, the most simple atom in nature. Whereas hydrogen consists of a positively charged proton circled by a negatively charged electron, in antihydrogen the proton's counterpart, a positively charged antiproton, is circled by an antielectron, otherwise known as a positron.

According to the standard theories of physics, the antimatter universe should look identical to our own. Antihydrogen and hydrogen atoms should have the same properties, emitting the exact same frequencies of light, for example. . . .

Antimatter has been part of physics since 1927 when its existence was predicted by the British physicist Paul Dirac. The antielectron, or positron, was discovered in 1932. According to the theory, matter can only be created in particle-antiparticle pairs. It is still a mystery, cosmologists say, why the universe seems to be overwhelmingly composed of normal matter.

Dennis Overbye, "More Sci- Than Fi, Physicists Create Antimatter,"
New York Times, Sept. 19, 2002

9. The author of the passage concerning antimatter incorrectly reported the findings of the experiment on antimatter. Which particle mentioned in the article has the charge incorrectly identified?

10. How should the emission spectrum of antihydrogen compare to the emission spectrum of hydrogen?

11. Identify *one* characteristic that antimatter particles must possess if clouds of them can be corralled by electric and magnetic fields.

12. According to the article, why is it a mystery that "the universe seems to be overwhelmingly composed of normal matter"?

13. If a proton were to combine with an antiproton, they would annihilate each other and become energy. Calculate the amount of energy that would be released by this annihilation. [Show all work, including the equation and substitution with units.]

14. After a uranium nucleus emits an alpha particle, the total mass of the new nucleus and the alpha particle is less than the mass of the original uranium nucleus. Explain what happens to the missing mass.

Answers to Part B–2 and C questions can be found on pages 544–546.

Glossary

absolute index of refraction The ratio of the speed of light in a vacuum to the speed of light in a medium.

absolute temperature The temperature as measured on the Kelvin scale; a measure of the average kinetic energy of the molecules of a body.

absolute zero The temperature at which the internal energy of an object is at a minimum (0 K or $-273°C$).

absorption spectrum A series of dark spectral lines or bands formed by the absorption of specific wavelengths of light by atoms or molecules.

acceleration The time rate of change in velocity. The SI unit is meters per second^2.

accuracy The agreement of a measured value with an accepted standard.

alpha decay A natural radioactive process that results in the emission of an alpha particle from a nuclide.

alpha particle A helium nucleus; a particle consisting of two protons and two neutrons.

alternating current An electric current that varies in magnitude and alternates in direction.

ammeter A device used to measure electric current. It is constructed by placing a low-resistance shunt across the coil of a galvanometer.

ampere (A) The SI unit of electric current, equivalent to the unit coulomb per second.

amplitude The maximum displacement in periodic phenomena such as wave motion, pendulum motion, and spring oscillation.

angle of incidence The angle made by the incident wave with the surface of a medium; the angle made by the incident ray with the normal to the surface of the medium.

angle of reflection The angle made by the reflected wave with the surface of a medium; the angle made by the reflected ray with the normal to the surface of the medium.

angle of refraction The angle made by the refracted wave with the surface of a medium; the angle made by the refracted ray with the normal to the surface of the medium.

anode The positive terminal of a DC source of potential difference.

antimatter One or more atoms composed entirely of antiparticles.

antinode The point or locus of points on an interference pattern (such as a standing wave or double slit pattern) that results in maximum constructive interference.

antiparticle The counterpart of a subatomic particle. An antiparticle has the same mass as its companion particle, but its electric charge is opposite in sign.

atomic mass unit (u) A unit of mass defined as one-twelfth the mass of an atom of carbon-12.

atomic number The number of protons in the nucleus of an atom. The atomic number defines the element.

Balmer series The visible-ultraviolet line spectrum of atomic hydrogen. It is the result of electrons falling from higher levels to the $n = 2$ state.

battery A combination of two or more electric cells.

beta (−) decay A natural radioactive process that results in the emission of a beta (−) particle from a nuclide.

beta (−) particle An electron formed in the nucleus by the disintegration of a neutron.

beta (+) decay A natural radioactive process that results in the emission of a beta (+) particle from a nuclide.

beta (+) particle A positron, the antiparticle of the electron, formed in the nucleus by the disintegration of a proton.

binding energy The energy equivalent of the mass defect of a nucleus.

cathode The negative terminal of a DC source of potential difference.

Celsius scale (°C) The temperature scale that fixes the (atmospheric) freezing point of water at 0° and the boiling point of water at 100°.

centripetal acceleration The acceleration that is directed along the radius and toward the center of a curved path in which an object is moving.

centripetal force The force that causes centripetal acceleration. It is responsible for changing an object's direction, not its speed.

circuit A closed loop formed by a source of potential difference connected to one or more resistances.

coefficient of kinetic friction The ratio of the force of kinetic friction on an object to the normal force on it.

coherent light A series of light waves that have a fixed phase relationship; the type of light produced by a laser. Lasers produce beams of monochromatic coherent light.

component One of the two or more vectors into which a given vector may be resolved.

concurrent forces Two or more forces acting at the same point.

conductivity The reciprocal of a material's resistivity. The SI unit of conductivity is the mho, which is equivalent to the $ohm^{-1} \cdot meter^{-1}$.

conductor A material that allows electrons to flow through it freely. Metals such as copper and silver are conductors.

constructive interference The combination of two in-phase wave disturbances to produce a single wave disturbance whose amplitude is the sum of the amplitudes of the individual disturbances.

coulomb (C) The SI unit of electric charge, approximately equal to 6.25×10^{18} elementary charges.

Coulomb's law The electrostatic force between two point charges is directly proportional to the product of the charges and inversely proportional to the square of the distance between the charges.

critical angle The angle of incidence for which the corresponding angle of refraction is 90°.

cycle One complete repetition of the pattern in any periodic phenomenon.

De Broglie wavelength The wavelength of a matter wave.

destructive interference The combination of two out-of-phase wave disturbances to produce a single wave disturbance whose amplitude is the difference of the amplitudes of the individual disturbances.

deuterium An isotope of hydrogen that contains one proton and one neutron in its nucleus.

diffraction The bending of a wave around a barrier.

diffuse reflection The reflection of parallel light rays by irregular surfaces.

direct current An electric current that flows in one direction only.

dispersion The separation of polychromatic light into its individual colors.

dispersive medium A medium in which the speed of a wave depends on its frequency.

displacement A change of position in a specific direction.

Doppler effect An apparent change in frequency that results when a wave source and an observer are in relative motion with respect to each other.

Einstein's postulates of special relativity (1) The laws of physics are valid in all inertial frames of reference. (2) The speed of light has the same value in all frames of reference.

elastic potential energy The energy stored in a spring when it is compressed or stretched.

electric current The time rate of flow of charged particles. The SI unit of electric current is the ampere (A).

electric field The region of space around a charged object that affects other charges.

electric field barrier The region in a semiconductor, established by the combination of holes and electrons, that prevents further migration of charge carriers across the *P-N* junction.

electric field intensity The ratio of the force that an electric field exerts on a charge to the magnitude of the charge.

electric motor A device that converts electrical energy into mechanical energy.

electric potential The total work done by an electric field in bringing 1 coulomb of positive charge from infinity to a specific point. The potential is a positive number if the charge is repelled by the field and a negative number if the charge is attracted by the field. At infinity, the potential is taken to be zero. Electric potential is measured in volts.

electromagnet A solenoid whose magnetic field is intensified by the insertion of certain ferromagnetic materials.

electromagnetic induction The process by which the magnetic field and the mechanical energy are used to generate a potential difference.

electromagnetic radiation The propagation of electromagnetic waves in space.

electromagnetic spectrum The entire range of electromagnetic waves from the lowest to the highest frequencies.

electromagnetic wave A periodic wave, consisting of mutually perpendicular electric and magnetic fields, that is radiated away from the vicinity of an accelerating charge.

electron A fundamental, negatively charged, subatomic particle.

electron capture A radioactive process in which a nucleus absorbs one of an atom's innermost electrons.

electron cloud In quantum theory, the region of space where an electron is most likely to be found.

electron-volt (eV) A unit of energy equal to the work needed to move an elementary charge across a potential difference of 1 volt.

electroscope A device used to detect the presence of electric charges.

elementary charge The magnitude of charge present on a proton or an electron. An elementary charge is approximately equal to 1.6×10^{-19} coulomb.

emission spectrum A series of bright spectral lines or bands formed by the emission of certain wavelengths of light by excited atoms falling to lower energy states.

energy A quantity related to work.

equilibrant A single balancing force that maintains the static equilibrium of an object.

equivalent resistance A single resistance that can be substituted for a group of resistances in series or in parallel.

ether A hypothetical medium whose existence was proposed as the carrier of all electromagnetic waves.

excited state A condition in which the energy of an atom is greater than its lowest energy state.

ferromagnetic Referring to a material, such as iron, that has the ability to strengthen greatly the magnetic field of a current-carrying coil.

field lines A series of lines used to represent the magnitude and direction of a field.

flux density The number of magnetic field lines per unit area. The flux density is one way of measuring the strength of a magnetic field.

force A push or a pull on an object. If the

force is unbalanced, an acceleration will result.

free fall A motion in the Earth's gravitational field without regard to air resistance.

frequency The number of repetitions produced per unit time by periodic phenomena.

friction The force present as the result of contact between two surfaces. The direction of a frictional force is opposite to the direction of motion.

Gallilean-Newtonian relativity principle The laws of mechanics are valid in all inertial frames of reference.

galvanometer A device, consisting of a coil-shaped wire placed between the opposite poles of a permanent magnet, that is used to detect small amounts of electric current.

gamma radiation Very high energy photons of electromagnetic radiation. Gamma photons have the highest frequencies in the electromagnetic spectrum.

generator A device that uses a magnetic field and mechanical energy to induce a source of electromotive force.

gravitational field The region of space around a mass that affects other masses.

gravitational field intensity The ratio of the force that a gravitational field exerts on a mass to the magnitude of the mass, numerically equal to the acceleration due to gravity.

gravitational force The universal attraction between two masses.

gravitational potential energy The energy that an object acquires as a result of the work done in moving the object against a gravitational field.

ground An extremely large source or reservoir of electrons, which can supply or accept electrons as the need arises.

ground state The lowest energy state of an atom.

hertz (Hz) The SI unit of frequency, equivalent to the unit second^{-1}.

ideal mechanical system a closed system upon which no friction or other external forces are acting.

impulse The product of the net force acting on an object and the time during which the force acts. The impulse delivered to an object is equal to its change in momentum. The direction of the impulse is the direction of the force. The SI unit of impulse is the newton · second, which is equivalent to the kilogram · meter per second.

incident ray A ray of a wave impinging on a surface.

incident wave A wave impinging on a surface.

induced current An electric current that is the result of an induced electromotive force.

induced emf A potential difference created when a magnetic field is interrupted over a time period.

induction (1) A method of charging a neutral object by using a charged object and a ground. The induced charge is always opposite to the charge on the charged object. (2) See *induced current* and *induced emf.*

inertia The property of matter that resists changes in motion. Mass is the quantitative measure of inertia.

inertial frame of reference A frame of reference in which Newton's first law of motion is valid.

instantaneous velocity The ratio of displacement to time at any given instant; the slope of a line tangent to a displacement-time graph at any given point.

insulator A material that is a very poor conductor because it has few conduction electrons. Wood and glass are examples of insulators.

interference pattern Regions of constructive and destructive interference that are present in a medium as a result of the combination of two or more waves.

internal energy The total kinetic and potential energy associated with the atoms and molecules of an object.

ionization energy See *ionization potential.*

ionization potential The quantity of energy needed to remove a single electron from an atom or ion.

isotopes Atoms with identical atomic numbers but different mass numbers. Two isotopes of the same element have identical numbers of protons but different numbers of neutrons.

joule (J) The SI unit of work and energy, equivalent to the unit newton · meter.

***k*-capture** Electron capture.

kelvin (K) The SI unit of temperature.

Kelvin scale (K) The absolute temperature scale. The single fixed point on the Kelvin scale is the triple point of water, which is set at 273.16K.

kilogram (kg) The SI unit of mass; a fundamental unit.

kinetic energy The energy that an object possesses because of its motion.

laser An acronym for light amplification by the stimulated emission of radiation. A laser is a device that emits extremely intense, monochromatic, coherent light.

longitudinal wave A wave in which the disturbance is parallel to the direction of the wave's motion. Sound waves are longitudinal.

magnet Any material that aligns itself, when free to do so, in an approximate north-south direction. Magnets exert forces on one another and on charged particles in motion.

magnetic field The region of space around a magnet or charge in motion that exerts a force on magnets or other moving charges.

magnification The ratio of image size to object size.

mass (1) The measure of an object's ability to obey Newton's second law of motion. (2) The measure of an object's ability to obey Newton's law of universal gravitation. The SI unit of mass is the kilogram.

mass defect The mass lost by a nucleus when it is assembled from its nucleons. (See also *binding energy.*)

mass number The sum of the number of protons and neutrons in a nucleus; the number of nucleons the nucleus contains.

medium A material through which a disturbance, such as a wave, travels.

momentum The product of mass and velocity. The direction of an object's momentum is the direction of its velocity. The SI unit of momentum is the kilogram·meter per second.

natural frequency A specific frequency with which an elastic body may vibrate if disturbed.

net force The unbalanced force present on an object; the accelerating force.

neutrino A subatomic particle with no charge and questionable mass. It and its antiparticle are products of beta-decay reactions.

newton (N) The SI unit of force, equivalent to the unit kilogram · meter per second^2.

Newton's first law of motion Objects remain in a state of uniform motion unless acted upon by an unbalanced force.

Newton's law of universal gravitation Any two bodies in the universe are attracted to each other with a force that is directly proportional to their masses and inversely proportional to the square of the distance between them.

Newton's second law of motion The unbalanced force on an object is equal to the product of its mass and acceleration.

Newton's third law of motion If object A exerts a force on object B, then object B exerts an equal and opposite force on object A.

node The point or locus of points on an interference pattern, such as a standing wave or double-slit pattern, that results in total destructive interference.

nondispersive medium A medium in which the speed of a wave does *not* depend on its frequency.

normal A line perpendicular to a surface.

normal force The force that keeps two surfaces in contact. If an object is on a *horizontal* surface, the normal force on the object is equal to its weight.

nuclear force The attractive, short-range force responsible for binding protons and neutrons in the nucleus.

nucleon A proton or a neutron.

nucleus The dense, positively charged core of an atom.

nuclide An atomic nucleus.

ohm (Ω) The SI unit of electrical resistance, equivalent to the unit volt per ampere.

Ohm's law A relationship in which the ratio of the potential difference across certain conductors to the current in them is constant at constant temperature.

parallel circuit An electric circuit with more than one current path.

period The time for one complete repetition of a periodic phenomenon. The SI unit of period is the second.

periodic wave A regularly repeating series of waves.

phase (1) A form in which matter can exist, including liquid, solid, gas, and plasma. (2) In wave motion, the points on the wave that have specific time and space relationships.

photoelectric effect A phenomenon in which light causes electrons to be ejected from certain materials. (See also *photoemissive*.)

photoelectrons Electrons that have been emitted as a result of the photoelectric effect.

photoemissive Referring to materials whose surfaces can eject electrons on exposure to light.

photon The fundamental particle of electromagnetic radiation.

Planck's constant A universal constant (h) relating the energy of a photon to its frequency; its approximate value is 6.62×10^{-34} joule · second.

point charge A charge with negligible physical dimensions.

polarization A process that produces transverse waves that vibrate in only one plane. Polarization is limited to transverse waves: light can be polarized; sound cannot.

polychromatic Referring to light waves of different colors (frequencies).

positron See *beta (+) particle*.

potential difference The ratio of the work required to move a test charge between two points in an electric field to the magnitude of the test charge. The unit of potential difference is the volt.

potential energy The energy that a system has because of its relative position or condition.

power The time rate at which work is done or energy is expended. The SI unit of power is the watt, which is equivalent to the unit joule per second

precision The limit of the ability of a measuring device to reproduce a measurement.

principal quantum number The integer that defines the main energy level of an atom.

proton A positively charged subatomic particle with a charge equal in magnitude to that of the electron.

pulse A nonperiodic disturbance in a medium.

quantum A discrete quantity of energy.

quarks The particles of which protons, neutrons, and certain other subatomic particles are composed. Quarks carry a charge of either one-third or two-thirds of an elementary charge and come in six "flavors": top, bottom, up, down, charm, and strange.

radioactive decay A spontaneous change in the nucleus of an atom.

radioactivity Changes in the nucleus of an atom that produce the emission of subatomic particles or photons.

ray A straight line indicating the direction of travel of a wave.

real image An image created by the actual convergence of light waves. Real images from single mirrors and single lenses are inverted and can be projected on a screen.

refraction The change in the direction of a wave when it passes obliquely from one medium to another in which it moves at different speed.

regular reflection The reflection of parallel light rays incident on a smooth plane surface.

resistance The opposition of a material to the flow of electrons through it; the ratio of potential difference to current.

resistivity A quantity that allows the resistance of substances to be compared. Numerically, it is the resistance of a 1-meter conductor with a cross-sectional area of 1 square meter. The SI unit of resistivity is the ohm·meter.

resistor A device that supplies resistance to a circuit.

resolution The process of determining the magnitude and direction of the components of a vector.

resonance The spontaneous vibration of an object at a frequency equal to that of the wave that initiates the resonant vibration.

resultant A vector sum.

satellite A body that revolves around a larger body as a result of a gravitational force.

scalar quantity A quantity, such as mass or work, that has magnitude but not direction.

series circuit A circuit with only one current path.

shunt A low-resistance device for divert-

ing electric current, used to convert a galvanometer into an ammeter.

significant digits The digits that are part of any measurement.

solenoid A coil of wire wound as a helix. When a current is passed through the solenoid, it becomes an electromagnet.

speed The time rate of change of distance; the magnitude of velocity. The SI unit of speed is the meter per second.

spring constant The ratio of the force required to stretch or compress a spring to the magnitude of the stretch or compression.

standard pressure One atmosphere; approximately 1.01×10^5 pascals.

standard temperature Approximately 273 kelvins.

standing wave A wave pattern created by the continual interference of an incident wave with its reflected counterpart. The standing wave does not travel, but oscillates about an equilibrium position.

static equilibrium The condition of a body when a net force of zero is acting on it.

stationary state A condition in which an atom is neither absorbing nor releasing energy.

superconductor A material with no electrical resistance.

tesla (T) The SI unit of flux density, equal to the units weber per square meter and newton per ampere · meter.

thermal equilibrium The point at which materials in contact reach the same temperature.

threshold frequency The lowest frequency at which photons striking a specific surface can produce a photoelectric effect.

torque A force, applied perpendicularly to a designated line, that tends to produce rotational motion.

total internal reflection The reflection of a wave inside a relatively dense medium produced when the angle of the wave with the boundary exceeds the critical angle.

total mechanical energy The sum of the potential and kinetic energies of a mechanical system.

transmutation The change of one radioactive nuclide into another, either by decay or by bombardment.

transverse wave A wave in which the disturbance is perpendicular to the direction of the wave's motion. Light waves are transverse.

uniform In the study of motion, a term that is equivalent to *constant*.

vaporization See *boiling*.

vector A representation of a vector quantity; an arrow in which the length represents the magnitude of the quantity and the arrowhead points in the direction of its orientation.

vector quantity A quantity, such as force or velocity, that has both magnitude and direction.

velocity The time rate of change of displacement.

virtual image An image formed by projecting diverging light behind a mirror or a lens.

volt (V) The SI unit of potential difference, equivalent to the unit joule per coulomb.

voltage Another term for *potential difference*.

voltmeter A device used to measure potential difference and constructed by placing a large resistor in series with the coil of a galvanometer.

watt (W) The SI unit of power, equivalent to the unit joule per second.

wave A series of periodic oscillations of a particle or a field both in time and in space.

wave front All points on a wave that are in phase with each other.

wavelength The length of one complete wave cycle.

weber (Wb) The SI unit of magnetic flux, equivalent to the unit joule per ampere.

weight The gravitational force present on an object.

work The product of the force on an object and its displacement. The SI unit of work is the joule.

work function The minimum radiant energy required to remove an electron from a photoemissive surface.

Answers to Questions for Review

Chapter One, pages 12–13

1.	2	4.	1	7.	3	10.	3	13.	2	16.	2
2.	3	5.	1	8.	1	11.	3	14.	3	17.	4
3.	3	6.	3	9.	3	12.	2	15.	2	18.	1

Chapter Two, pages 27–40

1.	3	14.	1	27.	4	40.	3	53.	3	66.	4
2.	3	15.	4	28.	3	41.	2	54.	2	67.	1
3.	3	16.	4	29.	2	42.	1	55.	2	68.	2
4.	3	17.	3	30.	2	43.	4	56.	4	69.	1
5.	2	18.	4	31.	3	44.	1	57.	2	70.	4
6.	2	19.	2	32.	2	45.	3	58.	1	71.	3
7.	2	20.	2	33.	3	46.	1	59.	1	72.	1
8.	1	21.	2	34.	2	47.	2	60.	2	73.	2
9.	3	22.	2	35.	3	48.	1	61.	4	74.	1
10.	3	23.	2	36.	2	49.	1	62.	3	75.	3
11.	2	24.	4	37.	2	50.	2	63.	2		
12.	3	25.	4	38.	4	51.	3	64.	1		
13.	1	26.	1	39.	3	52.	4	65.	3		

Chapter Three, pages 56–64

1.	1	10.	3	19.	3	28.	2	37.	1	46.	1
2.	4	11.	3	20.	4	29.	2	38.	4	47.	3
3.	3	12.	2	21.	2	30.	3	39.	3	48.	1
4.	2	13.	2	22.	1	31.	4	40.	4	49.	4
5.	4	14.	1	23.	1	32.	1	41.	2	50.	1
6.	3	15.	3	24.	4	33.	1	42.	3	51.	2
7.	3	16.	3	25.	1	34.	3	43.	1	52.	1
8.	2	17.	1	26.	4	35.	3	44.	2		
9.	1	18.	4	27.	2	36.	1	45.	1		

Chapter Four, pages 88–105

1. 4	16. 1	31. 2	46. 3	61. 2	76. 4
2. 2	17. 4	32. 2	47. 2	62. 1	77. 2
3. 3	18. 3	33. 2	48. 3	63. 3	78. 3
4. 2	19. 4	34. 3	49. 1	64. 4	79. 2
5. 2	20. 2	35. 3	50. 1	65. 2	80. 1
6. 1	21. 3	36. 2	51. 4	66. 2	81. 4
7. 2	22. 4	37. 2	52. 3	67. 4	82. 3
8. 2	23. 2	38. 1	53. 4	68. 1	83. 1
9. 1	24. 1	39. 1	54. 4	69. 1	84. 3
10. 3	25. 4	40. 4	55. 1	70. 2	85. 3
11. 3	26. 4	41. 1	56. 1	71. 2	86. 2
12. 2	27. 2	42. 2	57. 1	72. 2	87. 1
13. 3	28. 4	43. 2	58. 1	73. 3	88. 2
14. 1	29. 1	44. 1	59. 2	74. 2	89. 3
15. 3	30. 3	45. 2	60. 2	75. 1	

Chapter Five, pages 125–140

1. 4	15. 4	29. 3	43. 1	57. 3	71. 2
2. 1	16. 1	30. 3	44. 3	58. 3	72. 2
3. 1	17. 1	31. 1	45. 2	59. 2	73. 1
4. 4	18. 4	32. 4	46. 2	60. 4	74. 2
5. 2	19. 4	33. 3	47. 2	61. 3	75. 4
6. 4	20. 2	34. 1	48. 2	62. 2	76. 1
7. 3	21. 4	35. 3	49. 3	63. 3	77. 3
8. 2	22. 3	36. 1	50. 3	64. 1	78. 3
9. 1	23. 3	37. 2	51. 3	65. 2	79. 3
10. 4	24. 2	38. 2	52. 2	66. 3	80. 1
11. 4	25. 4	39. 3	53. 4	67. 4	
12. 3	26. 2	40. 4	54. 1	68. 4	
13. 4	27. 2	41. 2	55. 2	69. 1	
14. 3	28. 4	42. 4	56. 3	70. 3	

Chapter Six, pages 148–157

1. 4	11. 2	21. 2	31. 4	41. 1	51. 2
2. 3	12. 4	22. 4	32. 3	42. 3	52. 3
3. 2	13. 4	23. 3	33. 4	43. 4	53. 1
4. 2	14. 1	24. 4	34. 2	44. 2	54. 3
5. 1	15. 1	25. 2	35. 2	45. 4	55. 2
6. 2	16. 2	26. 3	36. 3	46. 4	56. 3
7. 4	17. 3	27. 4	37. 2	47. 4	57. 4
8. 3	18. 2	28. 2	38. 3	48. 2	58. 1
9. 3	19. 4	29. 3	39. 3	49. 4	59. 4
10. 1	20. 1	30. 4	40. 3	50. 4	

Chapter Seven, pages 172–189

1. 3	18. 2	35. 3	52. 3	69. 4	86. 4
2. 4	19. 1	36. 2	53. 2	70. 2	87. 4
3. 3	20. 1	37. 3	54. 1	71. 4	88. 3
4. 3	21. 1	38. 2	55. 3	72. 4	89. 2
5. 1	22. 3	39. 3	56. 3	73. 1	90. 4
6. 3	23. 4	40. 2	57. 3	74. 4	91. 1
7. 3	24. 1	41. 3	58. 2	75. 1	92. 1
8. 1	25. 3	42. 3	59. 3	76. 1	93. 4
9. 2	26. 2	43. 2	60. 3	77. 4	94. 1
10. 4	27. 3	44. 1	61. 3	78. 4	95. 1
11. 1	28. 2	45. 1	62. 1	79. 1	96. 3
12. 1	29. 2	46. 3	63. 4	80. 1	97. 1
13. 2	30. 4	47. 3	64. 1	81. 4	98. 4
14. 2	31. 2	48. 2	65. 3	82. 2	99. 3
15. 2	32. 3	49. 4	66. 1	83. 2	100. 4
16. 1	33. 2	50. 2	67. 1	84. 1	
17. 2	34. 2	51. 2	68. 1	85. 2	

Chapter Eight, pages 216–234

1. 4	18. 3	35. 1	52. 2	69. 2	86. 3
2. 4	19. 2	36. 4	53. 1	70. 1	87. 3
3. 2	20. 4	37. 3	54. 1	71. 4	88. 3
4. 2	21. 1	38. 3	55. 4	72. 2	89. 4
5. 2	22. 1	39. 1	56. 1	73. 2	90. 2
6. 3	23. 4	40. 1	57. 3	74. 4	91. 1
7. 3	24. 1	41. 3	58. 2	75. 3	92. 1
8. 2	25. 2	42. 1	59. 4	76. 4	93. 3
9. 1	26. 2	43. 4	60. 4	77. 2	94. 3
10. 3	27. 3	44. 4	61. 2	78. 3	95. 4
11. 3	28. 1	45. 1	62. 1	79. 4	96. 3
12. 3	29. 3	46. 3	63. 3	80. 2	97. 2
13. 1	30. 4	47. 1	64. 2	81. 1	98. 3
14. 4	31. 2	48. 2	65. 1	82. 4	99. 3
15. 1	32. 1	49. 3	66. 1	83. 1	100. 2
16. 1	33. 2	50. 2	67. 1	84. 3	
17. 3	34. 1	51. 1	68. 2	85. 4	

Chapter Nine, pages 255–272

1. 1	18. 3	35. 3	52. 3	69. 4	86. 4
2. 4	19. 4	36. 1	53. 3	70. 2	87. 4
3. 2	20. 1	37. 3	54. 1	71. 1	88. 4
4. 4	21. 4	38. 2	55. 2	72. 3	89. 2
5. 2	22. 2	39. 3	56. 4	73. 1	90. 3
6. 2	23. 1	40. 4	57. 1	74. 3	91. 4
7. 4	24. 1	41. 1	58. 4	75. 3	92. 1
8. 4	25. 4	42. 3	59. 3	76. 1	93. 4
9. 3	26. 3	43. 1	60. 3	77. 3	94. 3
10. 4	27. 3	44. 2	61. 1	78. 2	95. 4
11. 2	28. 3	45. 1	62. 3	79. 4	96. 2
12. 3	29. 4	46. 2	63. 4	80. 4	97. 3
13. 3	30. 2	47. 4	64. 1	81. 1	98. 4
14. 4	31. 1	48. 4	65. 3	82. 3	99. 2
15. 2	32. 1	49. 1	66. 1	83. 1	100. 1
16. 4	33. 1	50. 4	67. 2	84. 2	
17. 1	34. 1	51. 3	68. 2	85. 3	

Chapter Ten, pages 292–300

1. 1	6. 2	11. 3	16. 1	21. 2	26. 2
2. 4	7. 1	12. 4	17. 2	22. 2	27. 1
3. 2	8. 1	13. 2	18. 3	23. 2	28. 3
4. 2	9. 1	14. 4	19. 1	24. 3	29. 1
5. 1	10. 3	15. 1	20. 1	25. 3	30. 1

Chapter Eleven, pages 317–338

1. 4	20. 1	39. 2	58. 4	77. 1	96. 4
2. 3	21. 2	40. 4	59. 4	78. 1	97. 4
3. 1	22. 1	41. 1	60. 4	79. 1	98. 2
4. 3	23. 1	42. 2	61. 1	80. 2	99. 4
5. 3	24. 2	43. 4	62. 3	81. 4	100. 2
6. 4	25. 4	44. 3	63. 4	82. 2	101. 4
7. 2	26. 2	45. 2	64. 3	83. 2	102. 2
8. 1	27. 3	46. 1	65. 4	84. 1	103. 2
9. 2	28. 3	47. 2	66. 2	85. 3	104. 3
10. 3	29. 2	48. 1	67. 4	86. 2	105. 1
11. 3	30. 4	49. 4	68. 3	87. 3	106. 1
12. 1	31. 4	50. 2	69. 2	88. 4	107. 1
13. 4	32. 1	51. 3	70. 2	89. 2	108. 3
14. 1	33. 1	52. 4	71. 4	90. 1	109. 4
15. 2	34. 4	53. 1	72. 2	91. 1	110. 1
16. 2	35. 2	54. 3	73. 4	92. 4	
17. 2	36. 2	55. 1	74. 4	93. 2	
18. 1	37. 3	56. 3	75. 1	94. 3	
19. 2	38. 2	57. 4	76. 3	95. 4	

Chapter Twelve, pages 363–382

1.	4	20.	1	39.	1	58.	1	77.	1	96.	4
2.	1	21.	4	40.	1	59.	3	78.	1	97.	2
3.	1	22.	2	41.	3	60.	1	79.	3	98.	2
4.	1	23.	4	42.	1	61.	2	80.	1	99.	1
5.	1	24.	4	43.	4	62.	1	81.	4	100.	4
6.	3	25.	4	44.	2	63.	2	82.	3	101.	2
7.	3	26.	3	45.	3	64.	2	83.	1	102.	1
8.	1	27.	3	46.	1	65.	1	84.	3	103.	3
9.	3	28.	4	47.	2	66.	3	85.	2	104.	4
10.	4	29.	4	48.	4	67.	3	86.	1	105.	2
11.	1	30.	4	49.	2	68.	3	87.	3	106.	4
12.	3	31.	1	50.	2	69.	2	88.	4	107.	2
13.	1	32.	4	51.	3	70.	1	89.	4	108.	3
14.	2	33.	3	52.	2	71.	1	90.	1	109.	4
15.	2	34.	1	53.	1	72.	1	91.	1		
16.	3	35.	3	54.	3	73.	1	92.	1		
17.	1	36.	1	55.	2	74.	4	93.	2		
18.	3	37.	2	56.	2	75.	3	94.	4		
19.	3	38.	4	57.	2	76.	4	95.	4		

Chapter Thirteen, pages 403–415

1.	4	15.	3	29.	3	43.	1	57.	3	71.	3
2.	1	16.	2	30.	4	44.	2	58.	3	72.	3
3.	3	17.	4	31.	3	45.	3	59.	1	73.	2
4.	1	18.	1	32.	2	46.	2	60.	4	74.	2
5.	2	19.	4	33.	1	47.	3	61.	3	75.	3
6.	3	20.	1	34.	2	48.	3	62.	1	76.	3
7.	4	21.	2	35.	4	49.	2	63.	1	77.	1
8.	4	22.	3	36.	2	50.	3	64.	2	78.	2
9.	2	23.	2	37.	4	51.	3	65.	1	79.	4
10.	1	24.	2	38.	4	52.	1	66.	3	80.	2
11.	2	25.	1	39.	1	53.	1	67.	1	81.	1
12.	3	26.	1	40.	1	54.	3	68.	2	82.	1
13.	3	27.	1	41.	4	55.	2	69.	2		
14.	1	28.	1	42.	2	56.	2	70.	2		

Chapter Fourteen, pages 434–442

1. 3	13. 2	25. 2	37. 1	49. 4	61. 2
2. 1	14. 2	26. 2	38. 1	50. 2	62. 4
3. 2	15. 1	27. 4	39. 4	51. 4	63. 1
4. 3	16. 4	28. 2	40. 3	52. 4	64. 2
5. 3	17. 3	29. 3	41. 4	53. 3	65. 3
6. 2	18. 1	30. 2	42. 1	54. 1	66. 2
7. 3	19. 2	31. 1	43. 3	55. 3	67. 2
8. 3	20. 4	32. 1	44. 3	56. 3	
9. 3	21. 4	33. 1	45. 2	57. 4	
10. 2	22. 1	34. 2	46. 3	58. 2	
11. 3	23. 2	35. 3	47. 2	59. 4	
12. 2	24. 3	36. 3	48. 3	60. 3	

PART B–2 AND C QUESTIONS

Chapter Two, pages 40–44

1. Write out the problem.

Given: $v_l = 0\,\text{m/s}$ Find: $a = ?$
$v_f = 40.\,\text{m/s}$
$\Delta t = 20.\,\text{s}$

Refer to the *Mechanics* section in the *Reference Tables* to find an equation(s) that relates acceleration with velocity and time.

Solution: $\bar{a} = \dfrac{\Delta v}{\Delta t} = \dfrac{v_f - v_l}{\Delta t}$

$$= \frac{40.\ \text{m/s} - 0\ \text{m/s}}{20.\,\text{s}} = \frac{40.\,\text{m/s}}{20.\,\text{s}}$$

$$= 2.0\,\text{m/s}^2$$

Alternatively, you can solve this problem using the slope of line *AB* on the graph. The slope of a line is defined as the change in *y* over the change in *x*. In this graph, *y* is speed and *x* is time; thus, the slope on a speed versus time graph is equal to the change in speed over the change in time, which is equal to the acceleration.

$$\text{slope} = \frac{\Delta y}{\Delta x} = \frac{\Delta v}{\Delta t} = \bar{a}$$

$$= \frac{40. \text{ m/s}}{20. \text{ s}}$$

$$= 2.0 \text{ m/s}^2$$

Note: One point is awarded for writing the equation and for the substitution of values with units. If the equation and/or units are not shown, you do not receive this point. One point is awarded for the correct answer (number and units). If there are no units, you do not receive this point. Note also that significant figures and scientific notation are not required to obtain credit.

2. Write out the problem.

Given: $\bar{v}$ = 40. m/s   Find: Δs = ?
         Δt = 20. s

Refer to the *Mechanics* section in the *Reference Tables* to find an equation(s) that relates displacement with velocity and time.

Solution: $\bar{v} = \frac{\Delta s}{\Delta t}$

$\Delta s = \bar{v} \Delta t$

$$= \left(40. \frac{\text{m}}{\not{s}} \right) (20. \not{s})$$

$$= 800 \text{ m} \quad \text{or} \quad 8.0 \times 10^2 \text{ m}$$

Alternatively, you can solve this problem directly from the graph. The area of a square or rectangle is calculated by multiplying the base times the height. In this case the base is the time, and the height is the speed; therefore, the area under the curve is equal to the displacement.

$A = bh = \Delta t \bar{v}$

$$= (20. \not{s}) \left(40. \frac{\text{m}}{\not{s}} \right)$$

$$= 800 \text{ m} \quad \text{or} \quad 8.0 \times 10^2 \text{ m}$$

Note: One point is awarded for writing the equation and for the substitution of values with units. If the equation and/or units are not shown, you do not receive this point. One point is awarded for the correct answer (number and units). If there are no units, you do not receive this point. Note also that significant figures and scientific notation are not required to obtain credit.

3. Write out the problem.

Given: $v_i = 40.\,\text{m/s}$ Find: $v = ?$
$v_f = 0\,\text{m/s}$
$\Delta t = 10.\,\text{s}$

Refer to the *Mechanics* section in the *Reference Tables* to find an equation(s) that relates average velocity to the variables we are given.

$$\text{Solution: } \bar{v} = \frac{v_f + v_i}{2}$$

$$= \frac{40.\ \text{m/s} + 0\ \text{m/s}}{2} = \frac{40.\ \text{m/s}}{2}$$

$$= 20.\,\text{m/s} \quad \text{or} \quad 20\ \text{m/s}$$

Note that time is not required for this calculation.

Note: One point is awarded for the correct answer. To receive this point, you must include units. Note also that significant figures and scientific notation are not required to obtain credit.

4/5.

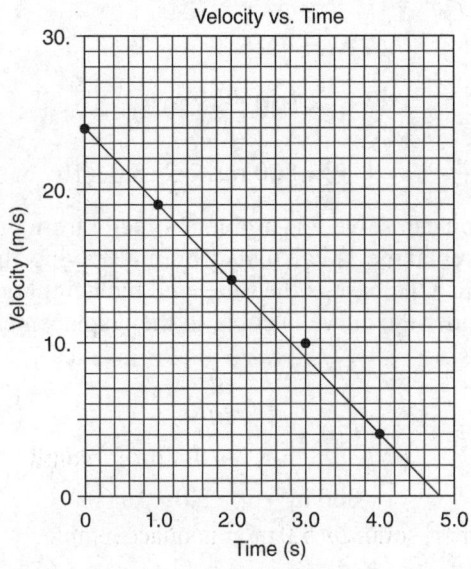

4. One credit is granted if all points are plotted accurately within ±0.3 grid space.

5. Credit is granted if the best-fit line is straight. Credit will also be granted if one or more points are plotted incorrectly in question 4, but the best-fit line is straight.

6. The slope of a velocity versus time graph is equal to the acceleration:

$$\text{slope} = \frac{\Delta y}{\Delta x}$$

For this problem, y = velocity and x = time, so

$$\text{slope} = \frac{\Delta v}{\Delta t}$$

or $= a$ (acceleration)

Examples of Acceptable Responses

$$\text{slope} = \frac{\Delta y}{\Delta x}$$
$$= \frac{-20\,\text{m/s}}{4\text{s}}$$
$$= -5\,\text{m/s}^2 \pm (0.3\,\text{m/s}^2)$$

or

$$5\,\text{m/s}^2 \pm (0.3\,\text{m/s}^2)\,\text{south}$$

or

$$\bar{a} = \frac{\Delta v}{\Delta t}$$
$$= \frac{1\,\text{m/s} - 21\,\text{m/s}}{4.6\text{s} - 0.6\text{s}}$$
$$= -5\,\text{m/s}^2 \pm (0.3\,\text{m/s}^2)$$

or

$$5\,\text{m/s}^2 \pm (0.3\,\text{m/s}^2)\,\text{south}$$

Note that $-5.0\,\text{m/s}^2$ south *or* $5.0\,\text{m/s}^2$ is unacceptable.

Note: One point is awarded for writing the equation and for the substitution of values with units. If the equation and/or units are not shown, you do not receive this point. One point is awarded for the correct answer (number and units). If there are no units, you do not receive this point. Note also that significant figures and scientific notation are not required to obtain this credit. Credit is granted as long as the answer is consistent with your graph (i.e., if the best-fit line passes through the data values used in the calculation). If no credit is granted for questions 4 and 5, then credit may be granted for this question if acceleration is calculated using values from the data table.

7–9. **Example of an appropriate graph:**

10. The acceleration of the object is 1.2 m/s^2.

11. Allow 1 credit for marking an appropriate scale.

12. Allow 1 credit for correctly plotting all points ± 0.3 grid space.

13. Allow 1 credit for drawing the line or curve of best fit.

Example of a 3-credit response for questions 11 through 13:

Speed vs. Height above Earth's Surface

14. Allow 1 credit for 15.7 m/s ± 0.3 m/s *or* an answer that is consistent with the student's graph.

15. Allow 1 credit for 2.7 cm ± 0.2 cm.

16. Allow a maximum of 2 credits. Refer to *Scoring Criteria for Calculations*.

Example of a 2-credit response:

$$d = v_i t + \tfrac{1}{2} a t^2$$

$$a = \frac{2d}{t^2}$$

$$a = \frac{2(2.7 \text{ cm})}{(0.30 \text{ s})^2}$$

$$a = 60 \text{ cm/s}^2 \text{ or } 6.0 \text{ m/s}^2$$

Note: Allow credit for an answer that is consistent with the student's response to question 15.

17. Allow a maximum of 2 credits. Refer to *Scoring Criteria for Calculations*.

    Example of a 2-credit response:

    $$\bar{v} = \frac{d}{t}$$

    $$\bar{v} = \frac{2.7 \text{ cm}}{0.30 \text{ s}}$$

    $$\bar{v} = 9.0 \text{ cm/s } or \ 0.090 \text{ m/s}$$

    Note: Allow credit for an answer that is consistent with the response to question 15 or 16.

18. Allow 1 credit for *at least four* dots that are equally spaced ± 0.2 cm.

    Example of a 1-credit response:

    Recording Tape

    (Drawn to scale)

Chapter Three, pages 65–67

1. Within a certain range the amount that a spring is stretched or compressed (x) is proportional to the force (F) applied. This relationship is known as *Hooke's law*.

    Refer to the *Energy* equations in the *Reference Tables*:

    $$F = kx$$

    where k is the spring constant. Substitute the given values:

    $$6.0 \text{ N} = k \, (0.040 \text{ m})$$

    Solve for k:

    $$\frac{6.0 \text{ N}}{0.040 \text{ m}} = k = 150 \text{ N/m}$$

    Note: Units must appear in the substitution and in the final answer for full credit.

2. One credit is granted if the scale is linear and the scale divisions are appropriate. A scale of 0.10 m per division is *not* acceptable.

   See the graph located after question 4 for examples of acceptable graphs.

3. One credit is granted if all points are plotted accurately (±0.3 grid space). This credit will be granted if you correctly use your response from question 2.

   See the graph located after question 4 for examples of acceptable graphs.

4. One credit is granted if the best-fit line is straight. If one or more points are plotted incorrectly in question 3 but a best-fit straight line is drawn, this credit will be granted.

   Examples of Acceptable Graphs

or

5. Write out the problem.

   Given: $F = 12.0 \, \text{N}$ Find: $k = ?$

   $x = 0.42 \, \text{m}$

   Refer to the *Energy* section in the *Reference Tables* to find an equation(s) that relates the spring constant with applied force and elongation of a spring.

Solution: $F = kx$

$$k = \frac{F}{x}$$

$$= \frac{12.0 \text{ N}}{0.42 \text{ m}}$$

$$= 29 \text{ N/m} (\pm 2 \text{ N/m})$$

(Note that this response is based on the assumption that the elongation of 0.42 m due to an applied force of 12.0 N lies on the best-fit line and that the line passes through the origin.)

Note: A total of two credits is allowed. One credit is awarded for writing the equation and for the substitution of values with units. If the equation and/or units are not shown, you do not receive this credit. One credit is awarded for the correct answer (number and units). If there are no units, you do not receive this credit. Significant figures and scientific notation are not required to obtain this credit.

Note: The slope *may* be determined by direct substitution into the equation $k = F/x$ *only* if the best-fit line passes through the *origin* and the data values used for substitution are on that line.

Note also that credit may be granted for an answer that is consistent with your graph, *unless* you receive no credits for questions 3 and 4. In that case, credit will be awarded if you correctly calculate the spring constant using data in the table.

6. Allow a maximum of 4 credits for finding the minimum coefficient of friction.

   Examples of acceptable responses and allocation of credits include, but are not limited to:

   Formulas: $F_f = \mu F_N$ $F_N = mg$ $F_c = \dfrac{mv^2}{r}$ [1]

   Rearrangement: $\mu = \dfrac{v^2}{rg}$ [1]

   Substitution: $\mu = \dfrac{(20. \text{ m/s})^2}{(80. \text{ m})(9.8 \text{ m/s}^2)}$ [1]

   Answer: $\mu = 0.51$ [1]

                    or

$$F_c = ma_c \quad a_c = \frac{v^2}{r}$$

$$F_c = \frac{mv^2}{r} = \frac{(1{,}600 \text{ kg})(20. \text{ m/s})^2}{80. \text{ m}} = 8.0 \times 10^3 \text{ N} \quad [1]$$

$$F_N = mg = (1{,}600 \text{ kg})(9.81 \text{ m/s}^2) = 1.6 \times 10^4 \text{ N} \quad [1]$$

$$F_f = F_c \quad [1]$$

$$F_f = \mu F_N \quad \mu = \frac{F_f}{F_N} = \frac{8.0 \times 10^3 \text{ N}}{1.6 \times 10^4 \text{ N}} = 0.50 \quad [1]$$

7. Allow 1 credit for indicating that changing the mass of the car would have no effect on the maximum speed at which it could round the curve.

8. Allow 1 credit for 7500 N.

9. Allow 1 credit for correctly plotting all the data points ± 0.3 grid space.

10. Allow 1 credit for drawing the line or curve of best fit.

Example of a 2-credit graph for questions 9 and 10:

Attached Weight vs. Total Length

11. [1] Allow 1 credit for 0.30 m ± 0.01 m *or* an answer that is consistent with the student's graph.

12. Allow 1 credit for indicating that a feather does not accelerate at 9.81 meters per second^2 when dropped near the surface of Earth because the net force is less than F_2. Acceptable responses include, but are not limited to:

    —Air friction acts on the feather.
    —The feather is not in free fall.

13. Allow a maximum of 2 credits. Refer to *Scoring Criteria for Calculations*.

    Example of a 2-credit response:

    $$F_f = \mu F_N$$

    $$F_N = \frac{F_f}{\mu}$$

    $$F_N = \frac{39\,\text{N}}{.05}$$

    $$F_N = 780\,\text{N}$$

Chapter Four, pages 105–116

1. *a.*

   b. Let's call the force vector F. To construct its horizontal component, we make a vector triangle by drawing a horizontal line from the tail of F and a vertical line down from the head of F. Place an arrowhead at the intersection of these two lines, and label the horizontal component H. See the diagram in part *a*.

c. Since your diagram was drawn to scale, measure the length of *H* and convert to newtons. The length turns out to be 4.1 cm. The scale specified in the question is 1.0 cm – 10. N. Therefore the magnitude of *H* is 41 N.

You could also use simple trigonometry to calculate the magnitude of *H*:

$$\cos 35° = \frac{H}{F}$$

$$.8192 = \frac{H}{50\ N}$$

$$H = 41\ n$$

2. A maximum of 4 credits may be given for explaining how to find the coefficient of kinetic friction between a wooden block of unknown mass and a table top in the laboratory. The response must include:

- Measurements needed: normal force (weight or mass) of block, friction force   [1]
- Equipment needed: spring scale (and balance if mass of block is used) or computer force sensor   [1]
- Procedure: The procedure must include a means of finding the normal force and the force of friction, *and* a means of using them to determine the coefficient of friction, e.g., using the equation *or* finding the slope of a graph.   [1]
- Equation: $F_f = \mu F_N$ (and $F_g = mg$ if mass is found first)

Examples of acceptable responses include, but are not limited to:

To determine the coefficient of friction between a block and the table, I would need to measure the normal force or weight of the block, and the force of friction. The equipment needed is a spring scale. First I would hang the block on the scale to find its weight. Then I would pull the block smoothly (or at constant speed) across the table with the spring scale to find the force of friction. Once I measured the weight and friction forces, I would use the formula $F_f = \mu F_N$ to calculate the coefficient of friction.

or

Use another device for measuring force (e.g., a computer force sensor).

Use a balance to find the mass of the block, then $g = \dfrac{F_g}{m}$ to find the weight. Then proceed as above.

or

Load various weights onto the block, find the friction for each weight, plot a graph of friction *vs.* weight, and find the slope of the graph.

3. *a.* (1) To receive full credit, each of the two vectors must include an arrowhead, have an appropriate label, and be drawn using the specified scale. Also, each vector must be drawn in the correct direction, the second displacement vector must originate at the arrowhead end of the first displacement vector, and the two vectors must be drawn at right angles (see diagrams below).

(2) Construct the resultant displacement by drawing a vector from the starting point (*P*) of the first vector (200.0 m) to the endpoint of the second vector (300. m). Label this vector "Resultant." See the diagram above.

b. Since your diagram was drawn to scale, measure the length of the resultant and convert to meters. The length turns out to be 3.61 centimeters.

The scale specified in the question is 1.0 cm = 50. m. Therefore the magnitude of the resultant is

$$3.61 \text{ cm} \times \frac{100. \text{ m}}{1.0 \text{ cm}} = 361 \text{ m}$$

(An answer that is within ±15 meters of 361 meters is also accepted.)

c. Place your protractor at point P and measure the angle between the first vector, which points north, and the resultant vector. This angle turns out to be 56°. (An answer that is within ±2° of 56° is also accepted.)

4.

Distance vs. Time

Note: One credit is awarded for plotting all points accurately (±0.3 grid space). One credit is awarded for drawing a best-fit straight line. If one or more points are plotted incorrectly but a best-fit straight line is drawn, this credit is still granted.

5. The slope of a graph is defined as the change in y over the change in x. In this graph, y is distance and x is time; thus, the slope of a distance-versus-time graph is equal to the change in distance over the change in time, which is equal to velocity.

Examples of acceptable responses therefore include:

The slope is the average speed of the safe.
The slope is the velocity.
The change in distance divided by the change in time is equal to the velocity.
The slope is the speed of the safe.

Note: One credit is granted for an acceptable response.

6. If you measure the 40.-N vector, you obtain a length of approximately 8.0 cm, which gives you the following scale: 1.0 cm is approximately equal to 5.0 N (±0.2 N).

   Note: One credit is granted for an acceptable response with units.

7. The vector that represents the resultant force is as follows:

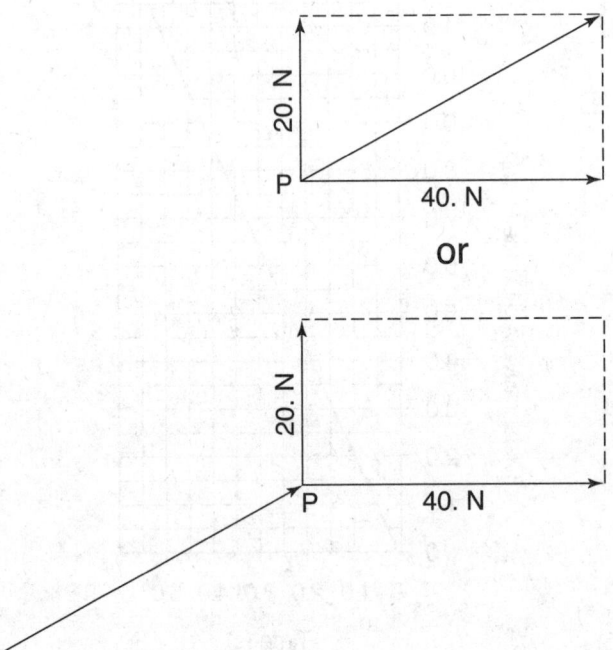

or

   Note: To receive one credit, the 8.9 cm ± 0.2-cm vector must include an arrowhead at the end. The resultant vector need not be labeled to receive this credit.

8. If you measure the resultant vector, you obtain a magnitude of 45 N (±2 N) for the resultant force. Alternatively, the magnitude may be obtained by calculation using the Pythagorean theorem or trigonometry.

   Note: To receive one credit the correct unit must be included. The credit is granted also if the answer is based correctly on the answers to questions 6 and 7.

9. Using a protractor, you obtain a degree measurement of 27° ± 2° for the angle between east and the resultant force.

   Note: The credit is granted also if the answer is based correctly on the answer to question 7 or is calculated using the tangent function $\tan \theta = 20\,N/40\,N$.

10. Write out the problem.

    Given: $F = 45\,N$ (from question 14)   Find: $a = ?$
    $m = 10\,kg$

    Refer to the *Mechanics* section in the *Reference Tables* to find an equation(s) that relates acceleration with force and mass.

    $$F = ma$$

    $$a = \frac{F}{m}$$

    $$= \frac{45\,N}{10.\,kg}$$

    $$= 4.5\ N/kg$$

    or

    $$= 4.5\ m/s^2$$

    Note: One credit is awarded for writing the equation and for the substitution of values with units. If the equation and/or units are not shown, you do not receive this credit. One credit is awarded for the correct answer (number and units). If there are no units in your answer, you do not receive this credit. Note also that significant figures and scientific notation are not required to obtain this credit. Credit is granted if you correctly use your response to question 8.

11. The three forces acting on the block are (1) the force of gravity (the block's weight), (2) the normal force, which acts perpendicular to the surface and in this situation is equal in magnitude but opposite in direction to the weight of the block, and (3) the applied horizontal force to the right.

    Example of Acceptable Response

475

If each of the three vectors meets *all three* of the following criteria, a total of three credits is awarded.

- A line originating at point C and having an arrowhead indicating the correct direction, such as those shown in the preceding diagram. (A vector with either of its ends in contact with an edge of the block is acceptable.)

  —applied force: parallel to the horizontal

  —weight: perpendicular to the horizontal

  —normal force: perpendicular to the horizontal

- The vector, including its arrowhead, is drawn to the appropriate length.

  —applied force: 2.0 cm ± 0.2 cm long

  —weight: 4.9 cm ± 0.2 cm long

  —normal force: 2.0 cm ± 0.2 cm long

- The vectors are labeled.

  —applied force: 20. N or 20 N

  —weight: 49 N

  —normal force: 49 N

If each of the three vectors meets *at least two* of the three criteria, a total of two credits is awarded.

If each of the three vectors meets *at least one* of the three criteria, a total of one credit is awarded.

12. Write out the problem.

    Given: $m = 5.0\ \text{kg}$ Find: $a = ?$

            $w = 49\ \text{N}$

    $F_{\text{applied}} = 20.\ \text{N}$

    Refer to the *Mechanics* section in the *Reference Tables* to find an equation(s) that relates acceleration with mass, weight, and force.

    Solution: $F = ma$

    $$a = \frac{F}{m}$$

    $$= \frac{20.\ \text{N}}{5.0\ \text{kg}}$$

    $$= 4.0\ \text{m/s}^2 \quad \text{or} \quad 4.0\ \text{N/kg}$$

(Note that weight was not needed in this calculation as the weight was balanced by the normal force and the net force was due solely to the applied horizontal force.)

Note: A total of two credits is allowed. One credit is awarded for writing the equation and for the substitution of values with units. If the equation and/or units are not shown, you do not receive this credit. One credit is awarded for the correct answer (number and units). If there are no units, you do not receive this credit. Significant figures and scientific notation are not required to obtain this credit.

13. Allow a maximum of 2 credits.
    - Allow 1 credit for a line originating at point B and having an arrowhead indicating the correct direction with a label.
    - Allow 1 credit if the vector, including its arrowhead, is drawn to the proper length (4.0 cm ± 0.2 cm).

    Example of Acceptable Response

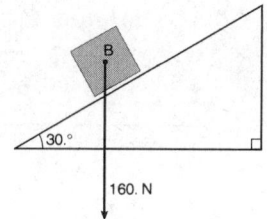

14. Allow a maximum of 2 credits.
    - Allow 1 credit for the substitutions $\frac{m}{s}$ for v and m or d.
    - Allow 1 credit for the answer $\frac{m}{s^2}$.

    Example of Acceptable Response
    $$\frac{v^2}{d} = \frac{(m/s)^2}{m} = \frac{m^2/s^2}{m} = \frac{m}{s^2}$$

    Note: Credit should *not* be allowed for merely giving acceleration units as m/s².

Examples of *Unacceptable* Responses

$$\frac{v^2}{d} = \frac{m}{s^2}$$

or

$$\frac{v^2}{d} = \frac{m/s}{s}$$

15. Allow a maximum of 2 credits for drawing and labeling the triangle.
    - Allow 1 credit for a straight line segment 10.0 cm (±0.2 cm) drawn from point B and a 30.° (±2°) angle at point B.
    - Allow 1 credit for a properly drawn right triangle with A and K labeled correctly.

Example of Acceptable Response

16. Allow 1 credit for determining the height, AK, of the kite to be a value consistent with the student's answer to question 15. The answer should be 58 m (±2 m) if the answer to question 15 is drawn correctly. Do *not* allow credit for an answer of 58 m if the answer to question 15 is drawn incorrectly or missing.

17. Allow a maximum of 2 credits for calculating the amount of time required for the sphere to fall to the ground. Refer to *Scoring Criteria for Calculations*.

Example of Acceptable Response

$$d = v_i t + \frac{1}{2} at^2 \text{ or } d = \frac{1}{2} at^2$$

$$t = \sqrt{\frac{2d}{a}}$$

$$t = \sqrt{\frac{2(58 \text{ m})}{9.81 \text{ m/s}^2}}$$

$$t = 3.4 \text{ s}$$

Allow credit for an answer that is consistent with the student's answer to question 16.

Note: The use of 9.8 m/s^2 in the equation is also acceptable.

18. 20. N or 20 N.

19. **Example of Acceptable Response**

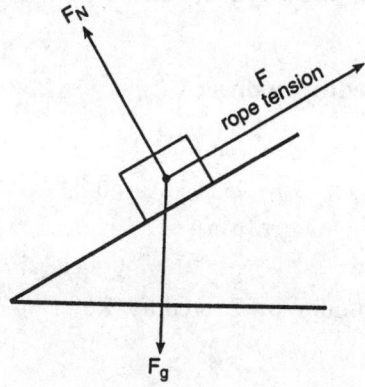

20. [1] Allow 1 credit for 1.0 cm = 2.0 m ± 0.2 m.

21. [1] Allow 1 credit for drawing a vector 5.0 cm ± 0.2 cm long, including an arrowhead at the end directed away from the starting point.

Example of a 1-credit response:

Note: The vectors need *not* be labeled to receive this credit.

22. [1] Allow 1 credit for 10. m ± 0.4 m.

Note: Allow credit for an answer that is consistent with the response to questions 20 and/or 21.

23. [2] Allow a maximum of 2 credits. Refer to *Scoring Criteria for Calculations*.

Example of a 2-credit response:

$$A_y = A \sin \theta$$
$$v_{iy} = (25 \text{ m/s })(\sin 40.°)$$
$$v_{iy} = 16 \text{ m/s}$$

24. [2] Allow a maximum of 2 credits. Refer to *Scoring Criteria for Calculations*.

Example of a 2-credit response:

$$v_f^2 = v_i^2 + 2ad$$
$$d = \frac{v_f^2 - v_i^2}{2a}$$
$$d = \frac{(16 \text{ m/s})^2}{2(9.81 \text{ m/s}^2)}$$
$$d = 13 \text{ m}$$

Note: Allow credit for an answer that is consistent with the response to question 22.

25. [1] Allow 1 credit for drawing a generally parabolic path.

   Example of a 1-credit response:

26. [1] Allow 1 credit for drawing an arrow at *X* toward the ground and perpendicular to the ground.

27. [1] Allow 1 credit for drawing an arrow at *Y* toward the ground and perpendicular to the ground.

   Example of a 2-credit response for questions 26 and 27:

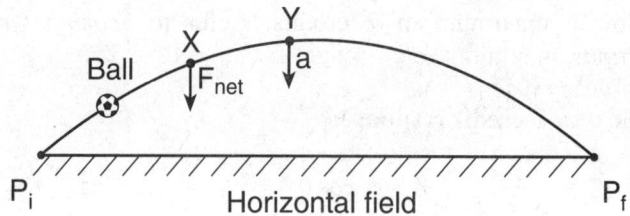

   Note: The arrows need *not* be labeled to receive credit.

28. [2] Allow a maximum of 2 credits. Refer to *Scoring Criteria for Calculations*.

   Example of a 2-credit response:

$$v_f^2 = v_i^2 + 2ad$$

$$d = \frac{v_f^2 - v_i^2}{2a}$$

$$d = \frac{(0 \text{ m/s})^2 - (70. \text{ m/s})^2}{2(-2.0 \text{ m/s}^2)}$$

$$d = 1200 \text{ m}$$

29. [2] Allow a maximum of 2 credits, allocated as follows.
    - Allow 1 credit for a direction south.
    - Allow 1 credit for a vector drawn 4.0 cm ± 0.2 cm long.

    Example of a 2-credit response:

    Note: The vector need *not* begin at point *P* to receive this credit.

30. [2] Allow a maximum of 2 credits. Refer to *Scoring Criteria for Calculations.*

    Example of a 2-credit response:

    $$A_x = A \ cos \ \theta$$
    $$F_x = (60. \ N) cos \ 30.°$$
    $$F_x = 52 \ N$$

31. [1] Allow 1 credit for 52 N *or* an answer that is consistent with your response to question 30.

32. [1] Allow 1 credit for a parabolic-shaped path.

    Example of a 1-credit response:

33. [1] Allow 1 credit for indicating that the projectile's maximum altitude will increase.

34. [1] Allow 1 credit for indicating that the total horizontal distance will increase.

35. [1] Allow 1 credit for drawing and labeling a vector 5.0 cm (± 0.2 cm) long, directed upward. Do *not* allow credit if the vector is not labeled or is missing the arrowhead.

Example of a 1-credit response:

Note: Allow credit if the student draws the correct vector from the box and *not* from point *P.*

36. [2] Allow a maximum of 2 credits. Refer to *Scoring Criteria for Calculations*.

Example of a 2-credit response:

$$F_f = \mu F_N$$
$$F_f = (0.30)(20.\,N)$$
$$F_f = 6.0\,N$$

37. [1] Allow 1 credit for 2.0 N *or* an answer that is consistent with your response to question 36.

38. [1] Allow 1 credit for 2.0 kg.

39. [2] Allow a maximum of 2 credits. Refer to *Scoring Criteria for Calculations*.

Example of a 2-credit response:

$$a = \frac{F_{net}}{m}$$

$$a = \frac{2.0\,\text{N}}{2.0\,\text{kg}}$$

$$a = 1.0 \ \text{m/s}^2$$

Note: Allow 1 credit for an answer that is consistent with the responses to questions 37 and 38.

40. [1] Allow 1 credit for 1 cm = 2.0 N ± 0.2 N.

41. [1] Allow 1 credit for construction the resultant 3.7 cm ± 0.2 cm long, at an angle of 36° ± 2° from vector B.

Example of 1-credit responses:

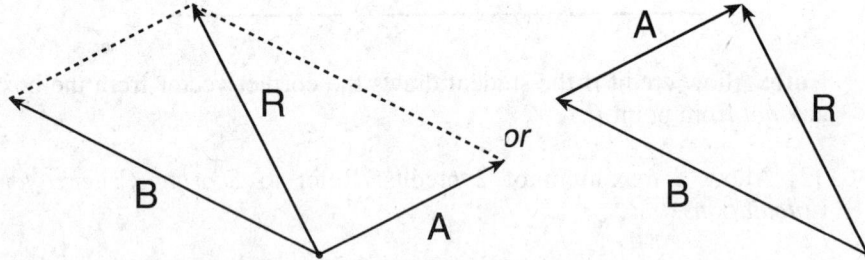

Note: Do *not* deduct credit if the resultant vector is not labeled.

42. [1] Allow 1 credit for 7.4 N ± 0.4 N *or* an answer that is consistent with the responses to questions 40 and 41.

43. Allow 1 credit for explaining the difference between a scalar and a vector quantity. Acceptable responses include, but are not limited to:

— A scalar quantity has magnitude only. A vector quantity has both magnitude and direction.
— A vector quantity has direction.
— A vector quantity has no direction.

44. Allow a maximum of 2 credits. Refer to *Scoring Criteria for Calculations.*

Example of a 2-credit response:

$$F_f = \mu F_N$$
$$F_f = (.15)(10.\,\text{kg})(9.81\,\text{m/s}^2)$$
$$F_f = 15\,\text{N} \;\; or \;\; 14.7\,\text{N}$$

45. Allow a maximum of 2 credits, allocated as follows:

• Allow 1 credit for a length of 2.5 cm (± 0.2 cm) and an arrow.
• Allow 1 credit for an angle above the horizontal of 60.° (± 2°).

Example of a 2-credit response:

46. Allow 1 credit for 125 m/s (± 10 m/s).

Allow credit for an answer that is consistent with the student's response to question 62.

47. Allow 1 credit for explaining why the projectile has no acceleration in the horizontal direction. Acceptable responses include, but are not limited to:

— No force on object in horizontal direction.
— The only force is vertical.
— Gravity acts only vertically.

Chapter Five, pages 141–143

1. *a.*

b. Applying Newton's second law ($F = ma$) to objects near the surface of a planet gives

$$w = mg$$

where w is the object's weight, m is its mass, and g is the gravitational acceleration of the planet. (Refer to the *Mechanics* equations in the *Reference Tables*.) Then

$$g = \frac{w}{m}$$
$$= \frac{170 \text{ N}}{24 \text{ kg}}$$
$$= 7.08 \text{ N/kg} \quad \text{or} \quad 7.08 \text{ m/s}^2$$

Note: The values substituted into the equation must come from a data point that lies on the best-fit straight line. The values from the given data table do not necessarily fall on this line. You must also be sure to include units when you make the substitution and when you give the final answer.

The planet's gravitational acceleration could also be determined by calculating the slope of the best-fit straight line:

$$g = \frac{\Delta y}{\Delta x}$$

$$= \frac{220 \text{ N} - 120 \text{ N}}{31 \text{ kg} - 17 \text{ kg}}$$

$$= \frac{100}{14}$$

$$= 7.1 \text{ N/kg} \quad \text{or} \quad 7.1 \text{ m/s}^2$$

Note: Again, the values substituted must come from data points that lie on the best-fit straight line.

2. Allow a maximum of 2 credits. Refer to *Scoring Criteria for Calculations.*

Examples of Acceptable Responses

$$v = \frac{d}{t}$$

$$t = \frac{d}{v}$$

$$t = \frac{2\pi r}{v}$$

$$t = \frac{2\pi(2.40\text{m})}{4.00 \text{ m/s}}$$

$$t = 3.77 \text{ s}$$

or

$$\bar{v} = \frac{d}{t}$$

$$4.00 \text{ m/s} = \frac{15.08\text{m}}{t}$$

$$t = 3.77 \text{ s}$$

3. Allow 1 credit for describing a change that would quadruple the magnitude of the centripetal force.

Examples of acceptable responses include, but are not limited to:

— double the speed of the car
— reduce the radius to 0.60 m
— quadruple the mass
— double the mass of the cart and halve the radius
— increase the speed of the cart to 5.66 m/s and double the mass of the cart
— increase the speed of the cart to 5.66 m/s and halve the radius

4. Allow 1 credit for drawing and labeling an arrow that represents the direction of the acceleration of the cart.

Example of Acceptable Response

Note: The label must be included to receive credit.

5. [2] Allow a maximum of 2 credits. Refer to *Scoring Criteria for Calculations.*

Example of a 2-credit response:

$$F_c = ma_c \quad \text{and} \quad a_c = \frac{v^2}{r}$$

$$F_c = \frac{mv^2}{r}$$

$$F_c = \frac{\left(5.98 \times 10^{24} \text{ kg}\right)\left(3.00 \times 10^4 \text{ m/s}\right)^2}{1.5 \times 10^{11} \text{ m}}$$

$$F_c = 3.59 \times 10^{22} \text{ N}$$

6. [2] Allow a maximum of 2 credits. Refer to *Scoring Criteria for Calculations.*

    Example of a 2-credit response:

    $$F_g = \frac{Gm_1m_2}{r^2}$$

    $$F_g = \frac{(6.67 \times 10^{-11} \text{ N} \cdot \text{m}^2/\text{kg}^2)(1.90 \times 10^{27} \text{ kg})(8.93 \times 10^{22} \text{ kg})}{(4.22 \times 10^8 \text{ m})^2}$$

    $$F_g = 6.35 \times 10^{22} \text{ N}$$

7. [2] Allow a maximum of 2 credits. Refer to *Scoring Criteria for Calculations.*

    Examples of 2-credit responses:

    $$a = \frac{F_{net}}{m}$$

    $$a = \frac{6.35 \times 10^{22} \text{ N}}{8.93 \times 10^{22} \text{ kg}}$$

    $a = 0.711 \text{ m/s}^2$ *or* $a = 0.711 \text{ N/kg}$

    or

    $$a = \frac{F_{net}}{m} \text{ and } F_g = \frac{Gm_1m_2}{r^2}$$

    $$a = G\frac{m_2}{r^2}$$

    $$a = \frac{(6.67 \times 10^{-11} \text{ N} \cdot \text{m}^2/\text{kg}^2)(1.90 \times 10^{27} \text{ kg})}{(4.22 \times 10^8 \text{ m})^2}$$

    $$a = 0.712 \text{ m/s}^2$$

    Note: Allow 1 credit for an answer that is consistent with the response to question 6.

8. Allow 1 credit for indicating that gravity is the fundamental force to which the author is referring.

9. Allow a maximum of 2 credits. Refer to *Scoring Criteria for Calculations.*

    Example of a 2-credit response:

    $$F = \frac{Gm_1m_2}{r^2}$$

    $$F = \frac{(6.67 \times 10^{-11} \text{ N} \cdot \text{m}^2/\text{kg}^2)(8.73 \times 10^{25} \text{ kg})(1.03 \times 10^{26} \text{ kg})}{(1.63 \times 10^{12} \text{ m})^2}$$

    $$F = 2.26 \times 10^{17} \text{ N}$$

10. Allow 1 credit for indicating that the Sun is larger in mass.

Note: Do *not* allow credit for just "larger."

Chapter Six, page 158

1. *a.* The *momentum* (*p*) of an object is equal to the product of its mass
(*m*) and its velocity (*v*). Refer to the *Mechanics* equations in the
Reference Tables:

$$p = mv$$
$$= (2.0 \times 10^3 \text{ kg})(3.0 \text{ m/s})$$
$$= 6.0 \times 10^3 \text{ kg·m/s (or 6000 kg·m/s)}$$

NOTE: Again, you must be sure to include units when you make the
substitution and when you give the final answer.

b.

c. For full credit, you must indicate that the momentum of the two carts
is zero after the collision, and give a reason based on the principle of
the *conservation of momentum*. *A complete sentence* begins with a
capital letter and ends with a period, and has a subject and a verb. The
following are sample correct answers:

*The two carts have zero momentum because the initial momentum of
the system is zero, so the final momentum of the system must be zero.*
    *The two carts will have zero momentum after the collision because
the initial momentum of cart A cancels out the initial momentum of
cart B.*
    *The momentum of the locked carts after the collision is zero. The
initial momenta of the carts are equal and opposite, totaling zero, and
momentum is conserved.*

2. Allow 1 credit for 6000 $\frac{\text{kg·m}}{\text{s}}$.

Chapter Seven, pages 189–199

1. *a.* The *gravitational potential energy* (ΔPE) of an object above the ground is equal to the product of the object's mass (*m*), the value of the acceleration due to gravity (*g*), and the vertical height (Δ*h*) to which the object was raised.

   Refer to the *Energy* equations in the *Reference Tables:*

$$\Delta PE = mg\Delta h$$
$$= (6.0\,\text{kg})(9.8\,\text{m/s}^2)(55\,\text{m})$$
$$= 3234\,\text{kg} \cdot \text{m}^2/\text{s}^2 \text{ or } \text{N} \cdot \text{m or J}$$

   or

$$= 3200\,\text{J to two significant figures}$$

Note: The value for *g*, the acceleration due to gravity, 9.8 m/s², is obtained from the *List of Physical Constants* in the *Reference Tables*.

   b. The *kinetic energy* (KE) of an object is equal to one-half the product of its mass (*m*) and the speed (*v*) squared.

   Refer to the *Energy* equations in the *Reference Tables:*

$$KE = \frac{1}{2}mv^2$$
$$= \frac{1}{2}(6.0)(30.\ \text{m/s})^2$$
$$= 2700\ \text{kg} \cdot \text{kg} \cdot \text{m}^2/\text{s}^2 \text{ or } \text{N} \cdot \text{m or J}$$

   c. *Mechanical energy* is the sum of the kinetic and potential energies of an object. At the instant the block was released from the top of the 55-m building, it had 0 kinetic energy (because it was at rest) plus 3200 J of gravitational potential energy. Therefore the block started with a total mechanical energy of 3200 J.

   When the block hit the ground, it had 0 gravitational potential energy (because it was on the ground level and Δ*h* = 0) plus 2700 J of kinetic energy. Therefore, when the block hit the ground it had a total mechanical energy of 2700 J.

   We conclude that the block lost 3200 J − 2700 J = *500 J* as it fell.

   d. For full credit, you must write one or more *complete* sentences that correctly explain what happens to the "lost" mechanical energy. A complete sentence begins with a capital letter, has a subject and verb, and ends with a period. The following are four sample correct answers:
   The energy was lost due to air friction.
   The energy was converted to heat energy.
   The energy was lost due to work done against friction.
   Work was done on the air by the block.

This is a sample of an *unacceptable* answer:
It was friction.

2. If you measure the first hill correctly, you should measure it to be 3.0 cm in height. By using the scale 1.0 cm = 8.0 m, this would correspond to 24 m ± 1 m.

Note: One point is awarded for the correct answer (number and units). If there are no units, you do not receive this point. Note also that significant figures and scientific notation are not required to obtain this credit.

3. Refer to the *Energy* section in the *Reference Tables* to find an equation for potential energy.

$$\Delta PE = mg\Delta h$$
$$= (650 \text{ kg})(9.8 \text{ m/s}^2)(24 \text{ m})$$
$$= 152{,}880 \text{ kg} \cdot \text{m}^2/\text{s}^2$$
$$\text{or } 1.5 \times 10^5 \text{ J}$$

Note: One point is awarded for writing the equation and for substituting values with units. If the equation and/or units are not shown, you do not receive this point. One point is awarded for the correct answer. Note also that significant figures and scientific notation are not required to obtain this credit. Credit may be granted if you correctly use your response to question 3.

4. The kinetic energy of the car at the top of the second hill is less than the kinetic energy of the car at the top of the third hill.

or

The car's KE is less.

or

The loss of potential energy from the second hill to the third hill results in an increase in kinetic energy from the second to the third hill.

Note: Credit is granted if the response is written in one or more complete sentences.

5. *a.*

b. The *slope* is the change in the ordinate or *y* value, divided by the change in the abscissa or *x* value, for any two points *on the best-fit line*:

$$\text{Slope} = \frac{\Delta y}{\Delta x}$$

or

$$\text{Slope} = \frac{\Delta W}{\Delta t}$$
$$= \frac{120\,\text{J} - 60\,\text{J}}{4\,\text{s} - 2\,\text{s}}$$
$$= 30\,\text{J/s (or W)}$$

The slope, $\Delta W/\Delta t$, represents the power developed in this problem.

Note: The values substituted must come from points that lie on the best-fit straight line. The values from the given data table do *not* necessarily fall on this line. You must also be sure to include units when you make the substitution and when you give the final answer.

c. Move up the vertical scale until the value for work = 75 J. At that level move along horizontally until the best-fit line is reached. Drop a vertical to the time axis from that point, and you see that *2.5 s* were needed to do 75 J of work.

493

6. Allow a maximum of 3 credits, 1 for each correct energy conversion.

   Examples of Acceptable Responses

   > work into
   > potential energy (spring) into
   > kinetic energy into
   > potential energy (gravity)

7. Allow a maximum of 2 credits.
   - Allow 1 credit for indicating the toy has less mass.
   - Allow 1 credit for indicating the toy has the same energy.

   Examples of acceptable responses include, but are not limited to:

   — The toy has less mass without the base but the same energy. Therefore it can go higher.
   — The work put into the toy is the same but the mass is less. With less mass the toy could go higher because it is moving faster.

8. Allow 1 credit for a scale that is linear and has appropriate divisions.

9. Allow 1 credit for plotting all points accurately (± 0.3 grid space). Allow credit if you correctly use your response to question 8. Do *not* penalize if no line is drawn.

8–9. Example of Acceptable Response

Kinetic Energy vs Time

10. Allow a maximum of 2 credits for calculating the speed of the mass. Refer to *Scoring Criteria for Calculations*.

    Example of Acceptable Response

    $$KE = \frac{1}{2}mv^2$$

    $$v = \sqrt{\frac{2\ KE}{m}}$$

    $$v = \sqrt{\frac{2\ (32\ J)}{4.0\ kg}}$$

    $$v = \sqrt{16\ \frac{m^2}{s^2}}$$

    $$v = 4.0\ m/s \ \ or \ \ v = 4.0\ \sqrt{\frac{J}{kg}}$$

11. Allow 1 credit for indicating that the speed of the mass at 6.0 seconds and the speed of the mass at 10.0 seconds are equal.

    Examples of Acceptable Responses

    The speeds are the same.

    or

    The speed of the mass at 6.0 seconds and the speed of the mass at 10.0 seconds are both 4.0 m/s.

    Allow credit for an answer that is consistent with the response to question 10.

12. Appropriate responses include, but are not limited to:

    Position *A*: kinetic or KE, or energy of motion
    Position *B*: elastic or potential, or energy of position

13. $mg\Delta h = \frac{1}{2}kx^2$ or $\Delta PE = PE_s$.

    Example of Acceptable Response

    $$\Delta PE = mg\Delta h$$

    $$PE_s = \frac{1}{2}kx^2$$

    $$\frac{1}{2}kx^2 = mg\Delta h$$

    $$k = \frac{2mg\Delta h}{x^2}$$

14. Allow a maximum of 2 credits for determining the amount of work done. Refer to *Scoring Criteria for Calculations*.

Examples of Acceptable Responses

$$F_y = F\sin\theta \text{ and } w = Fd$$
$$F_y = (160.\ N)(\sin 30.°) = 80.\ N$$
$$w = (80.\ N)(10.\ m) = 800\ J$$

or

$$w = Fd\sin\theta$$
$$w = (160.\ N)(10.\ m)(\sin 30.°)$$
$$w = 800\ J$$

or

$$w = Fd = \Delta E_T = mgh \text{ and } h = d\sin\theta$$
$$w = mgd\sin\theta$$
$$w = (160.\ N)(10.\ m)(\sin 30.°)$$
$$w = 800\ J$$

15. [1] Allow 1 credit for 50. m.

16. [2] Allow a maximum of 2 credits. Refer to *Scoring Criteria for Calculations*.

Example of a 2-credit response:

$$KE = \Delta PE = mg\Delta h$$
$$KE = (65\ kg)(9.81\ m/s^2)(5.5\ m)$$
$$KE = 3.5 \times 10^3\ J$$

17. [2] Allow a maximum of 2 credits. Refer to *Scoring Criteria for Calculations*.

Example of a 2-credit response:

$$KE = \frac{1}{2}mv^2$$
$$v = \sqrt{\frac{2KE}{m}}$$
$$v = \sqrt{\frac{2(3.5 \times 10^3\ J)}{65\ kg}}$$
$$v = 10.\ m/s$$

Note: Allow credit for an answer that is consistent with the response to question 16.

18. [1] Allow 1 credit for correctly plotting all data points ± 0.3 grid space.

19. [1] Allow 1 credit for drawing a line or curve of best fit.

Example of a 2-credit response for questions 18 and 19:

Dart's Maximum Vertical Displacement vs. Spring Compression

20. [2] Allow a maximum of 2 credits. Refer to *Scoring Criteria for Calculations*.

Example of a 2-credit response:

$$PE_s = \frac{1}{2}kx^2$$

$$PE_s = \frac{1}{2}(140 \text{ N/m})(0.070 \text{ m})^2$$

$$PE_s = 0.34 \text{ J}$$

Note: Allow credit for an answer that is consistent with the graph.

21. [1] Allow 1 credit for 5.6 N.

22. [2] Allow a maximum of 2 credits, allocated as follows:
    - Allow 1 credit for setting expressions for elastic potential energy and kinetic energy equal.
    - Allow 1 credit for correctly solving for k.

    Example of 2-credit responses:

    $$PE_s = KE$$
    $$\frac{1}{2}kx^2 = \frac{1}{2}mv^2$$
    $$k = \frac{mv^2}{x^2}$$

    or

    $$\frac{1}{2}kx^2 = \frac{1}{2}mv^2$$
    $$k = \frac{mv^2}{x^2}$$

23. [1] Allow 1 credit for 1.5 m/s.

24. [2] Allow a maximum of 2 credits. Refer to *Scoring Criteria for Calculations*.

    Example of a 2-credit response:

    $$KE = \frac{1}{2}mv^2$$

    $$KE = \frac{1}{2}(75\,\text{kg})(1.5\,\text{m/s})^2$$

    $$KE = 84 \text{ J}$$

    Note: Allow credit for an answer that is consistent with the response to question 23.

25. [1] Allow 1 credit for a straight line showing decreasing kinetic energy and increasing work.

    Example of a 1-credit response:

498

26. [2] Allow a maximum of 2 credits, 1 credit for indicating that the kinetic energy decreases and 1 credit for indicating that internal energy increases.

    Note: Do *not* allow credit for indicating that kinetic energy changes into potential energy.

27. [2] Allow a maximum of 2 credits. Refer to *Scoring Criteria for Calculations*.

    Example of a 2-credit response:

    $$PE_s = \tfrac{1}{2}kx^2$$
    $$PE_s = \tfrac{1}{2}(150\,\text{N/m})(0.050\,\text{m})^2$$
    $$PE_s = 0.19\,\text{J or } 1.9 \times 10^{-1}\,\text{J or } 0.1875\,\text{J}$$

28. [2] Allow a maximum of 2 credits. Refer to *Scoring Criteria for Calculations*.

    Example of a 2-credit response:

    $$\Delta PE = mg\Delta h$$
    $$\Delta h = \frac{\Delta PE}{mg}$$
    $$\Delta h = \frac{0.19\,\text{J}}{0.020\,\text{kg}\left(9.81\,\text{m/s}^2\right)}$$
    $$\Delta h = 0.97\,\text{m}$$

    Note: Allow credit for an answer that is consistent with the response to question 27.

29. Allow 1 credit for 8.0 ± 0.2 m.

30. Allow a maximum of 2 credits. Refer to *Scoring Criteria for Calculations*.

    Example of a 2-credit response:

    $$P = \frac{Fd}{t}$$
    $$P = \frac{(50.\,\text{N})(8.0\,\text{m})}{3.0\,\text{s}}$$
    $$P = 130\,\text{W} \ \ or \ \ 133\,\text{W}$$

Allow credit for an answer that is consistent with the response to question 29.

31. Allow a maximum of 2 credits. Refer to *Scoring Criteria for Calculations.*

    Example of a 2-credit response:

    $$\Delta PE = mg\Delta h$$
    $$\Delta PE = (3.0 \text{ kg})(9.81 \text{ m/s}^2)(3.0 \text{ m})$$
    $$\Delta PE = 88 \text{ J} \quad or \quad 88.3 \text{ kg} \cdot \text{m}^2/\text{s}^2$$

32. Allow a maximum of 2 credits. Refer to *Scoring Criteria for Calculations.*

    Example of a 2-credit response:

    $$E_T = PE + KE + Q$$
    $$KE = mg\Delta h$$
    $$KE = (3.0 \text{ kg})(9.81 \text{ m/s}^2)(3.0 \text{ m} - 1.0 \text{ m})$$
    $$KE = 59 \text{ J} \quad or \quad 58.9 \text{ J}$$

33. Allow 1 credit for G.

34. Allow 1 credit for 55 J.

35. Allow a maximum of 2 credits. Refer to *Scoring Criteria for Calculations.*

    Examples of 2-credit responses:

    $$PE = mgh$$
    $$m = \frac{PE}{gh}$$
    $$m = \frac{25 \text{ J}}{(9.81 \text{ m/s}^2)(1.0 \text{ m})}$$
    $$m = 2.5 \text{ kg} \quad or \quad 2.55 \text{ kg}$$

    or

    $$PE = mgh$$
    $$m = \frac{55 \text{ J}}{(2.25 \text{ m})(9.81 \text{ m/s}^2)}$$
    $$m = 2.5 \text{ kg}$$

    Allow credit for an answer that is consistent with the response to question 34.

36. Allow 1 credit. Acceptable responses include, but are not limited to:

    —weight of object *or* weight
    —*mg*
    —force
    —F_g

37. Allow 1 credit for drawing a line that would represent the relationship between gravitational potential energy and vertical height for an object having a greater mass. The line must be straight, with a slope steeper than that of the given line.

    Example of a 1-credit response:

Gravitational Potential Energy vs. Vertical Height

Chapter Eight, pages 235–237

1. Electric field lines are drawn in such a way that the arrows of the lines point in the direction in which a small positive test charge would move if it were placed in the field. Since like charges repel and opposite charges attract, the field lines will point from the positive plate toward the negative plate. See the diagram at the top of page 502.

Note: One point is given for drawing three parallel lines passing through points *A*, *B*, and *C* as long as the lines do not pass through either of the plates.

One point is awarded for drawing three arrowheads, one on each line, pointing in the correct direction. Additional field lines may be added with no penalty as long as they are drawn correctly.

2. Write out the problem.

   Given: $d = 2.0 \times 10^{-2}\,\text{m}$ Find: $E = ?$
   $V = 1.0 \times 10^{-2}\,\text{V}$

   Refer to the *Electricity* section in the *Reference Tables* to find an equation(s) that relates electric field strength to the variables given.

   Solution: $E = \dfrac{V}{d}$

   $= \dfrac{1.0 \times 10^{2}\,\text{V}}{2.0 \times 10^{-2}\,\text{m}}$

   $= 0.5 \times 10^{4}\,\dfrac{\text{V}}{\text{m}}$ or $5.0 \times 10^{3}\,\dfrac{\text{V}}{\text{m}}$

   Alternatively, $\dfrac{\text{V}}{\text{m}} = \dfrac{\dfrac{\text{J}}{\text{C}}}{\text{m}} = \dfrac{\dfrac{\text{N} \cdot \text{m}}{\text{C}}}{\text{m}},$

   $E = 5000\,\dfrac{\text{N}}{\text{C}}$

Note: One point is awarded for writing the equation and for the subsitution of values with units. If the equation and/or units are not shown, you do not receive this point. One point is awarded for the correct answer (number and units). If there are no units, you do not receive this point. Note also that significant figures and scientific notation are not required to obtain credit, and any correct units are accepted.

3. Write out the problem.

Given: $q = 25\,\text{C}$ Find: $W = ?$

$V = 1.8 \times 10^6\,\text{V}$

$I = 2.0 \times 10^4\,\text{A}$

Refer to the *Electricity* section in the *Reference Tables* to find an equation(s) that relates energy with charge, electric potential difference, and current.

Solution: $V = \dfrac{W}{q}$

$W = V_q$

$= (1.8 \times 10^6\,\text{V})(25\text{C})$

$= 4.5 \times 10^7\,\text{J or } 45 \times 10^6\,\text{V} \cdot \text{C}$

(Note that current was not needed in this calculation.)

Note: A total of two credits is allowed. One credit is awarded for writing the equation and for the substitution of values with units. If the equation and/or units are not shown, you do not receive this credit. One credit is awarded for the correct answer (number and units). If there are no units, you do not receive this credit. Significant figures and scientific notation are not required to obtain this credit.

4. By definition, electric field lines are drawn away from positive charges and toward negative charges.

Note: One point is awarded for three lines drawn from point A to point B. The lines may consist of one straight line and two curved lines or three curved lines.

One credit is awarded for three arrowheads, one on each line pointing in the direction of point B. There is no penalty for extra lines with arrowheads drawn correctly.

503

5. Refer to the *Electricity* section in the *Reference Tables* to find an equation(s) for the electrostatic force between two charges.

$$\text{Solution:} \quad F = \frac{kq_1q_2}{r^2}$$

Refer to the *List of Physical Constants* in the *Reference Tables* for the electrostatic constant.

$$F = \frac{\left(9.0 \times 10^9 \, \frac{N \cdot m^2}{C^2}\right)(2.4 \times 10^{-6} \, C)(-2.4 \times 10^{-6} \, C)}{(0.50 \, m)^2}$$

$$= -0.21 \, N$$

$$\text{or} - 2.1 \times 10^{-1} \, N$$

Note: One point is awarded for writing the equation and for substituting values with units. If the equation and/or units are not shown, you do not receive this point. One point is awarded for the correct answer (number and units). If no units are shown, you do not receive this point. Note also that significant figures and scientific notation are not required to obtain this credit. Note also that the negative sign indicates direction and is not necessary to receive credit.

6. The magnitude of the electrostatic force is inversely proportional to the distance squared.

Example of Accceptable Response

Magnitude of Electrostatic Force

0

Distance of Separation

Note: One point is awarded for an appropriately sketched graph. No credit is granted if the curve intersects either axis, and no credit is granted if the sketch is of a straight line.

7.

Sphere	Charge
R	neutral
T	positive
U	positive

8. [2] Allow a maximum of 2 credits. Refer to *Scoring Criteria for Calculations.*

   Examples of a 2-credit response:

   $$F_e = \frac{kq_1q_2}{r^2}$$

   $$F_e = \frac{\left(8.99 \times 10^9 \; \text{N} \cdot \text{m}^2 / \text{C}^2\right)\left(2.0 \times 10^{-6} \; \text{C}\right)\left(2.0 \times 10^{-6} \; \text{C}\right)}{\left(2.0 \times 10^{-1} \; \text{m}\right)^2}$$

   $$F_e = 9.0 \times 10^{-1} \; \text{N}$$

9. [1] Allow 1 credit for drawing *at least five* straight parallel lines perpendicular to the plates and pointing toward the negative plate. The lines must originate and end on the plates.

   Examples of a 1-credit response:

   Note: Curved lines beyond the edges of the plates are acceptable. Parallel lines need not be equally spaced.

10. [2] Allow a maximum of 2 credits. Refer to *Scoring Criteria for Calculations.*

Example of a 2-credit response:

$$E = \frac{F_e}{q}$$

$$F_e = Eq$$

$$F_e = (2.0 \times 10^3 \text{ N/C}) \ (1.6 \times 10^{-19} \text{ C})$$

$$F_e = 3.2 \times 10^{-16} \text{ N}$$

Chapter Nine, pages 272–282

1. *a.* 60. Ω   We reason as follows: If switch S_1 is kept open, no current can flow through the first resistor, which we have labeled R_1 in the diagram below. Since switch S_2 is closed, the circuit resistance consists of resistor R_2 only.

   b. If switch S_2 is kept open, no current will flow through resistor R_2 and we can ignore it. We then apply Ohm's law to the rest of the circuit. According to the *Electricity* equations in the *Reference Tables*,

$$R = \frac{V}{I}$$

$$60. \ \Omega = \frac{120 \text{ V}}{I}$$

$$I = \frac{120 \text{ V}}{60. \ \Omega}$$

$$= 2.0 \text{ A}$$

Note: It is also correct to rearrange the terms so that the unknown, I, stands alone, before substituting values:

$$R = \frac{V}{I}$$

$$I = \frac{V}{R}$$

$$I = \frac{120 \text{ V}}{60.\,\Omega}$$

$$= 2.0 \text{ A}$$

Units must appear in the substitution and in the final answer for full credit.

c. 4.0 A When both switches are closed, we have a parallel combination of two resistors and a power supply. The ammeter will read the total current in the circuit (I_t). There are several ways of determining this current.

Here's one way. First calculate the total resistance of the circuit (R_t). Refer to the *Electricity* equations in the *Reference Tables*:

$$\frac{1}{R_t} = \frac{1}{R_1} + \frac{1}{R_2}$$

$$\frac{1}{R_t} = \frac{1}{60.\,\Omega} + \frac{1}{60.\,\Omega}$$

$$R_t = 30.\,\Omega$$

The total current can now be found by applying Ohm's law to the whole circuit:

$$I_t = \frac{V_t}{R_t}$$

$$= \frac{120 \text{ V}}{30.\,\Omega}$$

$$= 4.0 \text{ A}$$

Another way of solving this problem is to realize that in a parallel circuit each resistor acts independently, drawing the same amount of current from the source whether or not other resistors are connected. The total current equals the sum of the branch currents. In part b, we found that 2.0 amperes flow through R_1. Since R_2 is a resistor of the same size, it will also draw 2.0 amperes. The total current then equals 2.0 A + 2.0 A = 4.0 A.

d. 120 V In a parallel circuit, the voltage across each branch is equal to the voltage of the source.

2. *a.*

b. Apply *Ohm's law* to determine the value of the current (*I*) in resistor R_1. Since R_1 is connected in parallel to the 12-volt battery, the potential difference (*V*) across R_1 is 12 volts. Refer to the *Electricity* equations in the *Reference Tables*:

$$R = \frac{V}{I}$$

$$18\,\Omega = \frac{12\,V}{I}$$

$$I = \frac{12\,V}{18\,\Omega}$$

$$= 0.67\ A$$

Note: It is also correct to rearrange the terms so that the unknown, *I*, stands alone, before substituting values:

$$R = \frac{V}{I}$$

$$I = \frac{V}{R}$$

$$= \frac{12\,V}{18\,\Omega}$$

$$= 0.67\ A$$

To obtain full credit, you must include units in the substitution and in the final answer.

3. *a.*

b. The *slope* is the change in the ordinate or *y* value, divided by the change in the abscissa or *x* value, for any two points *on the best-fit line*, such as the points (0.040 A, 10.0 V) and (0.010 A, 2.5 V):

$$\text{Slope} = \frac{\Delta y}{\Delta x}$$

or

$$\text{Slope} = \frac{\Delta V}{\Delta I}$$
$$= \frac{10.0\,\text{V} - 2.5\,\text{V}}{0.040\,\text{A} - 0.010\,\text{A}}$$
$$= 250\ \Omega$$

Note: The values substituted must come from points that lie on the best-fit straight line. The values from the given data table do *not* necessarily fall on this line. You must also be sure to include *units* when you make the substitution *and* when you give the final answer.

c. Resistance (R) is the opposition to the flow of current. It is defined as the ratio of the potential difference (*V*) across a conductor to the current (*I*) flowing through it.

Refer to the *Electricity* equations in the *Reference Tables*:

$$R = \frac{V}{I}$$

When the temperature is held constant, the ratio V/I remains constant for metallic conductors, a relationship known as *Ohm's law*. When the potential difference applied to the ends of a resistor is plotted against the current, as shown in the graph in part a, the graph is a straight line sloping upward.

The slope of this graph, $\Delta V/\Delta I$, represents the *resistance of the conductor*.

4. The circuit diagram is as follows:

Note: One credit is granted for a circuit that contains two resistors, labeled "5 ohms" and "20 ohms," connected in parallel with a source that may be labeled "source," "24 V," or "24-V source." One credit is granted for a single ammeter properly placed to measure the total current. This credit is disallowed if more than one ammeter is present in the circuit. If the circuit drawn is a series circuit, and the first credit is not granted, this credit may still be granted if the ammeter is properly placed to measure the total current.

5. Write out the problem.

Given: $R_1 = 50\text{-}\Omega$ Find: $R_t = ?$
         $R_2 = 20\text{-}\Omega$
         $V = 24\text{-V source}$

Refer to the *Electricity* section in the *Reference Tables* to find an equation(s) that relates total circuit resistance with individual resistance.

$$\frac{1}{R_t} = \frac{1}{R_1} + \frac{1}{R_2}$$

$$= \frac{1}{5.0\,\Omega} + \frac{1}{20.0\,\Omega}$$

$$= \frac{4}{20\,\Omega} + \frac{1}{200\,\Omega}$$

$$= \frac{5}{20\,\Omega}$$

$$R_t = \frac{20\,\Omega}{5}$$

$$= 4.0\,\Omega$$

Note: One credit is awarded for writing the equation and for the substitution of values with units. If the equation and/or units are not shown, you do not receive this credit. One credit is awarded for the correct answer (number and units). If there are no units, you do not receive this credit. Note also that significant figures and scientific notation are not required to obtain this credit. Credit is granted if you correctly use the response to question 4.

6. Write out the problem.

   Given: $V_T = 24\,V$ Find: $I_T = ?$

           $R_T = 4.0\text{-}\Omega$

   Refer to the *Electricity* section in the *Reference Tables* to find an equation(s) that relates current with potential difference and resistance.

   $$R_T = \frac{V_t}{I_T}$$

   $$I_T = \frac{V_T}{R_T}$$

   $$= \frac{24\,V}{4.0\,\Omega}$$

   $$= 6.0\,V/\Omega \text{ or } 6.0\,A$$

Note: One credit is awarded for writing the equation and for the substitution of values with units. If the equation and/or units are not shown, you do not receive this credit. One credit is awarded for the correct answer (number and units). If there are no units, you do not receive this credit. Note also that significant figures and scientific notation are not required to obtain this credit. Credit is granted if you correctly use the response to question 5.

7. Solve as follows:

   $$R = \frac{V}{I}$$

   $$W = IR$$

   $$= (0.50\,A)(5.0\,\Omega)$$

   $$= 2.5\,A \cdot \Omega \text{ or } 2.5\,V$$

Note: One point is awarded for writing the equation and for the substitution of values with units. If the equation and/or units are not shown, you do not receive this point. One point is awarded for the correct answer (number and units). If there are no units, you do not receive this point. Note also that significant figures and scientific notation are not required to obtain this credit.

8. Solve as follows:

$$W = VIt$$
$$= (15 \text{ V})(0.50 \text{ A})(600. \text{ s})$$
$$= 4.5 \times 10^3 \text{ J or } 4500 \text{ V} \cdot \text{A} \cdot \text{s}$$

Note: One point is awarded for writing the equation and for the substitution of values with units. If the equation and/or units are not shown, you do not receive this point. One point is awarded for the correct answer (number and units). If there are no units, you do not receive this point. Note also that significant figures and scientific notation are not required to obtain this credit.

9. Allow 1 credit for accurately plotting the data points (±0.3 grid space) for power vs. current.

10. Allow 1 credit for drawing a straight best-fit line. If one or more points are plotted incorrectly in question 9, but a best-fit line is drawn, allow this credit.

9–10. Example of Acceptable Response

Power vs. Current for a Variable Resistor

11. Allow 1 credit for determining the power delivered to the circuit.

    Example of Acceptable Response

    10.5 W ± 0.3 W

    Allow credit for an answer that is consistent with the student's answer to question 10.

12. Allow a maximum of 2 credits for determining the slope of the graph. Refer to *Scoring Criteria for Calculations*. Allow credit for an answer that is consistent with the graph, unless no credit is awarded for questions 10 and 11. In that case, credit may be awarded if you correctly calculate the slope using data in the table.

    Note: The slope may be determined by direct substitution of data points only if the data values are on the best-fit line.

    Examples of Acceptable Responses

    $$\text{slope} = \frac{\Delta P}{\Delta I}$$

    or

    $$\text{slope} = \frac{\Delta Y}{\Delta X}$$

    $$\text{slope} = \frac{10.5\ W - 0.0\ W}{3.5A - 0.0A}$$

    $$\text{slope} = 3.0\ V$$

    or

    $$\text{slope} = 3\frac{W}{A}$$

13. Allow 1 credit for stating the physical significance of the slope of the graph.

    Examples of acceptable responses include, but are not limited to:

    — voltage
    — potential difference

14. Allow a maximum of 5 credits for explaining how to find the resistance of an unknown resistor, allocated as follows:

   a Allow 1 credit for listing the necessary measurements (voltage and current).

   b Allow 1 credit for listing the necessary equipment (ammeter, voltmeter, battery or power supply, and wires).

   c Allow a maximum of 2 credits for completing the circuit diagram.
   - Allow 1 credit for drawing the ammeter in series with the resistor.
   - Allow 1 credit for drawing the voltmeter in parallel with the resistor.

   d Allow 1 credit for listing the necessary formula ($R = V/I$).

Example of Acceptable Response

   a To determine the resistance of an unknown resistor, I would need to measure the current and potential difference for the resistor in a circuit.

   b The equipment I would need would be the resistor, an ammeter, a voltmeter, a battery or power supply, and connecting wires.

   c The circuit would be connected as in the diagram below.

   d Once I measured the current and potential for the resistor, I would use the formula for Ohm's law ($R = V/I$) to calculate the resistance.

15. Allow a maximum of 2 credits.

   - Allow 1 credit for a circuit containing two resistors connected in parallel with a battery.
   - Allow 1 credit for a voltmeter connected in parallel with either resistor, or the battery. If the student has drawn a series circuit and the voltmeter is properly placed to measure the potential difference across either resistor, allow this credit.

Examples of Acceptable Responses

or

16. Allow a maximum of 2 credits. Refer to *Scoring Criteria for Calculations.*

Examples of Acceptable Responses

$$\frac{1}{R_{eq}} = \frac{1}{R_1} + \frac{1}{R_2}$$

$$\frac{1}{R_{eq}} = \frac{1}{3.0\ \Omega} + \frac{1}{3.0\ \Omega}$$

$$\frac{1}{R_{eq}} = \frac{2}{3.0\ \Omega}$$

$$R_{eq} = 1.5\ \Omega$$

or

$$R_{eq} = \frac{R_1 R_2}{R_1 + R_2}$$

$$R_{eq} = \frac{(3.0\ \Omega)(3.0\ \Omega)}{3.0\ \Omega + 3.0\ \Omega}$$

$$R_{eq} = 1.5\ \Omega$$

Allow credit if you correctly use your responses to question 15. That is, if you connected the resistors in series in question 15, then the following answer is acceptable.

$$R_{eq} = R_1 + R_2$$

$$R_{eq} = 3.0\ \Omega + 3.0\ \Omega$$

$$R_{eq} = 6.0\ \Omega$$

17. Allow a maximum of 2 credits. Refer to *Scoring Criteria for Calculations.*

    Examples of Acceptable Responses

    $$P = \frac{V^2}{R}$$

    $$R = \frac{V^2}{P}$$

    $$R = \frac{(120. \text{ V})^2}{1050 \text{ W}}$$

    $$R = 13.7 \ \Omega$$

    or

    $$P = VI$$

    $$I = \frac{P}{V}$$

    $$I = \frac{1050 \text{ W}}{120 \text{ V}}$$

    $$I = 8.75 \text{ A}$$

    $$R = \frac{V}{I}$$

    $$I = \frac{120 \text{ V}}{8.75 \text{ A}}$$

    $$R = 13.7 \ \Omega$$

18. Allow a maximum of 2 credits.

    • Allow 1 credit for correctly indicating the total current.

    • Allow 1 credit for stating whether it is possible to operate the toaster and the microwave simultaneously. If the current indicated is greater than 15 A, the answer should be no. If the current indicated is less than 15 A, the answer should be yes.

    Example of Acceptable Response

    $$I = \frac{P}{V} = \frac{1050 \text{ W}}{120 \text{ V}} = 8.75 \text{ A}$$

    $$I_{total} = 8.75 \text{ A} + 10.0 \text{ A} = 18.8 \text{ A}$$

Answer: No

No penalty for incorrect or missing units.

Note: Allow no credit for a yes or no answer with no mathematical justi-
fication.

19. **Example of Acceptable Response**

20. 2.0 V or 2 V.

21. **Examples of Acceptable Responses**

$$\frac{1}{R_1} + \frac{1}{R_2} = \frac{1}{R_{eq}}$$

$$\frac{1}{R_1} = \frac{1}{R_{eq}} \times \frac{1}{R_2}$$

$$\frac{1}{R_1} = \frac{1}{2.0 \ \Omega} \times \frac{1}{3.0 \ \Omega}$$

$$\frac{1}{R_1} = \frac{1}{6.0 \ \Omega}$$

$$R_1 = 6 \ \Omega$$

or

$$I_2 = \frac{V_2}{R_2} = \frac{12\text{ V}}{3.0\ \Omega} = 4.0\text{ A}$$

$$I_1 = 6.0\text{ A} - 4.0\text{ A} = 2.0\text{ A}$$

$$R_1\frac{V_1}{I_1} = \frac{12\text{ V}}{2.0\text{ A}} = 6\ \Omega$$

22. [1] Allow 1 credit for 3.1×10^{-6} m^2.

23. [2] Allow a maximum of 2 credits. Refer to *Scoring Criteria for Calculations.*

    Example of a 2-credit response:

    $$R = \frac{\rho L}{A}$$
    $$R = \frac{(1.72 \times 10^{-8}\ \Omega \bullet \text{m})(10.0\text{ m})}{3.1 \times 10^{-6}\text{ m}^2}$$
    $$R = 5.5 \times 10^{-2}\ \Omega$$

    Note: Allow credit for an answer that is consistent with the response to question 22.

24. [1] Allow 1 credit for drawing two resistors in parallel, completing the circuit.

    Example of a 1-credit response:

90. V

25. [1] Allow 1 credit for 90. V.

    Note: Allow credit for an answer that is consistent with the response to question 24.

518

26. [2] Allow a maximum of 2 credits. Refer to *Scoring Criteria for Calculations.*

   Example of a 2-credit response:

$$R = \frac{V}{I}$$

$$I = \frac{V}{R}$$

$$I = \frac{90. \text{ V}}{15 \ \Omega}$$

$$I = 6.0 \text{ A}$$

   Note: Allow credit for an answer that is consistent with the response to question 25.

27. [1] Allow 1 credit for indicating that the resistance increases.

28. [1] Allow 1 credit for indicating that the resistivity remains the same *or* there is no effect.

29. [2] Allow a maximum of 2 credits, allocated as follows:

   • Allow 1 credit for drawing a complete circuit, including a source of potential difference.
   • Allow 1 credit for connecting resistors with an equivalent resistance of 15 Ω.

   Example of a 2-credit response:

30. [1] Allow 1 credit for 2.4 Ω.

31. [2] Allow a maximum of 2 credits. Refer to *Scoring Criteria for Calculations.*

Example of a 2-credit response:

$$R = \frac{V}{I}$$

$$I = \frac{V}{R}$$

$$I = \frac{12\ V}{3.0\ \Omega}$$

$$I = 4.0\ A$$

32. [2] Allow a maximum of 2 credits. Refer to *Scoring Criteria for Calculations.*

Examples of 2-credit responses:

$$R = \frac{V}{I} \qquad\qquad \frac{1}{R_{eq}} = \frac{1}{R_1} + \frac{1}{R_2}$$

$$R = \frac{12\ V}{1.0\ A} \quad or \quad \frac{1}{R} + \frac{1}{3\ \Omega} = \frac{1}{2.4\ \Omega}$$

$$R = 12\ \Omega \qquad\qquad R = 12\ \Omega$$

Note: Allow credit for an answer that is consistent with the response to question 30 and/or question 31.

33. [2] Allow a maximum of 2 credits. Refer to *Scoring Criteria for Calculations.*

Example of a 2-credit response:

$$R = \frac{\rho L}{A}$$

$$R = \frac{(150. \times 10^{-8}\ \Omega \bullet m)\ (1.00 \times 10^3\ m)}{3.50 \times 10^{-6}\ m^2}$$

$$R = 429\ \Omega$$

34. [2] Allow a maximum of 2 credits. Refer to *Scoring Criteria for Calculations.*

Example of a 2-credit response:

$$W = VIt$$
$$W = (115\ V)\ (20.0\ A)\ (60.\ s)$$
$$W = 1.4 \times 10^5\ J\ or\ 138\ 000\ J$$

35. [2] Allow a maximum of 2 credits. Refer to *Scoring Criteria for Calculations.*

    Example of a 2-credit response:

    $$R = \frac{V}{I}$$
    $$R = \frac{120\,\text{V}}{0.50\,\text{A}}$$
    $$R = 240\ \Omega$$

36. [1] Allow 1 credit for 190 Ω *or* an answer that is consistent with the response to question 35.

37. [2] Allow a maximum of 2 credits. Refer to *Scoring Criteria for Calculations.*

    Example of a 2-credit response:

    $$P = I^2 R$$
    $$P = (0.50\,\text{A})^2 (50.\ \Omega)$$
    $$P = 12\ \text{W or } 12.5\ \text{W}$$

38. Allow a maximum of 2 credits. Refer to *Scoring Criteria for Calculations.*

    Example of a 2-credit response:

    $$I = \frac{\Delta q}{t}$$
    $$\Delta q = It$$
    $$\Delta q = (0.50\,\text{A})(60.\,\text{s})$$
    $$\Delta q = 30.\,\text{C}$$

39. Allow a maximum of 2 credits, allocated as follows:

    • Allow 1 credit for drawing a parallel circuit containing two resistors and a battery.
    Note: Do *not* allow this credit if the student draws a cell instead of a battery.
    • Allow 1 credit for correct placement of the ammeter.

Example of a 2-credit response:

40. Allow a maximum of 2 credits. Refer to *Scoring Criteria for Calculations.*

Example of a 2-credit response:

$$\frac{1}{R_{eq}} = \frac{1}{R_1} + \frac{1}{R_2}$$

$$\frac{1}{R_{eq}} = \frac{1}{18\,\Omega} + \frac{1}{36\,\Omega}$$

$$R_{eq} = 12\,\Omega$$

Allow credit for an answer that is consistent with the response to question 39.

41. Allow a maximum of 2 credits. Refer to *Scoring Criteria for Calculations.*

Examples of 2-credit responses:

$$P = \frac{V^2}{R}$$

$$P = \frac{(24\,\text{V})^2}{12\,\Omega}$$

$$P = 48\,\text{W}$$

or

$$I = \frac{V}{R} = \frac{24\,\text{V}}{12\,\Omega} = 2\,\text{A}$$

and

$$P = VI$$

$$P = (24\,\text{V})(2\,\text{A})$$

$$P = 48\,\text{W}$$

Allow credit for an answer that is consistent with the response to question 40.

Chapter Ten, page 301

1. Allow 1 credit for drawing the correct orientation of the needle of compass Y and labeling its polarity.

   Example of Acceptable Response

2. Allow a maximum of 2 credits, allocated as follows:

   • Allow 1 credit for drawing four field lines that do *not* cross and are closest together at the poles.
   • Allow 1 credit for four field lines drawn from N to S.

   Examples of 2-credit responses:

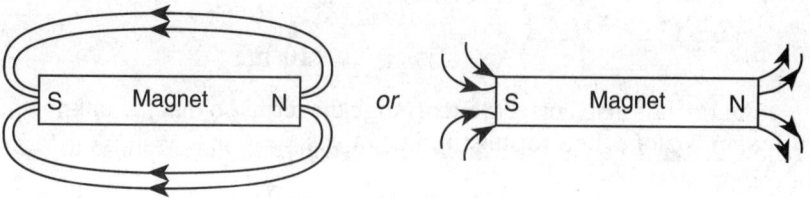

Chapter Eleven, pages 338–345

1. *a.* The speed (v) of a wave is equal to the product of its frequency (f) and its wavelength (λ). Refer to the *Waves and Optics* equations in the *Reference Tables*:

$$v = f\lambda$$

$$1.5 \times 10^3 \text{ m/s} = (5.0 \times 10^3 \text{ Hz})\lambda$$

$$\frac{1.5 \times 10^3 \text{ m/s}}{5.0 \times 10^3 \text{ Hz}} = \lambda$$

$$3.0 \times 10^{-1} \text{ m} = \lambda$$

Note: It is also correct to rearrange the terms so that the unknown, λ, stands alone, before substituting values:

$$v = f\lambda$$

Therefore

$$\lambda = \frac{v}{f}$$

$$= \frac{1.5 \times 10^3 \ \text{m/s}}{5.0 \times 10^3 \ \text{Hz}}$$

$$= 0.30 \ \text{m}$$

You must be sure to include units when you make the substitutions and when you give the final answer.

b. The average speed ($\bar{v}$) at which an object moves is equal to the distance traveled (Δs) divided by the time (Δt) required to travel that distance. The echo traveled from the ship to the ocean floor and then back to the ship again in 4.0 s. Therefore the time required for the sonar signal to reach the floor is half of 4.0 s, or 2.0 s. Refer to the *Mechanics* equations in the *Reference Tables*:

$$\bar{v} = \frac{\Delta s}{\Delta t}$$

$$1.5 \times 10^3 \ \text{m/s} = \frac{\Delta s}{2.0 \text{s}}$$

$$\Delta s = 3.0 \times 10^3 \text{m}$$

Again, it is also correct to rearrange the terms so that the unknown, Δs, stands alone, before substituting values:

$$\Delta s = \bar{v} \Delta t$$

$$= (1.5 \times 10^3 \text{m/s})(2.0 \text{s})$$

$$= 3.0 \times 10^3 \text{m}$$

Units must appear in the substitution and in the final answer for full credit.

2.

Note: One credit will be granted if the arrow forms an angle of 60.° ± 2° with the barrier and is directed away from the barrier as shown.

3. Write out the problem.

Given: $f = 12\,\text{Hz}$ Find: $v = ?$
            $\lambda = 0.9\,\text{cm}\ (0.009\,\text{m})$

Refer to the *Waves and Optics* section in the *Reference Tables* to find an equation that relates velocity with frequency and wavelength.

$$\text{Solution:}\quad v = f\lambda$$
$$= (12\,\text{Hz})(0.009\,\text{m})$$
$$= 0.1\,\text{m/s}$$

Other acceptable responses are 10.8 cm/s and 100 mm/s.

Note: One credit is awarded for writing the equation and for the substitution of values with units. If the equation and/or units are not shown, you do not receive this credit. One credit is awarded for the correct answer (number and units). If no units are shown, you do not receive this credit. Significant figures and scientific notation are not required to obtain this credit.

4. Examples of acceptable responses for one credit:
   The angle of incidence is equal to the angle of reflection.

$$\angle \text{ of incidence} = \angle \text{ of reflection}$$

   Examples of *unacceptable* responses:

   The angle that the incoming waves make with the barrier is equal to the angle that the reflected waves make with the barrier.

$$\angle i = \angle r$$

5. The resultant wave produced by the superposition of waves A and B is as follows:

Note: One credit is awarded for the correct answer.

6. From the grid you obtain 0.3 m (±0.02 m) or 30 cm (±2 cm) as the amplitude of the resultant wave.

   Note: To receive one credit, the correct unit must be included in your answer. Note also that one credit is granted if your answer is based correctly on the answer to question 5.

7. From the grid you obtain 1.0 m or 100 cm as the wavelength of the resultant wave.

   Note: To receive one credit, the correct unit must be included in your answer. Note also that credit is granted if your answer is correct based on the answer to question 5.

8. Allow a maximum of 2 credits for drawing at least one complete cycle of the periodic wave.

   - Allow 1 credit for a wavelength of 3 meters (± 0.2 grid space).
   - Allow 1 credit for an amplitude of .2 meter (± 0.2 grid space).

   Note: The waveform may be another shape (e.g., triangular), provided that it is periodic and has the required amplitude and wavelength (± 0.2 grid space).

   Example of Acceptable Response

9. 3

10. 0.6 cm ± 0.2 cm.

11. 2.3 cm ± 0.2 cm.

12. The wavelength would decrease.

13. 4.

14. [1] Allow 1 credit for marking an **X** 180° out of phase with point A.

Example of a 1-credit response:

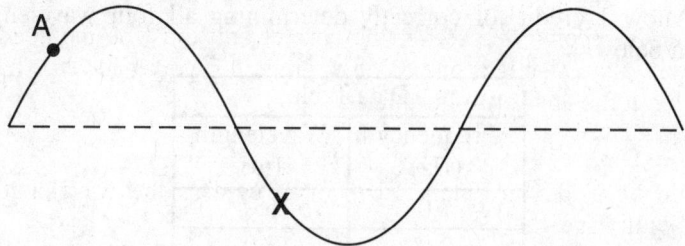

15. [2] Allow a maximum of 2 credits, allocated as follows:

   • Allow 1 credit for drawing a wave with smaller amplitude.
   • Allow 1 credit for drawing a wave with the same wavelength.

Example of a 2-credit response:

Note: Waves need *not* be in phase to receive credit.

16. [1] Allow 1 credit for drawing any two points horizontally across from each other and separated by one or two wavelengths.

Example of a 1-credit response:

Note: The points need *not* be labeled to receive credit.

17. [1] Allow 1 credit for visible light *or* green light.

18. [1] Allow 1 credit for correctly determining all four wavelengths, as shown below.

Data Table	
Frequency (Hz)	Wavelength (m)
1.0	6.0
2.0	3.0
3.0	2.0
6.0	1.0

19. [1] Allow 1 credit for correctly plotting all four data points (±0.3 grid space).

   Note: Allow credit for an answer that is consistent with the response to question 18.

20. [1] Allow 1 credit for drawing the best-fit curve.

   Note: Allow credit for an answer that is consistent with the response to question 19.

   Example of a 2-credit response for questions 19 and 20:

Wavelength vs. Frequency

21. [2] Allow a maximum of 2 credits. Refer to *Scoring Criteria for Calculations*.

Examples of 2-credit responses:

$$v = f\lambda$$
$$v = (5.0 \text{ Hz})(1.0 \text{ m}) \qquad or \qquad v = (2.0 \text{ Hz})(2.5 \text{ m})$$
$$v = 5.0 \text{ m/s} \qquad\qquad\qquad v = 5.0 \text{ m/s}$$

22. Allow a maximum of 2 credits, 1 credit for correct amplitude and 1 credit for correct frequency.

Example of a 2-credit response:

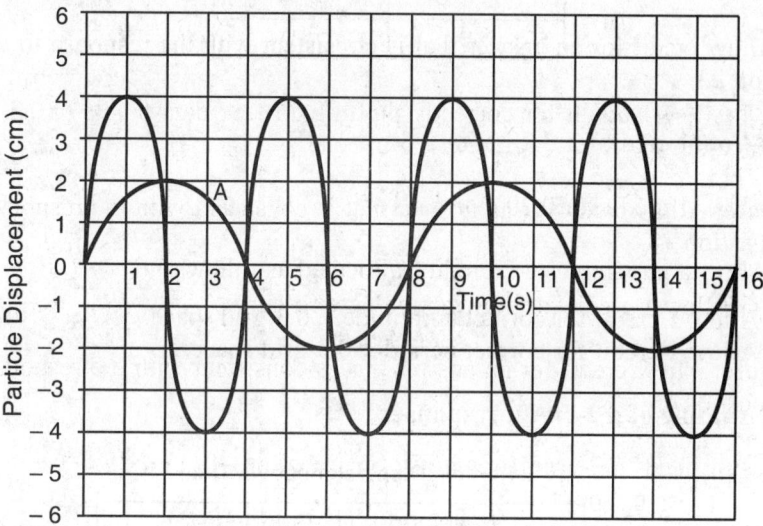

Note: If more than one cycle is drawn, rate only the first cycle.

23. Allow a maximum of 2 credits. Refer to *Scoring Criteria for Calculations.*

Example of a 2-credit response:

$$T = \frac{1}{f}$$

$$f = \frac{1}{T}$$

$$f = \frac{1}{5.0 \text{ s}}$$

$$f = 0.20 \text{ Hz}$$

24. Allow a maximum of 2 credits. Refer to *Scoring Criteria for Calculations.*

Examples of 2-credit responses:

$$v = f\lambda$$
$$v = (0.20 \text{ Hz})(2.0 \text{ m})$$
$$v = 0.40 \text{ m/s}$$

or

$$\bar{v} = \frac{d}{t}$$
$$\bar{v} = \frac{2.0 \text{ m}}{5.0 \text{ s}}$$
$$\bar{v} = 0.40 \text{ m/s}$$

Allow credit for an answer that is consistent with the response to question 23.

25. Allow 1 credit for 3.28 m *or* 3.3 m.

26. Allow 1 credit for 0.20 s *or* $\frac{1}{5}$ s.

27. Allow a maximum of 2 credits, allocated as follows:
    • Allow 1 credit for correct amplitude ± 0.3 grid space.
    • Allow 1 credit for correct period ± 0.3 grid space.

Example of a 2-credit response:

Particle Displacement vs. Time

Note: Allow credit for any periodic wave form (e.g., square or triangular) that meets these criteria.

Allow credit for an answer that is consistent with the response to question 26.

28. Allow a maximum of 2 credits. Refer to *Scoring Criteria for Calculations.*

    Example of 2-credit responses:

$$v = f\lambda$$
$$v = (5.0\,\text{Hz})(0.080\,\text{m})$$
$$v = 0.40\,\text{m/s}$$

or

$$v = \frac{d}{t}$$
$$v = \frac{0.080\,\text{m}}{0.2\,\text{s}}$$
$$v = .4\,\text{m/s}$$

Chapter Twelve, pages 382–389

1. *a.* The *angle of incidence* is the angle between the incident ray and the normal, or imaginary line drawn perpendicular to the surface. In this question, the angle of incidence for light ray *AO* measures 33°. (An answer that is within plus or minus of 2° from 33° is also acceptable.)

   b. According to the *law of reflection*, the angle of incidence is equal to the *angle of reflection*. Therefore the angle of reflection of the light ray also equals 33°.

   c.

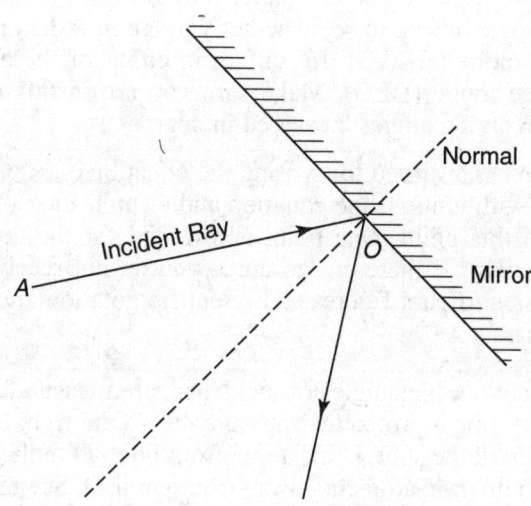

2. Write out the problem.

Given: $\theta_1 = 30°$ Find: $\theta_2 = ?$
medium 1 = Lucite
medium 2 = air

Refer to the *Waves and Optics* section in the *Reference Tables* to find an equation(s) that relates angle of refraction to angle of incidence.

Solution: $n_1 \sin \theta_1 = n_2 \sin \theta_2$

To calculate the angle of refraction from the angle of incidence, you need to know the refractive indices of the two media. You can easily obtain this information from the *Absolute Indices of Refraction* list in the *Reference Tables*. The refractive indices for Lucite and air are 1.50 and 1.00, respectively.

$$\sin\theta_2 = \frac{n_1 \sin\theta_1}{n_2}$$

$$\theta_2 = \sin^{-1}\left(\frac{n_1 \sin\theta_1}{n_2}\right)$$

$$= \sin^{-1}\left(\frac{1.50 \sin 30.°}{1.00}\right)$$

$$= \sin^{-1}(1.50 \times 0.5) = \sin^{-1}(0.750)$$

$$= 49°$$

(If you are using a scientific calculator and are not getting the correct value for the angle, check to see whether you are in radian mode (RAD) or in gradient mode (GRAD). If you are in either of these, change the mode to degree mode (DEG). Make sure you are in this mode for all calculations involving angles measured in degrees.)

Note: One point is awarded for writing the equation and for the subsitution of values with units. If the equation and/or units are not shown, you do not receive this point. One point is awarded for the correct answer (number and units). If there are no units, you do not receive this point. Note also that significant figures and scientific notation are not required to obtain credit.

3. Angles of refraction, like angles of incidence, are measured with respect to the normal. Using a protractor, measure 49° to the right of the normal in the air and mark the spot. Draw a line from point O to the spot marked off, and add an arrowhead facing away from point O. See the diagram at the top of page 533.

Note: One point will be awarded if the angle between the normal and the refracted ray is equal (±2°) to the value calculated in question 2. One point will be awarded if the refracted ray is drawn as shown (straight line, originating at point O, drawn to the right of the normal and having an arrow directed away from point O).

4. Refer to the *Waves and Optics* section of the *Reference Tables* to find an equation that will allow you to calculate the critical angle of incidence for Lucite relative to air.

$$\text{Solution:} \quad \sin\theta_c = \frac{1}{n}$$

$$\theta_c = \sin^{-1}\left(\frac{1}{n}\right)$$

$$= \sin^{-1}\left(\frac{1}{1.50}\right) = \sin^{-1}(0.667)$$

$$= 42°$$

Note: One point is awarded for writing the equation and for the substitution of values with units. If the equation and/or units are not shown, you do not receive this point. One point is awarded for the correct answer (number and units). If there are no units, you do not receive this point. Note also that significant figures and scientific notation are not required to obtain credit.

5. The angle of refraction is measured with respect to the normal. One credit is awarded for an answer of 40.° or 40° ± 2°.

6. The correct equation needed to solve for index of refraction is

$$n_1 \sin \theta_1 = n_2 \sin \theta_2$$

Refer to *Absolute Indices of Refraction* in the *Reference Tables* to obtain a value for water. Then:

$$n_2 = \frac{n_1 \sin \theta_1}{\sin \theta_2}$$

$$= \frac{(1.33)(\sin 60\,°)}{\sin 40\,°}$$

$$= 1.79 \text{ or } 1.8$$

Note: One point is awarded for writing the equation and for the substitution of values with units. If the equation and/or units are not shown, you do not receive this point. One point is awarded for the correct answer. Note also that significant figures and scientific notation are not required to obtain this credit. Credit may be granted if you correctly use the response to question 5.

7. Allow 1 credit for extending the line straight into the water (±2°) as shown below. Do *not* penalize if there is no arrowhead.

Example of Acceptable Response

8. Allow a maximum of 2 credits for calculating the angle of refraction. Refer to *Scoring Criteria for Calculations*.

Example of Acceptable Response

$$n_1 \sin \theta_1 = n_2 \sin \theta_2$$

$$\sin \theta_2 = \frac{n_1 \sin \theta_1}{n_2}$$

$$\sin \theta_2 = \frac{1.66 \sin 34.0°}{1.00}$$

$$= 68.2° \ or = 68°$$

9. Allow a maximum of 2 credits for constructing the refracted light ray.

   • Allow 1 credit if the angle between the normal and the ray is equal to the angle ($\pm 2°$) the student calculated in question 8.
   • Allow 1 credit for a straight line originating at the point where the ray inside the flint glass meets the right side of the prism drawn to the right of the normal in air.

Example of Acceptable Response

10. 2

11. Allow 1 credit for **orange**.

12. Allow a maximum of 2 credits. Refer to *Scoring Criteria for Calculations*.

   Examples of Acceptable Responses

$$n_1 \sin \theta_1 = n_2 \sin \theta_2$$

$$\sin \theta_2 = \frac{n_1 \sin \theta_1}{n_2}$$

$$\sin \theta_2 = \frac{1.00 \sin 33°}{1.50}$$

$$\sin \theta_2 = \frac{1.00(0.5446)}{1.50}$$

$$\sin \theta_2 = 0.363$$

$$\theta_2 = 21°$$

or

$$n_1 \sin i = n_2 \sin r$$
$$(1.00)(\sin 33°) = (1.50) \sin r$$
$$r = 21°$$

13. Allow a maximum of 2 credits.

    • Allow 1 credit if the angle between the normal and the ray is equal to the angle calculated in question 12 (±2°).

    • Allow 1 credit for an arrow originating at point *B*, drawn to the right of the normal in Lucite, and directed away from point *B*.

Example of Acceptable Response

14. Allow 1 credit for indicating that yellow light travels faster in air than in Lucite.

    Examples of acceptable responses include, but are not limited to:

    Light travels faster in air than in Lucite.

      or

    Yellow light travels slower in Lucite than in air.

    Allow credit if the student calculates and identifies the numerical values.

15. [2] Allow a maximum of 2 credits. Refer to *Scoring Criteria for Calculations*.

Example of a 2-credit response:

$$n_1 \sin \theta_1 = n_2 \sin \theta_2$$

$$\sin \theta_2 = \frac{n_1 \sin \theta_1}{n_2}$$

$$\sin \theta_2 = \frac{1.00 \sin 35°}{1.47}$$

$$\theta_2 = 23°$$

16. [1] Allow 1 credit. Acceptable responses include, but are not limited to:

—The light does not bend because light travels at the same speed in both layers.

—The absolute indices of refraction are the same.

17. [1] Allow 1 credit for drawing the refracted ray at an angle of 35° ± 2° to the normal.

Example of a 1-credit response:

18. [1] Allow 1 credit for 35° ± 2°.

19. [1] Allow 1 credit for drawing the reflected ray at an angle of reflection of 35° ± 2°.

Example of a 1-credit response:

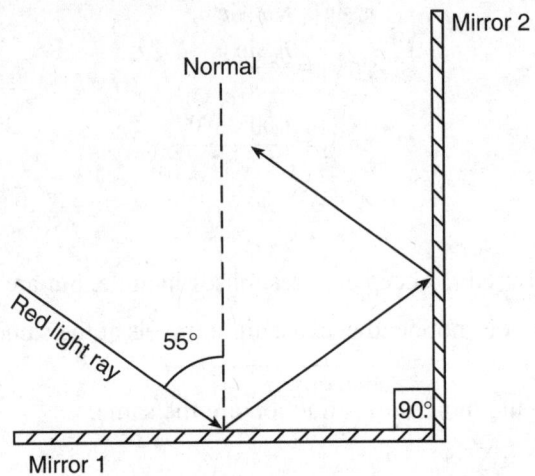

Note: The normal and the arrowhead need *not* be drawn to receive credit. Allow credit for an answer that is consistent with the response to question 18.

20. [1] Allow 1 credit for $17° \pm 2°$.

21. [2] Allow a maximum of 2 credits. Refer to *Scoring Criteria for Calculations*.

Example of a 2-credit response:

$$n_1 \sin \theta_1 = n_2 \sin \theta_2$$
$$\sin \theta_2 = \frac{n_1 \sin \theta_1}{n_2}$$
$$\sin \theta_2 = \frac{1.00 \sin 17°}{1.46}$$
$$\theta_2 = 12° \text{ or } 11.6°$$

Note: Allow credit for an answer that is consistent with the response to question 20.

22. [1] Allow 1 credit for drawing the refracted ray at an angle of $12° \pm 2°$, *or* an answer that is consistent with the response to question 21.

23. [1] Allow 1 credit for drawing the reflected ray at an angle of $17° \pm 2°$, *or* an answer that is consistent with the response to question 20.

Example of a 2-credit response for questions 22 and 23:

Normal

Incident light ray

Air
Fused quartz

Note: Rays do *not* have to have arrows to receive credit.

24. [1] Allow 1 credit for flint glass.

25. [1] Allow 1 credit for 1.81×10^8 m/s.

26. [2] Allow a maximum of 2 credits. Refer to *Scoring Criteria for Calculations*.

Examples of 2-credit responses:

$$n_1 \sin \theta_1 = n_2 \sin \theta_2$$

$$\sin \theta_2 = \frac{n_1 \sin \theta_1}{n_2}$$

$$\sin \theta_2 = \frac{1.00(\sin 55°)}{1.66}$$

$$\sin \theta_2 = 0.493$$

$$\theta_2 = 30.° \ or \ 29.6°$$

or

$$n = \frac{\sin i}{\sin r}$$

$$r = \sin^{-1}\left(\frac{\sin i}{n}\right)$$

$$r = \sin^{-1}\left(\frac{\sin 55°}{1.66}\right)$$

$$r = 30.°$$

27. [1] Allow 1 credit for drawing a ray in material X at an angle of 30.° ± 2° to the right of the normal. Allow this credit even if an arrowhead is *not* drawn on the ray.

Example of a 1-credit response:

Note: Allow credit for an answer that is consistent with the response to question 26.

28. Allow a maximum of 2 credits. Refer to *Scoring Criteria for Calculations*.

Example of a 2-credit response:

$$n_1 \sin \theta_1 = n_2 \sin \theta_2$$
$$n_2 = \frac{n_1 \sin \theta_1}{\sin \theta_2}$$
$$n_2 = \frac{1.33 \sin 45°}{\sin 29°}$$
$$n_2 = 1.94$$

29. Allow 1 credit for zircon. Allow credit for an answer that is consistent with the response to question 28.

30. Allow 1 credit for 55° (± 2°).

31. Allow a maximum of 2 credits. Refer to *Scoring Criteria for Calculations*.

Example of a 2-credit response:

$$n_1 \sin \theta_1 = n_2 \sin \theta_2$$

$$\sin \theta_2 = \frac{n_1 \sin \theta_1}{n_2}$$

$$\sin \theta_2 = \frac{(1.00)(\sin 55°)}{1.66} = 0.493$$

$$\theta_2 = 29.6° \quad or \quad 30.°$$

Allow credit for an answer that is consistent with the response to question 30.

32. Allow 1 credit for drawing the refracted ray at an angle of 30.° (± 2°).

Example of a 1-credit response:

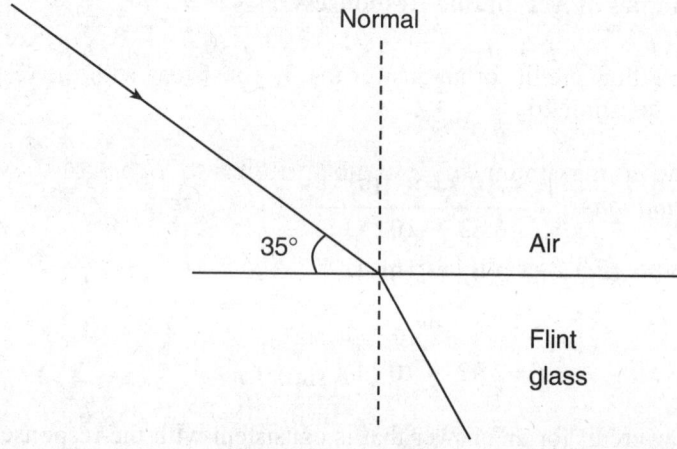

Allow credit for an answer that is consistent with the response to question 31.

33. Allow 1 credit for stating what would happen to light from the incident ray that is not refracted or absorbed. Acceptable responses include, but are not limited to:

—reflected
—scattered

Chapter Thirteen, pages 416–418

1. Allow 1 credit for determining the minimum energy necessary for an electron to change from the B energy level to the F energy level.

   Examples of Acceptable Responses

   $$19.34 \times 10^{-19} \text{ J}$$
   $$or$$
   $$1.934 \times 10^{-18} \text{ J}$$

   Allow this credit if the answer above is negative.

2. Allow a maximum of 2 credits. Refer to *Scoring Criteria for Calculations*.

   Examples of Acceptable Responses

   $$E = hf$$
   $$f = \frac{E}{h}$$

   $$f = \frac{19.34 \times 10^{-19} \text{ J}}{6.63 \times 10^{-34} \text{ J} \cdot \text{s}}$$

   $$f = 2.92 \times 10^{15} \text{ Hz}$$

   $$or$$

   $$f = 2.92 \times 10^{15} \text{ l/s}$$

   Allow credit for an answer that is consistent with the response to question 1.

3. **Example of Acceptable Response**

   $$E = \frac{hc}{l}$$

   $$E = \frac{6.63 \times 10^{-34} \text{ J} \cdot \text{s}(3.00 \times 10^{8} \text{m/s})}{6.58 \times 10^{-17} \text{ m}}$$

   $$E = 3.02 \times 10^{-19} \text{ J}$$

4. 1.89 eV.

5. 3 and 2.

6. It cannot be an X-ray because the wavelength is too long.

7. [1] Allow 1 credit for 8.82 eV.

8. [1] Allow 1 credit for 1.41×10^{-18} J *or* an answer that is consistent with the response to question 7.

9. [2] Allow a maximum of 2 credits. Refer to *Scoring Criteria for Calculations.*

   Example of a 2-credit response:

$$E_{photon} = hf$$
$$E_{photon} = (6.63 \times 10^{-34} \text{Jis}) (5.02 \times 10^{14} \text{ Hz})$$
$$E_{photon} = 3.33 \times 10^{-19} \text{ J}$$

10. [1] Allow 1 credit for 2.08 eV *or* an answer that is consistent with the response to question 9.

11. [1] Allow 1 credit for $n = 3$ *or* an answer that is consistent with the response to question 10.

12. [1] 2

13. [1] 4

14. Allow 1 credit for 4.52×10^{14} Hz.

15. Allow 1 credit for 3.02 eV.

16. Allow 1 credit for 4.83×10^{-19} J.

   Allow credit for a response that is consistent with the response to question 15.

17. Allow a maximum of 2 credits. Refer to *Scoring Criteria for Calculations.*

Example of a 2-credit response:

$$E = hf$$

$$f = \frac{E}{h}$$

$$f = \frac{4.83 \times 10^{-19} \text{ J}}{6.63 \times 10^{-34} \text{ J} \cdot \text{s}}$$

$$f = 7.29 \times 10^{14} \text{ Hz}$$

Allow credit for an answer that is consistent with the student's response to question 16.

18. Allow 1 credit for explaining why this is not the only energy and/or frequency that an electron in the $n = 6$ energy level of a hydrogen atom could emit. Acceptable responses include, but are not limited to:

— No, the $n = 6$ level can return to any of the 5 lower energy levels.
— No, the electron can drop to many different energy levels.
— The electron can fall from $n = 6$ to any other level between $n = 5$ and $n = 1$.
— $6 \rightarrow 5$ $6 \rightarrow 4$ $6 \rightarrow 3$ $6 \rightarrow 1$

Note: Do *not* allow credit for "no" without a correct explanation.

Chapter Fourteen, pages 442–445

1. Allow a maximum of 2 credits. Refer to *Scoring Criteria for Calculations*.

Example of Acceptable Response

$$E = mc^2$$
$$E = 2(9.11 \times 10^{-31} \text{ kg}) (3.00 \times 10^8 \text{ m/s})^2$$
$$E = 1.64 \times 10^{-13} \text{ J}$$

2. Allow 1 credit for **conservation of charge**.

3. 10^{-8}.

4. 10^{-47}.

5. Appropriate responses include, but are not limited to:
   - The electrostatic force is 10^{39} stronger than the gravitational force.
   - The gravitational force is smaller than the electromagnetic interaction.

6. [1] Allow 1 credit for 0.018 63u.

7. [1] Allow 1 credit for 17.3 MeV.

   Note: Allow credit for an answer that is consistent with the response to question 6.

8. [1] Allow 1 credit for indicating the correct charge on each particle.

   Example of a 1-credit response:

   $\underline{\;\;-1\;\;}$ e $\rightarrow$ $\underline{\;\;-1\;\;}$ e + $\underline{\;\;0\;\;}$ e + $\underline{\;\;0\;\;}$ e

9. [1] Allow 1 credit for antiproton.

10. [1] Allow 1 credit for stating how the emission spectrum of antihydrogen should compare to the emission spectrum of hydrogen. Acceptable responses include, but are not limited to:
    — identical
    — the same

11. [1] Allow 1 credit for identifying charge as one characteristic that antimatter particles must possess.

12. [1] Allow 1 credit for explaining why it is a mystery that "the universe seems to be overwhelmingly composed of normal matter." Acceptable responses include, but are not limited to:
    — Although matter is only created in matter-antimatter pairs, most matter is normal.
    — Matter, not ½ antimatter.
    — It should be balanced by antimatter.
    — Matter can only be created in particle-antiparticle pairs.

13. Allow a maximum of 2 credits. Refer to *Scoring Criteria for Calculations*.

Examples of 2-credit responses:

$E = mc^2$

$E = 2(1.67 \times 10^{-27} \text{ kg})(3.00 \times 10^8 \text{ m/s})^2$ or $\dfrac{1 \text{ u}}{931 \text{ MeV}} = \dfrac{2 \text{ u}}{x \text{ MeV}}$

$x = 1860 \text{ MeV}$

$E = 3.01 \times 10^{-10} \text{ J}$

14. Allow 1 credit for indicating that mass is converted into energy.

APPENDIX 1 _____

New York State Reference Tables for Physical Setting: Physics

If you take the New York State Regents Exam in Physical Setting: Physics, you will be provided with a set of reference tables to aid you in answering the questions on the examination. For students not taking this examination, the tables will provide a convenient reference for answering the questions and problems presented in this book. Each of these tables, along with a brief description of it, is given in this appendix.

List of Physical Constants

The most important physical constants and (where appropriate) their symbols are given in this table. Each constant is given to three significant digits along with its units.

For example, if you were asked to calculate the gravitational force between the Moon and the Earth, you would use this table to find the masses of the Moon and the Earth and the mean Earth-Moon distance. (See table on p. 548.)

Approximate Coefficients of Friction

The coefficients of friction (μ) for a number of pairs of surfaces are provided. The *kinetic* coefficients of friction are used when the surfaces are in relative motion. The *static* coefficients of friction are used when the surfaces are at rest with respect to each other.

Reference Tables for Physical Setting: Physics

List of Physical Constants

Name	Symbol	Value(s)
Universal gravitational constant	G	6.67×10^{-11} N•m^2/kg^2
Acceleration due to gravity	g	9.81 m/s^2
Speed of light in a vacuum	c	3.00×10^8 m/s
Speed of sound in air at STP		3.31×10^2 m/s
Mass of Earth		5.98×10^{24} kg
Mass of the Moon		7.35×10^{22} kg
Mean radius of Earth		6.37×10^6 m
Mean radius of the Moon		1.74×10^6 m
Mean distance—Earth to the Moon		3.84×10^8 m
Mean distance—Earth to the Sun		1.50×10^{11} m
Electrostatic constant	k	8.99×10^9 N•m^2/C^2
1 elementary charge	e	1.60×10^{-19} C
1 coulomb (C)		6.25×10^{18} elementary charges
1 electronvolt (eV)		1.60×10^{-19} J
Planck's constant	h	6.63×10^{-34} J•s
1 universal mass unit (u)		9.31×10^2 MeV
Rest mass of the electron	m_e	9.11×10^{-31} kg
Rest mass of the proton	m_p	1.67×10^{-27} kg
Rest mass of the neutron	m_n	1.67×10^{-27} kg

Approximate Coefficients of Friction

	Kinetic	Static
Rubber on concrete (dry)	0.68	0.90
Rubber on concrete (wet)	0.58	
Rubber on asphalt (dry)	0.67	0.85
Rubber on asphalt (wet)	0.53	
Rubber on ice	0.15	
Waxed ski on snow	0.05	0.14
Wood on wood	0.30	0.42
Steel on steel	0.57	0.74
Copper on steel	0.36	0.53
Teflon on Teflon	0.04	

The Electromagnetic Spectrum

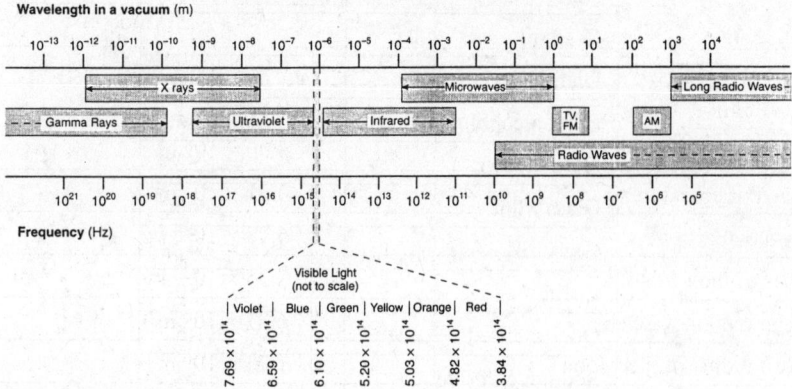

The wavelength and frequency ranges for the principal types of electromagnetic radiation are given in this chart. Note that the type of radiation is principally determined by its *source* and not necessarily by its wavelength (or frequency). As a result, there is some overlap between certain types of radiation. In addition, the frequency ranges for visible light are provided at the bottom of the chart.

Absolute Indices of Refraction

The *absolute index of refraction* is defined as the ratio of the speed of light in a vacuum to the speed of light in a medium. (Monochromatic yellow light of 5.09×10^{14} Hz (μ) was used to compute these indices.)

The larger the index of refraction, the slower light travels in the medium. Therefore, according to this table, light travels slowest in diamond and fastest in air. In a vacuum, the index of refraction would be exactly 1.

This table is useful in solving Snell's law and critical angle problems, as well as in comparing the speed of light in different media.

Absolute Indices of Refraction ($f = 5.09 \times 10^{14}$ Hz)	
Air	1.00
Corn oil	1.47
Diamond	2.42
Ethyl alcohol	1.36
Glass, crown	1.52
Glass, flint	1.66
Glycerol	1.47
Lucite	1.50
Quartz, fused	1.46
Sodium chloride	1.54
Water	1.33
Zircon	1.92

Prefixes for Powers of 10

This table lists a number of the more important metric prefixes, with their notations and symbols. A prefix placed in front of a base unit creates a multiple or submultiple of that unit. For example, a nanometer (nm) is a unit of length whose value is 10^{-9} meter.

Prefixes for Powers of 10		
Prefix	Symbol	Notation
tera	T	10^{12}
giga	G	10^{9}
mega	M	10^{6}
kilo	k	10^{3}
deci	d	10^{-1}
centi	c	10^{-2}
milli	m	10^{-3}
micro	μ	10^{-6}
nano	n	10^{-9}
pico	p	10^{-12}

Energy Level Diagrams for Mercury and Hydrogen

Each energy state is represented by a horizontal line. Its energy value (in electron-volts) is given at the right. Note that these values are *negative* numbers that increase to a maximum value of *zero*. The lowest energy state is called the *ground state*; every other state is an *excited state*. At a value of 0 electron-volt, the electron is no longer associated with the atom, a condition known as *ionization*.

At the left is the label corresponding to the energy state. For hydrogen, these states are integers known as *principal quantum numbers*. For mercury, the atom is much more complex and the various states are represented as letters.

To calculate the energy involved in a particular transition, one *subtracts* the final energy value from the initial value. If the difference is negative, energy is released by the atom as a photon; if the difference is positive, the energy is absorbed by the atom.

Energy-Level Diagrams

Hydrogen

Energy levels for the hydrogen atom

Mercury

A few energy levels for the mercury atom

Classification of Matter

This flowchart indicates the hierarchy of the different nuclear and subnuclear particles.

Classification of Matter

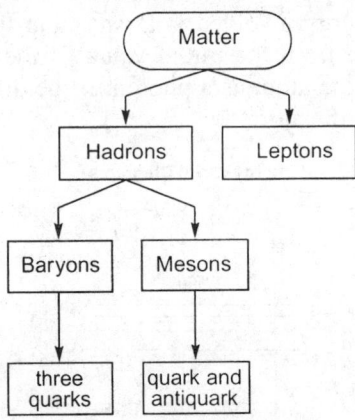

Particles of the Standard Model

The names, symbols, and charges of the six quarks and six leptons are given in this chart. Electric charges are given in terms of the elementary charge (e). Each particle has a corresponding *antiparticle* that has an electric charge opposite that of the particle. For example, the antimuon ($\bar{\mu}$) has an electric charge of $+1e$.

Particles of the Standard Model

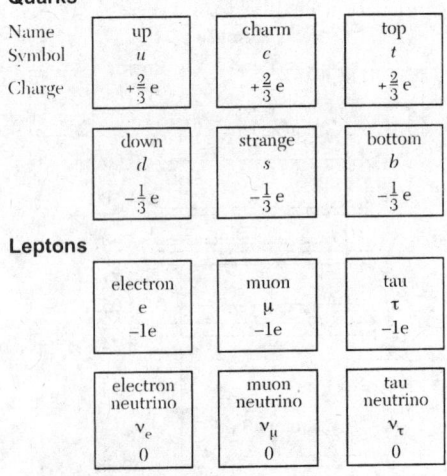

Note: For each particle there is a corresponding antiparticle with a charge opposite that of its associated particle.

Circuit Symbols

This table lists the various symbols used in circuit diagrams.

Resistivities at 20°C

The resistivities, in ohm • meters, are given for a number of conductors at 20°C. Resistivities are useful for comparing the relative abilities of various materials to conduct an electric current. They are also used to calculate the resistance of a conductor whose length and cross-sectional area are known.

Resistivities at 20°C	
Material	**Resistivity** $(\Omega \cdot m)$
Aluminum	2.82×10^{-8}
Copper	1.72×10^{-8}
Gold	2.44×10^{-8}
Nichrome	$150. \times 10^{-8}$
Silver	1.59×10^{-8}
Tungsten	5.60×10^{-8}

APPENDIX 2 _____

Summary of Equations for New York State Physical Setting: Physics Regents Examination

A summary of the equations used in the New York State Physical Setting: Physics Regents core curriculum is provided in this appendix.

The equations are categorized by areas:

> Electricity
> Waves and Optics
> Modern physics
> Geometry and trigonometry
> Mechanics

Notes:
The relevant symbols are provided at the right of each set of equations. Your teacher or textbook may use different symbols in connection with these equations.

Some of the equations used in this book do not appear on these summary tables; you will not be required to use these equations on the New York State Regents Examination.

Electricity

$$F_e = \frac{kq_1q_2}{r^2}$$

$$E = \frac{F_e}{q}$$

$$V = \frac{W}{q}$$

$$I = \frac{\Delta q}{t}$$

$$R = \frac{V}{I}$$

$$R = \frac{\rho L}{A}$$

$$P = VI = I^2R = \frac{V^2}{R}$$

$$W = Pt = VIt = I^2Rt = \frac{V^2t}{R}$$

A = cross-sectional area
E = electric field strength
F_e = electrostatic force
I = current
k = electrostatic constant
L = length of conductor
P = electrical power
q = charge
R = resistance
R_{eq} = equivalant resistance
r = distance between centers
t = time
V = potential difference
W = work (electrical energy)
Δ = change
ρ = resistivity

Series Circuits

$I = I_1 = I_2 = I_3 = \cdots$
$V = V_1 + V_2 + V_3 + \cdots$
$R_{eq} = R_1 + R_2 + R_3 + \cdots$

Parallel Circuits

$I = I_1 + I_2 + I_3 + \cdots$
$V = V_1 = V_2 = V_3 = \cdots$

$$\frac{1}{R_{eq}} = \frac{1}{R_1} + \frac{1}{R_2} + \frac{1}{R_3} + \cdots$$

Equations are provided for electrostatic forces, electric fields, potential difference, current, resistance, and power and energy in electric circuits. In addition, the current, potential difference, and resistance relationships are provided for series and parallel circuits.

Waves and Optics

$$v = f\lambda$$

$$T = \frac{1}{f}$$

$$n_1 \sin\theta_1 = n_2 \sin\theta_2$$

$$\frac{n_2}{n_1} = \frac{v_1}{v_2} = \frac{\lambda_1}{\lambda_2}$$

$$\theta_i = \theta_r$$

$$n = \frac{c}{v}$$

c = speed of light in a vacuum
f = frequency
n = absolute index of refraction
T = period
v = velocity
λ = wavelength
θ = angle
θ_i = incident angle
θ_r = reflected angle

Equations are provided for all types of waves (relationships among speed, wavelength, frequency, and period) as well as for reflection and refraction.

Modern Physics

$$E_{photon} = hf = \frac{hc}{\lambda}$$

$$E_{photon} = E_i - E_f$$

$$E = mc^2$$

c = speed of light in a vacuum
E = energy
f = frequency
h = Planck's constant
m = mass
λ = wavelength

Three energy equations are included in this table. The first is the Planck equation for relating the energy of a photon to its frequency (or wavelength). The second equation is the energy of a photon that is emitted or absorbed as the result of a transition between two energy levels in an atom. The third equation is the Einstein equation that relates energy and mass.

Geometry and Trigonometry

Rectangle

$$A = bh$$

Triangle

$$A = \frac{1}{2}bh$$

Circle

$$A = \pi r^2$$
$$C = 2\pi r$$

A = area
b = base
C = circumference
h = height
r = radius

Right Triangle

$$c^2 = a^2 + b^2$$

$$\sin\theta = \frac{a}{c}$$

$$\cos\theta = \frac{b}{c}$$

$$\tan\theta = \frac{a}{b}$$

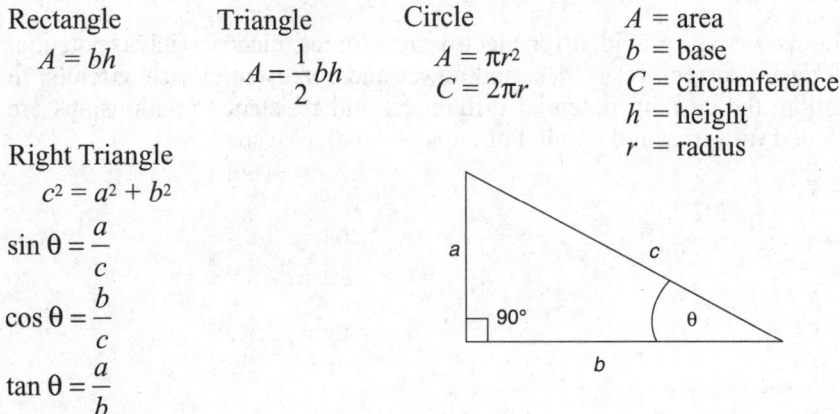

The geometric relationships for the areas of rectangles, triangles, and circles are given. In addition, the relationship for the circumference of a circle is provided. This table also gives the Pythagorean theorem and several trigonometric relationships for a right triangle.

Mechanics

$$\bar{v} = \frac{d}{t}$$

$$a = \frac{\Delta v}{t}$$

$$v_f = v_i + at$$

$$d = v_i t + \frac{1}{2}at^2$$

$$v_f^2 = v_i^2 + 2ad$$

$$A_y = A\sin\theta$$

$$A_x = A\cos\theta$$

$$a = \frac{F_{net}}{m}$$

$$F_f = \mu F_N$$

$$F_g = \frac{Gm_1 m_2}{r^2}$$

$$g = \frac{F_g}{m}$$

$$p = mv$$

$$p_{before} = p_{after}$$

$$J = Ft = \Delta p$$

$$F_s = kx$$

$$PE_s = \frac{1}{2}kx^2$$

$$F_c = ma_c$$

$$a_c = \frac{v^2}{r}$$

$$\Delta PE = mg\Delta h$$

$$KE = \frac{1}{2}mv^2$$

$$W = Fd = \Delta E_T$$

$$E_T = PE + KE + Q$$

$$P = \frac{W}{t} = \frac{Fd}{t} = F\bar{v}$$

a = acceleration
a_c = centripetal acceleration
A = any vector quantity
d = displacement/distance
E_T = total energy
F = force
F_c = centripetal force
F_f = force of friction
F_g = weight/force due to gravity
F_N = normal force
F_{net} = net force
F_s = force on a spring
g = acceleration due to gravity or gravitational field strength
G = universal gravitational constant
h = height
J = impulse
k = spring constant
KE = kinetic energy
m = mass
p = momentum
P = power
PE = potential energy
PE_s = potential energy stored in a spring
Q = internal energy
r = radius/distance between centers
t = time interval
v = velocity/speed
$\bar{v}$ = average velocity/average speed
W = work
x = change in spring length from the equilibrium position
Δ = change
θ = angle
μ = coefficient of friction

Equations are provided for motion, vectors, Newton's laws, momentum and impulse, work, energy, and power.

557

APPENDIX 3 _____

Answering Short-Constructed Response and Free-Response Questions

Techniques and Tips

A *short-constructed response* or *free-response question* is an examination question that requires the test taker to do more than to choose among several responses or to fill in a blank. You may need to perform numerical calculations, draw and interpret graphs, and provide short or extended written responses to a question or problem.

Parts B–2 and C of the New York State Physical Setting: Physics Regents Examination in Physics contain free-response questions. This appendix is designed to provide you with a number of general guidelines for answering them.

Solving Problems Involving Numerical Calculations

To receive full credit you must:

- Provide the appropriate equation(s).
- Substitute values and units into the equation(s).
- Display the answer, with appropriate units.
- If the answer is a vector quantity, include its direction.

Although SI units are used on the Regents examination, you are expected to have some familiarity with other metric units such as the gram and the kilometer.

You will *not* be penalized if your answer has an incorrect number of significant digits. However, it is always good practice to pay attention to this detail.

You should write as legibly as possible. Teachers are human, and nothing irks them more than trying to decipher a careless, messy scrawl. It is also a good idea to identify your answer clearly, either by placing it in a box or by writing the word "answer" next to it.

A final word: If you provide the correct answer but do not show any work, you will not receive any credit for the problem!

The following is a sample problem and its model solution.

PROBLEM
A 5.0-kilogram object has a velocity of 10. meters per second [east]. Calculate the momentum of this object.

SOLUTION

$$\mathbf{p} = m\mathbf{v}$$

$$\mathbf{p} = (5.0 \text{ kg})(10. \text{ m/s [east]})$$

$$\boxed{\mathbf{p} = 50. \text{ kg·m/s [east]}}$$

Graphing Experimental Data

To receive full credit you must:

- Label both axes with the appropriate variables and units.
- Divide the axes so that the data ranges fill the graph as nearly as possible.
- Plot all data points accurately.
- Draw a best-fit line carefully with a straightedge if the best-fit line is a straight line, and carefully freehand if the best-fit line is a curve. The line should pass through the origin *only if the data warrant it.*
- If a part of the question requires that the slope be calculated, calculate the slope *from the line,* not from individual data points.

Generally a graph should have a title, and the *independent variable* is usually drawn along the *x*-axis. However, you will not be penalized on the Regents examination if you do not follow these conventions.

The following is a sample problem and its model solution.

PROBLEM
The weights of various masses, measured on Planet *X*, are given in the table below.

Mass (kg)	Weight (N)
15	21
20.	32
25	35
30.	48
35	56

(1) Draw a graph that illustrates these data.
(2) Use the *graph* to calculate the acceleration due to gravity on Planet *X*.

559

SOLUTION

(1) The graph shown below incorporates the essential items that were listed in the table.

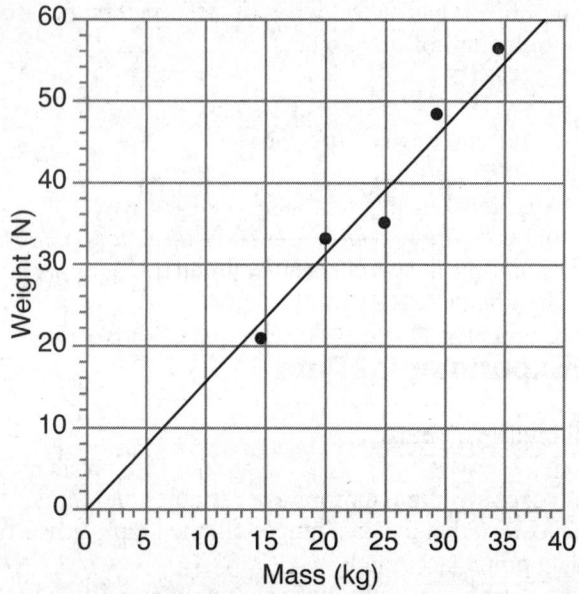

(2) Since the magnitude of the gravitational acceleration can be determined by calculating the ratio of weight to mass ($g = W/m$), we can calculate the value of g from the slope of the graph.

We choose two points on the line; we do not use the data points themselves:

$$g = \frac{\Delta W}{\Delta m} = \frac{50.\ N - 10.\ N}{32\ kg - 6\ kg} = 1.5\ \frac{N}{kg} = 1.5\ m/s^2$$

Drawing Diagrams

To receive full credit you must:

- Draw your diagrams neatly, and label them clearly.
- Draw vectors *to scale* and *in the correct direction* (i.e., must have an arrowhead pointing in the correct direction). If you are given a scale, you **must** draw your vectors to that scale.
- Bring a straightedge and a protractor with you so that you can draw neat, accurate diagrams.

Writing a Free-Response Answer

To receive full credit you must:

- Use complete, clear sentences that make sense to the reader.
- Use correct physics in your explanations.

A sample question and acceptable and unacceptable answers are given below.

QUESTION
If the data from two different photoemissive materials are graphed, which characteristic of the two graphs will be the same?

ACCEPTABLE ANSWERS
- The characteristic that will be the same is the slope of the line.
- It is the slope.
- The two lines have the same slope.

The following answers are unacceptable:

- The slope. (Incomplete sentence)
- Both lines will have the same *y*-intercept. (Incorrect physics)

APPENDIX 4 _____

Physics: The Physical Setting Regents Exam

The Physics Regents Exam is divided into four parts.

Part A consists of multiple-choice questions that count for approximately 41 percent of the total exam. These are content-based questions that assess a student's knowledge of core material. The ability to apply, analyze, synthesize, and evaluate core material may also be tested.

Parts B–1 and B–2 consist of multiple-choice or short constructed-response items. These items are content- and skills-based questions that assess a student's ability to apply, analyze, synthesize, and evaluate material primarily from Standard 4 (content) and Standards 1, 2, 6, and 7. Parts B1 and B2 count for approximately 35 percent of the exam.

Part C consists of free-response questions that count for approximately 24 percent of the exam. These items assess content and its applications by requiring students to apply knowledge of scientific concepts and skills to address real-world situations. These situations may come from newspapers or magazine articles, scientific journals, or current events. Students may be asked to apply scientific concepts, formulate hypotheses, make predictions, or use other scientific inquiry techniques.

Refer to the detailed outline that follows for a complete listing of the concepts to be mastered and skills needed to be demonstrated for the New York State core curriculum in Physics: The Physical Setting. Refer to the page numbers that follow each concept or skill to find where it is reviewed in the text.

I. MECHANICS

A. Concepts to be mastered.

II. ENERGY

A. Concepts to be mastered.

B. Skills to be demonstrated. The student will be able to:

III. ELECTRICITY AND MAGNETISM

A. Concepts to be mastered.

B. Skills to be demonstrated. The student will be able to:

IV. WAVES

A. Concepts to be mastered.

B. Skills to be demonstrated. The student will be able to:

V. MODERN PHYSICS

A. Concepts to be mastered.

7. The standard model of particle physics has evolved from previous attempts to explain the nature of the atom and states that (1) atomic particles are composed of subnuclear particles, (2) the nucleus is a conglomeration of quarks that manifest themselves as protons and neutrons, and (3) each elementary particle has a corresponding antiparticle.......... 396–402, 430–433

8. Behaviors and characteristics of matter, from the microscopic to the cosmic levels, are manifestations of its atomic structure. The macroscopic characteristics of matter, such as electrical and optical properties, are the result of microscopic interactions............ 203–204, 284

9. The total of the fundamental interactions is responsible for the appearance and behavior of the objects in the universe....................................... 430–433

10. The fundamental source of all energy in the universe is the conversion of mass to energy.* 422

B. Skills to be demonstrated. The student will be able to:

1. interpret energy-level diagrams. 399–402
2. correlate spectral lines with an energy-level diagram..................................... 398–401

NOTE: Items with asterisks* require quantitative treatment per the Reference Table for Physics. Asterisks following individual words refer to the preceding word or phrase only; asterisks appearing after the final period of a sentence refer to all concepts or ideas presented in the sentence.

Examination
June 2019

Physics—
Physical Setting

PART A
Answer all questions in this part.

Directions (1–35): For each statement or question, select the *number* of the word or expression that, of those given, best completes the statement or answers the question. Some questions may require the use of the *2006 Edition Reference Tables for Physical Setting/Physics*. Record your answers in the spaces provided.

1 Which pair of quantities represent scalar quantities?

   (1) displacement and velocity   (3) energy and velocity

   (2) displacement and time   (4) energy and time

   1 ____

2 A sailboat on a lake sails 40. meters north and then sails 40. meters due east. Compared to its starting position, the new position of the sailboat is

   (1) 40. m due east   (3) 57 m northeast

   (2) 40. m due north   (4) 80. m northeast

   2 ____

3 A ball is thrown straight upward from the surface of Earth. Which statement best describes the ball's velocity and acceleration at the top of its flight?

   (1) Both velocity and acceleration are zero.

   (2) Velocity is zero and acceleration is nonzero.

   (3) Velocity is nonzero and acceleration is zero.

   (4) Both velocity and acceleration are not zero.

   3 ____

4 As a student runs a plastic comb through her hair, the comb acquires a negative electric charge. This charge results from the transfer of

(1) protons from the comb to her hair
(2) protons from her hair to the comb
(3) electrons from the comb to her hair
(4) electrons from her hair to the comb

4 _____

5 How would the mass and weight of an object on the Moon compare to the mass and weight of the same object on Earth?

(1) Mass and weight would both be less on the Moon.
(2) Mass would be the same but its weight would be less on the Moon.
(3) Mass would be less on the Moon and its weight would be the same.
(4) Mass and weight would both be the same on the Moon.

5 _____

6 An object is moving with constant speed in a circular path. The object's centripetal acceleration remains constant in

(1) magnitude, only
(2) direction, only
(3) both magnitude and direction
(4) neither magnitude nor direction

6 _____

7 As shown in the diagram below, a rope attached to a 500.-kilogram crate is used to exert a force of 45 newtons at an angle of 65 degrees above the horizontal.

The horizontal component of the force acting on the crate is
(1) 19 N   (3) 210 N
(2) 41 N   (4) 450 N

7 _____

8 A spring with a spring constant of 68 newtons per meter hangs from a ceiling. When a 12-newton downward force is applied to the free end of the spring, the spring stretches a total distance of

(1) 0.18 m   (3) 5.7 m
(2) 0.59 m   (4) 820 m

8 _____

9 As a student walks downhill at constant speed, his gravitational potential energy

(1) increases and his kinetic energy increases
(2) increases and his kinetic energy remains the same
(3) decreases and his kinetic energy increases
(4) decreases and his kinetic energy remains the same

9 _____

10 When 150 joules of work is done on a system by an external force of 15 newtons in 20. seconds, the total energy of that system increases by

(1) 1.5×10^2 J　　　　　　(3) 3.0×10^2 J
(2) 2.0×10^2 J　　　　　　(4) 2.3×10^3 J

10 _____

11 A person on a ledge throws a ball vertically downward, striking the ground below the ledge with 200 joules of kinetic energy. The person then throws an identical ball vertically upward at the same initial speed from the same point. What is the kinetic energy of the second ball when it hits the ground? [Neglect friction.]

(1) 200 J　　　　　　(3) less than 200 J
(2) 400 J　　　　　　(4) more than 400 J

11 _____

12 Two construction cranes are used to lift identical 1200-kilogram loads of bricks the same vertical distance. The first crane lifts the bricks in 20. seconds and the second crane lifts the bricks in 40. seconds. Compared to the power developed by the first crane, the power developed by the second crane is

(1) the same　　　　　　(3) half as great
(2) twice as great　　　　　(4) four times as great

12 _____

13 An ionized calcium atom has a charge of +2 elementary charges. If this ion is accelerated through a potential difference of 2.0×10^3 volts, the ion's change in kinetic energy will be

(1) 1.0×10^3 eV　　　　　(3) 3.0×10^3 eV
(2) 2.0×10^3 eV　　　　　(4) 4.0×10^3 eV

13 _____

14 A total charge of 100. coulombs flows past a fixed point in a circuit every 500. seconds. What is the current at this point in the circuit?

(1) 0.200 A　　　　　　(3) 5.00×10^4 A
(2) 5.00 A　　　　　　(4) 1.25×10^{18} A

14 _____

15 An aluminum wire of length 1.0 meter has a resistance of 9.0×10^{-3} ohm. If the wire were cut into two equal lengths, each length would have a resistance of

(1) $2.8 \times 10^{-8}\ \Omega$ (3) $9.0 \times 10^{-3}\ \Omega$

(2) $4.5 \times 10^{-3}\ \Omega$ (4) $1.8 \times 10^{-2}\ \Omega$

15 _____

16 In an operating electrical circuit, the source of potential difference could be

(1) a voltmeter   (3) an ammeter

(2) a battery   (4) a resistor

16 _____

17 A lightbulb with a resistance of 2.9 ohms is operated using a 1.5-volt battery. At what rate is electrical energy transformed in the lightbulb?

(1) 0.52 W   (3) 4.4 W

(2) 0.78 W   (4) 6.5 W

17 _____

18 A 40.0-kilogram child exerts a 100.-newton force on a 50.0-kilogram object. The magnitude of the force that the object exerts on the child is

(1) 0.0 N   (3) 100. N

(2) 80.0 N   (4) 125 N

18 _____

19 Two identical stationary bar magnets are arranged as shown in the diagram below.

What is the direction of the magnetic field at point *P*?

(1)

(3)

(2)

(4)

19 _____

20 A student claps his hands once to produce a sudden loud sound that travels through the air. This sound is classified as a

(1) longitudinal mechanical wave
(2) longitudinal electromagnetic wave
(3) transverse mechanical wave
(4) transverse electromagnetic wave

20 _____

21 A student generates water waves in a pool of water. In order to increase the energy carried by the waves, the student should generate waves with a

(1) greater amplitude   (3) greater wavelength
(2) higher frequency   (4) longer period

21 _____

22 A wave generator produces straight, parallel wave fronts in a shallow tank of uniform-depth water. As the frequency of vibration of the generator increases, which characteristic of the wave will always decrease?

(1) amplitude   (3) wavelength
(2) phase   (4) speed

22 _____

23 A space probe produces a radio signal pulse. If the pulse reaches Earth 12.3 seconds after it is emitted by the probe, what is the distance from the probe to Earth?

(1) 3.71×10^2 m   (3) 4.10×10^8 m
(2) 4.07×10^3 m   (4) 3.69×10^9 m

23 _____

24 The diagram below represents a light ray reflecting from a plane mirror.

The angle of reflection for this light ray is

(1) 20°   (3) 140°
(2) 70°   (4) 160°

24 _____

25 A light wave travels from one medium into a second medium with a greater absolute index of refraction. Which characteristic of the wave can *not* change as the wave enters the second medium?

(1) frequency   (3) direction
(2) speed   (4) wavelength

25 _____

26 The speed of light ($f = 5.09 \times 10^{14}$ Hz) in glycerol is

(1) 1.70×10^6 m/s   (3) 3.00×10^8 m/s
(2) 2.04×10^8 m/s   (4) 4.41×10^8 m/s

26 _____

27 The diagram below represents a standing wave produced in a string by a vibrating wave generator.

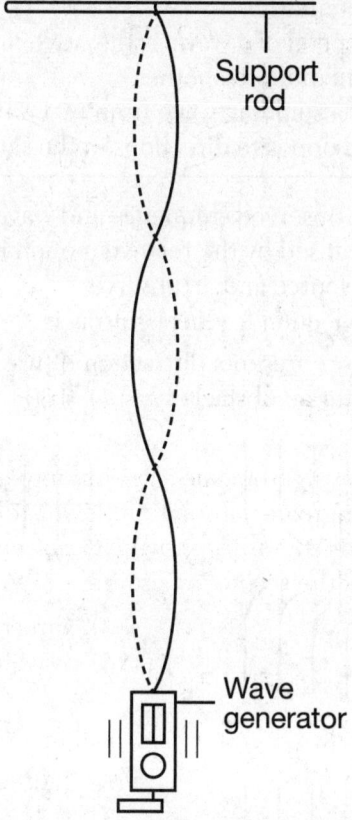

How many antinodes are shown in this standing wave?

(1) 6   (3) 3
(2) 2   (4) 4

27 _____

28 The Doppler effect is best described as the

(1) bending of waves as they pass by obstacles or through openings

(2) change in speed of a wave as the wave moves from one medium to another

(3) creation of a standing wave from two waves traveling in opposite directions in the same medium

(4) shift in the observed frequency and wavelength of a wave caused by the relative motion between the wave's source and an observer

28 _____

29 Which diagram represents diffraction of wave fronts as they encounter an obstacle?

(1)

(3)

(2)

(4)

29 _____

30 Which types of forces exist between the two protons in a helium nucleus?

  (1) a repulsive electrostatic force and a repulsive gravitational force
  (2) a repulsive electrostatic force and an attractive strong nuclear force
  (3) an attractive electrostatic force and an attractive gravitational force
  (4) an attractive electrostatic force and an attractive strong nuclear force

30 _____

31 A meson could be composed of

  (1) a top quark and a bottom quark
  (2) an electron and an antielectron
  (3) a strange quark and an anticharm quark
  (4) an up quark and a muon

31 _____

32 An electron in an excited mercury atom is in energy level g. What is the minimum energy required to ionize this atom?

  (1) 0.20 eV   (3) 2.48 eV
  (2) 0.91 eV   (4) 7.90 eV

32 _____

33 A student is standing in an elevator that travels from the first floor to the tenth floor of a building. The student exerts the greatest force on the floor of the elevator when the elevator is

  (1) accelerating upward as it leaves the first floor
  (2) slowing down as it approaches the tenth floor
  (3) moving upward at constant speed
  (4) at rest on the first floor

33 _____

34 At the bottom of a hill, a car has an initial velocity of +16.0 meters per second. The car is uniformly accelerated at –2.20 meters per second squared for 5.00 seconds as it moves up the hill. How far does the car travel during this 5.00-second interval?

  (1) 107 m   (3) 52.5 m
  (2) 74.5 m   (4) 25.0 m

34 _____

35 A particle enters the electric field between two oppositely charged parallel plates, as represented in the diagram below.

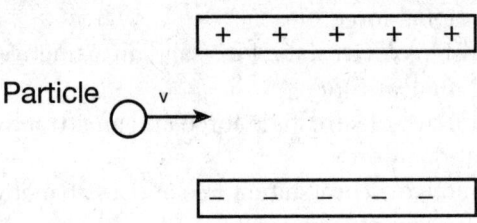

Which particle will be deflected toward the positive plate as it enters the electric field?

(1) photon   (3) electron

(2) proton   (4) neutrino

35 ____

PART B–1
Answer all questions in this part.

Directions (36–50): For *each* statement or question, select the number of the word or expression that, of those given, best completes the statement or answers the question. Some questions may require the use of the *2006 Edition Reference Tables for Physical Setting/Physics*. Record your answers in the spaces provided.

36 As represented in the diagram below, an object of mass m, located on the surface of the Moon, is attracted to the Moon with a gravitational force, F.

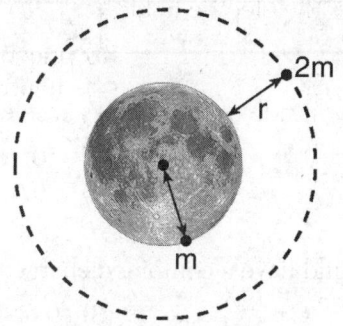

(Not drawn to scale)

An object of mass $2m$, at an altitude equal to the Moon's radius, r, above the surface of the Moon, is attracted to the Moon with a gravitational force of

(1) F (3) $F/2$
(2) $2F$ (4) $F/4$

36 _____

37 The graph below represents the relationship between velocity and time for an object moving along a straight line.

What is the magnitude of the object's acceleration?

(1) 5.0 m/s²
(2) 8.0 m/s²
(3) 10. m/s²
(4) 20. m/s²

37 _____

38 Two muons would have a combined charge of

(1) −3.2 × 10⁻¹⁹ C
(2) −1.6 × 10⁻¹⁹ C
(3) 0 C
(4) +3.2 × 10⁻¹⁹ C

38 _____

39 A 1.47-newton baseball is dropped from a height of 10.0 meters and falls through the air to the ground. The kinetic energy of the ball is 12.0 joules the instant before the ball strikes the ground. The maximum amount of mechanical energy converted to internal energy during the fall is

(1) 2.7 J
(2) 12.0 J
(3) 14.7 J
(4) 26.7 J

39 _____

40 A projectile lands at the same height from which it was launched. Which initial velocity will result in the greatest horizontal displacement of the projectile? [Neglect friction.]

(1)

(3)

(2)

(4)

40 _____

41 A 5.0-kilogram box is sliding across a level floor. The box is acted upon by a force of 27 newtons east and a frictional force of 17 newtons west. What is the magnitude of the acceleration of the box?

(1) 0.50 m/s^2 (3) 8.8 m/s^2
(2) 2.0 m/s^2 (4) $10. \text{ m/s}^2$

41 _____

42 The diagram below represents a 2.0-kilogram toy car moving at a constant speed of 3.0 meters per second counterclockwise in a circular path with a radius of 2.0 meters.

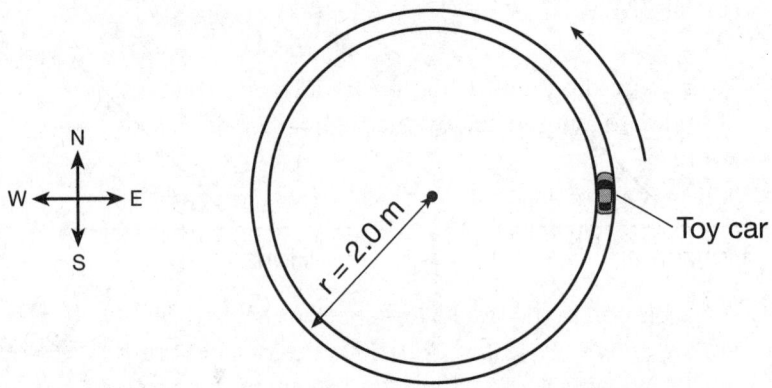

At the instant shown in the diagram, the centripetal force acting on the car is

(1) 4.5 N north   (3) 9.0 N north
(2) 4.5 N west   (4) 9.0 N west

42 _____

43 In which electric circuit would the voltmeter read 10 volts?

43 _____

44 The lambda baryon has the quark composition *uds*. Which particle has the same electric charge as the lambda baryon?

(1) neutron   (3) proton
(2) electron   (4) antimuon

44 _____

45 How many kilograms of matter would have to be converted into energy to produce 24.0 megajoules of energy?

(1) 2.67×10^{-16} kg   (3) 8.00×10^{-8} kg
(2) 2.67×10^{-10} kg   (4) 8.00×10^{-2} kg

45 _____

46 A red photon in the bright-line spectrum of hydrogen gas has an energy of 3.02×10^{-19} joule. What energy-level transition does an electron in a hydrogen atom undergo to produce this photon?

(1) $n = 3$ to $n = 2$ (3) $n = 5$ to $n = 2$
(2) $n = 4$ to $n = 2$ (4) $n = 6$ to $n = 2$

46 _____

47 In the diagram below, a negatively charged rod is placed between, but does not touch, identical small metal spheres R and S hanging from insulating threads.

What can be concluded if the rod repels sphere R but attracts sphere S?

(1) Sphere R must be negative and sphere S must be positive.

(2) Sphere R must be negative and sphere S may be positive or neutral.

(3) Sphere R must be positive and sphere S must be negative.

(4) Sphere R must be positive and sphere S may be negative or neutral.

47 _____

48 The amount of electric energy consumed by a 60.0-watt lightbulb for 1.00 minute could lift a 10.0 newton object to a maximum vertical height of

(1) 6.00 m   (3) 360. m
(2) 36.7 m   (4) 600. m

48 _____

49 Microwaves can have a wavelength closest to the

(1) radius of Earth   (3) length of a football field
(2) height of Mount Everest   (4) length of a physics student's thumb

49 _____

50 Two pulses approach each other in a uniform medium, as represented in the diagram below.

Which diagram best represents the superposition of the two pulses when the pulses overlap?

(1)   (2)   (3)   (4)

50 _____

PART B–2

Answer all questions in this part.

Directions (51–65): Record your answers on the answer sheet provided in the back. Some questions may require the use of the *2006 Edition Reference Tables for Physical Setting/Physics*.

Base your answers to questions 51 through 53 on the information below and on your knowledge of physics.

A toy launcher that is used to launch small plastic spheres horizontally contains a spring with a spring constant of 50. newtons per meter. The spring is compressed a distance of 0.10 meter when the launcher is ready to launch a plastic sphere.

51 Determine the elastic potential energy stored in the spring when the launcher is ready to launch a plastic sphere. [1]

52–53 The spring is released and a 0.10-kilogram plastic sphere is fired from the launcher. Calculate the maximum speed with which the plastic sphere will be launched. [Neglect friction.] [Show all work, including the equation and substitution with units.] [2]

54 Two 10.-ohm resistors have an equivalent resistance of 5.0 ohms when connected in an electric circuit with a source of potential difference. Using circuit symbols found in the *Reference Tables for Physical Setting/Physics*, draw a diagram of this circuit. [1]

55 The graph below shows the relationship between distance, d, and time, t, for a moving object.

On the axes *in your answer booklet*, sketch the general shape of the graph that shows the relationship between the magnitude of the velocity, v, and time, t, for the moving object. [1]

Base your answers to questions 56 through 58 on the information and diagram below and on your knowledge of physics.

A ray of monochromatic light ($f = 5.09 \times 10^{14}$ Hz) passes from medium X into air. The angle of incidence of the ray in medium X is 25°, as shown.

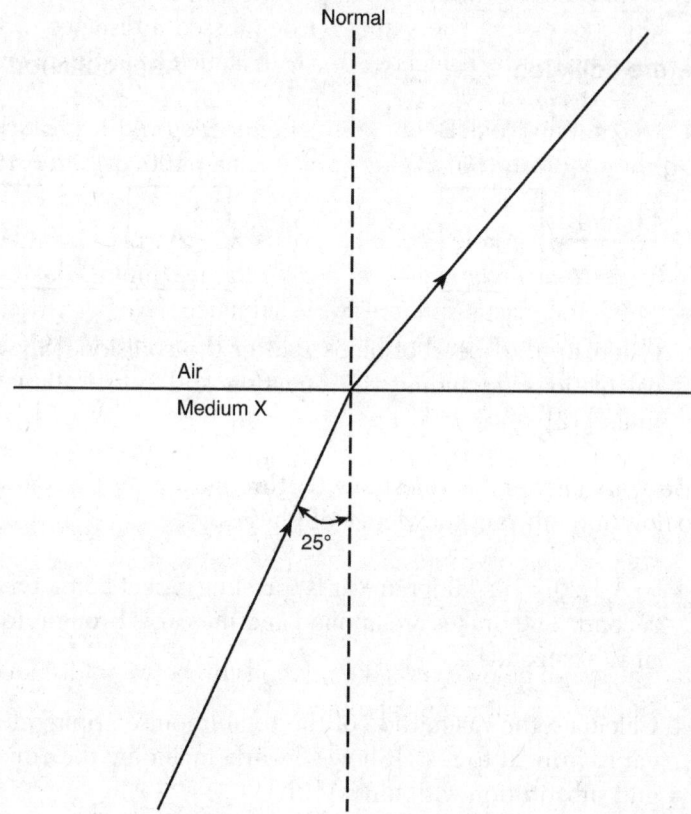

56 Using a protractor, measure and record the angle of refraction in the air, to the *nearest degree*. [1]

57–58 Calculate the absolute index of refraction of medium X. [Show all work, including the equation and substitution with units.] [2]

59–60 A student wishes to record a 7.5-kilogram watermelon colliding with the ground. Calculate how far the watermelon must fall freely from rest so it would be traveling at 29 meters per second the instant it hits the ground. [Show all work, including the equation and substitution with units.] [2]

61–62 As represented in the diagram below, block A with a mass of 100. grams slides to the right at 4.0 meters per second and hits stationary block B with a mass of 150. grams. After the collision, block B slides to the right and block A rebounds to the left at 1.5 meters per second. [Neglect friction.]

Before collision   **After collision**

Block A   Block B   Block A   Block B
m = 100. g   m = 150. g   m = 100. g   m = 150. g

[A] →4.0 m/s [B]   1.5 m/s← [A]   [B]

Calculate the speed of block B after the collision. [Show all calculations, including the equation and substitution with units.] [2]

Base your answers to questions 63 through 65 on the information below and on your knowledge of physics.

A 1.20×10^3-kilogram car is traveling east at 25 meters per second. The brakes are applied and the car is brought to rest in 5.00 seconds.

63–64 Calculate the magnitude of the total impulse applied to the car to bring it to rest. [Show all work, including the equation and substitution with units.] [2]

65 State the direction of the impulse applied to the car. [1]

PART C
Answer all questions in this part.

Directions (66–85): Record your answers in the spaces provided in the answer sheet provided in the back. Some questions may require the use of the 2006 *Edition Reference Tables for Physical Setting/Physics*.

Base your answers to questions 66 through 70 on the information and diagram below and on your knowledge of physics.

The diagram shows a negatively charged oil drop that is suspended motionless between two oppositely charged, parallel, horizontal metal plates. The electric field strength between the charged plates is 4.0×10^4 newtons per coulomb. The 1.96×10^{-15}-kilogram oil drop is being acted upon by a gravitational force, F_g, and an electrical force, F_e.

66–67 Calculate the magnitude of the gravitational force, F_g, acting on the oil drop. [Show all work, including the equation and substitution with units.] [2]

68 Determine the magnitude of the upward electrical force, F_e, acting on the oil drop suspended motionless between the charged metal plates. [1]

69–70 Calculate the net electric charge on the oil drop in coulombs. [Show all work, including the equation and substitution with units.] [2]

Base your answers to questions 71 through 75 on the information below and on your knowledge of physics.

In a circuit, a 100.-ohm resistor and a 200.-ohm resistor are connected in parallel to a 10.0-volt battery.

71–72 Calculate the equivalent resistance of the circuit. [Show all work, including the equation and substitution with units.] [2]

73–74 Calculate the current in the 200.-ohm resistor. [Show all work, including the equation and substitution with units.] [2]

75 Determine the power dissipated by the 100.-ohm resistor. [1]

Base your answers to questions 76 through 80 on the information below and on your knowledge of physics.

A wave traveling through a uniform medium has an amplitude of 0.20 meter, a wavelength of 0.40 meter, and a frequency of 10. hertz.

76–77 On the grid in your answer booklet, draw one complete cycle of the wave. [2]

78–79 Calculate the speed of the wave. [Show all work, including the equation and substitution with units.] [2]

80 Determine the period of this wave. [1]

Base your answers to questions 81 through 85 on the information and data table below and on your knowledge of physics.

In an experiment, the potential difference applied across an unmarked resistor was varied while the resistor was held at a constant temperature. The corresponding current through the resistor was measured. The data collected appear in the table below.

Potential Difference (volts)	Current (amperes)
1.5	0.0032
3.0	0.0059
6.0	0.0124
9.0	0.0177
12.0	0.0244

81 Mark an appropriate scale on the axis labeled "Current (A)." [1]

82 Plot the data points for current versus potential difference. [1]

83 Draw the line or curve of best fit. [1]

84–85 Using your graph, calculate the resistance of the resistor. [Show all work, including the equation and substitution with units.] [2]

Answer Sheet
June 2019

Physics—
Physical Setting

PART B–2

51 _____ J

52–53

54

55

56 _____°

57–58

59–60

61–62

63–64

65 _____

PART C

66–67

68 _____ **N**

69–70

71–72

73–74

75 _____ **W**

76–77

Particle
Displacement
(m)

Position
(m)

78–79

80 _____ **s**

81–83

Potential Difference vs. Current

Potential
Difference
(V)

15.0 —

10.0 —

5.0 —

0 —

Current (A)

84–85

Answers
June 2019
Physics—Physical Setting

Answer Key

PART A

1. 4	8. 1	15. 2	22. 3	29. 1
2. 3	9. 4	16. 2	23. 4	30. 2
3. 2	10. 1	17. 2	24. 2	31. 3
4. 4	11. 1	18. 3	25. 1	32. 3
5. 2	12. 3	19. 3	26. 2	33. 1
6. 1	13. 4	20. 1	27. 3	34. 3
7. 1	14. 1	21. 1	28. 4	35. 3

PART B–1

36. 3	39. 1	42. 4	45. 2	48. 3
37. 3	40. 3	43. 4	46. 1	49. 4
38. 1	41. 2	44. 1	47. 2	50. 1

PART B–2 and **PART C**—For sample methods of solutions, see Barron's *Regents Exams and Answers* book for Physics—Physical Setting.

Index